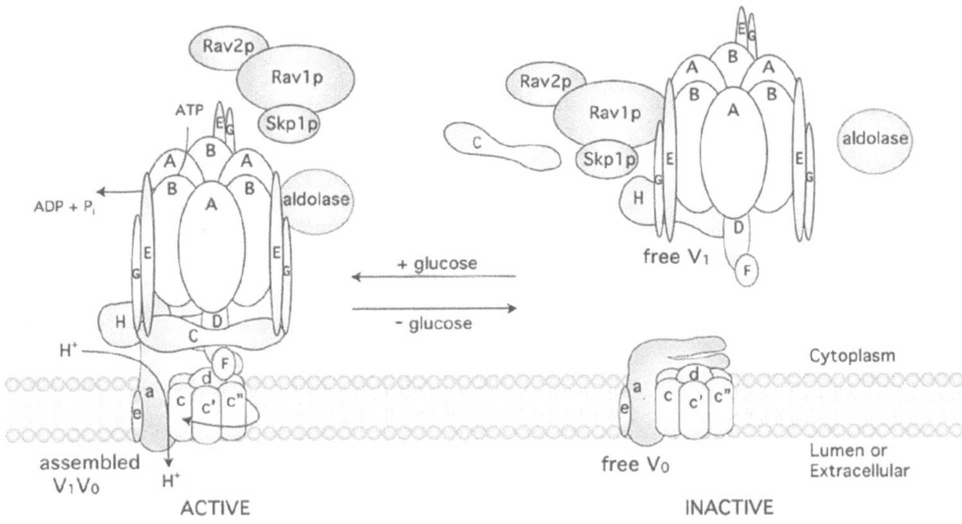

Handbook of
H^+-ATPases

Handbook of
H^+-ATPases

edited by

Suguru Nakamura

Pan Stanford Publishing

Published by

Pan Stanford Publishing Pte. Ltd.
Penthouse Level, Suntec Tower 3
8 Temasek Boulevard
Singapore 038988

Email: editorial@panstanford.com
Web: www.panstanford.com

British Library Cataloguing-in-Publication Data
A catalogue record for this book is available from the British Library.

Handbook of H^+-ATPases

ISBN 978-981-4411-91-2 (Hardcover)
ISBN 978-981-4411-92-9 (eBook)

Printed in the USA

Contents

Preface

Vacuolar-type H^+-ATPases (V-ATPases), which are found within the membranes of many organelles, such as endosomes, lysosomes, and secretory vesicles in eukaryotic cells, catalyze ATP hydrolysis to transport protons across intracellular and plasma membranes Plasma membrane H^+-ATPases, P-ATPases (E1E2-ATPases), which are found in bacteria, fungi eukaryotic plasma membranes and organelles, function to transport a variety of different ions across membranes. This book is dedicated to the scope of V-ATPases and P-ATPases by leading experts in the area of basic science and clinical medicine. This book presents recent findings on the structure and function of V-ATPase in glucose promoting assembly and in glucose signaling. It also describes the regulatory mechanisms of V-ATPases in yeast cells, neural stem cells, kidney cells, and cancer cells and under diabetic conditions. In addition, information on the role of V-ATPases on insulin secretion and cancer chemotherapy is also given in this book It also illustrates the activation of P-ATPases through glucose-induced calcium signaling in *Saccharomyces cerevisiae* yeast cells and the stimulation of proton–potassium pump (H^+–K^+-ATPase) by glucose in kidney cells.

V-ATPases are composed of two domains: the ATP hydrolysis V1 domain (subunits A, B, C, D, E, F, G, H) and the proton translocation V0 domain (subunits a, d, e, c, c', c" in yeast and subunits a, d, e, c, c" and Ac45 in mammals). The activity of V-ATPase is largely controlled by the assembly state of the complex, which consists of two domains. Glucose has been shown to play a critical role in both the functional structure and activity of V-ATPase and, as discussed in this book, has also been shown to promote V-ATPase assembly and activation. There is much evidence to suggest that the glucosemodulated, glycolytic enzyme aldolase mediates the assembly, expression, and activity of V-ATPase. Glucose signaling involves a number of different regulatory proteins including RAVE and protein kinase A, in order to control the assembly and disassembly of V-ATPase Glucose has also been shown to regulate cytosolic pH–promoted V-ATPase assembly in the field of nutrient sensing and signaling. Studies on cancer cells have shown that glucose withdrawal initiates reversible dissociation of V-ATPase, thereby shutting down its activity and leading to cell death; this suggests that glucose regulates V-ATPase in cancer cells.

It is also known that V-ATPase assembly can be regulated by other mechanisms such as angiotensin II or through certain extracellular factors. Angiotensin II is a peptide hormone that plays an endocrinological role in the regulation of blood pressure as well as fluid and electrolyte homeostasis. Angiotensin II is the major bioactive product of the renin–angiotensin system and it is involved in almost every pathophysiological process implicated in the development of diabetic nephropathy. Blocking the action of angiotensin II is a critical component in every therapeutic regimen designed to prevent and treat diabetic nephropathy. A study included in this book discusses how angiotensin II regulates the assembly of V0 and V1 domains through activation of P38 MAPKinase and PI3K pathways to form the active complex of V-ATPase. The activation of the PI3K/Akt pathway is responsible for glucose metabolism including glucose uptake and glycogenesis. It has been noted that V-ATPase assembly can also be regulated by other extracellular conditions through secondary messenger systems.

Several isoforms of the "a" subunit of V-ATPase have been identified including two "a" isoforms (Vph1p and Stv1p) that have been discovered in yeast. Vph1p is located in V-ATPase complexes of the vacuole while Stv1p is located in V-ATPase complexes of the Golgi and endosomes. Four different isoforms (a1–a4), encoded by different genes have been identified in mammalian cells; a1 (neural), a2 (endothelial and neural) a3 (osteoclasts, pancreatic β-cells and premature melanosomes), and a4 (renal and epididymis). The variants of the "a"subunit are generated by alternative splicing, with a1-I and IV being specifically expressed in the neurons of the brain while a3-III is expressed in the heart and lungs. Both a4-I and a4-II have been found to be expressed in the kidneys, lungs, and testis. Additionally, a4-I is also located in heart and skeletal muscle while a4-II can be located in liver. The diversity of the "a" subunit isoforms is not only important for tissue specificity and targeting different membrane compartments but it could also result in the generation of V-ATPases with different functional properties.

Three chapters of this book discuss the functional activities of the isoforms of the "a" subunit of V-ATPase in cancer-related inflammation, pancreatic β-cells, and yeast cells. The a2 isoform of V-ATPase has been shown to be important in tumor progression and metastases. The studies discussed in this book suggest that the a2 isoform has the capacity to redirect its activity and to function as either an ATPase or to control acid hydrolysis on the cell surface. Research conducted on endocrine tissues has shown that mutant mice lacking the a3 isoform of V-ATPase have a significantly lower level of plasma insulin than their wild-type groups. This suggests that the a3 isoform of V-ATPase has a regulatory function in the exocytosis of insulin secretion. In yeast, key gluconeogenic enzymes such as fructose-1,6-bisphosphatase (FBPase), phosphoenolpyruvatecarboxykinase, malate dehydrogenase, and isocitratelyase are degraded in the vacuole during glucose refeeding. This prevents energy futile cycles that are detrimental to cells. Vacuole import and degradation (Vid) vesicles are intermediate vesicles that carry gluconeogenic enzymes to the vacuole. Stv1p and Vph1p of the V-ATPase are required for FBPase degradation. Vph1p is required for both Vid vesicles and vacuoles, while Stv1p is required for the proper function of the Vid vesicles.

V-ATPase plays an important role in both the cell surface and vesicular trafficking signaling mechanisms for cancer cells and cells under diabetic conditions. V-ATPase is believed to be largely responsible for supporting cancer growth by controling related inflammatory processes and subsequent angiogenesis. V-ATPase acts as a modulator of chemokines and cytokine expression through a released peptide, which is the *N*-terminal portion of the a2 isoform of a V-ATPase V-ATPases have also been shown to be related to tumor pH control, metastasis, tumor cell growth and survival, and multidrug resistance (MDR), and also possess possible therapeutic applications associated with the use of specific V-ATPase inhibitors.

In the kidney, the V-ATPase E-subunit interacts directly with the glycolytic enzyme aldolase while the a-subunit interacts with phosphofructokinase-1, thereby providing a functional coupling of V-ATPase with the glycolytic pathway.This book discusses the localization and function of renal V-ATPases and their role in intracellular pH (pHi) regulation, transepithelial proton transport and acid–base homeostasis, in addition to providing an overview of V-ATPase in distal renal tubular acidosis and diabetes.

Diabetic nephropathy is a progressive kidney disease caused by angiopathy of capillaries in the kidney glomeruli, which leads to end-stage renal disease (ESRD) and finally to kidney failure. One of the many aspects of the late stages of diabetic nephropathy is the diminished protein reabsorption by proximal tubules via the megalin/cubilin-mediated endosomal/lysosomal protein degradative pathway that leads to proteinuria. V-ATPase, cytohesins and Arf-family GTP-binding proteins (Arfs) are essential for vesicular trafficking of receptors and their signaling along endocytic pathway of eukaryotic cells. A study in this book has demonstrated a novel specific interaction of cytohesin-2 with V-ATPase and aldolase in the V-ATPase/Arf6/cytohesin-2/aldolase complex on early endosomes. High glucose levels may regulate the activities of the components of this complex and trafficking of receptors in the protein degradative pathway, and thus, contribute to the development of early stages of diabetic nephropathy.

The role of V-ATPase has also been examined in the regulation of phosphate transporters in rodents. The type II sodium phosphate cotransporter, Npt2a (SLC34A1), is one of three known sodium-coupled phosphate transporters responsible for the reabsorption of filtered phosphate from the lumen of the proximal renal tubule. It has been found that V-ATPases regulate Npat2a at multiple sites. Apical membrane V-ATPase which is activated during metabolic acidosis may contribute to the phosphaturia associated with metabolic acidosis by decreasing Npt2a transport function. V-ATPases are critical for the forward trafficking of Npt2a as well as physiological regulation of the Npt2a degradation pathway via the transport of proteins from the endosome to the lysosome and ultimately the degradation by lysosomal enzymes. These findings provide a possible link between Npt2a in the regulation of serum phosphorus concentration and the risk of cardiovascular renal diseases.

Another study in this book shows that in neural stem cells of the mouse brain, V-ATPase is crucial for the transduction of Notch signaling and plays an important role for endosome acidification and endocytosis in signal transduction during neural stem cell differentiation and brain development.

The P-ATPases are a large group of transporter family–related ion and lipid pumps that can be divided into five subfamilies. H^+–K^+-ATPase in animal cells belongs to the type IIC subfamily while the P-ATPase is classified as type III in fungi. The relationship between glucose-induced calcium signaling and the activation of the P-ATPase in *S. cerevisiae* cells is discussed in this book We have also observed that glucose activates H^+–K^+-ATPase in kidney epithelial cells (our unpublished results).

We would like to thank our contributors and colleagues from around the world who have devoted a significant amount of time and effort into making this book as accurate and as useful as possible. Without their contributions, this project could not have possibly been as successful. Their expertise in biochemistry, cell biology, and pathophysiology has greatly added to our ability to bring the most recent results to our readers. I would also like to thank Ms. Shivani Sharma, Pan Stanford Publishing, for her professional assistance. We sincerely hope our readers will find this handbook beneficial and that the knowledge gained from the book will aid their future endeavors.

Suguru Nakamura
Winter 2013

Chapter 1

Vacuolar H$^+$-ATPase Assembly

Anne M. Smardon and Patricia M. Kane
Department of Biochemistry and Molecular Biology, SUNY Upstate Medical University, 750 East Adams Street, Syracuse, NY 13210, USA
smardona@upstate.edu

1.1 Introduction

The acidification of intracellular organelles is crucial for a wide range of cellular processes including protein sorting in biosynthetic and endocytic pathways; protein processing, activation and degradation; membrane transport; and coupled transport of small molecules [1–3]. Organelle acidification in all eukaryotic cells is mediated by the vacuolar proton-translocating ATPase known as V-ATPase. V-ATPases couple the energy released by hydrolysis of cytoplasmic ATP to active transport of protons from the cytosol into intracellular compartments. The proton gradient generated by the V-ATPase is required to maintain the pH of intracellular organelles and to regulate ion homeostasis [4]. V-ATPases have been shown to play a significant role in cytoplasmic pH homeostasis as well [5]. In addition to their role in intracellular pH regulation, some cells express V-ATPases at the plasma membrane where they pump

Handbook of H$^+$-ATPases
Edited by Suguru Nakamura

ISBN 978-981-4411-91-2 (Hardcover), 978-981-4411-92-9 (eBook)
www.panstanford.com

protons from the cytosol out of the cell mediating cell-specific functions, such as urine acidification, bone resorption, and sperm maturation [2, 6, 7].

V-ATPase activity is largely controlled by the assembly state of the holoenzyme complex or by targeting active complexes to different cellular sites where they serve a range of cellular functions. Much of our current knowledge of the structure and *in vivo* assembly of the eukaryotic V-ATPases comes from studies in the *Saccharomyces cerevisiae* yeast model system.

1.2 V-ATPase Subunit Composition and Mechanism

The structure and subunit composition of eukaryotic V-ATPases are highly conserved in all organisms from yeast to humans [1]. Figure 1.1 and Table 1.1 show a proposed structure and subunit composition of the yeast V-ATPase. They are large multisubunit complexes that are composed of two domains. The subunits of the V_1 and V_0 domains are denoted by capital and small letters, respectively. The V_1 domain (~650 kDa) is a peripheral cytoplasmic complex composed of eight different subunits (subunits A, B, C, D, E, F, G, and H), and it contains the sites of ATP hydrolysis. The V_0 domain (~300 kDa) is an integral membrane complex that is responsible for the translocation of protons across the membrane and is composed of six different subunits (subunits a, d, e, c, c′, and c″ in yeast and subunits a, d, e, c, c″, and Ac45 in mammals). The V_1 and V_0 domains must be structurally and functionally coupled for ATP-driven proton translocation to occur [8, 9].

V-ATPases are structurally and evolutionarily related to the F_1F_0-synthases (also known as F-ATPases) of mitochondria and chloroplasts [10, 11]. Both have descended from archaeal proton pumps and ATP synthases. However, the V-ATPases are dedicated proton pumps that, unlike the F-ATPases, do not synthesize ATP under physiological conditions. The catalytic rotary mechanism of V-ATPases and F-ATPases is also conserved [12–15]. When ATP is hydrolyzed, both enzymes couple the energy of ATP hydrolysis to proton transport.

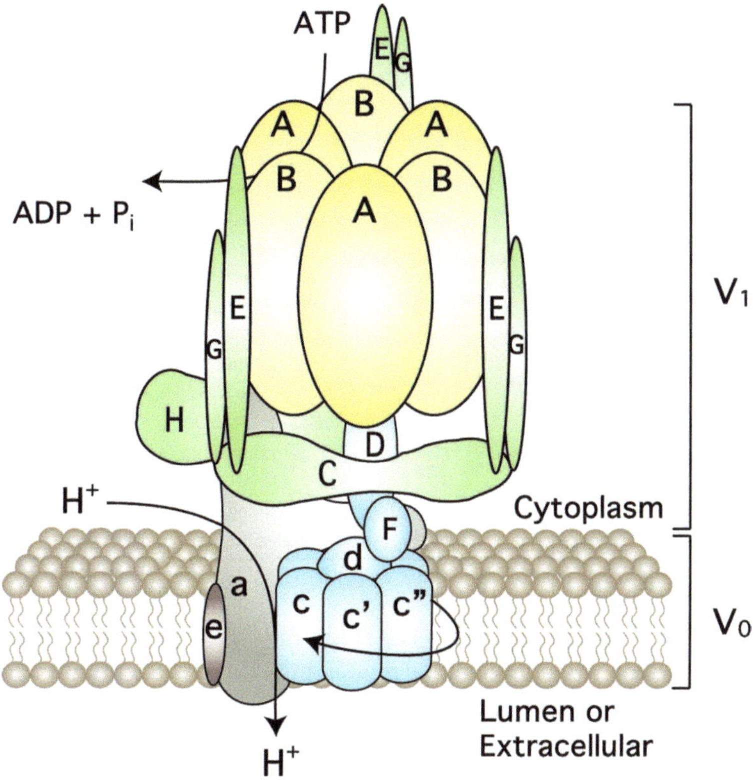

Figure 1.1 Schematic structure and subunit composition of the yeast V-ATPase. The vacuolar ATPase is composed of a V_1 peripheral domain (uppercase letters) responsible for ATP hydrolysis and a V_0 integral membrane domain (lowercase letters) involved in proton transport. The hexameric globular head domain composed of alternating A and B subunits (shaded yellow) is tethered via three stator stalks composed of subunits E and G that are anchored to V_0 through bridging subunits C and H (green). The rotor is composed of V_1 subunits D and F, V_0 subunit d, and proteolipid subunits c, c′, and c″ (blue). V_0 subunit a (gray) makes important contacts with the stator stalks to hold the a-subunit stationary relative to the rotation of the c-ring.

In the generally accepted model for V-ATPase function, the enzyme consists of two parts, a stator that remains fixed and a mobile rotor. Energy released from the hydrolysis of ATP at catalytic sites in V_1 causes rotation of a central stalk that is connected to a ring

Table 1.1 Yeast V-ATPase subunits

Domain	Subunit	Mass (kDa)	Gene	Subunit function
V_1	A	70	*VMA1*	ATP hydrolytic sites
	B	60	*VMA2*	ATP-binding (non-catalytic), binds aldolase
	C	44	*VMA5*	stator subunit, regulatory, binds RAVE
	D	34	*VMA8*	rotary subunit
	E	33	*VMA4*	stator subunit, binds RAVE, binds aldolase
	F	14	*VMA7*	rotary subunit
	G	13	*VMA10*	stator subunit, binds RAVE
	H	54	*VMA13*	stator subunit, regulatory
V_0	a	100	*VPH1/ STV1*	H^+ transport,stator subunit, targeting, binds aldolase
	d	38	*VMA6*	rotary subunit, coupling
	e	9	*VMA9*	unknown
	c	17	*VMA3*	H^+ transport, rotary subunit
	c'	17	*VMA11*	H^+ transport, rotary subunit, binds Vma21 assembly factor
	c"	21	*VMA16*	H^+ transport, rotary subunit

of proteolipid subunits in V_0. The rotation of the ring of V_0 subunits drives proton transport. In this model it is critical that subunits that are not part of the rotor remain stationary relative to the rotation.

The globular head of the V_1 domain contains three copies each of A and B subunits arranged in an alternating hexameric ring. ATP hydrolysis occurs at three catalytic ATP-binding sites located at alternating interfaces between subunits A and B, with most contact sites on subunit A [4, 16–18]. An additional three non-catalytic ATP-binding sites, which may have a regulatory function, are also present at A-B interfaces with most residues contributed by subunit B. The V_1 and V_0 domains are connected by multiple stalks [18–21]. The central rotary stalk fills the central cavity of the A_3B_3 ring and is composed of V_1 subunits D and F and V_0 subunit d [22]. Electron microscopy and cross-linking studies support a model in which the stator function is provided by three peripheral stalks composed of three E and G subunits that interact with the *N*-terminal domain of V_0 subunit a through bridging subunits H and C [20–26].

The V_0 proteolipid subunits c, c$'$, and c$''$ form a ring structure that spans the lipid bilayer. The c ring subunits are highly hydrophobic and each contains multiple transmembrane helices. There is one copy of each c′ and c″ and multiple copies of subunit c [2, 27]. Each proteolipid subunit contains a single acidic residue that is able to be reversibly protonated [27]. As the proteolipid ring rotates, each c subunit comes into contact with subunit a and is protonated. One complete 360° rotation of the c ring against the stationary a subunit drives the release of each proton to the other side of the membrane [2, 28]. The V_0 d subunit is proposed to rest on top of the proteolipid ring where it connects the V_1 central stalk subunits D and F with the V_0 proteolipid ring [22]. The rotor, then, consists of the subunits D, F, d, c, c′, and c″. V_0 subunit e is a small, integral membrane protein that is required for V_0 assembly [29].

Subunit a of V_0 is composed of an *N*-terminal cytoplasmic domain and a C-terminal domain that contains eight transmembrane helices [30]. Both of these domains are critical for proton transport. The *N*-terminal domain of subunit a makes important contacts with the stator stalks of V_1 and these contacts hold subunit a in a static position relative to the rotation of the proteolipid ring [31]. The C-terminal domain of subunit a contributes to a channel that allows protons to reach the acidic residues on the proteolipid ring. A single arginine residue in subunit a has been shown to be required for proton transport [2]. Subunit a is the only yeast V-ATPase subunit that is present in two isoforms, Vph1p and Stv1p, that target the pumps to the vacuole and Golgi/endosomes, respectively [32, 33].

1.3 Biosynthetic Assembly of the V-ATPase

Assembly and proper targeting of the V-ATPase is complicated by its complex multisubunit structure and its function in multiple organelles. V-ATPases are composed of a combination of membrane proteins that must be targeted to different cellular sites through the secretory pathway, and peripheral proteins that are synthesized in the cytoplasm. Numerous studies, mainly in yeast, have been able to identify some important factors required for assembly and targeting of V-ATPases. The first requirement for a functional V-ATPase is expression of a full set of V-ATPase subunits. In yeast, deletion of any one V-ATPase subunit (with the exception of *VPH1* and *STV1* where both isoforms must be deleted) results in loss of enzyme activity. Interestingly, however, loss of any V_1 subunit does not affect the levels of other V_1 or V_0 subunits [34]. An exception is destabilization of the E subunit when the G subunit is deleted [35]. V_0 domains in these V_1 mutants are assembled and targeted correctly [36]. Loss of a single V_0 subunit does affect expression and targeting of other V_0 subunits, particularly subunit a, but does not affect expression of the V_1 subunits [34]. Cells lacking a V_0 domain are able to assemble an inactive V_1 domain (minus subunit C) in the cytosol [37]. These studies served to demonstrate that V_1 and V_0 domains could be assembled independently. In fact, both mammalian cells and yeast contain free V_1 and V_0 domains *in vivo*. Yet it appears that the predominant biosynthetic assembly pathway does not involve association of preassembled V_1 and V_0 domains; rather, in yeast cells, there is evidence for a concerted assembly pathway in which individual V_1 and V_0 subunits associate at very early stages before full assembly of either complex [38].

Early assembly of the V_0 complex occurs in the endoplasmic reticulum (ER) and is dependent upon a set of five dedicated ER-associated assembly factors (Vma21p, Voa1p, Vma12p, Vma22p, and Pkr1p) that act to chaperone the proper assembly of the V_0 components (Fig. 1.2) [39–42]. Vma21p is an integral membrane protein that is proposed to interact with subunit c' and to direct assembly of the proteolipid ring and the binding of subunit d to the cytoplasmic side of the ring. The Vma21p/proteolipid ring complex has been shown to be a stable intermediate core complex in the V_0 assembly pathway [41]. In the absence of Vma21p, the V_0 subunits fail to assemble properly and are not able to exit the ER. Voa1p (an integral membrane protein) has been shown to associate with the partially assembled V_0/Vma21p core complex in the ER and to remain with the complex until all components of V_0 are present in the Vma21p/V_0 complex [42]. Homologues of Vma21p have been identified in higher organisms suggesting a universal V-ATPase assembly pathway.

Vma12p (an integral membrane protein) and Vma22p (a peripheral membrane protein) form a complex that binds transiently to subunit a and is required to stabilize and facilitate the assembly

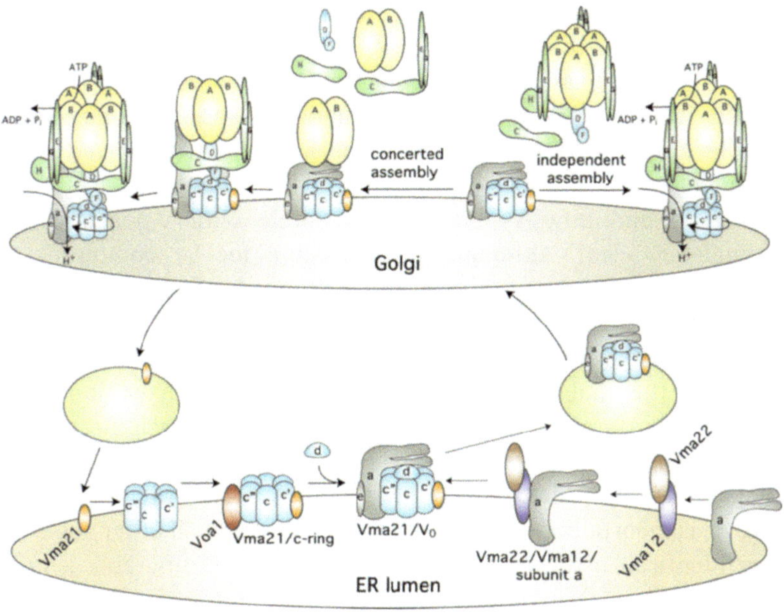

Figure 1.2 Biosynthetic assembly of the V-ATPase complex in yeast. The V_0 domain (blue and gray) is assembled in the endoplasmic reticulum (ER) and involves assembly factors Vma21p, Vma22p, Vma12p, and Voa1p. Vma21p associates with V_0 subunit c′ and directs the assembly of the proteolipid c-ring and the binding of subunit d to form the core Vma21p-proteolipid ring complex. Voa1p associates stably with the Vma21p-proteolipid complex. A complex of Vma22p and Vma12p binds to subunit a to form a second complex in the assembly pathway. The addition of subunits a and e to the core Vma21p/proteolipid complex results in the detachment of the Vma22p/Vma12p complex and Voa1p. Vma21p accompanies the V_0 complex to the Golgi in ER-derived vesicles and is recycled back to the ER. V_1 (yellow and green) is assembled with V_0 in the Golgi in either a concerted assembly pathway where V_0 assembles with individual V_1 subunits or V_1 subcomplexes, or in an independent assembly pathway where V_0 assembles with a mostly assembled V_1.

of subunit a with the core Vma21p/proteolipid ring complex [43]. If either protein of the Vma12p/Vma22p complex is not present, subunit a is rapidly degraded and V_0 is not able to assemble. Once subunit a binds to the Vma21p/proteolipid core complex, the Vma12p/Vma22p complex and the Voa1p protein dissociate and the fully assembled V_0/Vma21p complex is transported to the Golgi [41]. Vma21p is then released from V_0 and recycled back to the ER. Pkr1p appears to aid in the efficient assembly of V_0 and its export out of the ER. Cells lacking Pkr1p are only able to assemble extremely low levels (~5–10%) of active V-ATPases that are transported to the vacuole [39].

When and how V_1 subunits assemble with V_0 is still not completely clear. Although it is possible for V_1 complexes to completely assemble in the cytosol without V_0, there is evidence for an ER complex composed of V_1 subunit A and V_0 subunit a that suggests a more coordinated assembly of V_1 and V_0 [38]. In any case the final steps of V-ATPase assembly and activation occur in the Golgi apparatus, as there is evidence for acidification of this compartment [3]. There are two distinct V-ATPases in yeast, depending on which of the two subunit a isoforms, Vph1p or Stv1p, is incorporated into the complex. Although all other subunits are identical in these two complexes, their targeting and cellular functions are very different. Vph1p-containing complexes travel from the Golgi to the vacuole through the prevacuolar compartment. These compartments become progressively more acidic with the vacuole being the most highly acidic internal organelle [44]. Stv1p-containing complexes cycle between the prevacuolar compartment and the Golgi, and possibly the early endosome as well [45]. In addition to the difference in localization, Vph1p and Stv1p complexes have been shown to differ in transcript and protein levels, degree of assembly, and ability to couple ATP hydrolysis to proton transport [32, 33].

Stable association of the V_1 domain with the V_0 domain during biosynthesis in yeast also requires the RAVE complex (described in detail below) [46, 47]. In the absence of the RAVE complex, both V_1 and V_0 subcomplexes are able to assemble and there is some attachment of V_1 to V_0 at the vacuole, but these assembled complexes are unstable and there is very little V-ATPase activity in isolated vacuoles [47]. While the mechanistic role of RAVE in V-ATPase assembly is not known, it has been shown that V_1 subunit C is not able to stably assemble with the V-ATPase in the absence of the RAVE complex [48].

1.4 Assembly State Regulates V-ATPase Activity

1.4.1 *Reversible Dissociation of V_1V_0 Complex*

After its biosynthetic assembly, V-ATPase activity is regulated by adjusting the level of V_1–V_0 assembly. An important mechanism for regulation of activity is reversible disassembly of assembled complexes into free V_1 and V_0 domains in response to various extracellular conditions. During disassembly, V_1 is released into the cytosol as a complex, although subunit C is detached from both V_1 and V_0 in this process. Dissociation of V_1 from V_0 inhibits both the ATPase activity and proton transport of each respective domain [49, 50]. The V_1 complex loses all Mg^{2+} dependent ATP hydrolysis when it dissociates from the membrane, apparently to conserve cytoplasmic ATP stores in low energy conditions and to prevent non-productive ATP hydrolysis [50]. Reversible dissociation was first observed in response to glucose deprivation in yeast cells [49] (Fig. 1.3) and in the apical plasma membrane of the

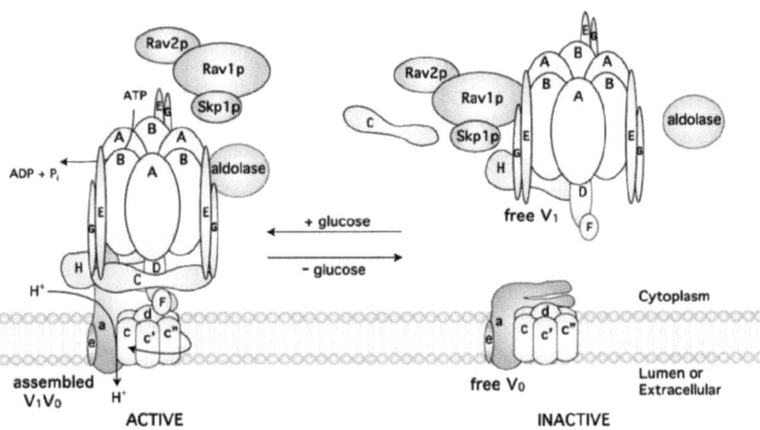

Figure 1.3 Regulation of V-ATPase activity by reversible dissociation. In yeast, intact, active V-ATPase complexes disassemble into inactive free V_1 and V_0 complexes in response to glucose depletion. Disassembly results in release of subunit C from both V_1 and V_0. Glucose readdition results in reassembly of the active enzyme. The RAVE complex (Rav1p, Rav2p, and Skp1p) and the glycolytic enzyme aldolase mediate active V-ATPase assembly of V_1 with V_0 at the membrane.

midgut in *Manduca sexta* larvae that have ceased feeding during molt [51]. This process proved to be completely reversible in both systems whereby detached V_1 and V_0 domains remain stable and are able to reassemble into fully active complexes when conditions are favorable. These results in organisms as diverse as yeast and insects suggested that reversible disassembly could be a universal regulatory mechanism of V-ATPases.

Reversible disassembly has been studied most extensively in yeast. Both disassembly and reassembly in response to glucose concentration are rapid and do not require new protein synthesis indicating that pre-existing V_1 and V_0 complexes are able to reassemble. In addition, intermediate levels of V-ATPase assembly are observed in yeast cells grown in a poor carbon source, such as raffinose, suggesting that disassembly is not an all or nothing event [49]. The dissociated V_1 and V_0 domains appear to be in a dynamic equilibrium with the fully assembled V-ATPase. These results suggest that the level of V-ATPase activity is modulated by the overall assembly state of the V-ATPase that is continually adjusted in response to extracellular glucose concentration. Subsequent work in yeast has demonstrated that V-ATPase assembly and activity can be regulated by signals other than glucose levels. A recent study shows that V-ATPase activity in yeast is upregulated in response to salt stress by increasing the assembly of V_1 with V_0 at the vacuolar membrane [52]. High extracellular pH has also been shown to increase V-ATPase activity and the level of assembly of V_1–V_0 at the vacuole [53]. Yeast cells subjected to high extracellular pH were unable to undergo reversible disassembly. These cells showed high levels of V-ATPase activity, assembly and proton pumping upon glucose depletion [53].

Subsequent work in mammalian cells supports the idea that regulation of V-ATPase activity by altering its assembly state may be more universal. In immature dendritic cells, antigen proteolysis is inhibited by reduced lysosomal acidification resulting in reduced processing of antigen. When maturation signals induce dendritic cells to process antigen, the level of assembly of the V-ATPase on the lysosomal membrane increases resulting in increased lysosomal acidification, which is required for antigen processing [54]. In another study, secretagogue stimulation of alveolar epithelial type II cells was shown to increase surfactant secretion by disassembling V_1 from V_0 thereby dissipating the pH gradient across lamellar bodies [55]. The V-ATPase assembly/disassembly signal in these two cases is not glucose related. The first evidence for glucose induced assembly of the V-ATPase in mammalian cells was an increase in assembly of the V-ATPase in human proximal tubular HK-2 cells and porcine renal epithelial LLC-PK1 cells in response to glucose [56, 57]. In both of these cell lines glucose stimulates V-ATPase-dependent acidification of intracellular compartments and rapid assembly of the V_1 and V_0 domains of the V-ATPase. Glucose deprivation induced the opposite effect, resulting in loss of organelle acidification and release of V_1 into the cytosol. In addition to glucose-dependent internal organelle acidification, glucose stimulation of HK-2 cells altered trafficking of assembled complexes between intracellular compartments and recruited V_1 to the apical plasma membrane [57]. Interestingly, V-ATPase activation in response to glucose in LLC-PK1 was found to be concentration dependent in a manner similar to that found in yeast cells [56, 58]. These experiments serve to illustrate the importance of the extent of V_1–V_0 assembly in the regulation of V-ATPase activity, and that different signals may be required to generate various degrees of enzyme assembly and activity in different cellular environments.

1.4.2 *Control of Reversible Disassembly*

1.4.2.1 The RAVE complex (regulator of the ATPase of vacuolar and endosomal membranes)

The V-ATPase assembly factors localized to the ER in yeast are specific for biosynthetic assembly. However, the overall assembly state of the V-ATPase is determined not only by initial synthetic assembly of the complex, but also by reversible disassembly of assembled complexes. The RAVE complex in yeast is the first assembly complex to be identified that is implicated in both biosynthetic assembly and in reassembly of dissociated V_1 and V_0 [46, 47]. RAVE is a heterotrimeric complex composed of Rav1p, Rav2p, and Skp1p. *SKP1* is an essential gene and Skp1p is an adaptor protein that is an important component in E3 ubiquitin ligases and other protein complexes. *rav1*Δ and *rav2*Δ mutants have growth phenotypes and biochemical defects that are characteristic of a loss of V-ATPase activity. RAVE has been shown to interact with V_1 in the cytosol and this interacton is required for the stable reassembly of disassembled V_1 and V_0 [46, 47]. The RAVE-V_1 interaction is, however, not glucose sensitive, suggesting that RAVE is not the glucose sensor for reassembly. Rav1p is the central component of the RAVE complex binding Rav2p and Skp1p, and RAVE interacts with V_1 through an interaction between Rav1p and V_1 peripheral stalk subunits E and G [47]. This suggests that RAVE may be important for docking the peripheral stalks to the membrane-bound V_0 complex during reassembly. This idea is supported by subsequent work that showed an additional interaction between RAVE and V_1 subunit C in the cytosol [48]. The C subunit is unique in that it is the only V_1 subunit that is released from both V_1 and V_0 during disassembly, suggesting that subunit C may control regulated assembly of the V-ATPase. It was shown that in the absence of the RAVE complex, subunit C was not able to stably assemble with V_1 and V_0 at the vacuolar membrane [48]. It appears that RAVE and subunit C are both stabilizing functional V-ATPase assembly, and the RAVE/subunit C physical interaction suggests that they work together in this process. The RAVE complex was first characterized in yeast; however, there are Rav1p homologues in virtually all eukaryotes [59]. Recently, Rav1p homologues in Drosophila and mice (rabconnectins) were shown to affect acidification of intracellular compartments by regulating V-ATPase function [60, 61]. This suggests that RAVE has a conserved role in regulating V-ATPase activity and assembly.

1.4.2.2 Glycolytic enzymes and V-ATPase assembly

In order for extracellular conditions to influence V-ATPase assembly on internal membranes there must exist a mechanism for signal transduction. In yeast the signal for extracellular glucose availability that regulates V-ATPase assembly state does not involve the usual glucose-dependent signal transduction pathways that sense the level of glucose 6-phosphate. Further metabolism of glucose is required to trigger V-ATPase assembly [58]. This finding suggested a possible link between glycolysis and V-ATPase assembly. Subsequent studies support this likelihood. A number of glycolytic enzymes including aldolase [62–64] and phosphofructokinase (PKF) [65] have been shown to bind to the V-ATPase. A physical interaction between aldolase and the V-ATPase was first described in kidney cells [62–64]. In addition to the direct interaction with aldolase, V-ATPase activity in kidney cells can be stimulated in the absence of glucose by pyruvate, an end product of glycolysis, and glucose activation of the V-ATPase can be inhibited by the glycolytic inhibitor 2-deoxy-D-glucose suggesting that metabolism of glucose through the glycolytic pathway is required [56]. The interaction between aldolase and V-ATPase activity was later explored more extensively in yeast [62–64]. This interaction has been shown to be glucose-sensitive and deletion of the aldolase gene in yeast causes disassembly of the V-ATPase complex. There is evidence that the interaction of the V-ATPase with aldolase is via subunits B and E of V_1 and subunit a of V_0 and this interaction increases in the presence of glucose [63]. A point mutation in aldolase that disrupts its interaction with subunit B, but does not interfere with aldolase enzymatic function, reduces assembly of V_1 and V_0, suggesting a role for aldolase in V-ATPase assembly in addition to its enzymatic role in glycolysis [64]. Removal of glucose disrupts the aldolase/V-ATPase interaction, and overexpression of aldolase inhibits V-ATPase dissociation upon glucose removal [64]. These studies led to the proposal that aldolase may be the glucose sensor that signals dissociation of the V-ATPase [63].

Another glycolytic enzyme, PFK, has been shown to bind the V_0 a subunit in human kidney and in yeast [65, 66]. Naturally occurring mutations in the human V_0 a4 subunit that cause distal renal tubular acidosis, disrupt its interaction with phosphofructokinase-1 [66]. These mutations were recreated in V_0 subunit a in the yeast model system where the PFK/V-ATPase interaction was also disrupted. V-ATPase expression, assembly, and functioning were all compromised in these mutants. Based on these studies, a model has been proposed in which interaction of the V-ATPase with glycolytic enzymes may provide a local source of ATP to the enzyme [62–66]. Whatever the mechanism of action, it appears that the regulatory connection between glycolysis and V-ATPase activity is conserved.

The PI3K (phosphatydylinositol-3-kinase) signaling pathway has also been implicated in control of V-ATPase signaling in mammalian cells [56, 57]. Both glucose-dependent V-ATPase assembly and trafficking were abolished in renal epithelial cells treated with a PI3K inhibitor and were restored by overexpression of a constitutively active catalytic subunit of PI3K [57]. Thus, glucose activation of V-ATPase activity requires both glycolysis and PI3K activity in this context.

1.4.2.3 PKA-dependent regulation of V-ATPase assembly

There is ample evidence for involvement of the Ras/cAMP/protein kinase A (PKA) pathway in the regulation of V-ATPase activity. Early work using the salivary glands of the blowfly *Calliphora vicina* as a model system has shown an increase in V-ATPase activity and assembly in the apical membrane of secretory epithelial cells in response to the hormone serotonin (5-hydroxytryptamine, 5-HT) [67]. It was later shown that V-ATPase activation in response to 5-HT was due to an increase in intracellular cAMP and that cAMP exerts its effects on the V-ATPase by activation of PKA [68]. Subsequent work showed that PKA binds to and phosphorylates recombinant *Manduca sexta* V-ATPase subunit C *in vitro,* and that in salivary glands of *C. vicina* transient subunit C phosphorylation correlates with the hormone induced reassembly and activation of the V-ATPase and with PKA activation by cAMP analogs [69]. Further support for this proposed mechanism was demonstrated in experiments that showed that treatment with 5-HT induces recruitment of the PKA catalytic subunit (PKA-C) to the V-ATPase enriched apical membrane [70]. These results suggest that phosphorylation of subunit C could be a key regulatory mechanism of V-ATPase activity in the insect system, although this has not been demonstrated in other organisms.

The Ras/cAMP/PKA pathway has also been shown to play a role in glucose-dependent V-ATPase assembly in yeast [71]. Two inhibitors of Ras, Ira1p and Ira2p, were identified in a screen for mutants that were unable to disassemble and inactivate the V-ATPase in response to glucose deprivation [71]. Both Ira1p and

Ira2p are negative regulators of Ras that are required to inactivate Ras in low glucose conditions. This suggested a possible role for the Ras/cAMP/PKA pathway in regulation of V-ATPase assembly in response to glucose. An increase in PKA activity by deletion of the gene encoding the PKA regulatory subunits or by constitutive activation of Ras does indeed result in assembly and/or blocked disassembly of the V-ATPase [71]. In fact this phenotype mimicked the one exhibited by the *ira1*Δ and *ira2*Δ mutants. Results from both insects and yeast support the glucose-sensitive Ras/cAMP/PKA pathway as playing an essential role in signaling V-ATPase assembly and activity.

While the argument presented above for PKA-dependent regulation of V-ATPase assembly is compelling, there is also evidence in yeast and pancreatic beta-cells that V-ATPase activity acts as an upstream activator in the PKA pathway. These studies suggested that increased cytosolic pH following initiation of glucose metabolism acts as a second messenger for V-ATPase assembly and that V-ATPase activity then activates the PKA pathway [72]. These results contradict the above hypothesis that PKA signals V-ATPase assembly and support cytosolic pH as the assembly signal. Recent work supporting pH as a determinant of V-ATPase assembly and activity has shown that disassembly of the V-ATPase is suppressed at high extracellular pH [53].

Some cells that have V-ATPases at the plasma membrane regulate activity by reversible exocytosis and endoctosis of vesicles containing assembled V-ATPases. It has been shown in both renal intercalated cells and epididymal clear cells that the recruitment of intact V-ATPases to the plasma membrane involves PKA-dependent phosphorylation of the V_1 A subunit [73–75]. In the case of epididymal clear cells the vesicular transport of V-ATPases from internal compartments to the apical membrane is mediated by a soluble adenylyl cyclase (sAC) that is sensitive to extracellular pH [76]. When sAC is activated, cAMP concentration increases causing exocytosis (or inhibiting endocytosis) of V-ATPases. Thus, while the evidence is substantial, these experiments highlight the need for more work to clarify the contributions of the cAMP/PKA pathway and pH changes in V-ATPase regulation.

1.4.2.4 Structural organization and V-ATPase reversible disassembly

The V- and F-type ATPases are evolutionarily related and share a similar rotary catalytic mechanism. However, the function, regulation, and targeting of the V-ATPase is considerably more complex. Where the F-ATPase is primarily an ATP synthase and exclusively localized to mitochondrial and chloroplast inner membranes in eukaryotic cells, the V-ATPase is a dedicated proton pump that functions in diverse endomembranes and plasma membranes of specialized cells. F-ATPases have no contact with the cell cytoplasm; in contrast, the V-ATPase is constantly responding to cellular signals, cytoplasmic environment and membrane composition. The structure of the F-ATPase has been well characterized from both bovine mitochondria and *E. coli* [77–80]. The general organization of the V-ATPase and the F-ATPase is similar. Both have a peripheral soluble domain (V_1 and F_1) and a membrane embedded proton pore domain (V_0 and F_0). They share a number of homologous subunits that are components of the rotational catalytic mechanism common to both enzymes. The V_1 ATP-binding subunits A and B and the proteolipid V_0 subunits c, c′, and c″ have significant homology to corresponding F-ATPase subunits. The principal structural difference between the V- and F-type ATPase lies in the region that connects the peripheral subunits to the proton pore subunits. Where the F_1 and F_0 are connected by one central rotary stalk and only one peripheral stator stalk, the region connecting V_1 and V_0 is more elaborate. The unique regulatory mechanism of reversible disassembly of the V-ATPase (F-ATPases do not disassemble) implies a complex structure capable of responding rapidly to various cellular conditions.

A consensus overall structure of the V-ATPase has emerged recently using three-dimensional image reconstruction generated primarily by negative stain and cryo-electron microscopy (EM) from several different species including bovine [81], insects [20, 82], plants [83, 84], and fungi [21, 85]. While there are still some discrepancies, these studies have been able to resolve a detailed arrangement of the subunits in the stalk region that connect V_1 and V_0. The eukaryotic V-ATPases appear to have three peripheral stator stalks comprised of V_1 subunits E and G that form tightly linked, long helical heterodimers that can be cross-linked to multiple B subunit sites [23, 26, 86–88]. It has been suggested that while the F-ATPase requires only one stator stalk to resist the torque of rotational catalysis, the V-ATPase has three weaker stalks that would allow for rapid release of V_1 during disassembly [20, 21, 24]. The three EG stator stalks are connected to a horizontal collar-like structure that extends $\sim 250^\circ$ around the central stalk at the V_1–V_0 interface and is made up of V_1 subunits C and H and the large cytoplasmic domain of V_0 subunit a [21, 25, 89]. As previously noted, subunit C is unique among V-ATPase subunits in that it is released from both V_1 and V_0in reversible disassembly. Subunit H is also unique as it is the only subunit that is not required for assembly of V_1 and V_0. However, subunit H is required for activity of the holoenzyme, and, importantly, is also responsible for silencing hydrolase activity of the disassembled V_1 subcomplex [50]. High-resolution structures are available for subunits H [90] and C [91] and studies have shown that both C and H form critical and reversible associations with V_0 subunit a [19, 24, 92]. This is consistent with a potential role for these subunits in the mechanism of reversible dissociation and silencing of V-ATPase activity.

The crystal structure of subunit C shows three distinct domains, a globular "head", an elongated "neck", and another globular "foot" [91]. In the EM reconstruction images, subunit C appears to bind to two of the three peripheral EG stator stalks [20, 21, 24,]. This is supported by previous cross-linking data that place subunits E and G in close proximity to both the head and the foot domain of subunit C [25]. Recent work with *in vitro* purified subunits shows that only the head domain of subunit C binds with high affinity to form of a stable EGC complex [93]. The lower affinity of the foot domain of subunit C for the EG heterodimer precludes the formation of a complex *in vitro*. The *in vivo* interaction between the foot domain of subunit C and EG may require the binding of an additional subunit such as the *N*-terminal domain of subunit a for this complex to form [93]. This is consistent with data that show cross-linking between the foot domain of subunit C and the *N*-terminal region of the V_0 a subunit [25]. Subunit H is also located at the V_1–V_0 interface in close proximity to V_0 subunit a, V_1 subunit B and the third EG stator stalk, and has been shown to bind to subunit a *in vitro* [21, 94, 95].

V_1 subunits C and H, and V_0 subunit a serve to interconnect and stabilize the three EG peripheral stalks. The interactions between these subunits provide the "stator" function necessary for rotational catalysis, but it is precisely these interactions that must be disrupted to cause destabilization of the stators for V_1 dissociation to occur. Conformational changes in both subunits C and H are believed to play a role in regulatory dissociation of the complex. The solution structure of the isolated EGC subcomplex was determined using small angle X-ray scattering (SAXS) and shown to be in a more open arrangement when compared to the EGC structure in the holoenzyme [24]. Crystal structures of subunit C show two distinct conformations that are similar in the foot and neck domains, but there is a large movement of the head domain [91]. Only one of these conformers provides a good fit to the three-dimensional V-ATPase model derived from electron microscopic images [21]. Interestingly, the region where the head domain of subunit C interacts with high affinity to an EG stator stalk is not well resolved in the EM-derived model suggesting that this region is flexible [21]. Moreover, structural changes have been observed in subunit C in both yeast and *Arabidopsis* that may depend upon the ATP/ADP ratio [96]. This ratio is affected by glucose availability. Taken together these results suggest that a change in subunit C conformation in response to glucose could affect the stability of the V_1–V_0 holoenzyme. There is also evidence that the H subunit alters its interaction with V_1 and V_0 during reversible disassembly. The H subunit is responsible for silencing ATPase activity in the free V_1 domain either by a conformational change that inhibits V_1 ATPase activity or by a bridging of the peripheral and central stalk by subunit H in the free V_1 complex [94, 97]. The current evidence, then, supports a key role for subunits C and H in the reversible dissociation mechanism. It appears that the first step in disassembly is a destabilization of subunit C, perhaps by phosphorylation [69, 98] and/or a conformational change, in response to cellular signals. The release of subunit C would destabilize the remaining link between subunit H and V_0 stator subunit a, and allow subunit H to assume a conformation that would inhibit rotor movement resulting in the silencing of the V-ATPase.

In addition to its role in dissociation, there is evidence supporting a role for subunit C in assembly of V_1 and V_0. As noted above, the RAVE complex is required for subunit C to stably assemble with the V-ATPase in both *de novo* assembly and reassembly upon glucose readdition. RAVE is able to bind V_1 subunits E, G, and C. The RAVE complex could bring subunit C to the V_1 subcomplex in a conformation that is able to bind the EG heterodimer allowing for the stable assembly of V_1 and V_0 at the membrane.

1.4.2.5 Isoform-specific regulation of V-ATPase assembly

V-ATPase reversible disassembly is regulated by subunit a isoforms and is determined by cellular targeting and membrane environment. The information required for correct cellular targeting is located in the *N*-terminal cytosolic domains of the subunit a isoforms [45]. As previously noted, yeast V-ATPases contain one of two a isoforms; a Vph1p-containing complex, targeted to the vacuole, that disassembles upon glucose deprivation, and a Stv1p-containing complex, targeted to the Golgi, that does not [32]. A chimeric a subunit containing the *N*-terminal domain of Vph1p and the C-terminal domain of Stv1p is targeted to vacuoles and can disassemble [45]. In order to compare the different properties of V-ATPases containing different isoforms, Stv1p was expressed at higher than normal levels in a strain disrupted for both *VPH1* and *STV1*. V-ATPases containing Stv1p are targeted to the vacuole in this strain, but are present at much lower levels and have lower assembly with the peripheral V_1 domain than a strain with Vph1p-containing complexes [32]. While both a-isoform containing complexes have similar kinetic properties, the Stv1p complexes were not able to couple proton transport to ATP hydrolysis as efficiently as Vph1p complexes [32]. Interestingly, Stv1p-containing complexes present at the vacuole were able to disassemble upon glucose removal and Vph1p-containing complexes mislocalized to post Golgi compartments in a *vps21* mutant or prevacuolar compartment in a *vps27* mutant were not able to disassemble [99]. These results suggest that activity and dissociation of the V-ATPase is controlled by both a-subunit isoform present and by the cellular environment. The environmental signals contributing to dissociation have not been identified. However, one study suggests the luminal pH of the compartment where the V-ATPase resides may play a role [100].

Subunit a isoforms are also implicated in intracellular localization in mammalian cells. Four a-subunit isoforms (a1-a4) have been identified in humans and mice [85, 101–103]. They are localized in different organelles and plasma membranes of specific tissues. In mammalian cells, V-ATPase targeting by different a-isoforms is a dynamic process [104]. For example, during osteoclast differentiation of cultured osteoclast cells the a3-isoform relocates from lysosomes to the plasma membrane, whereas most of the a1 and a2 isoforms are in the Golgi [6]. In pancreatic beta cells the a3-isoform is primarily, but not entirely, targeted to secretory granules [105]. All four a-isoforms are present in mouse kidney proximal tubule cells, however, while a1, a3, and a4-isoforms are all targeted to the plasma membrane, the a2-isoform is targeted to early endosomes [106]. Therefore, while it is generally accepted that the information for V-ATPase trafficking resides in the a-subunit isoforms, the mechanism is still not clear. Isoforms of other V-ATPase subunits are also found in higher organisms. The functions of these isoforms are not known, but there is some evidence that complexes can form that contain combinations of different isoforms and that localization of these complexes may be directed by their unique isoform configuration [107, 108].

1.5 Summary

Maintenance of optimal acidic pH in intracellular compartments is critical for numerous cellular processes. In eukaryotes this pH is primarily regulated by the V-ATPase. Tight regulation of V-ATPase activity is critical in order to maintain pH within narrow limits. The complicated multisubunit structure of the complex and different subunit isoforms allow for multiple possibilities for regulatory mechanisms. Regulation of the degree of functional assembly is accomplished through reversible dissociation of V_1V_0 complexes in response to extracellular conditions, control of cellular localization and changes in coupling efficiency. Numerous regulatory proteins, including RAVE, aldolase and other glycolytic enzymes, and PKA, have been identified that control V-ATPase activity by modulating V_1–V_0 assembly in response to not only glucose, but also other extracellular conditions.

References

1. Beyenbach, K.W. and Wieczorek, H. (2006). The V-type H^+-ATPase: molecular structure and function, physiological roles and regulation, *J Exp Biol*, 209, pp.577–589.
2. Forgac, M. (2007). Vacuolar ATPases: rotary proton pumps in physiology and pathophysiology, *Nat Rev Mol Cell Biol*, 8, pp. 917–929.
3. Kane, P.M. (2006). The where, when, and how of organelle acidification by the yeast vacuolar H^+-ATPase, *Microbiol Mol Biol Rev*, 70, pp. 177–191.
4. Stevens, T.H. and Forgac, M. (1997). Structure, function and regulation of the vacuolar (H+)-ATPase, *Annu Rev Cell Dev Biol*, 13, pp. 779–808.
5. Martinez-Munoz, G.A. and Kane, P. (2008). Vacuolar and plasma membrane proton pumps collaborate to achieve cytosolic pH homeostasis in yeast, *J Biol Chem*, 283, pp. 20309–20319.
6. Toyomura, T., Murata, Y., Yamamoto, A., Oka, T., Sun-Wada, G.H., Wada, Y., and Futai, M. (2003). From lysosomes to the plasma membrane: localization of vacuolar-type H^+-ATPase with the a3 isoform during osteoclast differentiation, *J Biol Chem*, 278, pp. 22023–22030.
7. Wagner, C.A., Finberg, K.E., Breton, S., Marshansky, V., Brown, D., and Geibel, J.P. (2004). Renal vacuolar H^+-ATPase, *Physiol Rev*, 84, pp. 1263–1314.
8. Beltran, C. and Nelson, N. (1992). The membrane sector of vacuolar H(+)-ATPase by itself is impermeable to protons, *Acta Physiol Scand Suppl*, 607, pp. 41–47.
9. Graf, R., Harvey, W.R., and Wieczorek, H. (1996). Purification and properties of a cytosolic V1-ATPase, *J Biol Chem*, 271, pp. 20908–20913.
10. Gruber, G. and Marshansky, V. (2008). New insights into structure-function relationships between archeal ATP synthase (A1A0) and vacuolar type ATPase (V1V0), *Bioessays*, 30, pp. 1096–1109.
11. Nelson, N. and Taiz, L. (1989). The evolution of H^+-ATPases, *Trends Biochem Sci*, 14, pp. 113–116.

12. Hirata, T., Iwamoto-Kihara, A., Sun-Wada, G.H., Okajima, T., Wada, Y., and Futai, M. (2003). Subunit rotation of vacuolar-type proton pumping ATPase: relative rotation of the G and C subunits, *J Biol Chem*, 278, pp. 23714–23719.

13. Imamura, H., Nakano, M., Noji, H., Muneyuki, E., Ohkuma, S., Yoshida, M., and Yokoyama, K. (2003). Evidence for rotation of V1-ATPase, *Proc Natl Acad Sci USA*, 100, pp. 2312–2315.

14. Nelson, N. (2003). A journey from mammals to yeast with vacuolar H^+-ATPase (V-ATPase), *J Bioenerg Biomembr*, 35, pp. 281–289.

15. Yoshida, M., Muneyuki, E., and Hisabori, T. (2001). ATP synthase: a marvellous rotary engine of the cell, *Nat Rev Mol Cell Biol*, 2, pp. 669–677.

16. Gruber, G., Radermacher, M., Ruiz, T., Godovac-Zimmermann, J., Canas, B., Kleine-Kohlbrecher, D., Huss, M., Harvey, W.R., and Wieczorek, H. (2000). Three-dimensional structure and subunit topology of the V(1) ATPase from Manduca sexta midgut, *Biochemistry*, 39, pp. 8609–8616.

17. Kawasaki-Nishi, S., Nishi, T., and Forgac, M. (2003). Proton translocation driven by ATP hydrolysis in V-ATPases, *FEBS Lett*, 545, pp. 76–85.

18. Wilkens, S., Vasilyeva, E., and Forgac, M. (1999). Structure of the vacuolar ATPase by electron microscopy, *J Biol Chem*, 274, pp. 31804–31810.

19. Hildenbrand, Z.L., Molugu, S.K., Stock, D., and Bernal, R.A. (2010). The C-H peripheral stalk base: a novel component in V1-ATPase assembly, *PLoS One*, 5, pp. e12588.

20. Muench, S.P., Huss, M., Song, C.F., Phillips, C., Wieczorek, H., Trinick, J., and Harrison, M.A. (2009). Cryo-electron microscopy of the vacuolar ATPase motor reveals its mechanical and regulatory complexity, *J Mol Biol*, 386, pp. 989–999.

21. Zhang, Z., Zheng, Y., Mazon, H., Milgrom, E., Kitagawa, N., Kish-Trier, E., Heck, A.J., Kane, P.M., and Wilkens, S. (2008). Structure of the yeast vacuolar ATPase, *J Biol Chem*, 283, pp. 35983–35995.

22. Smith, A.N., Francis, R.W., Sorrell, S.L., and Karet, F.E. (2008). The d subunit plays a central role in human vacuolar H(+)-ATPases, *J Bioenerg Biomembr*, 40, pp. 371–380.

23. Arata, Y., Baleja, J.D., and Forgac, M. (2002). Localization of subunits D, E, and G in the yeast V-ATPase complex using cysteine-mediated cross-linking to subunit B, *Biochemistry*, 41, pp. 11301–11307.

24. Diepholz, M., Venzke, D., Prinz, S., Batisse, C., Florchinger, B., Rossle, M., Svergun, D.I., Bottcher, B., and Fethiere, J. (2008). A different conforma-

tion for EGC stator subcomplex in solution and in the assembled yeast V-ATPase: possible implications for regulatory disassembly, *Structure*, 16, pp. 1789–1798.

25. Inoue, T. and Forgac, M. (2005). Cysteine-mediated cross-linking indicates that subunit C of the V-ATPase is in close proximity to subunits E and G of the V1 domain and subunit a of the V0 domain, *J Biol Chem*, 280, pp. 27896–27903.
26. Kitagawa, N., Mazon, H., Heck, A.J., and Wilkens, S. (2008). Stoichiometry of the peripheral stalk subunits E and G of yeast V1-ATPase determined by mass spectrometry, *J Biol Chem*, 283, pp. 3329–3337.
27. Hirata, R., Graham, L.A., Takatsuki, A., Stevens, T.H., and Anraku, Y. (1997). VMA11 and VMA16 encode second and third proteolipid subunits of the Saccharomyces cerevisiae vacuolar membrane H^+-ATPase, *J Biol Chem*, 272, pp. 4795–4803.
28. Fillingame, R.H., Angevine, C.M., and Dmitriev, O.Y. (2002). Coupling proton movements to c-ring rotation in F(1)F(o) ATP synthase: aqueous access channels and helix rotations at the a-c interface, *Biochim Biophys Acta*, 1555, pp. 29–36.
29. Sambade, M. and Kane, P.M. (2004). The yeast vacuolar proton-translocating ATPase contains a subunit homologous to the Manduca sexta and bovine e subunits that is essential for function, *J Biol Chem*, 279, pp. 17361–17365.
30. Toei, M., Toei, S., and Forgac, M. (2011). Definition of membrane topology and identification of residues important for transport in subunit a of the vacuolar ATPase, *J Biol Chem*, 286, pp. 35176–35186.
31. Qi, J. and Forgac, M. (2008). Function and subunit interactions of the *N*-terminal domain of subunit a (Vph1p) of the yeast V-ATPase, *J Biol Chem*, 283, pp. 19274–19282.
32. Kawasaki-Nishi, S., Nishi, T., and Forgac, M. (2001). Yeast V-ATPase complexes containing different isoforms of the 100-kDa a-subunit differ in coupling efficiency and in vivo dissociation, *J Biol Chem*, 276, pp. 17941–17948.
33. Manolson, M.F., Wu, B., Proteau, D., Taillon, B.E., Roberts, B.T., Hoyt, M.A., and Jones, E.W. (1994). STV1 gene encodes functional homologue of 95-kDa yeast vacuolar H(+)-ATPase subunit Vph1p, *J Biol Chem*, 269, pp. 14064–14074.
34. Kane, P.M., Kuehn, M.C., Howald-Stevenson, I., and Stevens, T.H. (1992). Assembly and targeting of peripheral and integral membrane subunits of the yeast vacuolar H(+)-ATPase, *J Biol Chem*, 267, pp. 447–454.

35. Tomashek, J.J., Graham, L.A., Hutchins, M.U., Stevens, T.H., and Klionsky, D.J. (1997). V1-situated stalk subunits of the yeast vacuolar proton-translocating ATPase, *J Biol Chem*, 272, pp. 26787–26793.

36. Doherty, R.D. and Kane, P.M. (1993). Partial assembly of the yeast vacuolar H(+)-ATPase in mutants lacking one subunit of the enzyme, *J Biol Chem*, 268, pp. 16845–16851.

37. Tomashek, J.J., Sonnenburg, J.L., Artimovich, J.M., and Klionsky, D.J. (1996). Resolution of subunit interactions and cytoplasmic subcomplexes of the yeast vacuolar proton-translocating ATPase, *J Biol Chem*, 271, pp. 10397–10404.

38. Kane, P.M., Tarsio, M., and Liu, J. (1999). Early steps in assembly of the yeast vacuolar H^+-ATPase, *J Biol Chem*, 274, pp. 17275–17283.

39. Davis-Kaplan, S.R., Compton, M.A., Flannery, A.R., Ward, D.M., Kaplan, J., Stevens, T.H., and Graham, L.A. (2006). PKR1 encodes an assembly factor for the yeast V-type ATPase, *J Biol Chem*, 281, pp. 32025–32035.

40. Graham, L.A., Flannery, A.R., and Stevens, T.H. (2003). Structure and assembly of the yeast V-ATPase, *J Bioenerg Biomembr*, 35, pp. 301–312.

41. Malkus, P., Graham, L.A., Stevens, T.H., and Schekman, R. (2004). Role of Vma21p in assembly and transport of the yeast vacuolar ATPase, *Mol Biol Cell*, 15, pp. 5075–5091.

42. Ryan, M., Graham, L.A., and Stevens, T.H. (2008). Voa1p functions in V-ATPase assembly in the yeast endoplasmic reticulum, *Mol Biol Cell*, 19, pp. 5131–5142.

43. Graham, L.A., Hill, K.J., and Stevens, T.H. (1998). Assembly of the yeast vacuolar H^+-ATPase occurs in the endoplasmic reticulum and requires a Vma12p/Vma22p assembly complex, *J Cell Biol*, 142, pp. 39–49.

44. Paroutis, P., Touret, N., and Grinstein, S. (2004). The pH of the secretory pathway: measurement, determinants, and regulation, *Physiology (Bethesda)*, 19, pp. 207–215.

45. Kawasaki-Nishi, S., Bowers, K., Nishi, T., Forgac, M., and Stevens, T.H. (2001). The amino-terminal domain of the vacuolar proton-translocating ATPase a subunit controls targeting and in vivo dissociation, and the carboxyl-terminal domain affects coupling of proton transport and ATP hydrolysis, *J Biol Chem*, 276, pp. 47411–47420.

46. Seol, J.H., Shevchenko, A., Shevchenko, A., and Deshaies, R.J. (2001). Skp1 forms multiple protein complexes, including RAVE, a regulator of V-ATPase assembly, *Nat Cell Biol*, 3, pp. 384–391.

47. Smardon, A.M., Tarsio, M., and Kane, P.M. (2002). The RAVE complex is essential for stable assembly of the yeast V-ATPase, *J Biol Chem*, 277, pp. 13831–13839.

48. Smardon, A.M. and Kane, P.M. (2007). RAVE is essential for the efficient assembly of the C subunit with the vacuolar H(+)-ATPase, *J Biol Chem*, 282, pp. 26185–26194.

49. Kane, P.M. (1995). Disassembly and reassembly of the yeast vacuolar H(+)-ATPase in vivo, *J Biol Chem*, 270, pp. 17025–17032.

50. Parra, K.J., Keenan, K.L., and Kane, P.M. (2000). The H subunit (Vma13p) of the yeast V-ATPase inhibits the ATPase activity of cytosolic V1 complexes, *J Biol Chem*, 275, pp. 21761–21767.

51. Sumner, J.P., Dow, J.A., Earley, F.G., Klein, U., Jager, D., and Wieczorek, H. (1995). Regulation of plasma membrane V-ATPase activity by dissociation of peripheral subunits, *J Biol Chem*, 270, pp. 5649–5653.

52. Li, S.C., Diakov, T.T., Rizzo, J.M., and Kane, P.M. (2012). The V-ATPase works in parallel with the HOG pathway to adapt yeast cells to osmotic stress, *Eukaryot Cell*, 11, pp. 282–291.

53. Diakov, T.T. and Kane, P.M. (2010). Regulation of vacuolar proton-translocating ATPase activity and assembly by extracellular pH, *J Biol Chem*, 285, pp. 23771–23778.

54. Trombetta, E.S., Ebersold, M., Garrett, W., Pypaert, M., and Mellman, I. (2003). Activation of lysosomal function during dendritic cell maturation, *Science*, 299, pp. 1400–1403.

55. Chintagari, N.R., Mishra, A., Su, L., Wang, Y., Ayalew, S., Hartson, S.D., and Liu, L. (2010). Vacuolar ATPase regulates surfactant secretion in rat alveolar type II cells by modulating lamellar body calcium, *PLoS One*, 5, pp. e9228.

56. Nakamura, S. (2004). Glucose activates H(+)-ATPase in kidney epithelial cells, *Am J Physiol Cell Physiol*, 287, pp. C97–105.

57. Sautin, Y.Y., Lu, M., Gaugler, A., Zhang, L., and Gluck, S.L. (2005). Phosphatidylinositol 3-kinase-mediated effects of glucose on vacuolar H^+-ATPase assembly, translocation, and acidification of intracellular compartments in renal epithelial cells, *Mol Cell Biol*, 25, pp. 575–589.

58. Parra, K.J. and Kane, P.M. (1998). Reversible association between the V1 and V0 domains of yeast vacuolar H^+-ATPase is an unconventional glucose-induced effect, *Mol Cell Biol*, 18, pp. 7064–7074.

59. Kane, P.M. and Smardon, A.M. (2003). Assembly and regulation of the yeast vacuolar H^+-ATPase, *J Bioenerg Biomembr*, 35, pp. 313–321.

60. Sethi, N., Yan, Y., Quek, D., Schupbach, T., and Kang, Y. (2010). Rabconnectin-3 is a functional regulator of mammalian Notch signaling, *J Biol Chem*, 285, pp. 34757–34764.

61. Yan, Y., Denef, N., and Schupbach, T. (2009). The vacuolar proton pump, V-ATPase, is required for notch signaling and endosomal trafficking in Drosophila, *Dev Cell*, 17, pp. 387–402.
62. Lu, M., Holliday, L.S., Zhang, L., Dunn, W.A., Jr., and Gluck, S.L. (2001). Interaction between aldolase and vacuolar H$^+$-ATPase: evidence for direct coupling of glycolysis to the ATP-hydrolyzing proton pump, *J Biol Chem*, 276, pp. 30407–30413.
63. Lu, M., Sautin, Y.Y., Holliday, L.S., and Gluck, S.L. (2004). The glycolytic enzyme aldolase mediates assembly, expression, and activity of vacuolar H$^+$-ATPase, *J Biol Chem*, 279, pp. 8732–8739.
64. Lu, M., Ammar, D., Ives, H., Albrecht, F., and Gluck, S.L. (2007). Physical interaction between aldolase and vacuolar H$^+$-ATPase is essential for the assembly and activity of the proton pump, *J Biol Chem*, 282, pp. 24495–24503.
65. Su, Y., Zhou, A., Al-Lamki, R.S., and Karet, F.E. (2003). The a-subunit of the V-type H$^+$-ATPase interacts with phosphofructokinase-1 in humans, *J Biol Chem*, 278, pp. 20013–20018.
66. Su, Y., Blake-Palmer, K.G., Sorrell, S., Javid, B., Bowers, K., Zhou, A., Chang, S.H., Qamar, S., and Karet, F.E. (2008). Human H$^+$-ATPase a4 subunit mutations causing renal tubular acidosis reveal a role for interaction with phosphofructokinase-1, *Am J Physiol Renal Physiol*, 295, pp. F950–958.
67. Zimmermann, B., Dames, P., Walz, B., and Baumann, O. (2003). Distribution and serotonin-induced activation of vacuolar-type H$^+$-ATPase in the salivary glands of the blowfly Calliphora vicina, *J Exp Biol*, 206, pp. 1867–1876.
68. Rein, J., Voss, M., Blenau, W., Walz, B., and Baumann, O. (2008). Hormone-induced assembly and activation of V-ATPase in blowfly salivary glands is mediated by protein kinase A, *Am J Physiol Cell Physiol*, 294, pp. C56–65.
69. Voss, M., Vitavska, O., Walz, B., Wieczorek, H., and Baumann, O. (2007). Stimulus-induced phosphorylation of vacuolar H(+)-ATPase by protein kinase A, *J Biol Chem*, 282, pp. 33735–33742.
70. Voss, M., Schmidt, R., Walz, B., and Baumann, O. (2009). Stimulus-induced translocation of the protein kinase A catalytic subunit to the apical membrane in blowfly salivary glands, *Cell Tissue Res*, 335, pp. 657–662.
71. Bond, S. and Forgac, M. (2008). The Ras/cAMP/protein kinase A pathway regulates glucose-dependent assembly of the vacuolar (H+)-ATPase in yeast, *J Biol Chem*, 283, pp. 36513–36521.

72. Dechant, R., Binda, M., Lee, S.S., Pelet, S., Winderickx, J., and Peter, M. (2010). Cytosolic pH is a second messenger for glucose and regulates the PKA pathway through V-ATPase, *EMBO J*, 29, pp. 2515–2526.

73. Alzamora, R., Thali, R.F., Gong, F., Smolak, C., Li, H., Baty, C.J., et al. (2010). PKA regulates vacuolar H^+-ATPase localization and activity via direct phosphorylation of the a subunit in kidney cells, *J Biol Chem*, 285, pp. 24676–24685.

74. Gong, F., Alzamora, R., Smolak, C., Li, H., Naveed, S., Neumann, D., Hallows, K.R., and Pastor-Soler, N.M. (2010). Vacuolar H^+-ATPase apical accumulation in kidney intercalated cells is regulated by PKA and AMP-activated protein kinase, *Am J Physiol Renal Physiol*, 298, pp. F1162–F1169.

75. Pastor-Soler, N.M., Hallows, K.R., Smolak, C., Gong, F., Brown, D., and Breton, S. (2008). Alkaline pH- and cAMP-induced V-ATPase membrane accumulation is mediated by protein kinase A in epididymal clear cells, *Am J Physiol Cell Physiol*, 294, pp. C488–494.

76. Pastor-Soler, N., Beaulieu, V., Litvin, T.N., Da Silva, N., Chen, Y., Brown, D., Buck, J., Levin, L.R., and Breton, S. (2003). Bicarbonate-regulated adenylyl cyclase (sAC) is a sensor that regulates pH-dependent V-ATPase recycling, *J Biol Chem*, 278, pp. 49523–49529.

77. Cingolani, G. and Duncan, T.M. (2011). Structure of the ATP synthase catalytic complex (F(1)) from Escherichia coli in an autoinhibited conformation, *Nat Struct Mol Biol*, 18, pp. 701–707.

78. Dickson, V.K., Silvester, J.A., Fearnley, I.M., Leslie, A.G., and Walker, J.E. (2006). On the structure of the stator of the mitochondrial ATP synthase, *EMBO J*, 25, pp. 2911–2918.

79. Rubinstein, J.L., Walker, J.E., and Henderson, R. (2003). Structure of the mitochondrial ATP synthase by electron cryomicroscopy, *EMBO J*, 22, pp. 6182–6192.

80. Wilkens, S. and Capaldi, R.A. (1998). Electron microscopic evidence of two stalks linking the F1 and F0 parts of the Escherichia coli ATP synthase, *Biochim Biophys Acta*, 1365, pp. 93–97.

81. Wilkens, S., Inoue, T., and Forgac, M. (2004). Three-dimensional structure of the vacuolar ATPase. Localization of subunit H by difference imaging and chemical cross-linking, *J Biol Chem*, 279, pp. 41942–41949.

82. Rizzo, V.F., Coskun, U., Radermacher, M., Ruiz, T., Armbruster, A., and Gruber, G. (2003). Resolution of the V1 ATPase from Manduca sexta into subcomplexes and visualization of an ATPase-active A3B3EG complex by electron microscopy, *J Biol Chem*, 278, pp. 270–275.

83. Domgall, I., Venzke, D., Luttge, U., Ratajczak, R., and Bottcher, B. (2002). Three-dimensional map of a plant V-ATPase based on electron microscopy, *J Biol Chem*, 277, pp. 13115–13121.
84. Li, Z. and Zhang, X. (2004). Electron-microscopic structure of the V-ATPase from mung bean, *Planta*, 219, pp. 948–954.
85. Toyomura, T., Oka, T., Yamaguchi, C., Wada, Y., and Futai, M. (2000). Three subunit a isoforms of mouse vacuolar H(+)-ATPase. Preferential expression of the a3 isoform during osteoclast differentiation, *J Biol Chem*, 275, pp. 8760–8765.
86. Fethiere, J., Venzke, D., Diepholz, M., Seybert, A., Geerlof, A., Gentzel, M., Wilm, M., and Bottcher, B. (2004). Building the stator of the yeast vacuolar-ATPase: specific interaction between subunits E and G, *J Biol Chem*, 279, pp. 40670–40676.
87. Kish-Trier, E., Briere, L.K., Dunn, S.D., and Wilkens, S. (2008). The stator complex of the A1A0-ATP synthase—structural characterization of the E and H subunits, *J Mol Biol*, 375, pp. 673–685.
88. Ohira, M., Smardon, A.M., Charsky, C.M., Liu, J., Tarsio, M., and Kane, P.M. (2006). The E and G subunits of the yeast V-ATPase interact tightly and are both present at more than one copy per V1 complex, *J Biol Chem*, 281, pp. 22752–22760.
89. Muench, S.P., Trinick, J. and Harrison, M.A. (2011). Structural divergence of the rotary ATPases, *Q Rev Biophys*, 44, pp. 311–356.
90. Sagermann, M., Stevens, T.H., and Matthews, B.W. (2001). Crystal structure of the regulatory subunit H of the V-type ATPase of Saccharomyces cerevisiae, *Proc Natl Acad Sci USA*, 98, pp. 7134–7139.
91. Drory, O., Frolow, F., and Nelson, N. (2004). Crystal structure of yeast V-ATPase subunit C reveals its stator function, *EMBO Rep*, 5, pp. 1148–1152.
92. Bernal, R.A. and Stock, D. (2004). Three-dimensional structure of the intact Thermus thermophilus H^+-ATPase/synthase by electron microscopy, *Structure*, 12, pp. 1789–1798.
93. Oot, R.A. and Wilkens, S. (2010). Domain characterization and interaction of the yeast vacuolar ATPase subunit C with the peripheral stator stalk subunits E and G, *J Biol Chem*, 285, pp. 24654–24664.
94. Diab, H., Ohira, M., Liu, M., Cobb, E., and Kane, P.M. (2009). Subunit interactions and requirements for inhibition of the yeast V1-ATPase, *J Biol Chem*, 284, pp. 13316–13325.

95. Venzke, D., Domgall, I., Kocher, T., Fethiere, J., Fischer, S., and Bottcher, B. (2005). Elucidation of the stator organization in the V-ATPase of Neurospora crassa, *J Mol Biol*, 349, pp. 659–669.

96. Armbruster, A., Hohn, C., Hermesdorf, A., Schumacher, K., Borsch, M., and Gruber, G. (2005). Evidence for major structural changes in subunit C of the vacuolar ATPase due to nucleotide binding, *FEBS Lett*, 579, pp. 1961–1967.

97. Jefferies, K.C. and Forgac, M. (2008). Subunit H of the vacuolar (H+) ATPase inhibits ATP hydrolysis by the free V1 domain by interaction with the rotary subunit F, *J Biol Chem*, 283, pp. 4512–4519.

98. Hong-Hermesdorf, A., Brux, A., Gruber, A., Gruber, G., and Schumacher, K. (2006). A WNK kinase binds and phosphorylates V-ATPase subunit C, *FEBS Lett*, 580, pp. 932–939.

99. Qi, J. and Forgac, M. (2007). Cellular environment is important in controlling V-ATPase dissociation and its dependence on activity, *J Biol Chem*, 282, pp. 24743–24751.

100. Shao, E. and Forgac, M. (2004). Involvement of the nonhomologous region of subunit A of the yeast V-ATPase in coupling and in vivo dissociation, *J Biol Chem*, 279, pp. 48663–48670.

101. Nishi, T. and Forgac, M. (2000). Molecular cloning and expression of three isoforms of the 100-kDa a subunit of the mouse vacuolar proton-translocating ATPase, *J Biol Chem*, 275, pp. 6824–6830.

102. Oka, T., Murata, Y., Namba, M., Yoshimizu, T., Toyomura, T., Yamamoto, A., Sun-Wada, G.H., Hamasaki, N., Wada, Y., and Futai, M. (2001). a4, a unique kidney-specific isoform of mouse vacuolar H^+-ATPase subunit a, *J Biol Chem*, 276, pp. 40050–40054.

103. Smith, A.N., Finberg, K.E., Wagner, C.A., Lifton, R.P., Devonald, M.A., Su, Y., and Karet, F.E. (2001). Molecular cloning and characterization of Atp6n1b: a novel fourth murine vacuolar H^+-ATPase a-subunit gene, *J Biol Chem*, 276, pp. 42382–42388.

104. Marshansky, V. and Futai, M. (2008). The V-type H^+-ATPase in vesicular trafficking: targeting, regulation and function, *Curr Opin Cell Biol*, 20, pp. 415–426.

105. Sun-Wada, G.H., Toyomura, T., Murata, Y., Yamamoto, A., Futai, M., and Wada, Y. (2006). The a3 isoform of V-ATPase regulates insulin secretion from pancreatic beta-cells, *J Cell Sci*, 119, pp. 4531–4540.

106. Hurtado-Lorenzo, A., Skinner, M., El Annan, J., Futai, M., Sun-Wada, G.H., Bourgoin, S., et al. (2006). V-ATPase interacts with ARNO and Arf6 in

early endosomes and regulates the protein degradative pathway, *Nat Cell Biol*, 8, pp. 124–136.

107. Norgett, E.E., Borthwick, K.J., Al-Lamki, R.S., Su, Y., Smith, A.N., and Karet, F.E. (2007). V1 and V0 domains of the human H^+-ATPase are linked by an interaction between the G and a subunits, *J Biol Chem*, 282, pp. 14421–14427.

108. Sun-Wada, G.H., Yoshimizu, T., Imai-Senga, Y., Wada, Y., and Futai, M. (2003). Diversity of mouse proton-translocating ATPase: presence of multiple isoforms of the C, d and G subunits, *Gene*, 302, pp. 147–153.

Chapter 2

Structure of Prokaryotic V-Type ATPase/Synthase

Ken Yokoyama and Jun-ichi Kishikawa
Department of Molecular Biosciences, Kyoto Sangyo University, Kamigamo-Motoyama, Kita-ku, Kyoto 603-8555, Japan
yokoken@cc.kyoto-su.ac.jp

The eukaryotic V-ATPases function as proton pumps that are involved in the acidification of cellular compartments such as the Golgi apparatus or lysosomes. In addition, some prokaryotes, such as the *Thermus thermophilus* or *Enterococcus hirae*, contain a member of the family (or homologue) of the V-ATPases in their membranes. The prokaryotic V-type ATPases/synthases (prokaryotic V-ATPases) have simpler subunit compositions than eukaryotic V-ATPases and, thus, are useful subjects for studying chemical, physical, and structural properties of V-ATPase. In this review, we focus on the results of recent studies on the structure of the V-ATPase from the eubacteria, both *T. thermophilus* and *E. hirae*.

Handbook of H^+-ATPases
Edited by Suguru Nakamura

ISBN 978-981-4411-91-2 (Hardcover), 978-981-4411-92-9 (eBook)
www.panstanford.com

2.1 Introduction: Model for Eukaryotic V-ATPases

The vacuole-type ATPases (V-ATPases) are commonly found in many organisms involved in a variety of physiological processes [1]. V-ATPases in eukaryotic cells (eukaryotic V-ATPases) pump protons across the membrane consuming ATP. They have complex architecture and consist of at least 14 kinds of different subunits. The homologues of eukaryotic V-ATPases are also found in archaea and some eubacteria (prokaryotic V-ATPases) [2–4]. The prokaryotic V-ATPases are sometimes called as A-type ATPase [5]. The prokaryotic V-ATPases have less complex architecture and consist of nine kinds of subunits. They function either as ATP synthases or sodium pumps. The V-type enzymes are composed of two functional domains: the hydrophilic V_1 domain and a membrane-embedded ion-translocating domain called V_o. These distinct domains are connected by the central and peripheral stalks (Fig. 2.1).

It is difficult to obtain large amounts of pure V-ATPase from vacuolar membranes. This has limited detailed investigations of the structure/function relationships of the eukaryotic enzymes. In contrast, stable V-ATPases can be obtained in large amounts from

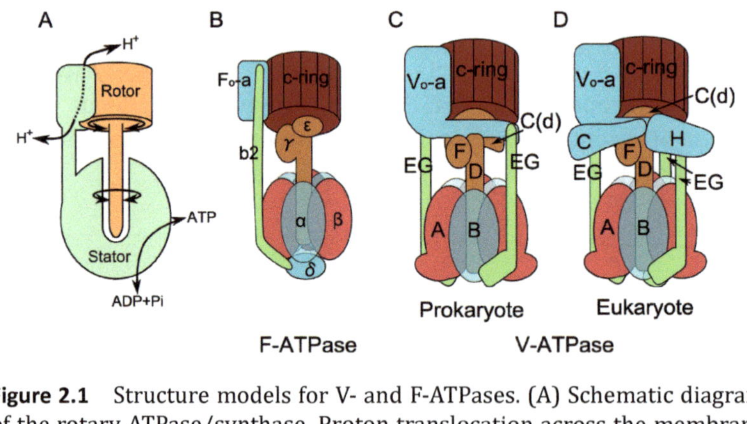

Figure 2.1 Structure models for V- and F-ATPases. (A) Schematic diagram of the rotary ATPase/synthase. Proton translocation across the membrane is coupled with ATP synthesis/hydrolysis by rotation of central rotor complex (yellow) against the surrounding stator apparatus (green). (B) Prokaryotic F-ATPase. (C) Prokaryotic V-ATPase. Eukaryotic V-ATPase contains additional stator composed of H and C subunit (D).

some prokaryotes [2, 6]. Thus, prokaryotic V-ATPases are useful subjects for studying chemical, physical, and structural properties of V-ATPase.

In this review, we summarize structural information of the prokaryotic V-ATPase, especially from both the eubacteria *T. thermophilus* and *E. hirae*.

2.2 Evolutionary Relationship between V-Type and F-Type ATPases

Rotary ATPases/synthases are energy converted enzymes by rotation of central rotor complex against surrounding stator apparatus (Fig. 2.1A) [1, 7]. There are two types of rotary enzymes: F- and V-ATPases in nature. F-ATPases are distributed in mitochondria and thylakoid membranes in plants and plasma membranes in bacteria [8]. F-ATPase genes in mitochondria are thought to be transferred from eubacterial kingdom [8]. That is, bacterial F-ATPase genes are evolutionary origin of mitochondrial enzymes in eukaryote. The origin of chloroplast F-ATPase is most likely cyanobacterium enzyme. On the other hand, V-ATPase in eukaryotic cells function as a proton pump [1]. They also utilize similar rotary catalytic mechanism to that of F-ATPase. However, V-ATPases constitute of different kinds of subunits and overall feature of V-ATPases are distinct from that of F-ATPases. Evolutionary-related rotary ATPases have been found in plasma membranes of archaea and some eubacteria [2, 9]. These are sometimes referred as archaeal-type ATP synthase (A-ATPase) [5]. However, the prokaryotic V/A-ATPase and eukaryotic V-ATPase are very similar, but are distinct from the F-ATPase family [4]. Each subunit of *T. thermophilus* V-ATPase shows a significant sequence similar to its eukaryotic counterpart. For example, amino acid homology between *T. thermophilus* V_1 subunits and yeast subunits are as follows: the A-subunit (51% identity/69% similarity), the B-subunit (54%/72%), the F-subunit (28%/48%), and the F-subunit (22%/38%) [10, 11]. In contrast, the subunits of *T. thermophilus* show no apparent sequence similarity to subunits of F-ATPases except the A-, B-, G-, and L-subunits, which are similar to β-, α-, b-, and c-subunits of F-ATPases, respectively (Table 2.1).

Table 2.1 Homologous and non-homologous subunits of the eukaryotic and prokaryotic V-ATPase, and prokaryotic F-ATPase

		V-ATPase		F-ATPase		
Yeast (MW)	Identity	*T. thermophilus* (MW)	Identity	*E. coli* (MW)	Stator or rotor	Function
A(70)	50%	A (64)	30%	α(50)	Stator	Catalysis
B (58)	55%	B (54)	25%	β(55)	Stator	Catalysis
D (28)	28%	D (25)	-	γ(32)	Rotor	Rotor shaft
F (13)	21%	F (12)	-	ε(15)	Rotor	Rotor shaft
E (26)	16%	E (21)	-	δ(19)	Stator	Peripheral stator
G (13)	22%	G (13)	33%	*b*(17)	Stator	Peripheral stator
a (96)	16%	I (72)	-	*a* (30)	Stator	Proton channel
c (16)	37%	L (8)	24%	*c* (8)	Rotor	Proton channel
d (40)	17%	C (36)	-		Rotor	Socket
c′ (17)					Rotor	Proton channel
c″(23)					Rotor	Proton channel
H (54)					Stator	Accessory stator
C (44)					Stator	Accessory stator

Conserve subunits are shown in the same row of the table.
Note: A dash indicates the absence of sequence similarity.

The phylogenic tree analysis for catalytic subunits of both F- and V-ATPases clearly indicated that prokaryotic V/A-ATPases are much closer to eukaryotic V-ATPases than to F-ATPases [8]. These clearly indicate that prokaryotic enzymes are evolutionary origin of eukaryotic V-ATPases. Thus, it is appropriate to classify prokaryotic V/A ATPases into a subclass of V-ATPase family. In this review, we call prokaryotic V-ATPase from *T. thermophilus* or *E. hirae* solely as V-ATPase.

2.3 Structure of Prokaryotic V-ATPases

2.3.1 *Subunit Structure of* T. Thermophilus *V-ATPase*

The *T. thermophilus* V-ATPase contains nine genes in the order of atpG-I-L-E-C-F-A-B-D, which encoded proteins with molecular masses of 13, 72, 10, 20, 35, 12, 64, 53, and 25 kDa, respectively [12]. The overall feature of *T. thermophilus* V-ATPase and yeast V-ATPase are depicted in Fig. 2.1. The product of atpL, the proteolipid subunit, contains two membrane-spanning domains rather than four [11]. The common operon encoding the *T. thermophilus* V-ATPase and other prokaryotic V-ATPases is well conserved [13]. Each subunit of *T. thermophilus* V-ATPase has significant sequence homology with its eukaryotic counterpart (Table 2.1). However, the molecular masses of some subunits are less than those of the corresponding subunits in the eukaryotic enzymes [10]. For instance, the 72-kDa I-subunit has an overall sequence similar to that of the eukaryotic 100-kDa a-subunit [14]. Although the molecular mass of subunit L is ~50% of eukaryotic c-subunit (16 k-Da proteolipid subunit), subunit L has an obvious sequence homology with the V_o-c subunit. The hydrophilic V_1 domain of *T. thermophilus* is made up of four kinds of subunits with a stoichiometry of $A_3B_3D_1F_1$. The peripheral stator is composed of G- and E-subunits, which are also hydrophilic. They are identified as V_o subunits together with hydrophobic I- and L-subunits, and hydrophilic C (d) subunit which sits on the L ring.

2.3.2 *Cryo EM Map of Intact* T. Thermophilus *V-ATPase*

Crystal structure of any intact rotary ATPases/synthases, F- and V-ATPases, has not been determined yet. Instead, EM images of *T. thermophilus* V-ATPases have been reported from several groups. Lau and Rubinstein recently presented a 9.7 Å resolution map of the intact *T. thermophilus* V-ATPase by electron cryomicroscopy (Fig. 2.2A) [15]. This resolution was sufficient to assign α-helices of each subunit in the complex. Two peripheral stators composed of E- and G-subunits are well observed. The *N*-terminal domain of E-subunit is attached onto B-subunit N terminal region. On the contrary, C terminal region of both E- and G-subunits is likely to interact with hydrophilic domain of I-subunit, which is a membrane-embedded protein. This indicated that one I-subunit contains two binding domains for EG stator. The EM map provided the insight of structure of V_o domain. The 12 distinct, well-resolved densities of membrane-embedded rotary ring (L_{12} ring) were observed in this cryo-EM map, even from an averaged map of single molecular intact V-ATPase (Fig. 2.2B). The detail structure of rotor rings will be described later. Density is visible within the center of the L_{12}

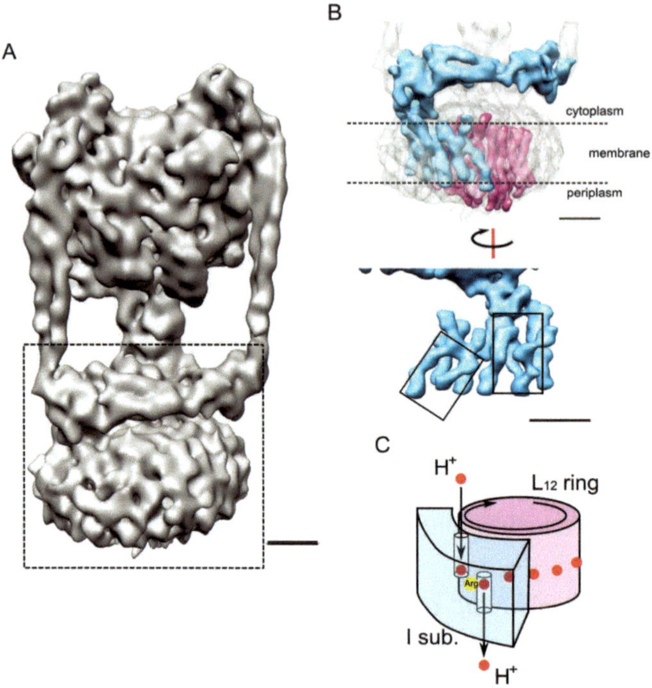

Figure 2.2 EM structure of the intact V-ATPase of *T. thermophilus.* (A) surface view of the three-dimensional map. (B) Map segments of the L ring (magenta) and I subunit (cyan). The proposed membrane-embedded domain is highlighted by two dot lines. (C) Proposal mechanism of rotation of L_{12} ring by proton transport through V_o. Schematic representation of the C-terminal domain of subunit I (sky blue) and the L_{12} ring (magenta) is shown. Protons (shown as red dots) enter the V_o complex through the cytoplasmic hemichannel on the subunit I and bind to the essential glutamic acid residue (Glu-63) on the L subunit that is aligned with the hemichannel. As the L ring rotates, the protonated form of the glutamic acid travels through the lipid bilayer and eventually returns to the L-subunit and aligns with the periplasmic hemichannel. Residue Arg-563 (shown in yellow) on I subunit promotes the deprotonation of the glutamate, and the proton is released through the hemichannel into the lumen. For ATP synthesis, a reverse reaction likely causes counterclockwise rotation of the L_{12} ring. Scale bar represents 25 Å.

ring and likely corresponds to a lipid plug. The 12-fold symmetry of the L_{12} ring is not broken significantly, even where the L_{12} ring contacts with stator I-subunit. These symmetric structures of L_{12} ring in the intact rotary ATPases/synthases are inconsistent with the models of proton translocation that require the significant swelling of helices of c-subunit in the ring [16] and previous model of hexagonal structure of L_{12} ring in the complex [17]. The EM map also indicates eight transmembrane densities that can be corresponded to α-helices in the C terminal hydrophobic domain of I-subunit. The number of transmembrane helices is consistent with an experimentally tested topology map of the yeast V-ATPase for V_o-a subunit [14]. The I-subunit sequence aligns well and several amino acid residues responsible for proton translocation are well conserved between prokaryotic I-subunit and eukaryotic V_o-a subunit [14]. The membrane domain of I-subunit is divided into two clusters of helices: one is mostly perpendicular to the membrane and another contains tilted helices adjacent to the L ring (Fig. 2.2). The idea of proton conducting hemichannels was first proposed as a mechanism of proton translocation through the integral F_o domain of F-ATPase [16]. For the V-ATPase, a proton is thought to first enter the membrane through a cytoplasmically oriented hemichannel that allows the proton to reach the buried carboxyl group on one of the L-subunit (Glu-63). Following protonation of this carboxylate, ATP-driven rotation of the c ring brings the protonated carboxyl group into contact with the luminally oriented hemichannel. Interaction between the carboxyl group on the L-subunit and a buried arginine residue in I-subunit releases the proton into the luminal hemichannel. This deprotonates Glu-63 available to bind another proton (Fig. 2.2C). The EM map structure of *T. thermophilus* V-ATPase supported these two hemichannel models. One cluster of transmembrane helices in the I-subunit could conduct protons from the periplasm to the Glu-63 residues of an L-subunit (Fig. 2.2C). The other cluster of transmembrane helices in I-subunit could conduct protons from the Glu-63 residue of the adjacent L-subunit to the cytoplasm. Subunit I contains a conserved and essential arginine residue (Arg-563) which has been postulated to cause a decrease in the pK_a of Glu-63 residue on the L-subunit in contact with the cytoplasmic hemichannel, thereby causing it to lose its proton to the channel. In this structure, Arg-563 (Arg-735 in yeast V_0-a subunit) has been proposed to reside on transmembrane helix and form part of the cytoplasmic hemichannel, which positions helix 7 as one of the two helices in contact with the L-subunit. At this resolution, however, it is not possible to assign the precise residues involved in proton translocation. In order to discuss the proton translocation mechanism or torque generation mechanism coupled with proton translocation, we need more detail structure of the intact *T. thermophilus* V-ATPase.

2.3.3 *Structure of Rotor Ring*

The membrane-embedded rotor ring constitutes of rotary proton translocation machinery with a stator subunit. The rotor ring of F-ATPase is composed of c-subunits with two membrane helices. In prokaryotic V-ATPases of *E. hirae*, K-subunits composed of four transmembrane helices are assembled into 10 ring (K ring) [18]. The C terminal helix of K-subunit contains a conserved carboxylate side chain that is essential for ion transport. The structure of K ring is shown in Fig. 2.3.

The 10 protomers of the K ring have very similar structures. Each protomer is folded into five a helices. Four helices, H1 to H4, are transmembrane segment. Thus, the 40 transmembrane helices are packed in a ring structure. The height of the K ring is similar to that of the F-ATPase c rings from *Saccharomyces cerevisiae* (58 Å) [19] and *Ilyobacter tartaricus* (70 Å) [20]. However, the outer diameter of K ring is ~85 Å, significantly larger than that of c ring of F-ATPase. External diameter at the top of ring in *S. cerevisiae* is 55 Å and in *I. tartaricus*, the value is 50 Å. These differences in diameter likely reflect significant differences in the association of the ring to central rotor complex in rotary motor. In F-ATPases, the interaction between the ring and central stalk subunits is direct. On the contrary, in V-ATPases, the C-subunit (V_0-d) is thought to be interposed between the equivalent central stalk subunits and the ring [21]. The details will be described later.

The crystal structure of K ring contains densities of sodium ion adjacent to the carboxylate of glutamic acid residue on the H4 (shown as red spheres). Each sodium ion is surrounded by five

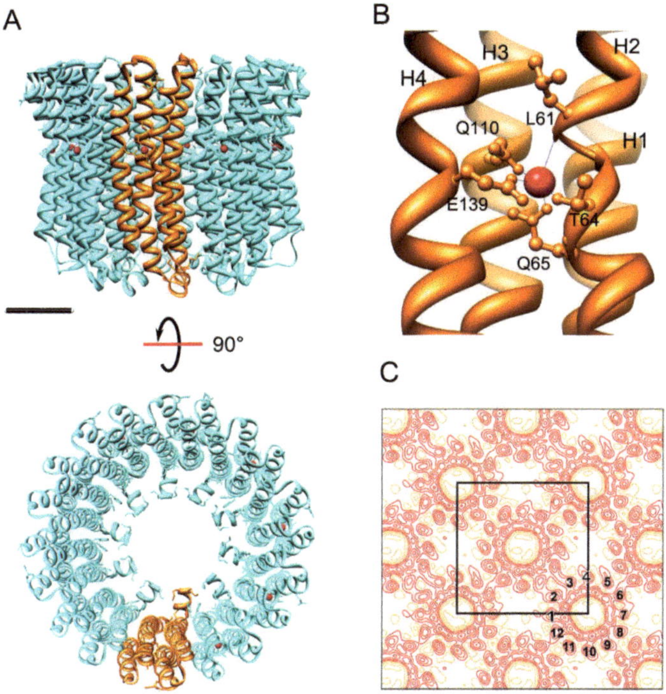

Figure 2.3 Structure of isolated rotor ring of prokaryotic V-ATPase of *E. hirae* and *T. thermophilus*. (A) Structure of K ring; a side view (upper) and a view from cytoplasmic side. A monomeric K-subunit composed of four membrane spanning helices is represented as gold. Bound sodium ions are represented as red spheres. Scale bar represents 25 Å. (B) The Na^+ binding site. Blue lines are proposed Na^+–oxygen bonds. Sodium ion is represented as red sphere. (C) The projection map, calculated from seven merged images, at 7.0 Å resolution, showing an orthogonal arrangement of L_{12} rings of *T. thermophilus*. One unit cell is outlined and has the dimensions of $a = 83.5 \pm 0.6$ Å, $b = 83.5 \pm 0.6$ Å, and $\gamma = 90.4 \pm 0.4°$.

oxygen atoms: four of them in the side chains of Thr64, Gln65, Gln110, and Glu139, and the fifth in the main-chain carbonyl of Leu61 (Fig. 2.3B). The *E. hirae* V-ATPase also contains I-subunit which shows high sequence and structure similar to I-subunit of *T. thermophilus* V-ATPase. Thus, the *E. hirae* V-ATPase contains similar sodium translocation channel in the interface with K ring and I-subunit.

The rotor ring of *T. thermophilus* V-ATPase constitutes of 12 L-subunits, which are composed of two membrane helices, not four [22]. The projection map of two-dimensional crystal of L_{12} ring is shown in Fig. 2.3C. The outer and inner rings are 58 Å and 22 Å, respectively. This size corresponds well to the corresponding measurement of the c ring of *I. tartaricus*, but significantly smaller than that of *E. hirae* K ring. The L_{12} ring structure is also identified in EM map of intact *T. thermophilus* V-ATPase as described in the former section. Single molecular analysis of rotation of *T. thermophilus* V-ATPase has revealed that V-ATPase exhibited 12 dwell position per revolutions, consistent with 12-fold symmetry of the L_{12} ring [23]. When the rotor ring is a decamer, the step size of unit rotation does not fit between V_1/F_1 and V_o/F_o. Because the coiled-coil structure of the rotor subunit allows some internal twisting, and the side stalk has extra flexibility, both can undergo elastic twisting or bending to enable the proper alignment of rotor–stator contacts at both $V_1/F_1$and V_o/F_o. This elastic energy due to the symmetry mismatch has been thought to be obligatory for the functioning of the rotary ATPase/synthase [19]. The presence of the dodecamer ring in the V/F-ATPases was previously undescribed. Although it was argued that the matching symmetry could be the consequence of the rotor ring's unusual size, it was noted that symmetry mismatch is not obligatory. Thus, independent of the rotor ring's size, symmetry mismatch is not obligatory for ATP hydrolysis or synthesis in V/F-ATPases.

2.3.4 *Structure of* V_1 *Domain* (A_3B_3DF)

The V_1 portion of *T. thermophilus*, which is ATPase-active and hence called V_1-ATPase, is made up of four subunits: A (64 kDa), B (54 kDa), D (25 kDa), and F (12 kDa) with a stoichiometry of $A_3B_3D_1F_1$ [2, 4, 12]. The A-subunit contains a catalytic site, and the A- and B-subunits are arranged alternately, forming a hexameric cylinder similar to the $\alpha_3\beta_3$ of F-ATPase. The structure of *T. thermophilus* A_3B_3 was solved by molecular replacement [24]. The structure refined using the data up to 2.8 Å resolution was shown in Fig. 2.4A. Both A- and B-subunits are nucleotide free in the crystal structure. This sixfold symmetry, however, breaks down because of protruding

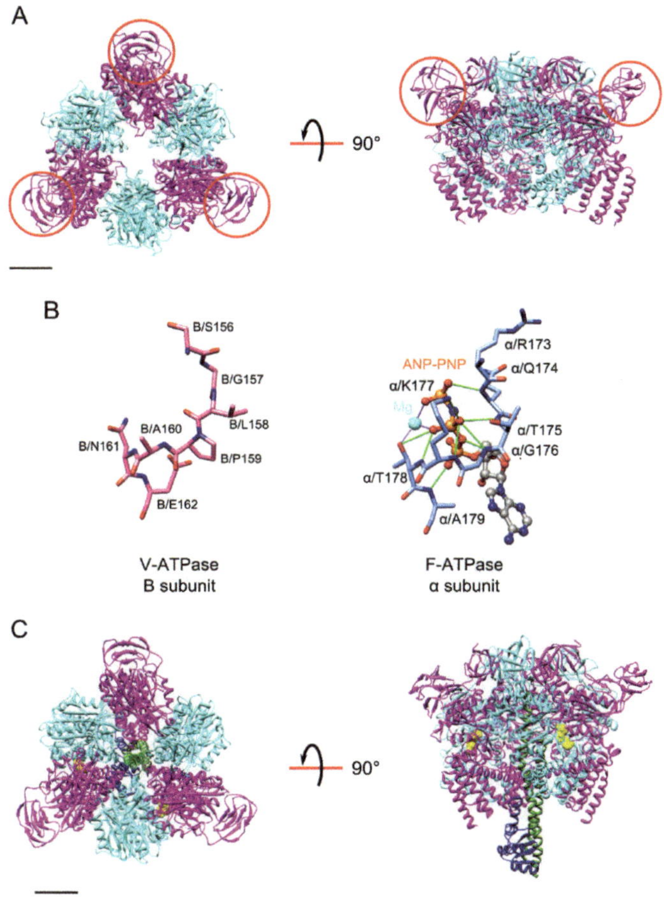

Figure 2.4 Structure of A_3B_3 of prokaryotic V-ATPase of *T. thermophilus* at 2.8 Å. (A) Crystal structure of the A_3B_3 complex. (A) Top view of the complex viewed towards the membrane (upper) and of the side view. The A and B subunits are represented by magenta and sky blue ribbons, respectively. (B) Structures of the P-loop equivalent region of B subunit of *T. thermophilus* V-ATPase (left) and of the nucleotide bound α subunit of F-ATPase. The bound AMP-PNP molecule and the magnesium ion are shown. Hydrogen bonds are shown in light green, whereas coordination links to the magnesium ion are shown in sky blue. (C) Structure of V_1 (A_3B_3DF) of prokaryotic V-ATPase of *T. thermophilus.* Crystal structure of V_1 of *T. thermophilus*; top view of the complex viewed towards the membrane (upper) and of the side view. Scale bar represents 25 Å. The A, B, D, and F subunit are represented as magenta, sky blue, green and blue ribbons, respectively. Bound nucleotides are represented as yellow spheres.

structures from the A-subunits (circled by red lines), which makes the A_3B_3 complex a triangular shape in the top view. The main structural differences between the V-ATPase A_3B_3 complex and the $\alpha_3\beta_3$ sub-domain in F_1 are observed in two regions in the A-subunit. One is the bulge mentioned above and the other is the presence of three longer helices in the C-terminal domain facing the V_o domain in the membrane. The structure of the potential nucleotide-binding site of the B-subunit in the A–B interface is very different from that observed in the β–α interface of F_1, which binds a non-catalytic nucleotide. Figure 2.4B compares the structures of the P-loop equivalent region in the B-subunit and of the nucleotide-bound P-loop in the yeast $F_1\alpha_E$ subunit. The P-loop has a unique main-chain structure, which allows five amide protons within the loop to form hydrogen-bonding interactions with the triphosphate group of bound ATP (Fig. 2.4B). The equivalent loop in the B-subunit shows a totally different main-chain conformation, which by analogy does not allow the formation of these hydrogen bonds. This is presumably because the P-loop equivalent region in the B-subunit lacks flexible glycine residues and includes an insertion of a proline residue, which does not have an amide group. In addition, several residues known to be involved in nucleotide binding at the β–α interface (β/Y368, β/R372, α/Q172, α/Q432, α/P367, the bovine F_1 residue numbering is used hereafter) are missing in the A_3B_3 complex. In conclusion, it is unlikely that the B-subunit binds a nucleotide at the A–B interface in a similar manner to that of the α-subunit. The catalytic nucleotide-binding sites are located in the interface between the B- and A-subunits. The site is mainly composed of the residues from the A-subunit, but some from the B-subunit are also essential. The nucleotide-binding domain of the A-subunit shows high sequence similarity to that of F_1, including the P-loop, which is responsible for coordination of phosphate moieties of ATP. However, there is one interesting and important difference in the nucleotide-binding sites of two ATPases. The residue α/S344 is well conserved in F_1 among species, and it lies in close proximity to the catalytic water interacting with β/E188 at the αD–βD interfaces [25]. Thus, the α/Ser344 residue has been thought to be critical for catalysis. However, the serine residue is not conserved at the identical position in the B-subunit of V_1. Instead, the conserved tyrosine residue (B/Y331) is located at the identical position. Mutation of study for B/Y331 indicates that an aromatic side chain in this position seems essential for the ATP-hydrolysis activity of V_1.

Numoto *et al.* have also reported crystal structure of *T. thermophilus* V_1 with bound nucleotides [26]. In this structure, the A_3B_3 hexamer is assembled asymmetrically around a central stalk (Fig. 2.4C). One AB pair forms a wide open conformation. On the contrary, the two AB pairs form a narrowly closed conformation. ADP or ATP was observed on the catalytic interface of the closed AB pairs. Long coiled-coil helices of the D-subunit are clearly seen in the electron density maps (Fig. 2.4C, green ribbon). The lengths of the two helices are about 80 Å for the *N*-terminal helix and 110 Å for the C-terminal helix. F-subunit was bound around the foot of the protruding region of the central stalk. The C-terminal helix of F-subunit has direct interactions with the C-terminal regions of A- and B-subunits, by which the F-subunit can modulate ATPase activity.

2.3.5 *Structure of Central Rotor Subunit C (V_o-d)*

The C-subunit, a homologue of Vma6p (or d-subunit) assigned to be V_o domain in yeast V-ATPase [27], is also part of V_o part in *T. thermophilus* V-ATPase. The crystal structure of the isolated C-subunit of *T. thermophilus* was determined at 1.95 Å resolution [21]. The molecule is highly α-helical. It is composed of six central helices and 12 peripheral helices arranged in a unique funnel-shaped structure (Fig. 2.5). There is a polar cavity inside (8 Å diameter, 25 Å depth), open to only one side. The peripheral helices form a rim, ~25 Å high around the central helices. The C-subunit is located in the center of central stalk of the holoenzyme complex obtained by single particle analysis using cryo-EM (Fig. 2.2A). That is, the C-subunit caps one end of the L_{12} ring, and the internal cavity of the γ-secretase open toward the upper V_1 side. The C-subunit is to interact with the DF central shaft. The C-subunit should recognize this complex, like "socket". This proposal model is finally established by cryo-EM map of the intact enzyme described above.

Despite the functional similarity, the central rotor apparatus of the F- and V-ATPases contain significant structural differences. The

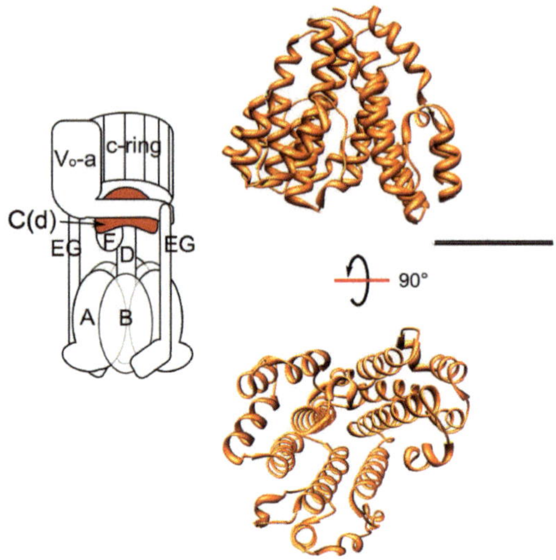

Figure 2.5 Structure of isolated subunit C (V_o-d) of *T. thermophilus.* Side view (upper) and top view from cytoplasmic side (lower). Scale bar represents 25 Å. Schematic model of the V-ATPase (left) represents the location of C subunit (colored by dark yellow).

V_o sector contains the funnelshaped V_o-d subunit, which serves as a socket for the DF complex in V_1. In contrast $F_1\gamma$ attaches directly onto the F_o-c ring, while the F_1-ε forms contacts with both the F_1-γ and F_o-c ring (see Fig. 2.1). The F_1-ε, composed of an N-terminal β sandwich and a short C terminal helix shows neither sequence nor structural similarity to V_o-d. Assuming that V-ATPases have conserved the ancestral form in the rotor apparatus, V_o-d has been replaced with F_1-ε during the evolutionary process to F_oF_1. The central rotor apparatus including V_o-d and the DF complex structure in V-ATPase is a landmark feature clearly distinguishing V-ATPases from F_oF_1

2.3.6 *Structure of DF Rotor*

The crystal structure of the F-subunit of *T. thermophilus* V-ATPase was also determined to 2.2 Å resolution [28]. Although the F-subunit had been proposed to have a function and structure similar to that of the F_1-ε-subunit, the structure of the F-subunit of the V-

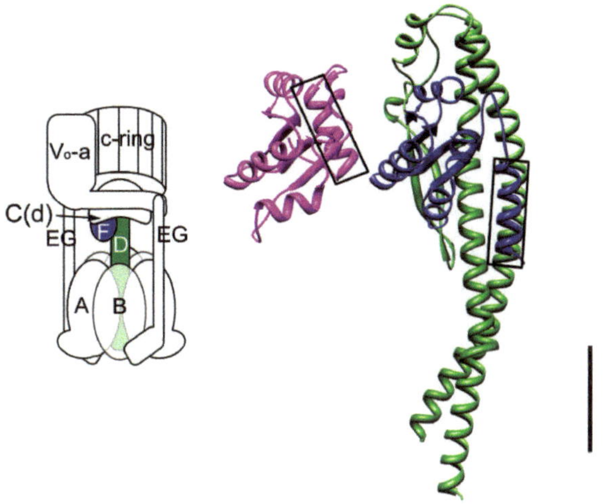

Figure 2.6 Structure of isolated F subunit of *T. thermophilus* (middle) and DF complex of *E. hirae* (right). D and F subunits of *E. hirae* are represented by green and blue ribbons, respectively. Scale bar represents 25 Å. Schematic model of the V-ATPase (left) represents the location of D- and F-subunits (colored by green and blue, respectively).

ATPase shows no structural similarity to the ε-subunit. Structure of *T. thermophilus* F-subunit is shown in Fig. 2.6. The subunit F has a flat spherical shape with dimensions of 30 × 35 × 20 Å. It has a four-stranded parallel β-sheets, a core that is surrounded by four long and one short α-helices. The structure of the retracted subunit F shows a striking similarity to the structure of CheY, which is the chemotaxis regulator protein from *E. coli* [29]. Saijo *et al.* reported crystal structure of DF complex of *E. hirae* V-ATPase [30]. In the complex, *E. hirae* F-subunit shows similar fold to that of *T. thermophilus* F-subunit, but C terminal helix is elongated forward to D-subunit. The C terminal helix is responsible for binding to D-subunit. The overall structure of DF complex is shown in Fig. 2.6. The D-subunit constitutes of a long coiled-coil helices and a short β-hairpin region. The D does not contain a globular domain between C- and N-helices such as F_1-γ subunit. Instead, *E. hirae* F-subunit binds to the middle portion of *E. hirae* D. Thus, F-subunit is most likely evolutionary-related proteins to globular domain in F_1-γ. In

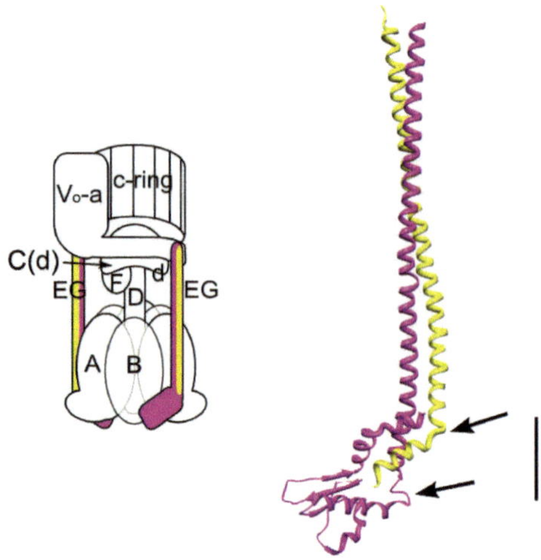

Figure 2.7 Structure of isolated EG heterodimer of *T. thermophilus* E- and G-subunits of *T. thermophilus* is represented by magenta and yellow ribbons, respectively. The kink of G-subunit and proline loop of E-subunit are indicated by arrows. Scale bar represents 25 Å. Schematic model of the V-ATPase (left) represents the location of E- and G-subunits (colored by magenta and yellow, respectively).

fact, fused DF-subunit can also function as an intact rotor. The detail will be described elsewhere.

2.3.7 *Structure of EG Peripheral Stalk*

The peripheral stalk is the conserved component of rotary ATPases/synthases. In F-type ATPase, there is only one peripheral stalk, whereas prokaryotic V-ATPase displays two peripheral stalks, each made of an EG heterodimers [12, 31] (Fig. 2.1). The crystal structure of the isolated EG heterodimers is shown in Fig. 2.7. The heterodimer contains two distinct domains: long coiled coil and a globular domain. The coiled coil is formed by the long *N*-terminal helices from E- and G-subunits. Subunit G is entirely α-helices with a long *N*-terminal helix and a short C-terminal helix that are separated by a kink (Fig. 2.7). The C-terminal half of subunit E forms the majority of the globular domain. In E-subunit, *N*-terminal helix is joined with C-terminal globular domain by proline loop. The loop in

E-subunit and the kink in G-subunit are thought to be two flexible joints, that is, a hinge that could facilitate movement of the globular domain. The EM map of the intact V-ATPase indicates that the globular domain of EG heterodimer binds to β-barrel in B-subunit (Fig. 2.2). The end of coiled coil likely binds to the hydrophilic domain of I-subunit. That is, A_3B_3 domain is connected with stator I-subunit in membrane-embedded domain by two EG heterodimers.

2.4 Perspective

Crystal structure of an intact F-ATPase has not been determined yet even though lots of group had made significant effort to determine it. On the contrary, Lau and Rubinstein determined subnanometer-resolution structure of the intact V-ATPase of *T. thermophilus* by cryo-EM analysis [15]. Their EM structure provides us several information about subunit–subunit interaction in the complex. Especially the map provides eight transmembrane densities that can be attributed to α-helices in the hydrophobic C-terminal domain of I-subunit. This is the first report of structural information of stator subunits in F_o/V_o of any rotary motors. Furthermore, atomic resolution models of V_1, EG stator, and rotor C-subunit of *T. thermophilus* V-ATPase have been already determined. Thus, *T. thermophilus* V-ATPase is the most suitable model enzyme for structural study of rotary ATPase/synthase including both F- and V-ATPase. Our goal of structural study of prokaryotic V-ATPase is to understand how rotary ATPase/synthase efficiently couple proton translocation with ATP hydrolysis or synthesis. Crystal structure of the intact rotary ATPase/synthase will be necessary for the precise knowledge of rotary catalytic mechanism. For this we need an atomic resolution model of the V-ATPase of *T. thermophilus*.

Acknowledgment

I thank Toei, M., and Murata, T., for useful discussion, and Nakanishi, A., for her technical assistant.

References

1. Forgac, M. (2007). Vacuolar ATPases: rotary proton pumps in physiology and pathophysiology, *Nat. Rev. Mol. Cell Biol.*, 8, pp. 917–929.
2. Yokoyama, K., Oshima, T. and Yoshida, M. (1990). *Thermus thermophilus* membrane-associated ATPase. Indication of a eubacterial V-type ATPase, *J. Biol. Chem.*, 265, pp. 21946–21950.
3. Kakinuma, Y., Yamato, I. and Murata, T. (1999). Structure and function of vacuolar Na^+-translocating ATPase in *Enterococcus hirae, J. Bioenerg. Biomembr.*, 31, pp. 7–14.
4. Yokoyama, K. and Imamura, H. (2005). Rotation, structure, and classification of prokaryotic V-ATPase, *J. Bioenerg. Biomembr.*, 37, pp. 405–410.
5. Muench, S. P., Trinick, J. and Harrison, M. A. (2011). Structural divergence of the rotary ATPases, *Q. Rev. Biophys.*, 44, pp. 311–356.
6. Yokoyama, K. *et al.* (1994). Isolation of prokaryotic V_oV_1-ATPase from a thermophilic eubacterium *Thermus thermophilus, J. Biol. Chem.*, 269, pp. 12248–12253.
7. Yoshida, M., Muneyuki, E. and Hisabori, T. (2001). ATP synthase—a marvellous rotary engine of the cell, *Nat. Rev. Mol. Cell Biol.*, 2, pp. 669–677.
8. Mulkidjanian, A. Y. *et al.* (2007). Inventing the dynamo machine: the evolution of the F-type and V-type ATPases, *Nat. Rev. Microbiol.*, 5, pp. 892–899.
9. Konishi, J. *et al.* (1987). Purification and properties of the ATPase solubilized from membranes of an acidothermophilic archaebacterium, *Sulfolobus acidocaldarius, J. Biochem. (Tokyo)*, 102, pp. 1379–1387.
10. Tsutsumi, S. *et al.* (1991). Molecular cloning of genes encoding major two subunits of a eubacterial V-type ATPase from *Thermus thermophilus, Biochim. Biophys. Acta*, 1098, pp. 13–20.
11. Yokoyama, K. *et al.* (2003). Subunit arrangement in V-ATPase from *Thermus thermophilus, J. Biol. Chem.*, 278, pp. 42686–42691.
12. Yokoyama, K. *et al.* (2000). V-Type H^+-ATPase/synthase from a thermophilic eubacterium, *Thermus thermophilus*. Subunit structure and operon, *J. Biol. Chem.*, 275, pp. 13955–13961.
13. Takase, K. *et al.* (1994). Sequencing and characterization of the ntp gene cluster for vacuolar-type Na(+)-translocating ATPase of *Enterococcus hirae, J. Biol. Chem.*, 269, pp. 11037–11044.

14. Toei, M., Toei, S. and Forgac, M. (2011). Definition of membrane topology and identification of residues important for transport in subunit a of the vacuolar ATPase, *J. Biol. Chem.*, 286, pp. 35176–35186.

15. Lau, W. C. and Rubinstein J. L. (2011). Subnanometre-resolution structure of the intact *Thermus thermophilus* H^+-driven ATP synthase, *Nature*, 481, pp. 214–218.

16. Fillingame, R. H., Angevine, C. M. and Dmitriev, O. Y. (2003). Mechanics of coupling proton movements to c-ring rotation in ATP synthase, *FEBS Lett.*, 555, pp. 29–34.

17. Zhou, M. *et al.* (2011). Mass spectrometry of intact V-type ATPases reveals bound lipids and the effects of nucleotide binding, *Science*, 334, pp. 380–385.

18. Murata, T. *et al.* (2005). Structure of the rotor of the V-type Na^+-ATPase from *Enterococcus hirae, Science*, 308, pp. 654–659.

19. Stock, D., Leslie, A. G. and Walker, J. E. (1999). Molecular architecture of the rotary motor in ATP synthase, *Science*, 286, pp. 1700–1705.

20. Meier, T. *et al.* (2005). Structure of the rotor ring of F-type Na^+-ATPase from *Ilyobacter tartaricus, Science*, 308, pp. 659–662.

21. Iwata, M. *et al.* (2004). Crystal structure of a central stalk subunit C and reversible association/dissociation of vacuole-type ATPase, *Proc. Natl. Acad. Sci. USA*, 101, pp. 59–64.

22. Toei, M. *et al.* (2007). Dodecamer rotor ring defines H^+/ATP ratio for ATP synthesis of prokaryotic V-ATPase from *Thermus thermophilus, Proc. Natl. Acad. Sci. USA*, 104, pp. 20256–20261.

23. Furuike, S. *et al.* (2011). Resolving stepping rotation in *Thermus thermophilus* H(+)-ATPase/synthase with an essentially drag-free probe, *Nat. Commun.*, 2, p. 233.

24. Maher, M. J. *et al.* (2009). Crystal structure of A_3B_3 complex of V-ATPase from *Thermus thermophilus, EMBO J.*, 28, pp. 3771–3779.

25. Abrahams, J. P. *et al.* (1994). Structure at 2.8 A resolution of F_1-ATPase from bovine heart mitochondria, *Nature*, 370, pp. 621–628.

26. Numoto, N. *et al.* (2009). Inter-subunit interaction and quaternary rearrangement defined by the central stalk of prokaryotic V_1-ATPase, *EMBO Rep.*, 10, pp. 1228–1234.

27. Bauerle, C. *et al.* (1993). The *Saccharomyces cerevisiae* VMA6 gene encodes the 36-kDa subunit of the vacuolar H(+)-ATPase membrane sector, *J. Biol. Chem.*, 268, pp. 12749–12757.

28. Makyio, H. *et al.* (2005). Structure of a central stalk subunit F of prokaryotic V-type ATPase/synthase from *Thermus thermophilus*, *EMBO J.*, 24, pp. 3974–3983.
29. Stock, A. M. *et al.* (1989). Three-dimensional structure of CheY, the response regulator of bacterial chemotaxis, *Nature*, 337, pp. 745–749.
30. Saijo, S. *et al.* (2011). Crystal structure of the central axis DF complex of the prokaryotic V-ATPase, *Proc. Natl. Acad. Sci. USA*, 108, pp. 19955–19960.
31. Lee, L. K. *et al.* (2010). The structure of the peripheral stalk of *Thermus thermophilus* H^+-ATPase/synthase, *Nat. Struct. Mol. Biol.*, 17, pp. 373–378.

Chapter 3

The Function of V-ATPase in the Degradation of Gluconeogenic Enzymes in Yeast Vacuoles

Jingjing Liu[a,b] **and Hui-Ling Chiang**[a,b]
[a] *Department of Surgery, University of Arizona, Tucson, AZ 85719, USA*
[b] *Department of Physiology, Pennsylvania State University, Hershey, PA 17033, USA*
jliu4@email.arizona.edu, hxc32@psu.edu

3.1 Introduction

Vacuolar H^+-ATPase (V-ATPase) is a multi-subunit proton pump that utilizes ATP hydrolysis to pump protons into the vacuole [1–3]. V-ATPase is essential for maintaining homeostasis in acid and base balances [4–6]. Acid and base homeostasis is critical for a number of cellular functions including endocytosis, the trafficking of lysosomal enzymes to lysosomes and the degradation of signaling molecules in lysosomes [4, 7]. V-ATPase is also important for the storage of glycogen and the sorting of the vacuole-resident protein, carboxypeptidase Y but is not required for the delivery of autophagosomes to the vacuole [8–11]. In addition to the well-established role of V-ATPase in acidifying organelles, it also forms a

Handbook of H^+-ATPases
Edited by Suguru Nakamura

ISBN 978-981-4411-91-2 (Hardcover), 978-981-4411-92-9 (eBook)
www.panstanford.com

trans-V0 complex fusion pore during homotypic vacuolar fusion in yeast [12, 13]. While V-ATPase has many important functions, the possible link of this complex with diabetes has recently emerged [14, 15]. For example, V-ATPase is necessary for exocytosis of GLUT4-containing vesicles (glucose transporter 4) during insulin stimulation [16]. Therefore, a better understanding of how V-ATPase functions may shed light on how insulin stimulates the transport of glucose and how defects in V-ATPase contribute to insulin resistance and the development of diabetes.

Fructose-1,6-bisphosphatase (FBPase) is the key gluconeogenic enzyme that is induced in patients with type 2 diabetes [17]. Specific FBPase inhibitors have been used successfully to reduce the production of glucose in animals with diabetes [18]. These inhibitors are in clinical trials to reduce glucose production in patients suffering from type 2 diabetes [19]. In *Saccharomyces cerevisiae*, FBPase is induced when cells starved glucose for a prolonged period of time [20–22]. However, when glucose is added to glucose-starved cells, this protein is first inactivated and then degraded in the vacuole *via* the vacuole import and degradation (Vid) pathway [20, 23]. Recent evidence indicated that this pathway is also utilized by other gluconeogenic enzymes, including malate dehydrogenase, phosphoenolpyruvate carboxykinase, and isocitrate lyase 1 [23–25]. Degradation of gluconeogenic enzymes during glucose re-feeding prevents energy futile cycles that are detrimental to cells. Genetic studies revealed that V-ATPase is required for the degradation of FBPase in the vacuole but is not required for the degradation of FBPase in the proteasome [26]. Here, we focus on the mechanistic insights into the function of V-ATPase in the regulation of the degradation of gluconeogenic enzymes in the vacuole using *S. cerevisiae* as the model system.

3.2 The Structure and Function of Vacuolar ATPase

3.2.1 *The Structure of V-ATPase*

V-ATPase is composed of 14 subunits that combine to form proton-translocation machinery to acidify intracellular compartments

Table 3.1 The yeast V-ATPase

	V-ATPase subunit	Yeast gene
V1 domain	A	*VMA1*
	B	*VMA2*
	C	*VMA5*
	D	*VMA8*
	E	*VMA4*
	F	*VMA7*
	G	*VMA10*
	H	*VMA13*
V0 domain	a	*VPH1 or STV1*
	c	VMA3
	c′	*VMA11*
	c″	*VMA16*
	d	*VMA6*
	e	*VMA9*

(Table 3.1) [1–3]. V-ATPase can be structurally divided into two major functional domains (Fig. 3.1) [1, 3]. A peripheral V1 domain consists of eight subunits (A–H) required for ATP hydrolysis (A–H). A V0 integral membrane domain consists of a ring of proteolipid subunits (a, c, c′, c″, d, and e) required for proton translocation [1, 27, 28]. V1 and V0 can exist separately but must combine to pump protons [1, 29]. In contrast to the mitochondrial F0F1-ATPase that synthesizes ATP *via* translocating protons from the mitochondrial lumen to the cytosol, V-ATPase uses the energy of ATP hydrolysis to translocate protons from the cytosol into the lumen of the vacuole [30, 31].

3.2.1.1 The V1 domain

The V1 domain consists of eight subunits (Table 3.1). Subunit A and B are encoded by the *VMA1* and *VMA2* genes [32, 33]. Subunits A and B each have three copies that form a hexameric structure for ATP binding and hydrolysis [28]. ATP hydrolysis may induce conformational changes in subunit A, and the torque is transduced to the proton-generating V0 domain [1, 34].

Subunits C, D, E, F, G, and H are encoded by the *VMA5, VMA8, VMA4, VMA7, VMA10,* and *VMA13* genes, respectively (Table 3.1).

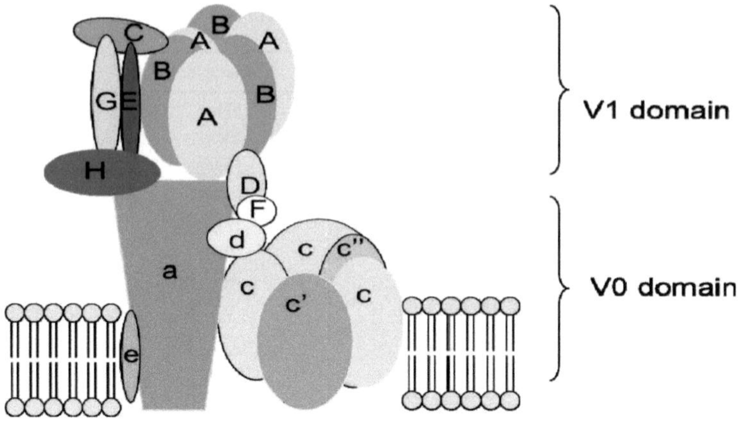

Figure 3.1 Diagram of V-ATPase. V-ATPase contains a membrane-bound V0 domain and a peripheral V1 domain. The V1 domain consists of eight subunits involved in ATP hydrolysis, and the V0 domain consists of six subunits participating in proton translocation.

They are involved in coupling V1 and V0 domains for V-ATPase activity [35–37].

3.2.1.2 The V0 domain

The yeast V0 domain consists of six subunits. The *VPH1* and *STV1* genes encode two 100 kDa subunit an isoforms in yeast and are highly homologous [38]. Subunits c, c′ and c″ are encoded by *VMA3*, *VMA11*, and *VMA16* in yeast [39]. Subunit d is encoded by the *VMA6* gene and associates with the V0 domain on the cytosolic side of the membrane [40]. Subunit e is encoded by *VMA9* and is required for V0 assembly [41]. The V0 subunits participate in proton translocation [42, 43].

3.2.2 *The Assembly of V-ATPase*

Genetic studies in yeast have identified three mutants that exhibit the *vma* phenotype [44–46]. The protein products of these three genes, Vma12p, Vma21p, and Vma22p are localized to the ER in yeast and are involved in the assembly of V-ATPase in the ER [47, 48]. The absence of any one of these assembly factors results in the rapid degradation of the 100 kDa V-ATPase subunit a in the ER [49, 50]. The stable assembly of V-ATPase also requires the proteins of the regulator of the H^+-ATPase of the vacuolar and endosomal membranes (RAVE) complex [51, 52].

V-ATPase activity can be regulated by the assembly and disassembly of the V1 and V0 domains in a glucose-dependent manner [49, 50, 53]. When cells are grown in low-glucose medium, a high percentage of the V1 domain is released into the cytosol. The V0 and V1 domains re-assembled when glucose is added to glucose-depleted cells [49, 53]. Glycolytic enzymes such as aldolase, glyceraldehyde 3-phosphate dehydrogenase (GAPDH), and phosphofructokinase-1 (PFK1) associate with V-ATPase [54]. Aldolase binds to V-ATPase in a glucose-dependent manner and is required for V-ATPase activity. Moreover, aldolase is also required for the glucose-induced assembly of V-ATPase [55]. The association of glycolytic enzymes with V-ATPase may provide a local source of ATP to acidify the vacuole. Alternatively, these enzymes may act as fusion molecules to facilitate the fusion event [55–57].

3.2.3 *The Function of V-ATPase in Organelle Acidification and Homotypic Vacuolar Fusion*

The maintenance of pH in organelles is essential for diverse cellular functions [58, 59]. V-ATPase is responsible for the acidification of several intracellular organelles [60, 61]. The differences in pH are often used to associate and dissociate cargo proteins from their receptors during transport along the endocytic pathway [59, 62, 63]. For example, a lower pH in early endosomes (pH 6–6.5) causes the dissociation of receptors from their cargoes, leading to the sorting of receptors into the recycling endosomes [64, 65].

Although V-ATPase was initially identified as an essential component to acidify the vacuole, V-ATPase has been implicated in homotypic vacuolar fusion [13, 66]. Vacuolar fusion depends on an electrochemical potential across the vacuolar membrane [67]. The early phase of fusion requires the acidification of the vacuole [67]. However, the V0 sector of V-ATPase forms a *trans*-V0 complex from the opposing membranes during a very late step of homotypic vacuolar fusion [13].

A study by Mayer *et al.* suggests that V-ATPase functions as a fusion pore in terminal homotypic vacuolar fusion [13]. According to this model, channels similar to those found in gap junctions connect the two opposing membranes during the fusion event [68]. The terminal phase of fusion depends on the release of Ca^{2+} from the vacuolar lumen to trigger bilayer fusion [69]. Vma3p is associated with calmodulin, which may induce conformational changes in the proteolipid channels, leading to their radial expansion, and lateral propagation [69]. This may create a dilated hydrophilic lumen allowing lipids to flow between the membranes [69]. There is a concomitant coupling of the V0 domain with the release of lumenal Ca^{2+} [69]. In support of this hypothesis, antibodies that react with Vph1p caused a defect in the release of Ca^{2+} [12]. Vph1p antibodies may block the interaction with the Ca^{2+} channel and impair the release of Ca^{2+} from a V0-coupled channel.

3.3 Catabolite Inactivation

S. cerevisiae can grow in various carbon sources, including galactose, maltose, acetate, lactate, and glucose [20, 70]. Glucose is converted to pyruvate anaerobically by glycolysis. Under glucose-deprived conditions, glucose can be produced from precursors such as acetate, glycerol, and amino acids by gluconeogenesis. Glycolysis and gluconeogenesis must be strictly controlled to prevent futile energy cycles. Failure in the regulation of glycolysis or gluconeogenesis results in metabolic disorders in humans. For instance, excessive glucose produced by gluconeogenesis can cause elevated levels of glucose in patients with type 2 diabetes [17]. As such, FBPase inhibitors such as managlinat dialanetil show promise in the treatment of patients with type 2 diabetes [17, 19].

Gluconeogenic enzymes are induced when *S. cerevisiae* is grown on poor carbon sources in acetate, oleate, galactose, or pyruvate [20]. In the presence of glucose, gluconeogenic genes are transcriptionally repressed (called catabolite repression). Furthermore, the existing enzymes are inactivated (called catabolite inactivation) [21, 22, 71, 72]. Four gluconeogenic enzymes, FBPase, phosphoenolpyruvate carboxykinase, isocitrate lyase 1, and malate dehydrogenase (MDH2) are all inactivated by glucose [73–77]. High-affinity hexose transporters such as Hxt6p and Hxt7p and the galactose transporter are also inactivated by glucose [76, 78]. In response to glucose, these transporters are delivered to the vacuole for degradation by the endocytic pathway [76, 77]. Peroxisomes are important for the metabolism of fatty acids and are also degraded by glucose in the vacuole by pexophagy [79]. Thus, catabolite inactivation has been described for a number of proteins including gluconeogenic enzymes, sugar transporters, and peroxisomes.

3.3.1 *FBPase Can Be Degraded in the Proteasome or in the Vacuole Depending on Growth Conditions*

Catabolite inactivation of FBPase has been studied extensively [20, 22, 80]. FBPase is degraded in the proteasome [80] or in the vacuole [20] depending on growth conditions. For instance, when glucose is added to cells that are starved of glucose for one day, FBPase is degraded in the proteasome [24]. In contrast, when glucose is added to cells that starved glucose for three days, FBPase is degraded in the vacuole [24]. MDH2 also exhibits similar patterns of degradation as FBPase. MDH2 is degraded in the proteasome when glucose is added to cells starved for one day. In contrast, MDH2 is degraded in the vacuole when glucose is added to cells starved for three days [24]. Other gluconeogenic enzymes, phosphoenolpyruvate carboxykinase, and isocitrate lyase 1 are also targeted to the vacuole for degradation when prolonged-starved cells are re-fed with glucose [24, 25].

A number of *VID* genes have been identified that are required for the vacuole-dependent pathway [23]. Some genes such as *VID24*, *VID28*, and *VID30* are essential for FBPase degradation in both the vacuole-dependent and the proteasome-dependent pathways [24]. In contrast, the *PRE9* gene encoding a subunit of the proteasome and the *BLM10* gene encoding a regulator of the proteasome, are required for the proteasomal pathway but not the vacuolar pathway [81].

3.3.2 *The Vacuole-Dependent Pathway*

The targeting of FBPase to the vacuole in response to glucose has been demonstrated using the Δ*pep4* strain [20]. The *PEP4* gene encodes vacuolar protease A (Pep4p) required for the maturation of proteinase B and proteinase C. In the wild-type strain, FBPase is degraded by these vacuolar proteinases after it is delivered to the vacuole. However, in the Δ*pep4* strain deficient in vacuolar proteinases, FBPase accumulates in the vacuole and can be visualized by immunoelectron microscopy and immunofluorescence microscopy [20, 77]. Ultra-structural studies indicated that FBPase was in invaginations of the vacuolar membrane and also within small vesicles inside the vacuole [77]. An *in vitro* system has also been developed to reconstitute the targeting of FBPase to the vacuole [82]. In this system, purified FBPase was radiolabeled and added to semi-intact cells in which the endogenous *FBP1* gene was deleted. In the presence of an ATP-regenerating system and cytosol, radiolabeled FBPase was detected in vacuoles after the addition of glucose for 40 min. The FBPase signal in the vacuole was significantly reduced at 60 min following the addition of glucose, indicating that most of the FBPase had been degraded by vacuolar proteinases by the 60 min time point [82].

3.3.3 *The Discovery of Vid Vesicles*

A number of *Vid* mutants defective in the degradation of FBPase in the vacuole have been isolated [83]. When the distribution of FBPase was examined by immunofluorescence microscopy, two distinct phenotypes were observed [83]. In some mutants, FBPase exhibited cytosolic staining, while in others, FBPase showed staining in punctate structures [83], suggesting that FBPase is associated with membranous structures. Using differential centrifugation, substantial amounts of FBPase were found in the 100,000 g pellet, supporting the idea that FBPase is associated with membrane-bound structures.

To capture and purify these intermediate vesicles, wild-type cells were replenished with glucose at 22°C to delay the process of FBPase degradation [84]. FBPase-containing peaks were further

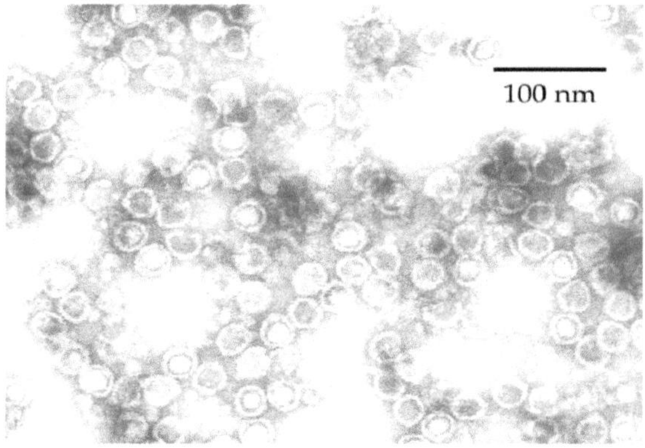

Figure 3.2 Vid vesicles are 30–50 nm in diameter as observed by negative staining and electron microscopy.

purified from the S-1000 column and sucrose density gradients. Novel vesicles that were 30–50 nm in diameter were observed by negative staining and electron microscopy (Fig. 3.2). These small vesicles have smooth surfaces and are called Vid vesicles. These Vid vesicles do not contain markers such as CPY (vacuole), Sec62p (ER), Mnn1p (Golgi), 3-oxoacyl CoA thiolase (peroxisome), or cytochrome C (mitochondria). The biogenesis of Vid vesicles requires the formation of ubiquitin chains and the ubiquitin conjugation enzyme Ubc1p [85]. The association of FBPase with Vid vesicles is impaired in the K48R/K63R ubiquitin mutant that prevents the formation of multi-ubiquitin chains [85].

3.3.4 *Vid24p and the COPI Coatomer Proteins are Components of Vid Vesicles*

Vid24p was the first Vid vesicle protein identified [86]. Vid24p is a novel 41 kDa protein peripherally associated with Vid vesicles. This protein is expressed at low levels during glucose starvation and increases its expression after glucose re-feeding. Vid24p co-localizes with FBPase in the fraction enriched with Vid vesicles. Vid24p is not required for growth, sporulation, osmolarity sensitivity, or CPY processing. The *vid24-1* mutant accumulates FBPase in punctate structures in the cytoplasm [86], suggesting that *VID24* is involved in a step after FBPase is associated with Vid vesicles or clusters of Vid vesicles.

When purified Vid vesicles were subjected to matrix-assisted laser desorption/ionization (MALDI) analysis, COPI coatomer proteins, including Sec28p, were identified [87]. COPI vesicles have well-established roles in retrograde transport from the Golgi to the ER and in endocytic trafficking in mammalian cells and yeast [88–91]. When coatomer genes were mutated, the association of FBPase with the Vid vesicles was defective. Coatomer proteins are localized to Vid vesicles and form a large protein complex with Vid24p [87]. Importantly, coatomer proteins are required to recruit Vid24p to Vid vesicles [87].

3.4 V-ATPase and the Vid Pathway

3.4.1 *An* in vitro *System to Study the Fusion of Vid Vesicles and Vacuoles*

We have developed an *in vitro* system to study the fusion of vesicles with vacuoles [92]. In this system, FBPase was fused to a truncated form of alkaline phosphatase (Δ60pho8p). The first 60 amino acids of alkaline phosphatase contain the signal sequence for translocation into the ER. When the first 60 amino acids were deleted, the resulting protein accumulated in the cytosol. We produced the FBPase–Δ60pho8p fusion protein and expressed it in various cells that also contained the Δpho8 deletion so that FBPase–Δ60pho8p can only be activated by Pep4p in the vacuole. This fusion protein was induced by glucose starvation and was targeted to the vacuole by the Vid pathway [92]. When Vid vesicles and vacuoles were isolated from wild-type cells and incubated in the presence of an ATP-regenerating system and cytosol, alkaline phosphatase activity was observed as quantified by spectrophotometry [92]. Using this *in vitro* system, we demonstrated that the fusion of Vid vesicles and vacuoles requires the general fusion machinery such as the *v*-SNARE on Vid vesicles and the *t*-SNARE on the vacuole [92]. Because Vid vesicles aggregate to form large clusters of vesicles when purified at low temperatures, this *in vitro* system may reconstitute the fusion of clusters of Vid vesicles with vacuoles. The fusion of small vesicles with the vacuole can also lead to the activation of alkaline phosphatase in the *in vitro* system.

3.4.2 *V-ATPase is Involved in the Fusion of Vesicles with Vacuoles*

V-ATPase is involved in the vacuole-dependent degradation of FBPase but not the proteasomal-dependent degradation [26]. All V-ATPase subunits except Vma1p and Vma2p are required for FBPase degradation in prolonged (three days) starved cells but are not required in short-term (one day) starved cells [26].

Stv1p and Vph1p are two isoforms of the large a-subunit in the V0 domain [93]. Previous immunofluorescence studies indicated that Vph1p is localized to the vacuole, whereas Stv1p is distributed to the Golgi or endosomes [38, 93]. FBPase was found in the fractions enriched with Vid vesicles in the Δ*stv1* and Δ*vph1* mutants, suggesting that the association of FBPase with Vid vesicles is not affected in these mutants. Using an *in vitro* fusion assay, we demonstrated that the Δ*stv1* mutant contained defective Vid vesicles but competent vacuoles. In contrast, the Δ*vph1* mutant contained partially defective vesicles and partially defective vacuoles [26].

It is also shown that all the V0 subunits of V-ATPase are required for FBPase degradation [26]. In the Δ*vma3* mutant, the majority of FBPase was sequestered in Vid vesicles, suggesting that FBPase is imported into Vid vesicles in the absence of this gene [85]. Using the *in vitro* fusion assay, we demonstrated that this mutant contained defective vesicles and partially defective vacuoles [26].

Interestingly, most of the V1 subunits are also required for the Vid pathway [26]. Vma5p is a subunit of the V1 domain. The Δ*vma5* mutant also contained defective vesicles and partially defective vacuoles in the *in vitro* fusion assay [26]. Other V1 subunits may play similar roles, as most of the V1 mutants accumulated FBPase in the fraction enriched with Vid vesicles and blocked the fusion of Vid vesicles with vacuoles [26]. The V1 sector may function as part of the fusion molecules during the formation of the trans-V0 complex.

Alternatively, V1 subunits may be required for the acidification of the vacuole for efficient fusion of vesicles with vacuoles.

3.4.3 *The Assembly of V-ATPase in Prolonged-Starved Cells*

In yeast cells lacking a V1 subunit, the V0 domain is assembled and is targeted to the vacuolar membrane correctly [13]. In contrast, in the absence of a single V0 subunit, the remaining V0 subunits are destabilized [26]. In prolonged-starved cells, Vph1p could be co-immunoprecipitated with the V1 subunit A (Vma1p). The RAVE complex proteins Rav1p and Rav2p that stimulate the assembly of V-ATPase in response to glucose [51] are not involved in the Vid pathway. V-ATPase was assembled properly in prolonged-starved Δ*rav1* and Δ*rav2* mutants [26]. Likewise, in prolonged-starved Δ*vph1* and Δ*stv1* cells, the expression of other V0 or V1 proteins was normal [26]. Furthermore, the majority of Vph1p and Stv1p do not interact under our experimental conditions [26].

In cells grown in rich media, *VMA3* is important to regulate expression of other V0 domain proteins, such as Vma6p and Vph1p [40, 50, 94]. As such, cells lacking *VMA3* had very low levels of Vma6p [40] and Vph1p [26, 40, 50]. In prolonged-starved cells, levels of Vma6p and Vph1p were low, but sub-complexes of V-ATPase were still formed in the absence of the *VMA3* and *VMA6* genes [26].

3.4.4 *The Acidification of the Vacuole is Required for the Vid Pathway*

The vacuole can be acidified by a mechanism independent of V-ATPase. For example, significant ATP-dependent proton-uptake activity was still observed in the Δ*vma2* mutant cells [12] or in cells treated with the V-ATPase-specific inhibitor concanamycin A [12]. Furthermore, concanamycin A did not block homotypic vacuolar fusion [13]. In contrast, drugs that dissipate both V-ATPase-dependent and V-ATPase-independent proton gradients-inhibited homotypic vacuolar fusion [12]. Under our conditions, concanamycin A and bafilomycin B1 did not affect the fusion of Vid vesicles and vacuoles (unpublished data). However, chloroquine and ammonium chloride that neutralize the vacuolar pH impaired the fusion of Vid vesicles with vacuoles [26]. Thus, an acidification independent of V-ATPase is also required for the fusion of Vid vesicles and vacuoles [26].

Vma1p and Vma2p are V1 subunits that are essential for V-ATPase activity [1]. Surprisingly, FBPase degradation was normal in cells lacking the *VMA1* and *VMA2* genes [26]. In prolonged-starved Δ*vma1* and Δ*vma2* mutants, the pH of the media was approximately 6.0. However, when these mutants were grown in media buffered at pH 6.5, FBPase degradation was defective. Therefore, other mechanisms independent of V-ATPase may acidify Vid vesicles and vacuoles in the absence of functional V-ATPase activity in these mutants. For instance, in the absence of V-ATPase, protons can still be taken up from the medium by endocytosis. This is supported by the observation that growing *vma* mutants in media of low pH rescued the growth defects in these mutants [95]. Since the degradation of FBPase is impaired at a higher pH, the Vid pathway is sensitive to changes in the pH [26].

3.4.5 *Distinct Functions of Stv1p and Vph1p in the Vid Pathway*

Stv1p is homologous to Vph1p, but they have distinct functions in the Vid pathway. The Δ*stv1* mutant contained defective vesicles but normal vacuoles [26]. In contrast, the Δ*vph1* mutant contained defective vesicles and defective vacuoles [26]. In the Δ*vph1*, Δ*vma3*, and Δ*vma5* mutants, FBPase was in the fraction enriched with Vid vesicles and was resistant to digestion with proteinase K. In the Δ*stv1* mutant, FBPase was in the Vid-vesicle fraction but was digested by proteinase K. These findings suggest that FBPase is enclosed in membrane-bound Vid vesicles in cells lacking the *VPH1*, *VMA2*, and *VMA5* genes, whereas FBPase is outside Vid vesicles in the Δ*stv1* mutant.

3.4.6 *The Vid Pathway Converges with the Endocytic Pathway*

The Vid pathway utilizes multiple compartments prior to being targeted to the vacuole. At least four different FBPase-containing

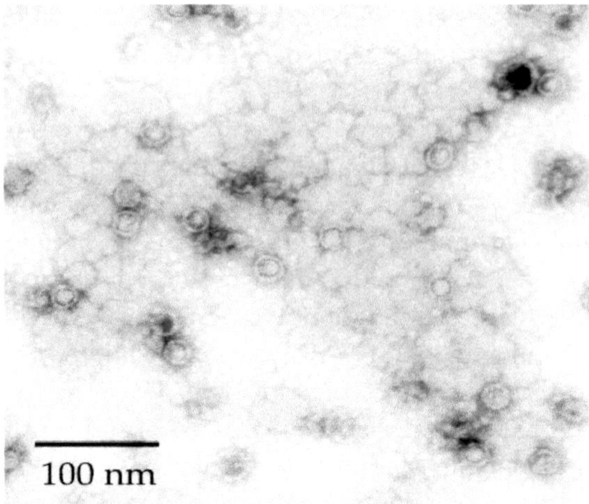

Figure 3.3 Vid/endosomes contain Vid24p, FBPase, and Pep12p. Vid/endosomes are heterogeneous in shape and in size. Vid vesicles are located both inside and outside these organelles.

peaks have been identified using lysates from Δ*pep4* cells that were subjected to differential centrifugation and fractionation on an S-1000 column [83, 87]. The first FBPase-containing peak co-localized with the plasma membrane marker Pma1p. The second and third peaks contained FBPase, Vid24p and the endosomal marker Pep12p. The last peak contained Vid vesicles [87]. Vid/endosomes have irregular shapes and vary in size (Fig. 3.3). Vid24p is inside these Vid/endosomes and also on the surfaces of these compartments [87, 96]. In the Δ*vph1* mutant, FBPase accumulated in Vid/endosomes and Vid vesicles [87], suggesting that the *VPH1* gene has a role in the trafficking of Vid/endosomes to the vacuole [87].

When FBPase distribution was examined at the ultra-structural level, FBPase is found in areas underneath the plasma membrane, suggesting that FBPase is targeted to the plasma membrane prior to being delivered to the vacuole [87]. In yeast, polymerization of actin is required for endocytosis [96, 97]. Actin polymerizes on endocytic vesicles forming on the plasma membrane [97]. Mutants that block the polymerization of actin and endocytosis showed defects in FBPase degradation [96, 98–100]. Moreover, cargo proteins, FBPase and MDH2 associate with actin patches transiently. The Vid-vesicle proteins Sec28p and Vid24p also associate with actin patches following the addition of glucose to prolonged-starved cells [96]. Thus, the Vid pathway merges with the endocytic pathway to deliver cargo proteins to the vacuole for degradation.

Recent studies showed that the Vid pathway also utilizes retrograde transport from the vacuolar membrane [87]. When endocytosis is induced in the absence of retrograde transport, the vacuole will expand. In order to maintain the size of the vacuole, retrograde transport should also be induced. In the *ret2-1* mutant that contains a defective coatomer subunit, Sec28p was targeted to the vacuole during anterograde transport. Following the addition of glucose for a prolonged period of time, Sec28p was in punctate structures that formed from the vacuolar membrane during retrograde transport [87]. These Sec28p-containing retrograde vesicles may become Vid vesicles later. Alternatively, they may travel to the plasma membrane and form Vid vesicles near the plasma membrane.

3.4.7 In vivo *Studies of VPH1 Functions in the Vid Pathway*

To understand how the endocytic pathway converges with the Vid pathway, the uptake of FM 4-64 was examined [87]. FM 4-64 is a lipophilic dye that is internalized from the plasma membrane, targeted to endosomes and then to the vacuole [101]. Following the addition of glucose to prolonged-starved wild-type cells, FM dye was found in endosome-like punctate structures at early time points and then in the vacuoles at 1–2 h. The Vid-vesicle protein, Sec28p was also found in FM-containing endosomes [87].

To track the trafficking of Vid vesicles, the distribution of Sec28p-GFP and FM was investigated in various known mutant strains that block the Vid pathway [87]. Two different patterns of FM staining were observed. Cells lacking Ypt7p (a small GTPase) or Vam3p (vacuolar *t*-SNARE) accumulated numerous small FM-containing vesicles, while the Δ*vid24* or Δ*vph1* strains accumulated FM in large compartments. In the Δ*vam3* mutant, no obvious co-localization of Sec28p-GFP or FBPase-GFP with the FM dye was found. The Δ*vph1* strain accumulated large FM-containing circles at later time points following the addition of glucose. When the distribution of FBPase-GFP was examined in the Δ*vph1* mutant, a significant

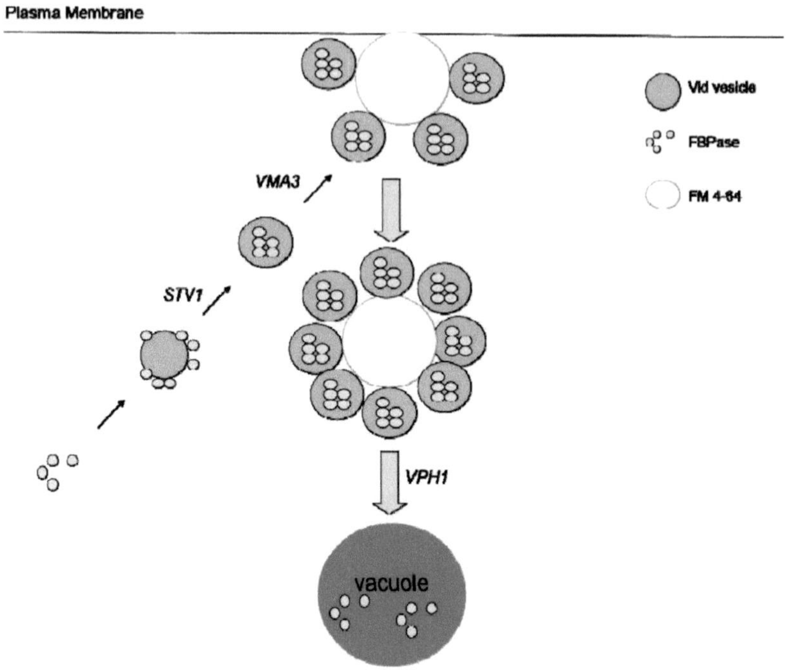

Figure 3.4 V-ATPase is required for the Vid pathway. FBPase is degraded in the Vid pathway when glucose is added to prolonged-starved cells. The *STV1* gene is involved in the sequestration of FBPase into Vid vesicles. The *VAM3* gene controls the convergence of the Vid pathway with the endocytic pathway, whereas the *VPH1* gene is required for the fusion of Vid/endosomes with the vacuole. Vid vesicles are located at the periphery of Vid/endosomes and are responsible for the fusion of Vid/endosomes with vacuoles. In contrast, endosomes are located inside Vid/endosomes and should not participate in the fusion process.

amount of FBPase was in punctate structures stained with FM dye. Our results suggest that *VAM3* is required for the convergence of the Vid pathway with the endocytic pathway, whereas *VPH1* is needed at a later step prior to the fusion of Vid clusters or Vid/endosomes with vacuoles (Fig. 3.4).

Purified Vid/endosomes are reminiscent of multivesicular bodies observed in yeast and other types of cells. Based on our immuno-electron microscopic studies using purified Vid/endosomes [87], Vid vesicles appear to constitute a large portion of

Vid/endosomes, as these vesicles are observed both inside and outside Vid/endosomes. Because Vid vesicles are located outside Vid/endosomes, the fusion of Vid/endosomes with vacuoles is likely to be mediated by Vid vesicles. Endosomes are located inside Vid/endosomes and should not contribute much to the fusion with vacuoles.

3.5 Conclusions and Future Perspectives

V-ATPase is the major protein complex responsible for the acidification of the vacuole [1, 30]. Most of the V0 and V1 subunits of V-ATPase play critical roles in the Vid pathway and are required for the fusion of Vid vesicles with vacuoles [26]. Both V-ATPase-dependent and V-ATPase-independent mechanisms of acidification are responsible for the fusion of vesicles with vacuoles [26].

Vph1p and Stv1p are isoforms of the a-subunit of the V0 domain [38]. In prolonged-starved cells, Stv1p and Vph1p do not form a complex with each other and have distinct roles in the Vid pathway [26, 93]. FBPase was sensitive to digestion with proteinase K in the Δ*stv1* mutant, whereas FBPase was resistant to digestion with proteinase K in the Δ*vph1* mutant. Furthermore, vesicles were the defective compartments in the Δ*stv1* mutant, whereas both Vid vesicles and vacuoles were defective in the Δ*vph1* mutant [26]. Large FM-containing vesicles accumulated in cells lacking *VPH1*, indicating that this gene is involved in a late step of the Vid pathway prior to the fusion of Vid/endosomes with the vacuole [87].

Endocytosis is essential for eukaryotic cells to adapt to their environment in response to extracellular stimuli. Multiple endocytic pathways have been described including fluid-phase endocytosis and receptor-mediated endocytosis [102]. The Δ*stv1* mutant strain is defective in fluid-phase endocytosis [93] and blocks the degradation of FBPase. Therefore, the Vid pathway may merge with the fluid-phase endocytic pathway. Alternatively, the Vid pathway may converge with the endocytic pathway that delivers the galactose receptor Gal2p to the vacuole in response to glucose. In contrast to the vacuolar protein sorting (Vps) pathway where late endosomes are the merging points for the secretory and endocytic pathways, the

Vid pathway converges with the endocytic pathway at earlier steps near the plasma membrane [11].

The convergence of the Vid and endocytic pathways allows cells to remove intracellular and extracellular proteins simultaneously. There are many questions remaining to be answered. Are cargo proteins secreted during glucose starvation? Are they internalized during glucose re-feeding? Where do Vid vesicles come from? Are Vid/endosomes similar to multivesicular bodies? How are Vid/endosomes organized? How do actin patches and Vid vesicles associate and dissociate? A better understanding of the Vid pathway may have important clinical implications. For example, we can activate the Vid pathway to reduce abnormal proteins that accumulate under pathological conditions. Alternatively, we can inactivate the Vid pathway to increase beneficial proteins that are needed in cells. Induced degradation of gluconeogenic enzymes may also be useful to decrease glucose levels in patients with type 2 diabetes.

References

1. Forgac, M. (2000) Structure, mechanism and regulation of the clathrin-coated vesicle and yeast vacuolar H(+)-ATPases. *J. Exp. Biol.* 203(Pt 1): pp. 71–80.
2. Kibak, H., *et al.* (1992) Evolution of structure and function of V-ATPases. *J. Bioenerg. Biomembr.* 24(4): pp. 415–24.
3. Graham, L.A., Flannery, A.R. and Stevens T.H. (2003) Structure and assembly of the yeast V-ATPase. *J. Bioenerg. Biomembr.* 35(4): pp. 301–312.
4. Brown, D., *et al.* (2009) Regulation of the V-ATPase in kidney epithelial cells: Dual role in acid-base homeostasis and vesicle trafficking. *J. Exp. Biol.* 212(Pt 11): pp. 1762–1772.
5. Shum, W.W., *et al.* (2009) Regulation of luminal acidification in the male reproductive tract via cell-cell crosstalk. *J. Exp. Biol.* 212(Pt 11): pp. 1753–1761.
6. Nanda, A., *et al.* (1996) Activation of proton pumping in human neutrophils occurs by exocytosis of vesicles bearing vacuolar-type H^+-ATPases. *J. Biol. Chem.* 271(27): pp. 15963–15970.

7. Kinouchi, K., Ichihara, A., and Itoh. H. (2012) Functional characterization of (pro)renin receptor in association with V-ATPase. *Front. Biosci.* 17: pp. 3216–3223.

8. Seeley, E.S., *et al.* (2002) Genomic analysis of homotypic vacuole fusion. *Mol. Biol. Cell* 13(3): pp. 782–794.

9. Wilson, W.A., Wang, Z., and Roach, PJ. (2002) Systematic identification of the genes affecting glycogen storage in the yeast Saccharomyces cerevisiae: Implication of the vacuole as a determinant of glycogen level. *Mol. Cell. Proteomics* 1(3): pp. 232–242.

10. Bonangelino, C.J. Chavez, E.M. and Bonifacino, J.S. (2002) Genomic screen for vacuolar protein sorting genes in Saccharomyces cerevisiae. *Mol. Biol. Cell* 13(7): pp. 2486–2501.

11. Gerrard, S.R., Bryant, N.J., and Stevens, TH. (2000) VPS21 controls entry of endocytosed and biosynthetic proteins into the yeast prevacuolar compartment. *Mol. Biol. Cell* 11(2): pp. 613–626.

12. Bayer, M.J., *et al.* (2003) Vacuole membrane fusion: V0 functions after trans-SNARE pairing and is coupled to the Ca^{2+}-releasing channel. *J. Cell. Biol.* 162(2): pp. 211–222.

13. Peters, C., *et al.* (2001) Trans-complex formation by proteolipid channels in the terminal phase of membrane fusion. *Nature* 409(6820): pp. 581–588.

14. Dechant, R., *et al.* (2010) Cytosolic pH is a second messenger for glucose and regulates the PKA pathway through V-ATPase. *EMBO J.* 29(15): pp. 2515–2526.

15. Hettiarachchi, K.D., *et al.* (2008) Transplacental exposure to the vacuolar-ATPase inhibitor bafilomycin disrupts survival signaling in beta cells and delays neonatal remodeling of the endocrine pancreas. *Exp. Toxicol. Pathol.* 60(4–5): pp. 295–306.

16. Choi, Y.O., *et al.* (2007) Involvement of vesicular H^+-ATPase in insulin-stimulated glucose transport in 3T3-F442A adipocytes. *Endocr. J.* 54(5): pp. 733–743.

17. Perriello, G., *et al.* (1997) Evidence of increased systemic glucose production and gluconeogenesis in an early stage of NIDDM. *Diabetes* 46(6): pp. 1010–1016.

18. Erion, M.D., *et al.* (2005) MB06322 (CS-917): A potent and selective inhibitor of fructose 1,6-bisphosphatase for controlling gluconeogenesis in type 2 diabetes. *Proc. Natl. Acad. Sci. USA.* 102(22): pp. 7970–7975.

19. Wang, Y. and Tomlinson, B. (2007) Managlinat dialanetil, a fructose-1,6-bisphosphatase inhibitor for the treatment of type 2 diabetes. *Curr. Opin. Investig. Drugs* 8(10): pp. 849–858.

20. Chiang, H.L. and Schekman, R. (1991) Regulated import and degradation of a cytosolic protein in the yeast vacuole. *Nature* 350(6316): pp. 313–318.
21. Gancedo, J.M. (1998) Yeast carbon catabolite repression. *Microbiol. Mol. Biol. Rev.* 62(2): pp. 334–361.
22. Gancedo, C. (1971) Inactivation of fructose-1,6-diphosphatase by glucose in yeast. *J. Bacteriol.* 107(2): pp. 401–405.
23. Alibhoy, A.A. and Chiang, H.L. (2011) Vacuole import and degradation pathway: Insights into a specialized autophagy pathway. *World J. Biol. Chem.* 2(11): pp. 239–245.
24. Hung, G.C., *et al.* (2004) Degradation of the gluconeogenic enzymes fructose-1,6-bisphosphatase and malate dehydrogenase is mediated by distinct proteolytic pathways and signaling events. *J. Biol. Chem.* 279(47): pp. 49138–49150.
25. Brown, C.R., *et al.* (2010) The TOR complex 1 is distributed in endosomes and in retrograde vesicles that form from the vacuole membrane and plays an important role in the vacuole import and degradation pathway. *J. Biol. Chem.* 285(30): pp. 23359–23370.
26. Liu, J., Brown, C.R., and Chiang, H.L. (2005) Degradation of the gluconeogenic enzyme fructose-1, 6-bisphosphatase is dependent on the vacuolar ATPase. *Autophagy* 1(3): pp. 146–56.
27. Nakanishi-Matsui, M., *et al.* (2010) The mechanism of rotating proton pumping ATPases. *Biochim. Biophys. Acta.* 1797(8): pp. 1343–1352.
28. Feng, Y. and Forgac, M. (1994) Inhibition of vacuolar H(+)-ATPase by disulfide bond formation between cysteine 254 and cysteine 532 in subunit A. *J. Biol. Chem.* 269(18): pp. 13224–13230.
29. Zhang, J., Feng, Y., and Forgac, M. (1994) Proton conduction and bafilomycin binding by the V0 domain of the coated vesicle V-ATPase. *J. Biol. Chem.* 269(38): pp. 23518–23523.
30. Stevens, T.H. (1992) The structure and function of the fungal V-ATPase. *J. Exp. Biol.* 172: pp. 47–55.
31. Jefferies, K.C., Cipriano, D.J. and Forgac, M. (2008) Function, structure and regulation of the vacuolar (H+)-ATPases. *Arch. Biochem. Biophys.* 476(1): pp. 33–42.
32. Hirata, R., *et al.* (1990) Molecular structure of a gene, VMA1, encoding the catalytic subunit of H(+)-translocating adenosine triphosphatase from vacuolar membranes of Saccharomyces cerevisiae. *J. Biol. Chem.* 265(12): pp. 6726–3733.

33. Ghislain, M. and Bowman, E.J. (1992) Sequence of the genes encoding subunits A and B of the vacuolar H(+)-ATPase of Schizosaccharomyces pombe. *Yeast* 8(9): pp. 791–799.

34. Kawasaki-Nishi, S., Nishi, T., and Forgac, M. (2003) Proton translocation driven by ATP hydrolysis in V-ATPases. *FEBS Lett.* 545(1): pp. 76–85.

35. Ohya, Y., *et al.* (1991) Calcium-sensitive cls mutants of Saccharomyces cerevisiae showing a Pet-phenotype are ascribable to defects of vacuolar membrane H(+)-ATPase activity. *J. Biol. Chem.* 266(21): pp. 13971–13977.

36. Ho, M.N., *et al.* (1993) VMA13 encodes a 54-kDa vacuolar H(+)-ATPase subunit required for activity but not assembly of the enzyme complex in Saccharomyces cerevisiae. *J. Biol. Chem.* 268(24): pp. 18286–18292.

37. Supekova, L., *et al.* (1996) A novel subunit of vacuolar H(+)-ATPase related to the b subunit of F-ATPases. *J. Exp. Biol.* 199(Pt 5): pp. 1147–1156.

38. Manolson, M.F., *et al.* (1994) STV1 gene encodes functional homologue of 95-kDa yeast vacuolar H(+)-ATPase subunit Vph1p. *J. Biol. Chem.* 269(19): pp. 14064–14074.

39. Hirata, R., *et al.* (1997) VMA11 and VMA16 encode second and third proteolipid subunits of the Saccharomyces cerevisiae vacuolar membrane H^+-ATPase. *J. Biol. Chem.* 272(8): pp. 4795–47803.

40. Bauerle, C., *et al.* (1993) The Saccharomyces cerevisiae VMA6 gene encodes the 36-kDa subunit of the vacuolar H(+)-ATPase membrane sector. *J. Biol. Chem.* 268(17): pp. 12749–12757.

41. Compton, M.A., Graham, L.A., and Stevens, T.H. (2006) Vma9p (subunit e) is an integral membrane V0 subunit of the yeast V-ATPase. *J. Biol. Chem.* 281(22): pp. 15312–15319.

42. Leng, X.H., *et al.* (1996) Site-directed mutagenesis of the 100-kDa subunit (Vph1p) of the yeast vacuolar (H+)-ATPase. *J. Biol. Chem.* 271(37): pp. 22487–22493.

43. Arai, H., Berne, M., and M. Forgac. (1987) Inhibition of the coated vesicle proton pump and labeling of a 17,000-dalton polypeptide by N,N'-dicyclohexylcarbodiimide. *J. Biol. Chem.* 262(23): pp. 11006–11011.

44. Hill, K.J. and Stevens, T.H. (1994) Vma21p is a yeast membrane protein retained in the endoplasmic reticulum by a di-lysine motif and is required for the assembly of the vacuolar H(+)-ATPase complex. *Mol. Biol. Cell* 5(9): pp. 1039–1050.

45. Hill, K.J. and Stevens, T.H. (1995) Vma22p is a novel endoplasmic reticulum-associated protein required for assembly of the yeast vacuolar H(+)-ATPase complex. *J. Biol. Chem.* 270(38): pp. 22329–22336.
46. Hirata, R., *et al.* (1993) VMA12 is essential for assembly of the vacuolar H(+)-ATPase subunits onto the vacuolar membrane in Saccharomyces cerevisiae. *J. Biol. Chem.* 268(2): pp. 961–967.
47. Jackson, D.D. and Stevens, TH. (1997) VMA12 encodes a yeast endoplasmic reticulum protein required for vacuolar H^+-ATPase assembly. *J. Biol. Chem.* 272(41): pp. 25928–25934.
48. Malkus, P., *et al.* (2004) Role of Vma21p in assembly and transport of the yeast vacuolar ATPase. *Mol. Biol. Cell* 15(11): pp. 5075–5091.
49. Kane, P.M. (1995) Disassembly and reassembly of the yeast vacuolar H(+)-ATPase *in vivo*. *J. Biol. Chem.* 270(28): pp. 17025–17032.
50. Kane, P.M., *et al.* (1992) Assembly and targeting of peripheral and integral membrane subunits of the yeast vacuolar H(+)-ATPase. *J. Biol. Chem.* 267(1): pp. 447–454.
51. Smardon, A.M., Tarsio, M., and Kane, PM. (2002) The RAVE complex is essential for stable assembly of the yeast V-ATPase. *J. Biol. Chem.* 277(16): pp. 13831–13839.
52. Seol, J.H., Shevchenko, A., and Deshaies, RJ. (2001) Skp1 forms multiple protein complexes, including RAVE, a regulator of V-ATPase assembly. *Nat. Cell. Biol.* 3(4): pp. 384–391.
53. Parra, K.J. and Kane P.M. (1998) Reversible association between the V1 and V0 domains of yeast vacuolar H^+-ATPase is an unconventional glucose-induced effect. *Mol. Cell. Biol.* 18(12): pp. 7064–7074.
54. Su, Y., *et al.* (2003) The a-subunit of the V-type H^+-ATPase interacts with phosphofructokinase-1 in humans. *J. Biol. Chem.* 278(22): pp. 20013–20018.
55. Lu, M., *et al.* (2004) The glycolytic enzyme aldolase mediates assembly, expression, and activity of vacuolar H^+-ATPase. *J. Biol. Chem.* 279(10): pp. 8732–8739.
56. Tisdale, E.J., Kelly, C., and Artalejo, C.R. (2004) Glyceraldehyde-3-phosphate dehydrogenase interacts with Rab2 and plays an essential role in endoplasmic reticulum to Golgi transport exclusive of its glycolytic activity. *J. Biol. Chem.* 279(52): pp. 54046–54052.
57. Sirover, M.A. (1999). New insights into an old protein: the functional diversity of mammalian glyceraldehyde-3-phosphate dehydrogenase. *Biochim. Biophys. Acta.* 1432(2): pp. 159–184.

58. Sun-Wada, G.H., *et al.* (2006) The a3 isoform of V-ATPase regulates insulin secretion from pancreatic beta-cells. *J. Cell. Sci.* 119(Pt 21): pp. 4531–4540.
59. Sorkin, A.D., Teslenko, L.V., and Nikolsky, N.N. (1988) The endocytosis of epidermal growth factor in A431 cells: A pH of microenvironment and the dynamics of receptor complex dissociation. *Exp. Cell. Res.* 175(1): pp. 192–205.
60. Nishi, T. and Forgac, M. (2002) The vacuolar (H+)-ATPases: Nature's most versatile proton pumps. *Nat. Rev. Mol. Cell. Biol.* 3(2): pp. 94–103.
61. Moriyama, Y., Maeda, M., and Futai, M. (1992) The role of V-ATPase in neuronal and endocrine systems. *J. Exp. Biol.* 172: pp. 171–178.
62. Sorkin, A. and Duex, J.E. (2010) Quantitative analysis of endocytosis and turnover of epidermal growth factor (EGF) and EGF receptor. *Curr. Protoc. Cell. Biol.*, Chapter 15: Unit 15.14.
63. Pfeffer, S.R. (1988) Mannose 6-phosphate receptors and their role in targeting proteins to lysosomes. *J. Membr. Biol.* 103(1): pp. 7–16.
64. Cortese, K., *et al.* (2008) Clathrin and LRP-1-independent constitutive endocytosis and recycling of uPAR. *PLoS One.* 3(11): p. e3730.
65. Goldstein, J.L. and Brown, M.S. (2009) The LDL receptor. *Arterioscler Thromb. Vasc. Biol.* 29(4): pp. 431–438.
66. Hiesinger, P.R., *et al.* (2005) The v-ATPase V0 subunit a1 is required for a late step in synaptic vesicle exocytosis in Drosophila. *Cell* 121(4): pp. 607–620.
67. Ungermann, C., Wickner, W., and Xu, Z. (1999) Vacuole acidification is required for trans-SNARE pairing, LMA1 release, and homotypic fusion. *Proc. Natl. Acad. Sci. USA* 96(20): pp. 11194–11199.
68. Lindau, M. and Alvarez de Toledo, G. (2003) The fusion pore. *Biochim. Biophys. Acta.* 1641(2–3): pp. 167–173.
69. Peters, C. and Mayer, A. (1998) Ca^{2+}/calmodulin signals the completion of docking and triggers a late step of vacuole fusion. *Nature* 396(6711): pp. 575–580.
70. Sanz, P. (2007) Yeast as a model system to study glucose-mediated signalling and response. *Front. Biosci.* 12: pp. 2358–2371.
71. Holzer, H. (1989) Proteolytic catabolite inactivation in Saccharomyces cerevisiae. *Revis. Biol. Celular.* 21: pp. 305–319.
72. Gamo, F.J., *et al.* (1994) Catabolite inactivation of heterologous fructose-1,6-bisphosphatases and fructose-1,6-bisphosphatase-beta-galactosidase fusion proteins in Saccharomyces cerevisiae. *Eur. J. Biochem.* 222(3): pp. 879–884.

73. Zubenko, G.S. and Jones, E.W. (1979) Catabolite inactivation of gluconeogenic enzymes in mutants of yeast deficient in proteinase B. *Proc. Natl. Acad. Sci. USA* 76(9): pp. 4581–4585.
74. Minard, K.I. and McAlister-Henn, L. (1992) Glucose-induced degradation of the MDH2 isozyme of malate dehydrogenase in yeast. *J. Biol. Chem.* 267(24): pp. 17458–17464.
75. Ordiz, I., *et al.* (1995) Glucose-induced inactivation of isocitrate lyase in Saccharomyces cerevisiae is mediated by an internal decapeptide sequence. *FEBS Lett.* 367(3): pp. 219–222.
76. Krampe, S., *et al.* (1998) Catabolite inactivation of the high-affinity hexose transporters Hxt6 and Hxt7 of Saccharomyces cerevisiae occurs in the vacuole after internalization by endocytosis. *FEBS Lett.* 441(3): pp. 343–347.
77. Chiang, H.L., Schekman, R., and Hamamoto, S. (1996) Selective uptake of cytosolic, peroxisomal, and plasma membrane proteins into the yeast lysosome for degradation. *J. Biol. Chem.* 271(17): pp. 9934–9941.
78. Horak, J., Regelmann, J., and Wolf, D.H. (2002) Two distinct proteolytic systems responsible for glucose-induced degradation of fructose-1,6-bisphosphatase and the Gal2p transporter in the yeast Saccharomyces cerevisiae share the same protein components of the glucose signaling pathway. *J. Biol. Chem.* 277(10): pp. 8248–8254.
79. Hutchins, M.U., Veenhuis, M., and Klionsky, D.J. (1999) Peroxisome degradation in Saccharomyces cerevisiae is dependent on machinery of macroautophagy and the Cvt pathway. *J. Cell. Sci.* 112(Pt 22): pp. 4079–4087.
80. Schork, S.M., Thumm, M., and Wolf, D.H. (1995) Catabolite inactivation of fructose-1,6-bisphosphatase of Saccharomyces cerevisiae. Degradation occurs via the ubiquitin pathway. *J. Biol. Chem.* 270(44): pp. 26446–26450.
81. Alibhoy, A.A., *et al.* (2012) Vid30 is required for the association of Vid vesicles and actin patches in the vacuole import and degradation pathway. *Autophagy* 8(1): 29–46.
82. Shieh, H.L. and Chiang, H.L. (1998) *In vitro* reconstitution of glucose-induced targeting of fructose-1, 6-bisphosphatase into the vacuole in semi-intact yeast cells. *J. Biol. Chem.* 273(6): pp. 3381–3387.
83. Hoffman, M. and Chiang, H.L. (1996) Isolation of degradation-deficient mutants defective in the targeting of fructose-1,6-bisphosphatase into the vacuole for degradation in *Saccharomyces cerevisiae. Genetics* 143(4): pp. 1555–1566.

84. Huang, P.H. and Chiang, H.L. (1997) Identification of novel vesicles in the cytosol to vacuole protein degradation pathway. *J. Cell. Biol.* 136(4): pp. 803–810.

85. Shieh, H.L., *et al.* (2001) Biochemical analysis of fructose-1,6-bisphosphatase import into vacuole import and degradation vesicles reveals a role for UBC1 in vesicle biogenesis. *J. Biol. Chem.* 276(13): pp. 10398–10406.

86. Chiang, M.C. and Chiang, H.L. (1998) Vid24p, a novel protein localized to the fructose-1,6-bisphosphatase-containing vesicles, regulates targeting of fructose-1,6-bisphosphatase from the vesicles to the vacuole for degradation. *J. Cell. Biol.* 140(6): pp. 1347–1356.

87. Brown, C.R., *et al.* (2008) The vacuolar import and degradation pathway merges with the endocytic pathway to deliver fructose-1,6-bisphosphatase to the vacuole for degradation. *J. Biol. Chem.* 283(38): pp. 26116–26127.

88. Duden, R., *et al.* (1991) Beta-COP, a 110 kD protein associated with non-clathrin-coated vesicles and the Golgi complex, shows homology to beta-adaptin. *Cell* 64(3): pp. 649–665.

89. Orci, L., *et al.* (1994) Coatomer-rich endoplasmic reticulum. *Proc. Natl. Acad. Sci. USA* 91(25): pp. 11924–11928.

90. Whitney, J.A., *et al.* (1995) Cytoplasmic coat proteins involved in endosome function. *Cell* 83(5): pp. 703–713.

91. Piguet, V., *et al.* (1999) Nef-induced CD4 degradation: a diacidic-based motif in Nef functions as a lysosomal targeting signal through the binding of beta-COP in endosomes. *Cell* 97(1): pp. 63–73.

92. Brown, C.R., *et al.* (2003) The Vid vesicle to vacuole trafficking event requires components of the SNARE membrane fusion machinery. *J. Biol. Chem.* 278(28): pp. 25688–25699.

93. Perzov, N., *et al.* (2002) Characterization of yeast V-ATPase mutants lacking Vph1p or Stv1p and the effect on endocytosis. *J. Exp. Biol.* 205(Pt 9): pp. 1209–1219.

94. Doherty, R.D. and Kane, P.M. (1993) Partial assembly of the yeast vacuolar H(+)-ATPase in mutants lacking one subunit of the enzyme. *J. Biol. Chem.* 268(22): pp. 16845–16851.

95. Sambade, M., *et al.* (2005) A genomic screen for yeast vacuolar membrane ATPase mutants. *Genetics* 170(4): pp. 1539–1551.

96. Brown, C.R., Dunton, D. and Chiang, H.L. (2010) The vacuole import and degradation pathway utilizes early steps of endocytosis and actin polymerization to deliver cargo proteins to the vacuole for degradation. *J. Biol. Chem.* 285(2): pp. 1516–1528.

97. Kaksonen, M., Sun, Y and Drubin, D.G. (2003) A pathway for association of receptors, adaptors, and actin during endocytic internalization. *Cell* 115(4): pp. 475–487.
98. Carreno, S., *et al.* (2004) Actin dynamics coupled to clathrin-coated vesicle formation at the trans-Golgi network. *J. Cell. Biol.* 165(6): pp. 781–788.
99. Merrifield, C.J., Perrais, D., and Zenisek, D. (2005) Coupling between clathrin-coated pit invagination, cortactin recruitment, and membrane scission observed in live cells. *Cell* 121(4): pp. 593–606.
100. Munn, A.L. and Riezman, H. (1994) Endocytosis is required for the growth of vacuolar H(+)-ATPase-defective yeast: Identification of six new END genes. *J. Cell. Biol.* 127(2): pp. 373–386.
101. Vida, T.A. and Emr, S.D. (1995) A new vital stain for visualizing vacuolar membrane dynamics and endocytosis in yeast. *J. Cell. Biol.* 128(5): pp. 779–792.
102. Mukherjee, S., Ghosh, R.N. and Maxfield, F.R. (1997) Endocytosis. *Physiol. Rev.* 77(3): pp. 759–803.

Chapter 4

The Role of Vacuolar ATPase in the Regulation of Npt2a Trafficking

Eleanor Lederer[a,b,c] and Syed Jalal Khundmiri[a,b,c]
[a] *Robley Rex VA Medical Center, University of Louisville School of Medicine, Louisville, KY 40202, USA*
[b] *Department of Medicine, University of Louisville School of Medicine, Louisville, KY 40202, USA*
[c] *Department of Physiology and Biophysics, University of Louisville School of Medicine, Louisville, KY 40202, USA*
e.lederer@louisville.edu, syed.khundmiri@louisville.edu

4.1 Introduction

The type 2 sodium phosphate co-transporter, Npt2a (SLC34A1) is one among three known sodium coupled phosphate transporters responsible for the re-absorption of filtered phosphate from the lumen of the proximal renal tubule [1–5]. This highly regulated protein is expressed in the apical membrane of renal proximal tubule cells and is thought to be the major protein responsible for maintenance of total body phosphate homeostasis. In rodents, Npt2a accounts for over 70% of proximal tubule phosphate re-absorption [1, 6–9]. Although similar analysis has not been possible in human subjects, several studies point toward a role for Npt2a in regulation

Handbook of H^+-ATPases
Edited by Suguru Nakamura

ISBN 978-981-4411-91-2 (Hardcover), 978-981-4411-92-9 (eBook)
www.panstanford.com

of serum phosphorus concentration, risk of chronic kidney disease, and risk of cardiovascular disease [10–13]. Published literature over the past few decades has outlined a role for pH, and specifically for vacuolar ATPase, in the regulation of the function, expression, and trafficking of Npt2a. This chapter will review the current status of knowledge on the relationship between vacuolar H^+-ATPase and Npt2a in renal proximal tubule.

4.2 Npt2a: Structure, Function, and Regulation

Npt2a is an intrinsic membrane protein exhibiting 10 or 12 transmembrane domains, predicted by hydropathy analysis [14–16]. Both amino and carboxyl termini are intracellular and express multiple protein binding domains including a terminal PDZ binding domain, SH3 domains, and consensus sequences for kinase phosphorylation. Post-translational modifications that have been identified for this protein include phosphorylation, glycosylation, and disulfide bond formation. Mutation analysis has identified regions of the protein responsible for response to parathyroid hormone (PTH), a major regulator of Npt2a, transmembrane ion transport, and binding to anchoring and scaffolding proteins. Interestingly, the amino acid residues responsible for apical membrane localization have not been identified. Although the protein can function as a monomer, studies suggest it may exist in multimeric form, the significance of which is not clear. Under experimental conditions where the protein can be forced into either monomeric or multimeric configurations, phosphate transport does not appear to be affected [17]. The mature protein exists as a heavily glycosylated and phosphorylated protein [18–21]. The roles that these two modifications play in the function of the protein remain controversial. Studies suggest that glycosylation facilitates apical membrane trafficking and localization but removal of the putative glycosylation sites by site specific mutagenesis and expression in oocytes showed that glycosylation is not necessary for apical membrane expression of Npt2a. Likewise, the role of phosphorylation of Npt2a remains unclear and largely unstudied.

In contrast, the role of associated proteins in the expression and function of Npt2a is well established [21–23]. A central protein is the Na–H exchanger regulatory factor isoform 1 (NHERF1), a complex scaffolding protein exhibiting two PDZ domains and a C-terminal ezrin binding domain [24–26]. A series of studies in animal and cell culture models have demonstrated that NHERF1 anchors Npt2a in the apical membrane and that interference with NHERF1–Npt2a association results in diminished apical membrane expression of Npt2a and response to physiologic regulators such as dietary or ambient phosphate and PTH. The NHERF1-deficient mouse shows chronic hypophosphatemia, phosphaturia, and decreased apical membrane expression of Npt2a when compared to wild-type mice. Treatment of these mice with either alteration in dietary phosphate or with PTH has no effect on Npt2a expression. Blocking Npt2a–NHERF1 association by introduction of a competing peptide into cells blocks apical membrane localization of Npt2a. Cell culture studies have also suggested a role for NHERF1 in regulation of Npt2a transcription and apical membrane trafficking of Npt2a. Recent studies from our laboratory suggest that Npt2a and NHERF1 may associate at the level of the Golgi apparatus through the C-terminal PDZ domain of Npt2a. An Npt2a construct lacking the PDZ domain does not associate with NHERF1 and shows faulty apical membrane localization (unpublished data).

Npt2a is an electrogenic transporter, coupling the transport of three sodium ions with one phosphate (HPO_4^{2-}) [14, 15]. An initial binding of a sodium ion allows binding of the phosphate ion followed by the binding and transmembrane transport of two additional sodium ions with the phosphate ion. The divalent form of phosphate is the preferred substrate. pH has a significant effect on the transport function [27–33]. Rising hydrogen ion concentrations in the presence of a decrease in pH results in titration of phosphate to the monovalent form, which is transported far less efficiently than the divalent form. Although large fluctuations in sodium concentration have demonstrable effects on the transporter, these changes are unlikely to have significant clinical relevance as proximal tubule fluid sodium concentration will be nearly identical to plasma sodium, that is, $>$100 mM with little variation.

The function of Npt2a is largely determined by expression level, that is, the greater number of transporters, the greater phosphate reabsorption [2, 9]. Potassium balance and cholesterol content of the membrane are two recognized regulators of Npt2a activity, distinct from transporter number; however, transporter expression is the predominating mechanism for regulation of phosphate transport [34]. The major regulators of Npt2a are dietary phosphate, PTH, fibroblast growth factor 23 (FGF23), and dopamine (DA) [1, 19, 35–38, 39–47]. High dietary phosphate, PTH, FGF23, and DA all decrease Npt2a expression and thus phosphate uptake, while low dietary phosphate has the opposite effect. A number of other circulating agents such as angiotensin II, insulin, and glucocorticoids; sex steroids and gender; and age also contribute to the regulation of renal proximal tubule transport but generally exert relatively small modulatory effects [29, 48–52].

The factors controlling synthesis and apical trafficking of Npt2a remain largely unknown. Under usual conditions, little protein is identifiable in intracellular compartments, thus the precise pathway taken by Npt2a to traffic from its initial translation to apical membrane expression has not been defined. The extensive glycosylation of the protein strongly suggests that it traverses the well-described ER–TGN process prior to apical trafficking [18, 19, 21]. The protein exists in the apical membrane in a protein complex that includes NHERF1, ezrin, the PTH receptor, protein kinase A regulatory and catalytic subunits, an A-kinase anchoring protein, protein phosphatase 2b, and receptor for activated C-kinase, isoform 1 (RACK1) [21–23]. Each of these protein interactions represents a potential site for regulation of the expression and function of Npt2a. Preliminary data from our laboratory (unpublished) suggests that the association between NHERF1 and Npt2a occurs prior to insertion into the apical membrane. Such pre-assembly of protein complexes has been described for a number of other transporters and signaling molecules [53]. Npt2a that has been removed from the apical membrane is not recycled; thus, these proteins that are internalized are subject to degradation in lysosomes [40]. Inhibitors of proteasomal degradation have little to no effect on Npt2a degradation, suggesting that lysosomes are the major degradation pathway for Npt2a (Fig. 4.1).

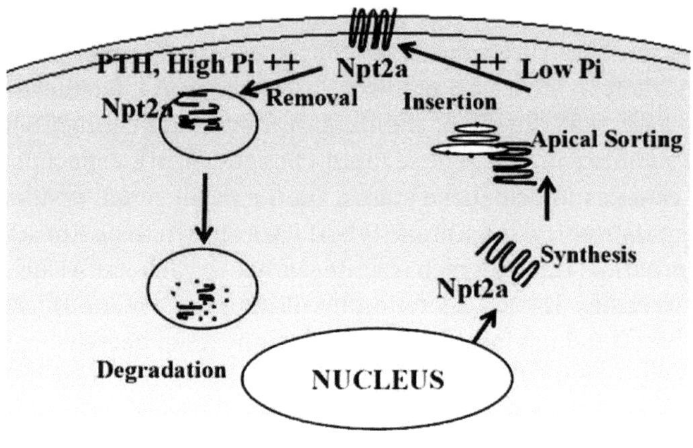

Figure 4.1 Trafficking of Npt2a.

4.2.1 *Role of pH in Regulation of Npt2a Function*

Regulation of protein function, specifically transporter function, by pH is well documented [1, 9, 27–33]. The transport activity of Npt2a increases with increase in pH up to a pH of 8.5 and decreases with decrease in pH down to a pH of 6. As stated previously, several mechanisms contribute to the pH effect including the alterations in the concentration of the preferred substrate, alterations in the affinity of the transporter for phosphate, and possibly voltage changes. Seminal studies by the Murer laboratory and others showed that alteration of ambient pH produced a dramatic effect on the transport activity of Npt2a, while the effect on expression of Npt2a was modest. Interestingly, decreasing the pH of medium surrounding cultured proximal renal tubule cells or acidifying filtered proximal tubule fluid by administration of NH_4Cl did not substantially alter the upregulation of Npt2a stimulated by incubation in low phosphate medium or by ingestion of a low phosphate diet. However, the increase in phosphate transport that would be expected from Npt2a upregulation was significantly blunted.

These findings correlate well with the clinical observation that metabolic acidosis causes phosphaturia. However, the mechanism for the phosphaturic effect of systemic pH on apical membrane Npt2a function is not really clear. First, although Npt2a-mediated phosphate transport varies through the above-described range of pH, systemic pH rarely varies by that degree, meaning that filtered fluid is likely to have a pH within 0.2 pH units of 7.4. Furthermore, it is not clear that the pH of the fluid in immediate contact with the apical membrane will reflect rapid changes in pH, especially with acute changes in acid–base status. On the other hand, systemic pH may regulate apical membrane H^+-ATPase known to be expressed in renal proximal tubule which is quite similar to vacuolar ATPase [54–61]. Increasing H^+ ion secretion by plasma membrane H^+-ATPase could result in high local H^+ concentrations, thus decreasing Npt2a transport activity leading to phosphaturia. Interestingly, a recent study reported that metabolic acidosis increases the release of FGF23 from bone, suggesting that some of the decrease in phosphate transport with metabolic acidosis may be related to FGF23-induced decrease in apical membrane Npt2a expression [62]. As FGF23 regulation of Npt2a expression becomes apparent after a relatively prolonged exposure, this mechanism likely reflects a long-term mechanism for regulation. In contrast, Sun *et al.* [63] were unable to show phosphaturia in the GPR4 knock out mouse model, lacking a known proton receptor. These considerations highlight a significant gap in our understanding of ion transport, in that while many studies of acute regulation have been performed, few studies of chronic regulation have been published. Whether plasma membrane H^+-ATPase plays a role in acute regulation of Npt2a function, chronic regulation of Npt2a function, or both remains unknown.

The pH of intracellular organelles also has effects on transporter expression and function [36, 64–80]. Progressive acidification of intracellular organelle compartments is well described. Following endocytosis of plasma membrane containing proteins such as Npt2a, the vesicles undergo progressive acidification through the lysosomal compartment. Lysosomal degradation of internalized proteins is dependent on acidification of the lysosomes by vacuolar ATPase which, by secreting a H^+ into the lysosomal vesicle, activates the lysosomal enzymes responsible for protein degradation. Similarly to what is seen after endocytosis, acidification of intracellular vesicles also regulates forward trafficking from the trans-Golgi network to the plasma membrane. Studies from a number of laboratories have demonstrated that members of the CLC family of chloride channels serve as Cl/H^+ exchangers, altering intravesicle pH, thus facilitating exocytosis of vesicular contents [53, 67, 74, 77, 81–91].

4.2.2 *The Role of Vacuolar H^+-ATPase in Npt2a Trafficking*

Although the precise pathways for both exocytosis and endocytosis of Npt2a remain undefined, vacuolar ATPases play a documented role in the trafficking of Npt2a. Inhibition of vacuolar ATPases blocks degradation of Npt2a induced by treatment of kidney tubules or proximal tubule cells with PTH [1, 40, 43, 92]. It is clear that the H^+ secreting role of vacuolar ATPase is required for this process as maneuvers that alkalinize the lysosomes such as chloroquine treatment also inhibit lysosomal degradation of Npt2a. Experiments where lysosomal degradation has been inhibited demonstrate that PTH or high phosphate still are able to inhibit apical membrane expression of Npt2a and phosphate transport but result in intracellular accumulation of Npt2a. Vacuolar ATPase has been identified on early and late endosomes [56, 58, 65, 93]. While inhibition of vacuolar ATPases has no effect on endocytosis of Npt2a stimulated by either PTH or high phosphate, loss of vacuolar ATPase function results in faulty transport of proteins from endosomes to lysosomes [40, 92]. Thus, vacuolar ATPases are critical for physiologic regulation of the Npt2a degradation pathway at two sites, transport of proteins from endosome to lysosome and ultimate degradation by lysosomal enzymes.

Vacuolar ATPases also play a role in the regulation of forward trafficking of several proteins including Npt2a. Our laboratory in collaboration with the laboratories of Dr. Jon Klein and Dr. Moshe Levi demonstrated that changes in dietary phosphate content in rats resulted in dramatic changes in brush border membrane-associated proteins, including specific chains of vacuolar ATPase [92]. When rats fed a chronically high phosphate diet were abruptly fed with a low phosphate diet, there was an increase in the brush border membrane association with the A chain subunit of vacuolar ATPase. Because ingestion of a low phosphate diet stimulates insertion of Npt2a into the brush border membrane, this finding suggested that vacuolar ATPases may be involved in the regulation of forward trafficking of Npt2a (Table 4.1). Subsequent studies showed that

Table 4.1 Differentially expressed proteins in BBMs of rats shifted from chronic high Pi to acute low Pi diet, as identified by proteomic analysis

Ref No.	Protein	GenInfo	Fold decrease (chronic high to acute low Pi)	pI	M_W
1	H^+-transporting ATPase (EC 3.6.3.6) chain A, vacuolar (VA68 type)	gi|1082432	0.4	5.3	68.63
2	DnaK-type molecular chaperone HSP72-ps1	gi|347019	0.4	5.4	71.14
3	Class I beta-tubulin	gi|3551054	0.58	4.8	50.11
4	ChainB, F1-ATPase	gi|6729935	0.57	4.9	51.33
5	Mid-1-related chloride channel 4	gi|17736806	0.52	5.6	42.21
6	NHERF-1 Mouse Isoform	gi|6755566	0.45	5.6	38.87
7	Apolipoprotein A-I	gi|6978515	0.46	5.5	30.10
8	Fructose-1,6-bisphosphatase	gi|119740	0.47	5.5	40.05
9	Serine/threonine kinase receptor associated protein	gi|6755682	0.53	5.0	38.78
10	Lipocortin V (annexin V)	gi|2981437	0.53	5.0	33.95
11	Chloride intracellular channel 1	gi|15617203	0.46	5.1	27.34
12	Actin, beta	gi|71620	0.33	5.3	42.08
13	Blood vessel epicardial substance	gi|3195598	0.61	6.9	41.51
14	Intracellular chloride ion channel protein p64Hl	gi|13929166	0.4	5.9	28.84
15	Sodium channel protein pCSC-1 (fragment)	gi|321336	0.55	5.3	37.24
16	NHERF-3 rat isoform (PDZK1, Diphor-1)	gi|2331224	0.31	5.3	52.63
17	HSP60 protein (chaperonin 60)	gi|1334284	0.56	5.3	58.08
18	H^+-transporting ATPase (EC 3.6.1.14) beta chain, mitochondrial (fragment)	gi|92350	0.38	4.9	50.75
19	ERM-binding phosphoprotein	gi|11024674	0.39	5.7	39.15
20	Tropomyosin 5; TM-5	gi|9653293	0.32	4.7	29.16
21	Solute carrier family 5 (Slc5a6) sodium-dependent multivitamin transporter	gi|18543353	0.47	8.9	69.54

Source: Reproduced with permission from Ref. [92], Copyright © 2011, Karger Publishers.
Note: Rats were maintained on different dietary conditions, as described in the Section Methods. The reference number assigned to each protein is the same that is used in Fig. 4.1 to identify the corresponding spot on a representative 2-d gel. Only significantly matched proteins by peptide mass fingerprinting are included. The identifier of each protein is the GenInfo number in the NCBI protein database. M_w and pI indicate the theoretical position of each protein spot on two-dimensional gel based on the protein database.

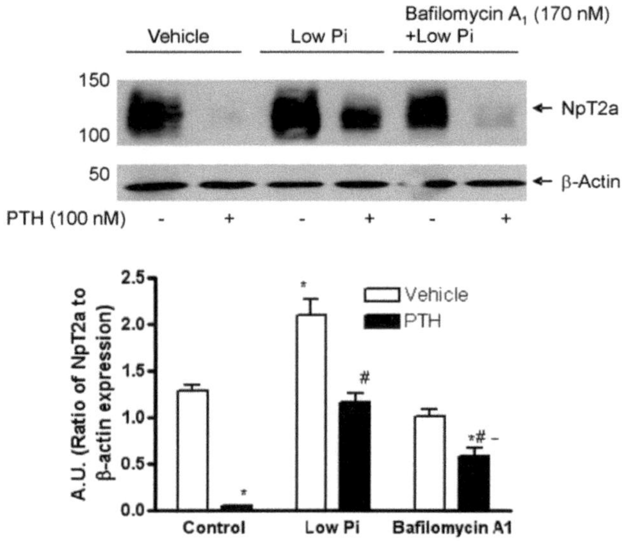

Figure 4.2 Effect of Bafilomycin A1 on low phosphate-induced increase in apical membrane expression of Npt2a. OK cell grown on transwell filters were treated with PTH for 6 h followed by incubation for 24 h in low phosphate medium in the presence or absence of Bafilomycin A1. Cells were lysed and BBM was prepared as described in Methods. 20 μg of BBM proteins were separated by 10% SDS PAGE, transferred to solid matrix, and blotted for Npt2a. Blot shown is representative of five separate experiments. The densitometric data shown are mean ± SE of the analysis on five different blots and is represented as ratio of Npt2a band density to β-actin band density. $^{*}p < 0.01$ vs. vehicle, $^{\#}p < 0.01$ vs. low phosphate medium alone, $^{+}p < 0.01$ vs. Bafilomycin A1 alone by ANOVA followed by Bonferroni analysis. Reproduced with permission from Ref. [92], Copyright © 2011, Karger Publishers.

treatment of proximal tubule cells with either of two inhibitors of vacuolar ATPase, Bafilomycin or Concanamycin A, blunted the increase in brush border membrane expression of Npt2a stimulated by low phosphate medium (Fig. 4.2). Neither agent had an effect on endocytosis of Npt2a stimulated by PTH but predictably did block lysosomal degradation of Npt2a (Fig. 4.3). Treatment of cells with Bafilomycin alone, Brefeldin A (an inhibitor of Golgi function), or Brefeldin A plus Bafilomycin resulted in similar decrease in low phosphate stimulated insertion of Npt2a into the apical membrane.

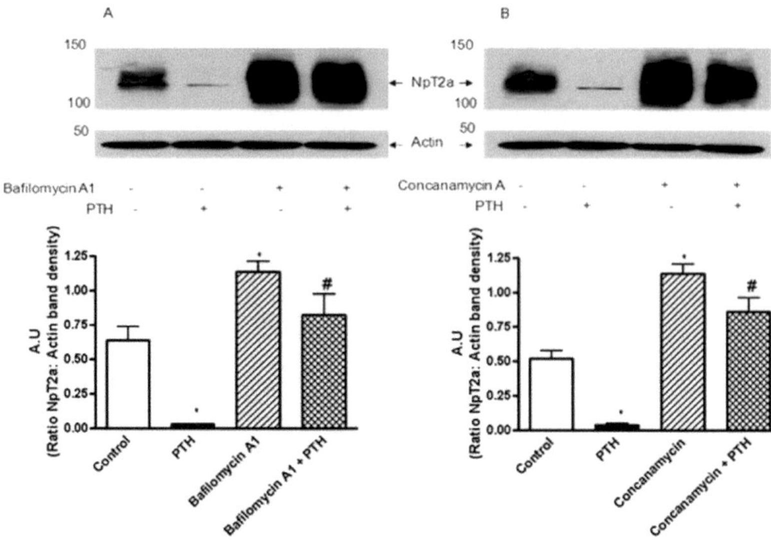

Figure 4.3 Role of Vacuolar ATPase in PTH inhibition of phosphate uptake in OK cells. OK cell monolayers were incubated (A) for 30 min with 170 nM bafilomycin A1 or (B) for 15 min with 50 nM concanamycin A, followed by treatment with 100 nM PTH 4 h. Phosphate uptake was determined using radiolabeled [32P] phosphoric acid, as described under Section Methods. Results shown are mean ± SE of three different experiments and are expressed as phosphate uptake (pmoles/mg protein/min). *$p < 0.001$ vs. control, $^{\dagger}p < 0.05$ vs. control, $^{\#}p < 0.05$ vs. bafilomycin A1 or concamamycin A alone by ANOVA followed by Bonferroni analysis. Reproduced with permission from Ref. [92], Copyright © 2011, Karger Publishers.

These findings are consistent with the conclusion that inhibition of vacuolar ATPase blocked forward trafficking of Npt2a similarly to Brefeldin A. The mechanisms for this effect remain unknown.

The next obvious question is how does vacuolar ATPase regulate forward trafficking of Npt2a? The exocytic pathway for Npt2a is not understood. Whether Npt2a is transported to the apical membrane through a vesicular or non-vesicular pathway has not been determined. Several lines of evidence suggest; however, that Npt2a is packaged into membrane vesicles for transport to the surface. Npt2a has been identified virtually exclusively in membrane fractions of cell preparations and not in cytosolic fractions [1, 2, 14, 23, 43]. Npt2a is an integral membrane protein with projected

10 or 12 transmembrane domains. Factors that regulate surface expression of Npt2a such as glycosylation also are associated with the trafficking of other proteins through vesicular pathways [18, 20, 21]. Our data, demonstrating that inhibition of vacuolar ATPases and treatment with Brefeldin, a Golgi poison, result in similar inhibition of Npt2a insertion into the apical membrane is also consistent with the hypothesis that Npt2a is packaged into membrane vesicles at the level of the Golgi for exocytosis to the apical membrane. Finally, the previously described study demonstrating changes in the expression of membrane-associated proteins after changing from a high to a low phosphate diet in a rodent model identified other proteins commonly associated with vacuolar transport including intracellular chloride channels, guanine nucleotide binding proteins, Brefeldin inhibitable guanine nucleotide exchange factors, and HSP (chaperone) proteins [92, 94].

If, in fact, Npt2a is trafficked through membrane vesicles, then the mechanism for regulation of Npt2a via vacuolar ATPases may be deduced from studies of the trafficking of proteins other than Npt2a. Progressive acidification of the vacuoles is critical for forward trafficking. Chloride channels that secrete H^+ and Cl^- into the vesicle play a major role in this process [84, 86, 90]. Vacuolar ATPases expressed on the vesicle secrete H^+ out of the vesicle, thus allowing continued function of the chloride channel which would otherwise be limited by the low pH. It has also been suggested that vacuolar ATPases may be important for recruitment of accessory proteins to the vesicles, required for normal trafficking. A role for small GTPases in the trafficking of Npt2a has been observed. Small GTPases such as cdc42 and vacuolar ATPases play mutually regulatory roles [66, 68, 70–80, 94–96]. Studies in yeast have demonstrated regulation of H^+-ATPase activity through a cdc42 dependent process. Whether H^+-ATPase is involved in Npt2a trafficking through a small G protein dependent mechanism has not been determined. Glycosylation of Npt2a enhances surface expression of the protein, though is probably not absolutely essential. Our demonstration that inhibition of vacuolar ATPase by either Bafilomycin or Concanamycin A had no effect on the glycosylation status of Npt2a suggests that vacuolar ATPases are not involved in this process [92]. In summary, significant circumstantial evidence points toward multiple roles of vacuolar ATPases in regulation of Npt2a trafficking but little hard data exist to corroborate these hypotheses.

A final consideration in examining the interactions between Npt2a and vacuolar ATPases is in the regulation of proximal renal tubule metabolism. Acidosis and limitation of essential nutrients such as phosphate result in substantial alterations in proximal tubule metabolic pathways [29, 35, 36, 68, 81, 97–100]. Aldolase, an enzyme of the glycolytic pathway, can interact directly with vacuolar ATPase subunits, suggesting a role for vacuolar ATPase in regulation of proximal tubule cellular metabolism [93, 101]. Deletion of phosphate from fluid perfusing proximal tubules blocks fluid transport if glucose is also present in the perfusate but inhibition of glucose uptake restores fluid transport even in the absence of luminal phosphate [97]. In view of the importance of phosphate for normal cellular energetic and signaling mechanisms, it seems obvious that states of altered phosphate homeostasis would have an impact on cellular metabolism [35, 36, 41, 42, 81, 97, 102–104]. How alterations in cellular phosphate homeostasis, phosphate transport by Npt2a, or Npt2a trafficking would be affected by vacuolar ATPase under these conditions has not been investigated; however, the increasing role of vacuolar ATPases in proximal tubule metabolism suggests that vacuolar ATPase could have an impact on Npt2a function or trafficking under these circumstances.

4.3 Conclusion

Studies from our laboratory and from others have both suggested and defined numerous roles for vacuolar ATPase in the regulation of function and expression of Npt2a. H^+-ATPase at the apical membrane, by altering local pH, can regulate Npt2a function without altering surface expression. Vacuolar ATPase is essential for endosomal transport to lysosomes after endocytosis of Npt2a and is essential for the terminal degradation of Npt2a in lysosomes. Vacuolar ATPases also play a significant role in trafficking of Npt2a to the apical membrane, likely through vacuolar acidification, recruitment of important accessory proteins, support for intracellular chloride channel function, and small G protein function. The involvement of

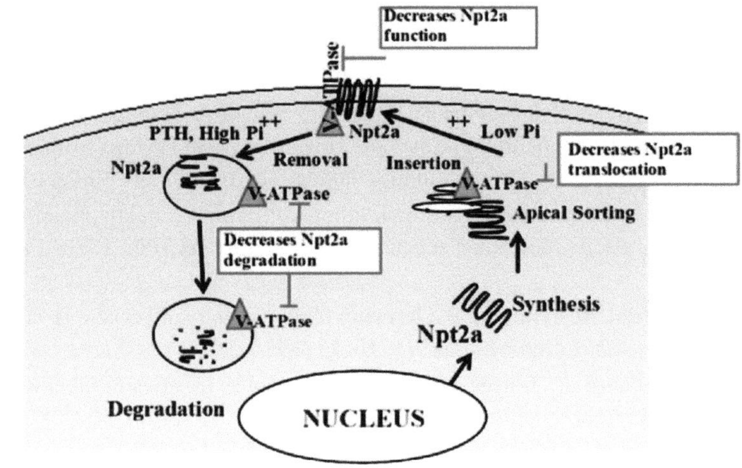

Figure 4.4 Proposed role of V-H$^+$-ATPase in Npt2a trafficking.

vacuolar ATPases and phosphate transport in regulation of renal proximal tubule metabolism points toward another area of mutual regulation between these two ion transporting proteins (Fig. 4.4).

References

1. Blaine J, Weinman EJ, Cunningham R. The regulation of renal phosphate transport. *Adv. Chronic. Kidney Dis.* 18: 77–84, 2011.
2. Forster IC, Hernando N, Biber J, Murer H. Proximal tubular handling of phosphate: A molecular perspective. *Kidney Int.* 70: 1548–1559, 2006.
3. Lederer E, Miyamoto K. Clinical consequences of mutations in sodium phosphate cotransporters. *Clin. J. Am. Soc. Nephrol.* 7: 1179–1187, 2012.
4. Tenenhouse HS. Regulation of phosphorus homeostasis by the Type IIa Na/phosphate cotransporter *Annu. Rev. Nutr.* 25: 197–214, 2005.
5. Tenenhouse HS, Roy S, Martel J, Gauthier C. Differential expression, abundance, and regulation of Na$^+$-phosphate cotransporter genes in murine kidney. *Am. J. Physiol.* 275: F527–F534, 1998.
6. Beck L, Karapalis AC, Amizuka N, Hewson AS, Ozawa H, Tenenhouse HS. Targeted inactivation of Npt2 in mice leads to severe renal phosphate wasting, hypercalciuria, and skeletal abnormalities. *Proc. Natl. Acad. Sci. USA* 95: 5372–5377, 1998.

7. Iwaki T, Sandoval-Cooper MJ, Tenenhouse HS, Castellino FJ. A missense mutation in the sodium phosphate co-transporter Slc34a1 impairs phosphate homeostasis. *J. Am. Soc. Nephrol.* 19: 1753–1762, 2008.
8. Miyamoto K, HaitoSugino S, Kuwahara S, Ohi A, Nomura K, Ito M, *et al.* Sodium dependent phosphate cotransporters: Lesions from gene knock out and mutation studies. *J. Pharm. Sci.* 100: 3719–3730, 2011.
9. Murer H, Hernando N, Forster I, Biber J. Proximal tubular phosphate reabsorption: Molecular mechanisms. *Physiol. Rev.* 80: 1373–1409, 2000.
10. Kendrick J, Kestenbaum B, Chonchol M. Phosphate and cardiovascular disease. *Adv. Chronic Kidney Dis.* 18: 113–119, 2011.
11. Kestenbaum B, Glazer NL, Kottgen A, Felix JF, Hwang SJ, Liu Y, *et al.* Common genetic variants associate with serum phosphorus concentration. *J. Am. Soc. Nephrol.* 21: 1223–1232, 2010.
12. Kottgen A, Pattaro C, Boger CA, Fuchsberger C, Olden M, Glazer NL, *et al.* New loci associated with kidney function and chronic kidney disease. *Nat. Genet.* 42: 376–384, 2010.
13. Magen D, Berger L, Coady MJ, Ilivitzki A, Militianu D, Tieder M, *et al.* A loss of function mutation in NaPi-IIa and renal Fanconi's syndrome. *N. Engl. J. Med.* 361: 1102–1109, 2010.
14. Forster IC, Kohler K, Biber J, Murer H. Forging the link between structure and function of electrogenic cotransporters: The renal Type IIa Na/Pi cotransporter as a case study. *Prog. Biophys. Mol. Biol.* 80: 69–108, 2002.
15. Forster IC, Hernando N, Biber J, Murer H. Phosphate transport kinetics and structure function relationships of SLC34 and SLC20 proteins. *Curr. Top. Membr.* 70: 313–356, 2012.
16. Radanovic T, Gisler SM, Biber J, Murer H. Topology of the Type IIa Na/Pi cotransporter *J. Membr. Biol.* 212(1): 41–49, 2006.
17. Köhler K, Forster IC, Lambert G, Biber J, Murer H. The functional unit of the Type IIa Na/Pi transporter is a monomer *J. Biol. Chem.* 275(34): 26113–26120, 2000.
18. Hayes G, Busch A, Lötscher M, Waldegger S, Lang F, Verrey F, *et al.* Role of N-linked glycosylation in regulation of rat renal NaPi cotransport. *J. Biol. Chem.* 269: 24143–24149, 1994.
19. Hu MC, Shi M, Zhang J, Pastor J, Nakatani T, Lanske B, *et al.* Klotho: A novel phosphaturic substance acting as an autocrine enzyme in the proximal tubule. *FASEB J.* 24: 3438–3450, 2010.

20. Jankowski M, Hilfiker H, Biber J, Murer H. The opossum kidney cell Type IIa Na/Pi cotransporter is a phosphoprotein. *Kidney Blood Press. Res.* 24(1): 1–4, 2001.

21. Khundmiri SJ, Ahmad A, Bennett RE, Weinman EJ, Steplock D, Cole J, *et al.* Novel regulatory function for NHERF1 in Npt2a transcription. *Am. J. Physiol. Renal. Physiol.* 294: F840–F849, 2008.

22. Hernando N, Gisler SM, Pribanic S, Deliot N, Capuano P, Wagner CA, *et al.* NaPi-IIa and interacting partners. *Am. J. Physiol.* 567: 21–26, 2005.

23. Khundmiri SJ, Rane MJ, Lederer ED. Parathyroid hormone regulation of Type II sodium phosphate cotransporters is dependent on an A kinase anchoring protein. *J. Biol. Chem.* 278: 10134–10141, 2003.

24. Capuano P, Bacic D, Roos M, Gisler SM, Stange G, Biber J, *et al.* Defective coupling of apical PTH receptors to phospholipase C prevents internalization of the Na-phosphate cotransproter NaPi-IIa in NHERF1 deficient mice. *Am. J. Physiol.* 292: C927–C934, 2007.

25. Hernando N, Déliot N, Gisler SM, Lederer E, Weinman EJ, Biber J, *et al.* PDZ domain interactions and apical expression of Type IIa Na/Pi transporters. *Proc. Natl. Acad. Sci. USA* 99: 11957–11962, 2002.

26. Shenolikar S, Voltz JW, Minkoff CM, Wade J, Weinman EJ. Targeted disruption of the mouse gene encoding a PDZ domain containing protein adaptor NHERF-1, promotes Npt2 internalization and renal phosphate wasting. *Proc. Natl. Acad. Sci. USA* 99: 11470–11475, 2002.

27. Ambuhl PM, Zajicek HK, Wang H, Puttaparthi K, Levi M. Regulation of renal phosphate transport by acute and chronic metabolic acidosis in the rat. *Kidney Int.* 53: 1288–1298, 1988.

28. Amstutz M, Mohrmann M, Gmaj P, Murer H. Effect of pH on phosphate transport in rat renal brush border membrane vesicles. *Am. J. Physiol.* 248: F705–F710, 1985.

29. Boross M, Kinsella J, Cheng L, Sacktor B. Glucocorticoids and metabolic acidosis-induced renal transports of phosphate calcium and NH4. *Am. J. Physiol.* 250: F827–F833, 1986.

30. Guntupalli J, Eby B, Lau K. Mechanism for the phosphaturia of NH_4Cl: Dependence on acidemia but not on diet PO_4 or PTH. *Am. J. Physiol.* 242: F552–F560, 1982.

31. Jehle AW, Hilfiker H, Pfister MF, Biber J, Lederer E, Krapf R, *et al.* Type II NaPi cotransport is regulated transcriptionally by ambient bicarbonate/carbon dioxide tension in OK cells. *Am. J. Physiol.* 276: F46–F53, 1999.

32. Nowik M, Picard N, Stange G, Capuano P, Tenenhouse HS, Biber J, *et al.* Renal phosphaturia during metabolic acidosis revisited: Molecular mechanisms for decreased renal phosphate reabsorption. *Pflugers Arch.-Eur J. Physiol.* 457: 539–549, 2008.
33. Quamme GA. Effect of pH on Na^+-dependent phosphate transport in renal outer cortical and outer medullary BBMV. *Am. J. Physiol.* 258: F356–F363, 1990.
34. Levi M, McDonald LA, Preisig PA, Alpern RJ. Chronic K depletion stimulates rat renal brush-border membrane Na-citrate cotransproter. *Am. J. Physiol.* 261: F767–F773, 1991.
35. Bergwitz C, Juppner H. Phosphate sensing. *Adv. Chronic Kidney Dis.* 18: 132–144, 2011.
36. Bevington A, Kemp GJ, Graham R, Russell G. Phosphate-sensitive enzymes: A possible molecular basis for cellular disorders of phosphate metabolism. *Clin. Chem. Enzym. Comms.* 4: 235–237, 1992.
37. Cheng L, Liang CT, Sacktor B. Phosphate uptake by renal membrane vesicles of rabbits adapted to high and low phosphorous diets. *Am. J. Physiol.* 245: F175–F180, 1983.
38. Gloor HJ, Bonjour JP, Caverzasio J, Fleisch H. Resistance to the phosphaturic and calcemic actions of parathyroid hormone during phosphate depletion. *J. Clin. Invest.* 63: 371–377, 1979.
39. Kempson SA, Shah SV, Werness PG, Berndt T, Lee PH, Smith LH, *et al.* Renal brush border membrane adaptation to phosphorous deprivation: Effects of fasting versus low-phosphorous diet. *Kidney Int.* 18: 36–47, 1980.
40. Keusch I, Traebert M, Lotscher M, Kaisling B, Murer H, Biber J. Parathyroid hormone and dietary phosphate provoke a lysosomal routing of the proximal tubular Na/Pi-cotransporter Type II. *Kidney Int.* 54: 1224–1232, 1998.
41. Kreusser WJ, Kurokawa K, Aznar E, Massry SG. Phosphate depletion: Effect on inorganic phosphorous and adenine nucleotides, urinary phosphate and calcium balance. *Electrolyte Metab.* 1: 30–42, 1978.
42. Kumar R. Phosphate sensing. *Curr Opin. Nephrol. Hypertens.* 18: 281–284, 2009.
43. Lederer ED, Sohi SS, Mathiesen JM, Klein JB. Regulation of Type II sodium phosphate cotransporters by protein kinases A and C. *Am. J. Physiol.* 275(2 Pt 2): F270–F277, 1998.
44. Levi M, Lotscher M, Sorribas V, Custer M, Arar M, Kaissling B, *et al.* Cellular mechanisms of acute and chronic adaptation of rat renal Pi

transporter to alterations in dietary Pi. *Am. J. Physiol.* 267: F900–F908, 1994.

45. Lotscher M, Wilson P, Nguyen S, Kaissling B, Biber J, Murer H, *et al.* New aspects of adaptation of rat renal Na-Pi cotransporter to alterations in dietary phosphate. *Kidney Int.* 49: 1012–1018, 1996.
46. Stoll R, Kinne R, Murer H. Effect of dietary phosphate intake on phosphate transport by isolated rat renal brush-border vesicles. *Biochem. J.* 180: 465–470, 1979.
47. Traebert M, Volkl H, Biber J, Murer H, Kaissling B. Luminal and contraluminal action of 1–34 and 3-34 PTH peptides on renal Type IIa Na-Pi cotransporter *Am. J. Physiol.* 278: F792–F798, 2000.
48. DeFronzo RA, Cooke CR, Andres R, Faloona R, Davis PJ. The effect of insulin on renal handling of sodium, potassium, calcium, and phosphate in man. *J. Clin. Invest.* 55: 845–855, 1975.
49. Freiberg JM, Kinsella J, Sacktor B. Glucocorticoids increase the Na-H exchange and decrease the Na gradient-dependent phosphate uptake systems in renal brush-border membrane vesicles. *Proc. Natl. Acad. Sci. USA* 79: 4932–4936, 1982.
50. Hammerman MR, Rogers S, Hansen VA, Gavin JR. Insulin stimulates Pi transport in brush border vesicles from proximal tubular segments. *Am. J. Physiol.* 247: E616–E624, 1984.
51. Hammerman MR. The growth hormone-insulin-like growth factor axis in the kidney. *Am. J. Physiol.* 246: F503–F514, 1989.
52. Loffing J, Lotscher M, Kaissling B, Biber J, Murer H, Seikaly M, *et al.* Renal Na/H exchanger NHE-3 and Na-PO_4 cotransporter protein expression in glucocorticoid excess and deficient states. *J. Am. Soc. Nephrol.* 9: 1560–1567, 1998.
53. Gentzsch M, Cui L, Mengos A, Chang XB, Chen JH, Riordan JR. The PDZ-binding chloride channel ClC-3B localizes to the Golgi and associates with CFTR-interacting PDZ proteins. *J. Biol. Chem.* 278: 6440–6449, 2003.
54. Alpern RJ. Cell mechanisms of proximal tubule acidification. *Physiol. Rev.* 70: 79–114, 1990.
55. Brown D, Hirsch S, Gluck S. Localization of a proton-pumping ATPase in rat kidney. *J. Clin. Invest.* 82: 2114–2126, 1988.
56. Brown D, Marshansky V. Renal VATPases: Physiology and pathophysiology. In: Futai M, Wada Y Kaplan J, editors. *Handbook of ATPases.* Weinheim: WileyVCH. pp. 413–442, 2004.

57. Brown D, Paunescu TG, Breton S, Marshansky V, Regulation of the V ATPase in kidney epithelial cells: Dual role in acid–base homeostasis and vesicle trafficking. *J. Exp. Biol.* 212: 1762–1772, 2009.
58. Chambrey R, Paillard M, Podevin RA. Enzymatic and functional evidence for adaptation of the vacuolar H(+)-ATPase in proximal tubule apical membranes from rats with chronic metabolic acidosis. *J. Biol. Chem.* 269: 3243–3250, 1994.
59. Gluck SL, Underhill DM, Iyori M, Holliday LS, Kostrominova TY, Lee BS. Physiology and biochemistry of the kidney vacuolar H^+-ATPase. *Ann. Rev. Physiol.* 58: 427–445, 1996.
60. Nakhoul, NL, Hamm LL. Vacuolar H^+-ATPase in the kidney. *J. Nephrol.* 12 (Suppl. 5): S22–S31, 2002.
61. CA, Finberg KE, Breton S, Marshansky V, Brown D, Geibel JP. Renal vacuolar H^+-ATPase. *Physiol. Rev.* 84(4): 1263–1314, 2004.
62. Lemann J, Bushinsky DA, Hamm LL. Bone buffering of acid and base in humans. *Am. J. Physiol.* 285: F811–F832, 2003.
63. Sun X, Yang LV, Tiegs BC, Arend LJ, McGraw DW, Penn RB, *et al.* Deletion of the pH sensor GPR4 decreases renal acid excretion. *J. Am. Soc. Nephrol.* 21: 1745–1755, 2010.
64. Clague MJ, Urbe S, Aniento F, Gruenberg J. Vacuolar ATPase activity is required for endosomal carrier vesicle formation. *J. Biol. Chem.* 269: 21–24, 1994.
65. Gu F, Gruenberg J. ARF1 regulates pH-dependent COP functions in the early endocytic pathway. *J. Biol. Chem.* 275: 8154–8160, 2000.
66. Hara-Chikuma M, Wang Y, Guggino SE, Guggino WB, Verkman AS. Impaired acidification in early endosomes of ClC-5 deficient proximal tubule. *Biochem. Biophys. Res. Comun.* 329: 941–946, 2005.
67. Hurtado-Lorenzo A, Skinner M, El Annan J, Futai M, SunWada GH, Bourgoin S, *et al.* VATPase interacts with ARNO and Arf6 in early endosomes and regulates the protein degradative pathway. *Nat. Cell Biol.* 8: 124–136, 2006.
68. Maede Y, Ide T, Koike M, Uchiyama Y, Kinoshita T. GPHR is a novel anion channel critical for acidification and functions of the Golgi apparatus. *Nat. Cell Biol.* 10: 1135–1145, 2008.
69. Maranda B, Brown D, Bourgoin S, Casanova JE, Vinay P, Ausiello DA, *et al.* Intra-endosomal pH-sensitive recruitment of the Arf-nucleotide exchange factor ARNO and Arf6 from cytoplasm to proximal tubule endosomes. *J. Biol. Chem.* 276: 18540–18550, 2001.

70. Marshansky V, Ausiello DA, Brown D. Physiological importance of endosomal acidification: Potential role in proximal tubulopathies. *Curr. Opin. Nephrol. Hypertens.* 11: 527–537, 2002.
71. Marshansky V, Futai M. The V-type H^+-ATPase in vesicular trafficking: Targeting, regulation and function. *Curr Opin. Cell Biol.* 20: 415–426, 2008.
72. Mellman I. The importance of being acid: The role of acidification in intracellular membrane traffic. *J. Exp. Biol.* 172: 39–45, 1992
73. Mindell JA. Lysosomal acidification mechanisms. *Annu. Rev. Physiol.* 74: 69–86, 2012.
74. Novarino G, Weinert S, Rickheit G, Jentsch TJ. Endosomal chloride-proton exchange rather than chloride conductance is crucial for renal endocytosis. *Science* 328: 1398–1401, 2010.
75. Paroutis P, Touret N, Grinstein S. The pH of the secretory pathway: Measurement determinants, and regulation. *Physiology* 19: 207–215, 2004.
76. Steinberg BE, Huynh KK, Brodovitch A, Jabs S, Stauber T, Jentsch TJ, *et al.* A cation counterflux supports lysosomal acidification. *J. Cell Biol.* 289: 1171–1186, 2010.
77. Weisz OA. Acidification and protein traffic. *Int. Rev. Cytol.* 226: 259–319, 2003.
78. Zeuzem S, Feick P, Zimmerman P, Haase W, Kahn RA, Schulz I. Intravesicular acidification correlates with binding of ADP-ribosylation factor to microsomal membranes. *Proc. Natl. Acad. Sci. USA* 89: 6619–6623, 1992.
79. Zeuzem S, Zimmerman P, Schulz I. Association of a 19- and a 21-kDa GTP-binding protein to pancreatic microsomal vesicles is regulated by the intravesicular pH estabilished by a vacuolar-type H^+-ATPase. *J. Membr. Biol.* 125: 231–241, 1992.
80. Aniento F, Gu F, Parton RG, Gruenberg J. An endosomal beta COP is involved in the pH-dependent formation of transport vesicles destined for late endosomes. *J. Cell Biol.* 133: 29–41, 1996.
81. Braun NA, Morgan B, Dick TP, Schwappach B. The yeast CLC protein counteracts vesicular acidification during iron starvation. *J. Cell Sci.* 123: 2342–2350, 2010.
82. Edwards JC, Kahl CR. Chloride channels of intracellular membranes. *FEBS Lett.* 584: 2102–2111, 2010.
83. Gunther W, Luchow A, Cluzeaud F, Vandewalle A, Jentsch TJ. ClC-5, the chloride channel mutated in Dent's disease, colocalizes with the proton

pump in endocytotically active kidney cells. *Proc. Natl. Acad. Sci. USA* 95: 8075–8080, 1998.

84. Jentsch TJ. Chloride and the endosomal-lysosomal pathway: Emerging roles of CLC chloride transporters. *J. Physiol.* 578: 633–640, 2007.
85. Mohammad-Panah R, Harrison R, Dhani S, Ackerley C, Huan LJ, Wang Y, *et al.* The chloride channel ClC-4 contributes to endosomal acidification and trafficking. *J. Biol. Chem.* 278: 29267–29277, 2003.
86. Piwon N, Gunther W, Schwake M, Bosl MR, Jentsch TJ. ClC-5 Cl-channel disruption impairs endocytosis in mouse model for Dent's disease. *Nature* 408: 369–373, 2000.
87. Sakamoto H, Sado Y, Naito I, Kwon TH, Inoue S, Endo K, *et al.* Cellular and subcellular immunolocalization of ClC-5 channel in mouse kidney: Colocalization with H^+-ATPase. *Am. J. Physiol.* 277: F957–F965, 1999.
88. Sonawane ND, Thiagaajah JR, Verkman AS. Chloride concentration in endosomes measured using a ratioable fluorescent Cl indicator: Evidence for chloride accumulation during acidification. *J. Biol. Chem.* 277: 5506–5513, 2002.
89. Stauber T, Jentsch TJ. Chloride in vesicular trafficking and function. *Annu. Rev. Physiol.* 75: 11.1–11.25, 2013.
90. Steinmeyer K, Schwappach B, Bens M, Vandewalle A, Jentsch TJ. Cloning and functional expression of rat CLC-5, a chloride channel related to kidney disease. *J. Biol. Chem.* 270: 31172–31177, 1995.
91. Wartosch L, Fuhrmann JC, Schweizer M, Stauber T, Jentsch TJ. Lysosomal degradation of endocytosed proteins depends on the chloride transport protein Clc-7. *FASEB J.* 23: 4056–4068, 2009.
92. Ahmad A, Khundmiri SJ, Pribble F, Merchant ML, Ameen M, Klein JB, *et al.* Role of the vacuolar ATPase in the trafficking of the Type IIa sodium phosphate cotransporter. *Cell Physiol. Biochem.* 27: 703–714, 2011.
93. Brown D, Wagner CA. Molecular mechanisms of acid sensing by the kidney. *J. Am. Soc. Nephrol.* 23: 774–780, 2012.
94. Ishizaki R, Shin HW, Mitsuhashi H, Nakayama K. Redundant roles of BIG2 and BIG1, guanine nucleotide exchange factors for ADP ribosylation factors in membrane traffic between the transgolgi network and endosomes. *Mol. Biol. Cell* 19(6): 2650–2660, 2008.
95. Musch A, Cohen D, Kreitzer G, Rodriguez-Boulain. Cdc42 regulates the exit of apical and basolateral proteins from the trans-Golgi network. *EMBO J.* 20: 2171–2179, 2001.

96. SantiagoTirado FH, Bretscher A. Membrane-trafficking sorting hubs: Cooperation between PI4P and small GTPases at the trans-Golgi Network. *Trends. Cell Biol.* 21: 515–525, 2011.
97. Brazy PC, Gullans SR, Mandel LJ, Dennis VW. Metabolic requirement for inorganic phosphate by the rabbit proximal tubule: Evidence for a Crabtree effect. *J. Clin. Invest.* 70: 53–62, 1982.
98. Chang C, Hu Y, Sun S, Zhu Y, Ma G, Xu G. Proton pump OsA8 is linked to phosphorus uptake and translocation in rice. *J. Exp. Bot.* 60: 557–565, 2009.
99. Gullans SR, Brazy PC, Mandel LJ, Dennis VW. Stimulation of phosphate transport in the proximal tubule by metabolic substrates. *Am. J. Physiol.* 247: F582–F587, 1984.
100. McKeown JW, Brazy PC, Dennis VW. Intrarenal heterogeneity for fluid, phosphate and glucose absorption in the rabbit. *Am. J. Physiol.* 237: F312–F318, 1979.
101. Merkulova M, Hurtado-Lorenzo A, Hosokawa H, Zheung Z, Brown D, Aussiello DA, *et al.* Aldolase directly interacts with ARNO and modulates cell morphology and acidic vesicle distribution. *Am. J. Physiol. Cell Physiol.* 300(6): C1442–C1455, 2011.
102. Brazy PC, Mandel LJ, Gullans SR, Soltoff SP. Interactions between phosphate and oxidative metabolism in proximal tubules. *Am. J. Physiol.* 247: F575–F581, 1984.
103. Brazy PC, Mandel LJ. Does availability of inorganic phosphate regulate cellular oxidative metabolism? *News Physiol. Sci.* 1: 100–103, 1986.
104. Sestoft L. Is the relationship between the plasma concentration of inorganic phosphate and the rate of oxygen consumption of significance in regulating energy metabolism in mammals? *J. Clin. Invest.* 39: 191–197, 1979.

Chapter 5

Cytosolic pH Regulated by Glucose Promotes V-ATPase Assembly

Reinhard Dechant and Matthias Peter
Institute of Biochemistry, ETH Zurich, 8093 Zurich, Switzerland
Competence Center for Systems Physiology and Metabolic Diseases, 8093 Zurich, Switzerland
reinhard.dechant@bc.biol.ethz.ch

Over the past few years, the view of V-ATPases has changed considerably. These proton pumps are no longer only considered as important housekeeping enzymes required for regulation and homeostasis of organelle pH, but have also emerged as important players in cellular signaling. Indeed, V-ATPase assembly and, consequently, enzymatic activity are tightly regulated in response to environmental conditions. Recent evidence primarily from yeast, and also from mammalian cells, indicates that V-ATPase assembly is regulated by cytosolic pH. This suggests that V-ATPases are potentially conserved pH sensors that couple cytosolic pH to the luminal pH of vacuoles/lysosomes and link cellular physiology to signal transduction.

Handbook of H^+-ATPases
Edited by Suguru Nakamura

ISBN 978-981-4411-91-2 (Hardcover), 978-981-4411-92-9 (eBook)
www.panstanford.com

5.1 Introduction

The V-ATPase is a multi-subunit protein complex that consists of a membrane-inserted V_0 sector comprising the proton pore and an associated V_1 sector containing the ATP hydrolyzing subunits required for proton pumping. While these subcomplexes assemble independently of each other, activity of V-ATPases requires association of the V_1 and V_0 sectors into the functional holocomplex. Once assembled, V-ATPase pumping activity promotes intraluminal acidification of the endomembrane system [1].

The main mechanism to modulate V-ATPase activity is regulation of V-ATPase assembly. First observed in budding yeast *Saccharomyces cerevisiae,* the V_1 sector was shown to dissociate from the V_0 sector after glucose starvation, or upon a shift to less optimal carbon sources [1, 2]. Re-addition of glucose leads to rapid reassembly of the V-ATPase independent of new protein synthesis, demonstrating reversible regulation [1]. Due to the fast kinetics of glucose-dependent V-ATPase assembly, it seemed likely that V-ATPase assembly, and hence activity, was closely linked to glucose-sensitive signaling pathways. Moreover, disassembly of V-ATPase into the V_1 and V_0 sectors was found to not only inhibit proton pumping, but also to efficiently shut off ATPase activity [3, 4], possibly to prevent unnecessary ATP turnover during starvation [1, 2].

As glucose is the preferred carbon source (C-source) for most eukaryotic cells and provides the necessary energy for all biosynthetic processes, glucose availability is a major determinant of cell growth. Consequently, cells have developed elaborate signaling networks that control the cellular response to changes in nutrient supply [5]. In yeast, glucose activates the Ras/PKA and TORC1 pathways, which stimulate cell growth by promoting ribosome biogenesis and general translation, while at the same time suppressing growth inhibitory processes, such autophagy and the general stress response [5–7].

Interestingly, in yeast, regulation of V-ATPase assembly was shown to be independent of glucose-sensitive signaling pathways [2, 8], although some data suggest otherwise [9] (see later in this chapter). Thus, V-ATPase assembly and glucose-sensitive cell signaling may be independently regulated. Alternatively, V-ATPase

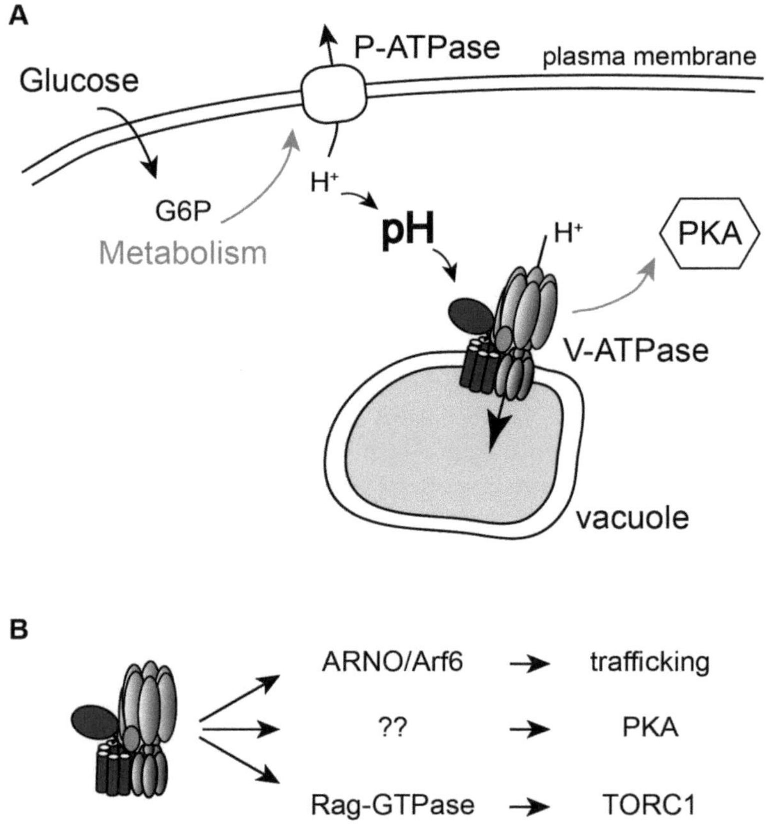

Figure 5.1 (A) proposed mechanism how cytosolic pH links glucose availability to the activation of V-ATPase assembly and activation of PKA. G6P: glucose-6-phosphate; (B) examples of signaling pathways regulated by V-ATPase. See text for details.

assembly itself might contribute to the activation of glucose-sensitive signaling pathways. Indeed, a number of recent studies indicate that V-ATPase is involved in cellular signaling in response to nutrient sensing, at least in part through regulating the activity of growth-promoting pathways [8, 10–16] (Fig. 5.1). Specifically, V-ATPase activity is required to regulate insulin secretion in pancreatic beta cells [8, 14, 15] and cell survival [11]. Moreover, V-ATPase contributes to the full activation of cAMP-dependent protein kinase A (PKA) in yeast and mammalian cells [8], and it is required for sensing amino acids and activation of the TORC1 complex [10]. Thus, elucidating the regulation of V-ATPase assembly is crucial to understand cellular signaling pathways that are involved in the regulation of cell growth, both in normal and pathological conditions. Interestingly, recent evidence demonstrates that V-ATPase is directly regulated by cytosolic pH and we will therefore summarize the available evidence supporting this notion.

5.2 Factors Regulating V-ATPase Assembly

Regulation of V-ATPase assembly has been observed in numerous species [1, 3, 4, 17, 18]. Yet, most mechanistic insight stems from work in budding yeast, from which several critical regulators of V-ATPase assembly, were identified [19–21].

5.2.1 *Signaling Pathways*

5.2.1.1 Yeast

As V-ATPase assembly is exquisitely sensitive to glucose availability, it was suggested that a glucose-sensitive signaling pathway might regulate V-ATPase assembly [1]. Interestingly, however, V-ATPase regulation is independent of glucose signaling pathways [2], but itself an important regulator of glucose signaling, thus establishing a crucial role for V-ATPase in the regulation of cellular signaling [8].

In yeast, three major signaling pathways are regulated by glucose availability, namely the Ras/PKA, TORC1, and Snf1 pathways [5, 22, 23]. While TORC1 and the Ras/PKA pathways are activated in the presence of glucose, Snf1, the yeast homologue of the mammalian AMP-activated protein kinase (AMPK) [24], is mostly active upon glucose starvation and upon growth on alternative C-sources. However, deletion of Snf1 does not affect the regulation of V-ATPase assembly by glucose [2], demonstrating that V-ATPase activity is independent of Snf1. Similarly, addition of Rapamycin, a potent inhibitor of TORC1 activity, does not interfere with V-ATPase assembly, ruling out a requirement of TORC1 activity in the assembly of V-ATPase [25]. Thus, only the Ras/PKA pathway remains as a candidate for V-ATPase assembly regulation in yeast.

The Ras/PKA pathway is activated by glucose through two different mechanisms [26]. First, PKA activity can be directly stimulated by glucose upon binding to a G-protein coupled receptor, Gpr1 [27–29]. Second, PKA activity also critically depends on glucose uptake and phosphorylation, demonstrating that PKA activity can be indirectly stimulated by glucose through a metabolic signal derived from glycolysis or a closely associated metabolite [30]. Both glucose-dependent activation mechanisms are mediated by cAMP; the direct and indirect glucose-triggered signals stimulate adenylate cyclase activity to produce cAMP, which eventually binds to the regulatory subunit of PKA and triggers the release of active catalytic subunits.

To test the influence of PKA activity on V-ATPase assembly, Parra and Kane studied the regulation of V-ATPase assembly in a strain deleted for the phosphodiesterase *PDE2*, the major enzyme degrading cAMP [2]. In this strain, PKA activity can be efficiently activated upon addition of exogenous cAMP. Interestingly, V-ATPase still disassembled upon starvation even in the presence of high levels of exogenously added cAMP, suggesting that high PKA activity does not interfere with disassembly upon starvation. Similarly, in this strain, addition of cAMP did not reverse disassembly of V-ATPase upon starvation. Finally, V-ATPase assembly was unaffected in a strain carrying a temperature sensitive allele of adenylate cyclase, $cdc35^{ts}$, which displays very low levels of cAMP [2], demonstrating that V-ATPase assembly is independent of PKA activity, and thus independent of the Ras/PKA glucose-sensitive signaling pathway.

However, this view has recently been challenged. By screening for mutations that would enhance V-ATPase function on a poor C-source, Bond and Forgac have identified the Ras-GAP Ira2 as negative regulators of V-ATPase function in response to glucose starvation [9]. In strains deleted for *IRA2*, the small GTPase Ras remains in its active, GTP-bound form and therefore also promotes high PKA activity during starvation [31, 32]. Indeed, while deleting *IRA2* did not affect overall proton pumping or ATPase activity on isolated vacuoles, levels of V-ATPase assembly remained high *in vitro* in *ira2*Δ cells even after 20 minutes of glucose starvation, suggesting that reduction of PKA activity is required to allow V-ATPase disassembly upon starvation [9]. Interestingly, several other mutants leading to constitutively high PKA activity, including strains deleted for *IRA1*, a homologue of *IRA2* [31, 32], or overexpressing of active Ras2 (Ras2-V19), also lead to sustained V-ATPase assembly during starvation, therefore contradicting previous data [2].

However, no differences in V-ATPase assembly were detected upon starvation in WT and *ira2*Δ*ra2*Δ *in* at the same time point [8]. Similarly, using their *in vitro* assay Bond and Forgac did not detect reduced V-ATPase assembly upon (partial) inactivation of Ras/PKA activity, suggesting that Ras/PKA activity is not essential for V-ATPase assembly [9]. Yet, PKA activity might feedback on V-ATPase assembly to stabilize V-ATPase assembly after glucose re-addition. The activity of this feedback loop might be readily detectable using the specific conditions in the *in vitro* assay, but might not be strong enough to detect differences in V-ATPase assembly *in vivo*. However, as V-ATPase assembly *in vivo* is exquisitely sensitive to cytosolic pH ([8], see later in this chapter), any *in vitro* assays for V-ATPase assembly or activity are unlikely to accurately account for this fact, but might be more sensitive to other influences, such as PKA activity.

Interestingly, as both V-ATPase assembly and activation of the Ras/PKA pathway are dependent on an intracellular, metabolic signal, it remains possible that V-ATPase might function upstream of the Ras/PKA pathway. Indeed, recent evidence suggests that full activation of PKA in yeast and mammalian cells requires V-ATPase, suggesting that V-ATPase is part of the glucose sensing mechanism responsible for activation of the Ras/PKA pathway [8]. V-ATPase was identified in genetic screens for potential activators of PKA activity [25], and this finding was confirmed by assaying various readouts for PKA pathway activity [8]. For example, V-ATPase mutants display severe negative genetic interactions with Gpr1 and Gpa2, which are required for activation of PKA in response to extracellular glucose [8, 33]. Moreover, V-ATPase and Gpr1/Gpa2 cooperatively regulate cAMP production. Similarly, V-ATPase-associated changes in PKA-dependent transcription can be suppressed by addition of exogenous cAMP, demonstrating that V-ATPase indeed acts upstream of the Ras/PKA pathway. Thus, V-ATPase acts in parallel to Gpr1/Gpa2 to mediate activation of the PKA pathway in response to metabolic cues. Although the molecular mechanisms on how V-ATPase activates the Ras/PKA pathway remain unclear, understanding the regulation of V-ATPase assembly *in vivo* should identify mechanisms underlying glucose sensing [8].

5.2.1.2 Higher eukaryotes

In contrast to yeast, regulation of cell growth in multicellular organisms is governed by nutrients and growth factors, which serve to coordinate cell growth in entire tissues or organisms. Therefore, it is not surprising that additional mechanisms exist, which regulate V-ATPase assembly in these systems. Indeed, there is ample evidence suggesting that V-ATPase assembly and activity is tightly regulated by nutrient-sensitive kinases, including PKA and AMPK [17, 34–38], at least in part through hormonal signals. For example, V-ATPase assembly and luminal acidification are regulated by the hormone serotonin in blowfly salivary glands [34]. Interestingly, this effect was dependent on cAMP and activation of PKA activity [35]. The molecular mechanisms by which PKA and/or AMPK activity would promote activation of V-ATPase are still largely unknown, but might involve direct phosphorylation of the V-ATPase *A* subunit [37]. Alternatively, PKA might regulate V-ATPase activation indirectly by promoting the interaction of V-ATPase with aldolase [9], which might contribute to the coupling of V-ATPase activity and cellular metabolism [39–42].

5.2.2 *RAVE Complex*

Another factor that is required for the assembly of the V-ATPase in yeast is the RAVE complex. The RAVE complex was first defined biochemically as a complex associated with, and required for, the activation of V-ATPase [21]. However, genetic evidence suggests that RAVE is not essential for V-ATPase assembly [20, 21, 43]. Loss of RAVE activity leads to partial loss of V-ATPase assembly, and only mild growth defects under growth conditions that require active V-ATPase for survival of yeast cells (i.e., media buffered to high pH or in the presence of high Ca^{2+} concentration) [20, 40]. Dynamic data of V-ATPase assembly upon glucose re-addition to starved cells revealed that the RAVE complex is not essential for assembly *per se*, but that it enhances assembly kinetics [21] by promoting the assembly of subunit *C* (Vma5) into the V_1 sector [45]. Whether RAVE contributes to the regulated V-ATPase assembly by mediating the glucose-dependent signal or whether it is only a permissive factor for V-ATPase assembly remains unclear.

5.2.3 *Regulation of V-ATPase by Metabolic Signals*

One attractive model for V-ATPase assembly suggests the process is mediated by a metabolic signal closely associated to glucose metabolism. Similar to the activation of the Ras/PKA pathway [30], assembly of V-ATPase requires glucose metabolism beyond the formation of glucose-6-phosphate [2]. Moreover, V-ATPase assembly is dependent on glucose concentration, suggesting that V-ATPase assembly is linked to glycolytic activity. Since V-ATPase binds to ATP and requires ATP hydrolysis for catalytic activity, the simplest hypothesis would be ATP binding to V-ATPase also triggers its assembly. Yet, somewhat surprisingly in yeast the levels of ATP do not correlate tightly with glucose concentration. Rather, both glucose starvation and re-addition lead to a transient change of ATP levels that quickly recover to pre-treatment values [2, 46, 47]. In contrast, V-ATPase remains disassembled even during prolonged periods of starvation, thus ruling out that ATP levels are directly responsible for promoting V-ATPase assembly.

Strikingly, V-ATPase stably interacts with the glycolytic enzyme aldolase in a glucose-dependent manner [9, 39–41], suggesting that this interaction mediates glucose sensing by V-ATPase. Indeed, genetic analysis showed that aldolase is required for growth in media buffered to increase pH and might, therefore, act upstream of V-ATPase assembly. The aldolase-V-ATPase interaction is independent of the substrate or products of aldolase, suggesting that the aldolase-V-ATPase interaction is mediated by a distinct binding surface on aldolase. However, how the binding of V-ATPase to aldolase couples V-ATPase assembly with glucose levels is still unclear. It is conceivable that the association contributes to the generation of a local micro-environment that triggers V-ATPase assembly.

Mechanistically it is hard to explain how high aldolase activity could result in increased interaction with V-ATPase. Generally, although often proposed, molecular mechanisms allowing for sensing metabolic fluxes—if at all possible—still remain to be identified [33]. Specifically, in the case of the aldolase–V-ATPase interaction, it is not clear how the extremely transient interactions with substrates and/or products could trigger a conformational change in aldolase sufficient for promoting V-ATPase assembly. Alternatively, aldolase might be directly modified by a signaling pathway that in turn is responsible for mediating V-ATPase assembly in an aldolase-dependent manner [9].

5.2.4 *Regulation of V-ATPase by pH*

Studies in yeast and mammalian cells have recently suggested the exciting possibility that V-ATPase assembly might be directly regulated by pH, both *in vivo* and *in vitro* [8, 16, 48, 49].

5.2.4.1 Sensing of luminal pH by V-ATPase

The first indication that V-ATPases might be pH sensors came from studies in kidney epithelial cells in which Marshansky andco-workers studied the regulation of the small GTPase Arf6 and its cognate activator, the guanine exchange factor ARNO. In this system, both Arf6 and ARNO are recruited to endosomes in a V-ATPase-dependent manner *in vivo* and *in vitro* [16, 50]. Specifically, recruitment of ARNO to isolated endosomes *in vitro* was strongly impaired upon treatment with a V-ATPase inhibitor or the uncoupling agent FCCP, suggesting that ARNO recruitment depends on intraluminal acidification [50].

Arf6 directly interacts with the c-subunit of V-ATPase. ARNO binds to the *N*-terminal domain of the a2-subunit of V-ATPase in a V-ATPase activity-dependent manner and this interaction is required to regulate endocytosis. Thus, V-ATPase itself was suggested as the pH sensing protein regulating endosomal recruitment of ARNO, possibly through pH-dependent regulation of the *a*-subunit [16].

The *a*-subunit of V-ATPase is an integral membrane protein with a large *N*-terminal cytoplasmic domain and up to 10 transmembrane domains in the C-terminus, but the exact membrane topology is still somewhat controversial. Endosomal acidification might trigger the protonation of a pH-sensitive residue facing the lumen of the endosome, which might cause a conformational change that is transduced to the cytoplasmic side of V-ATPase to trigger ARNO recruitment. Indeed, several histidine residues of the *a* subunit, which could be protonated at low pH, are facing the lumen of the endosome, but direct evidence for a requirement of these residues in the recruitment of ARNO is still lacking [49].

5.2.4.2 V-ATPase assembly is regulated by cytosolic pH in yeast

As discussed above in detail, V-ATPase assembly is dependent on glucose metabolism and required for the activation of the Ras/PKA pathway, suggesting that V-ATPase may act as a glucose sensor for the PKA pathway.

Interestingly, quantitative live cell imaging assays of yeast V-ATPase assembly *in vivo* were recently developed using a microfluidic setup that allows precise and rapid control of glucose starvation and re-addition. V-ATPase assembly was scored by monitoring accumulation of a Vma5-RFP construct at the vacuolar membrane [8]. Consistent with previous studies using both *in vitro* and *in vivo* assays [2, 21], these measurements revealed high dynamics of V-ATPase assembly and disassembly [8].

Similarly, cytosolic pH was highly dynamic and paralleled the regulation of V-ATPase assembly. In the presence of glucose, cytosolic pH was high and dropped rapidly upon starvation, suggesting that cytosolic pH might be a cellular signal mediating glucose sensing. Surprisingly, under these conditions, the regulation of cytosolic pH was independent of V-ATPase activity, demonstrating that changes in cytosolic pH are not the consequence of V-ATPase regulation. Rather, these data suggested that V-ATPase might itself be regulated by cytosolic pH.

Indeed, reduction of cytosolic pH using the protonophore 2,4-dinitrophenol (2,4-DNP), which acts to equilibrate pH gradients across biological membranes, efficiently triggered V-ATPase disassembly even in the presence of glucose, but did not affect overall vacuolar morphology. Importantly, this effect was only observed when the medium was buffered to low pH and was completely alleviated when the medium was buffered to near neutral pH, suggesting that this effect was indeed due to the effect of 2,4-DNP on intracellular pH [8].

Acidification of the cytosol upon glucose starvation is caused by the inability of cells to maintain high proton pumping across the plasma membrane activity necessary to maintain high cytosolic pH and subsequent proton influx from the medium [8, 51]. Consequently, buffering the medium to slightly alkaline pH efficiently prevents cytosolic acidification upon glucose starvation [8] and retained V-ATPase pumping activity and vacuolar acidification even in the absence of glucose [52], further demonstrating that reduction of cytosolic pH is both sufficient and required for V-ATPase disassembly.

In yeast, each of the subunits of V-ATPase is encoded by a single gene, the only exception being the *a*-subunit, which is encoded by either Stv1 or Vph1. Complexes containing Stv1 and Vph1 are targeted to the Golgi and vacuoles, respectively, yet only Vph1 containing vacuolar complexes are subject to glucose regulation [53]. Thus, the microenvironment at the different locations might influence V-ATPase assembly, or, alternatively, the *a*-subunit might act as a regulatory subunit that might mediate pH-dependent regulation of V-ATPase assembly. For example, protonation of the *a*-subunit could cause a conformational change in this subunit, which eventually triggers V-ATPase disassembly (Fig. 5.2).

Amino acid residues with side chains that can undergo reversible protonation/deprotonation at near neutral pH are histidines [49] and, depending on the structural context, aspartic acids [54]. Indeed, mutating one of the conserved aspartic acids in the cytoplasmic domain of Vph1 at least partially inhibited V-ATPase disassembly upon starvation. Thus, it is tempting to speculate that Asp329 mediates pH regulation of V-ATPase assembly, either by direct protonation or by participating in the conformational change triggering V-ATPase disassembly in response to protonation of another, yet to be identified, residue.

5.2.4.3 V-ATPase subunit 'a' is regulated by pH *in vitro*

Studies on the structure of V-ATPase by electron microscopy have suggested that the putatively pH sensitive *N*-terminal domain of

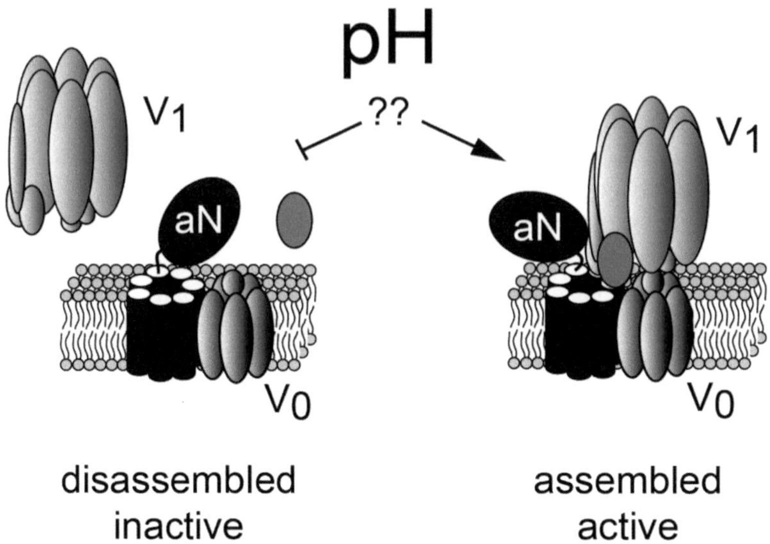

Figure 5.2 Schematic representation of V-ATPase assembly regulated by cytosolic pH. High cytosolic pH promotes V-ATPase assembly, possibly through deprotonation of the *N*-terminal cytoplasmic domain of the '*a*'-subunit (aN). See text for details.

subunit *a* (hereafter referred to as aN) is connected by a short linker to the C-terminal membrane-inserted domain [55]. Moreover, the aN domain undergoes a structural rearrangement upon dissociation of the V_1 sector [56], further supporting that the aN domain might be critical for pH-dependent regulation of V-ATPase.

Interestingly, recombinantly expressed and purified aN domain from the vacuolar *a*-subunit Vph1 (Vph1N) oligomerizes in a pH-dependent manner. Purified Vph1N elutes in two major peaks, corresponding to both the monomeric form and a high molecular weight fraction, most consistently with an octameric assembly of the protein. Incubating the protein in buffer of reduced pH shifted the elution profile toward more octameric form, suggesting that protonation of Vph1N induces its oligomerization [48].

Strikingly, expression of a slightly shorter protein fragment that lacks the putative linker region connecting Vph1N to the first transmembrane domain of Vph1, yielded mostly monomeric protein over the pH range tested, demonstrating that the predicted linker region is required for oligomerization *in vitro* and likely mediates pH-dependent regulation of Vph1. Consistent with these *in vitro* data, Vph1N expressed in yeast cells showed a similar pH-dependent behavior. When total yeast extracts from cells expressing Vph1N were separated by gel filtration chromatography, Vph1N was again detected in two major peaks, corresponding to monomeric and high molecular weight forms. Moreover, preparing the extracts with reduced pH yielded a greater fraction of the high molecular weight complexes, while a truncated version of Vph1N mostly eluted with low molecular weight independent of pH [48]. Thus, direct protonation of the aN domain of V-ATPase might cause pH-dependent regulation of V-ATPase *in vivo*. However, it remains unclear how oligomerization of the aN subunit could regulate V-ATPase assembly, especially as structural studies demonstrated only a single copy of the *a*-subunit per holocomplex. It is nevertheless possible that homo-oligomerization of the *a*-subunit would occur *in trans*, leading to the clustering of at least two V-ATPase complexes. Indeed, interactions between V sectors of V-ATPases have been proposed to underlie the role of V-ATPase in mediating homotypic vesicle fusions [57, 58]

Alternatively, the observed oligomerization *in vitro* might indicate that Vph1N *in vivo* forms topologically similar heterodimers with another subunit of V-ATPase. Interestingly, *ab initio* predictions of the structure of Vph1N suggested that it assumes a similar fold to Vma5, which encodes the Csubunit. These predictions were largely confirmed by solving the crystal structure of the Isubunit of V-ATPase from bacteria, which is homologous to the eukaryotic *a*-subunit [59]. Indeed, Vph1N and Vma5 directly interact *in vivo* [60]. Thus, further characterization of the interaction of Vph1N and Vma5 might provide important new insights into the regulation of V-ATPases.

5.3 Cytosolic pH Is Regulated by Glucose

Thus, although V-ATPase is a proton pump required for the regulation of luminal pH of organelles, it is not required for the regulation of cytosolic pH [8] (which might, however, depend on the specific growth conditions [61, 62]). Rather, cytosolic pH regulates V-ATPase. So then, what does regulate cytosolic pH?

In budding yeast, regulation of cytosolic pH largely depends on P-ATPase, encoded by *PMA1*, which regulates cytosolic pH in response to glucose levels [51]. Pma1 is a highly abundant protein localized to the plasma membrane and it actively pumps protons from the cytosol into the medium in an ATP-dependent manner. It has been suggested that Pma1 uses up to 20% of the cellular ATP to maintain a high cytosolic pH and to generate a steep pH gradient across the plasma membrane that is subsequently used for ion-homeostasis and nutrient uptake [63]. However, the regulation of Pma1 by glucose levels is still not completely understood.

Several lines of evidence in yeast suggest that cytosolic pH is tightly linked to glucose metabolism. For example, conditional inactivation of pyruvate kinase, which catalyzes the last step in glycolysis significantly reduces cytosolic pH. Similarly, addition of 2-deoxy glucose, a glucose analog that blocks glycolysis at the level of phosphoglucoisomerase, triggers a rapid drop of cytoplasmic pH and concomitant V-ATPase disassembly even in the presence of glucose [8]. Therefore, it seems likely that cellular metabolism directly modulates Pma1 activity, but the molecular mechanisms underlying this regulation remain elusive.

5.4 V-ATPase as a Cytosolic pH Sensor

Taken together, the available data suggest that cytosolic pH acts as a cellular signal that links V-ATPase assembly to cellular metabolism in yeast and in higher eukaryotes. Such a model has some intriguing implications.

5.4.1 *Coupling of Cytosolic and Luminal pH*

One striking aspect of regulating V-ATPase activity by cytosolic pH is that it readily allows for the coupling of cytosolic pH and luminal pH of acidified organelles.

Upon activation of V-ATPase, the pH gradient across the yeast vacuolar membrane is used for the active transport of ions, such

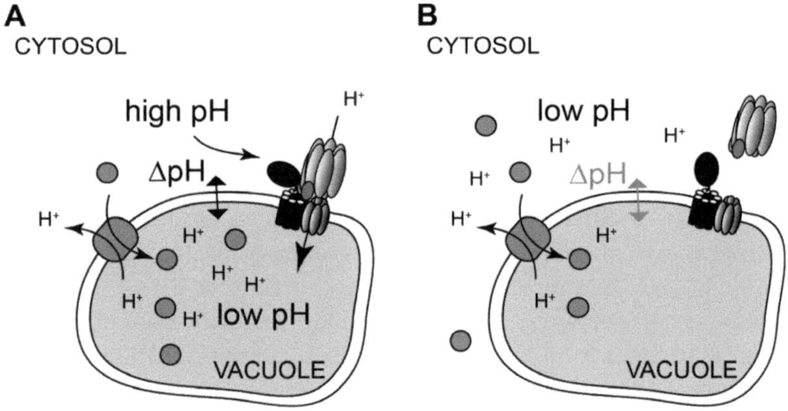

Figure 5.3 pH-dependent regulation of V-ATPase links regulation of cytosolic pH with intraluminal pH. When V-ATPase is active due to high cytosolic pH (A) the pumping activity of V-ATPase leads to acidification of vacuoles and, consequently, the establishment of a high proton gradient across the vacuolar membrane, which can be used by proton antiporters or symporters to redistribute other small molecules (dark circles). (B) Inactivation of V-ATPase activity at low cytoplasmic pH leads to dissipation of the proton gradient and therefore to lack of vacuolar acidification.

as Ca^{2+}, and small molecules (e.g., amino acids) into or out of the vacuole through proton antiporters or symporters, respectively (Fig. 5.3). Thus, V-ATPase is a major regulator of ion homeostasis and contributes to nutrient sensing [10, 44]. However, conditions that reduce cytosolic pH, such as nutrient limitation, not only directly affect the proton gradient across the vacuolar membrane due to decreasing pH at the cytoplasmic face of the membrane, but they also trigger V-ATPase disassembly and inactivation of pumping activity. As a consequence, the pH gradient dissipates, leading to the equilibration of the pH across the two compartments and to a redistribution of vacuolar ions.

Thus, pH-dependent regulation of V-ATPase leads to an amplification of the pH gradient across the vacuolar membrane and it is therefore tempting to speculate that the establishment of this pH gradient is responsible for the many physiological functions of V-ATPase. Indeed, addition of the V-ATPase inhibitor Concanamycin A, which stops pumping activity without affecting V-ATPase assembly

[2], is sufficient to trigger multiple V-ATPase-dependent cellular processes [8, 10, 16, 64].

In this context, it is also worth mentioning that increasing vacuolar pH during glucose starvation will also interfere with proteolytic activation of vacuolar enzymes that degrade cellular components targeted to the vacuole by autophagy [65]. Although autophagy is efficiently induced upon nitrogen limitation in yeast, less is known about the induction of autophagy by glucose limitation [66]. It will therefore be interesting to examine if the proposed coupling of cytosolic pH and vacuolar pH will also affect the degradative function of the vacuole, which might reveal new insights into why V-ATPase activity is so tightly controlled by glucose.

5.4.2 *Linking Cellular Physiology to Signal Transduction*

Coupling V-ATPase activity to cytosolic pH also provides an intuitive model of how glucose availability regulates cellular physiology and cellular signaling. As regulation of cytosolic pH is dependent on glucose metabolism, but not glucose per se, cytosolic pH is in principle able to integrate different environmental signals.

Indeed, yeast cells can utilize many different sugars, and all these sugars are funneled into cellular metabolism through glycolysis, albeit with different efficiencies. Therefore, differential regulation of cytosolic pH could readily explain quantitative differences in cell growth under different environmental conditions. In addition, cytosolic pH might also be influenced by several stresses. For example, oxidizing agents lead to a block in glycolytic activity and therefore, potentially, a drop in cytosolic pH [67]. Similarly, weak acid stress could cause growth arrest due to transient reduction of cytosolic pH [68]. Thus, cytosolic pH is ideally suited to integrate different kinds of growth and stress signals for cellular signaling.

Multiple lines of evidence suggest that cytosolic pH might also act as a cellular signal in mammalian systems. For example, a rather alkaline cytosolic pH is generally found in cancer cells [69–71] and might be causal for cellular transformation [72–74]. Moreover, cytosolic pH increases in pancreatic beta cells upon glucose stimulation [75]. Interestingly, V-ATPase is also regulated by glucose in this cell type and regulates insulin secretion [8, 14, 15].

Thus, it is tempting to speculate that cytosolic pH is a conserved cellular signal mediating glucose sensing, at least in part, through V-ATPase.

However, the mechanisms by which V-ATPase might regulate cellular signaling are still incompletely understood. Recently published evidence demonstrates that V-ATPase regulates some of its downstream processes through direct interaction with GTPases. As discussed, V-ATPase recruits the GTPase Arl6 and its GEF ARNO to endosomal membranes to regulate vesicular trafficking [16]. Similarly, V-ATPase interacts with and activates the Rag GTPases to activate TORC1 [10]. Future studies will be required to test if GTPase activation is a more general mechanism through which V-ATPases impinge on the regulation of various aspects of cellular physiology. In any case, studying the multiple roles of V-ATPase will undoubtedly reveal interesting new insights into the regulation of cellular signaling and physiology in both healthy and disease settings.

Acknowledgments

We would like to thank Alicia Smith, Serge Pelet, and Shady Saad for critical reading of the manuscript and helpful discussions. Financial support by the ETH Zurich, the Competence Center for Systems Physiology and Metabolic Diseases of the ETH and UNI Zürich, the Swiss Initiative in Systems Biology (SystemsX.ch, RTD grant YeastX), the Swiss National Science Foundation (SNF), and the European Research Council (ERC) is gratefully acknowledged.

References

1. Kane, P.M., Disassembly and reassembly of the yeast vacuolar H(+)-ATPase in vivo. *J. Biol. Chem.*, 1995, 270(28): pp. 17025–17032.
2. Parra, K.J. and P.M. Kane, Reversible association between the V1 and V0 domains of yeast vacuolar H^+-ATPase is an unconventional glucose-induced effect. *Mol. Cell Biol.*, 1998, 18(12): pp. 7064–7074.
3. Graf, R., W.R. Harvey, and H. Wieczorek, Purification and properties of a cytosolic V1-ATPase. *J. Biol. Chem.*, 1996, 271(34): pp. 20908–20913.

4. Kane, P.M. and A.M. Smardon, Assembly and regulation of the yeast vacuolar H^+-ATPase. *J. Bioenerg. Biomembr.*, 2003, 35(4): pp. 313–321.
5. Zaman, S., et al., How Saccharomyces responds to nutrients. *Annu. Rev. Genet.*, 2008, 42: pp. 27–81.
6. De Virgilio, C. and R. Loewith, Cell growth control: little eukaryotes make big contributions. *Oncogene*, 2006, 25(48): pp. 6392–6415.
7. Urban, J., *et al.*, Sch9 is a major target of TORC1 in *Saccharomyces cerevisiae. Mol. Cell*, 2007, 26(5): pp. 663–674.
8. Dechant, R., *et al.*, Cytosolic pH is a second messenger for glucose and regulates the PKA pathway through V-ATPase. *EMBO J.*, 2010, 29(15): pp. 2515–2526.
9. Bond, S. and M. Forgac, The Ras/cAMP/protein kinase A pathway regulates glucose-dependent assembly of the vacuolar (H+)-ATPase in yeast. *J. Biol. Chem.*, 2008, 283(52): pp. 36513–36521.
10. Zoncu, R., *et al.*, mTORC1 senses lysosomal amino acids through an inside-out mechanism that requires the vacuolar H(+)-ATPase. *Science*, 2011, 334(6056): pp. 678–683.
11. Li, G., *et al.*, A novel cellular survival factor–the B2 subunit of vacuolar H^+-ATPase inhibits apoptosis. *Cell Death Differ.*, 2006, 13(12): pp. 2109–2117.
12. Yan, Y., N. Denef, and T. Schupbach, The vacuolar proton pump, V-ATPase, is required for notch signaling and endosomal trafficking in Drosophila. *Dev. Cell*, 2009, 17(3): pp. 387–402.
13. Cruciat, C.M., *et al.*, Requirement of prorenin receptor and vacuolar H^+-ATPase-mediated acidification for Wnt signaling. *Science*, 2010, 327(5964): pp. 459–463.
14. Stiernet, P., *et al.*, Glucose acutely decreases pH of secretory granules in mouse pancreatic islets. Mechanisms and influence on insulin secretion. *J. Biol. Chem.*, 2006, 281(31): pp. 22142–22151.
15. Sun-Wada, G.H., *et al.*, The a3 isoform of V-ATPase regulates insulin secretion from pancreatic beta-cells. *J. Cell Sci.*, 2006, 119(Pt. 21): pp. 4531–4540.
16. Hurtado-Lorenzo, A., *et al.*, V-ATPase interacts with ARNO and Arf6 in early endosomes and regulates the protein degradative pathway. *Nat. Cell Biol.*, 2006, 8(2): pp. 124–136.
17. Sautin, Y.Y., *et al.*, Phosphatidylinositol-3-kinase-mediated effects of glucose on vacuolar H^+-ATPase assembly, translocation, and acidification of intracellular compartments in renal epithelial cells. *Mol. Cell. Biol.*, 2005, 25(2): pp. 575–589.

18. Zimmermann, B., *et al.*, Distribution and serotonin-induced activation of vacuolar-type H^+-ATPase in the salivary glands of the blowfly *Calliphora vicina*. *J. Exp. Biol.*, 2003, 206(Pt. 11): pp. 1867–1876.
19. Kane, P.M., Close-up and genomic views of the yeast vacuolar H^+-ATPase. *J. Bioenerg. Biomembr.*, 2005, 37(6): pp. 399–403.
20. Sambade, M., *et al.*, A genomic screen for yeast vacuolar membrane ATPase mutants. *Genetics*, 2005, 170(4): pp. 1539–1551.
21. Seol, J.H., A. Shevchenko, and R.J. Deshaies, Skp1 forms multiple protein complexes, including RAVE, a regulator of V-ATPase assembly. *Nat. Cell Biol.*, 2001, 3(4): pp. 384–391.
22. Hedbacker, K. and M. Carlson, SNF1/AMPK pathways in yeast. *Front Biosci.*, 2008, 13: pp. 2408–2420.
23. Dechant, R. and M. Peter, Nutrient signals driving cell growth. *Curr. Opin. Cell Biol.*, 2008, 20(6): pp. 678–687.
24. Hedbacker, K. and M. Carlson, SNF1/AMPK pathways in yeast. *Front. Biosci.*, 2008, 13: pp. 2408–2420.
25. Dechant, R. and M. Peter, *unpublished data.*
26. Rolland, F., *et al.*, Glucose-induced cAMP signalling in yeast requires both a G-protein coupled receptor system for extracellular glucose detection and a separable hexose kinase-dependent sensing process. *Mol. Microbiol.*, 2000, 38(2): pp. 348–358.
27. Kraakman, L., *et al.*, A *Saccharomyces cerevisiae* G-protein coupled receptor, Gpr1, is specifically required for glucose activation of the cAMP pathway during the transition to growth on glucose. *Mol. Microbiol.*, 1999, 32(5): pp. 1002–1012.
28. Xue, Y., M. Batlle, and J.P. Hirsch, GPR1 encodes a putative G protein-coupled receptor that associates with the Gpa2p Galpha subunit and functions in a Ras-independent pathway. *EMBO J.*, 1998, 17(7): pp. 1996–2007.
29. Yun, C.W., *et al.*, Gpr1p, a putative G-protein coupled receptor, regulates glucose-dependent cellular cAMP level in yeast *Saccharomyces cerevisiae*. *Biochem. Biophys. Res. Commun.*, 1998, 252(1): pp. 29–33.
30. Rolland, F., *et al.*, The role of hexose transport and phosphorylation in cAMP signalling in the yeast *Saccharomyces cerevisiae*. *FEMS Yeast Res.*, 2001, 1(1): pp. 33–45.
31. Tanaka, K., *et al.*, *S. cerevisiae* genes IRA1 and IRA2 encode proteins that may be functionally equivalent to mammalian ras GTPase activating protein. *Cell*, 1990, 60(5): pp. 803–807.

32. Tanaka, K., *et al.*, IRA2, a second gene of *Saccharomyces cerevisiae* that encodes a protein with a domain homologous to mammalian ras GTPase-activating protein. *Mol. Cell. Biol.*, 1990, 10(8): pp. 4303–4313.
33. Huberts, D.H., B. Niebel, and M. Heinemann, A flux-sensing mechanism could regulate the switch between respiration and fermentation. *FEMS Yeast Res.*, 2012, 12(2): pp. 118–128.
34. Dames, P., *et al.*, cAMP regulates plasma membrane vacuolar-type H^+-ATPase assembly and activity in blowfly salivary glands. *Proc. Natl. Acad. Sci. USA*, 2006. 103(10): pp. 3926–3931.
35. Rein, J., *et al.*, Hormone-induced assembly and activation of V-ATPase in blowfly salivary glands is mediated by protein kinase A. *Am. J. Physiol. Cell. Physiol*, 2008. 294(1): pp. C56–C65.
36. Gong, F., *et al.*, Vacuolar H^+-ATPase apical accumulation in kidney intercalated cells is regulated by PKA and AMP-activated protein kinase. *Am. J. Physiol. Renal. Physiol.* (February 10, 2010). doi:10.1152/ajprenal.00645.2009, 2010.
37. Hallows, K.R., *et al.*, AMP-activated protein kinase inhibits alkaline pH- and PKA-induced apical vacuolar H^+-ATPase accumulation in epididymal clear cells. *Am. J. Physiol. Cell Physiol.*, 2009, 296(4): pp. C672– C681.
38. Paunescu, T.G., *et al.*, cAMP stimulates apical V-ATPase accumulation, microvillar elongation, and proton extrusion in kidney collecting duct A-intercalated cells. *Am. J. Physiol. Renal Physiol.*, 2010, 298(3): pp. F643–F654.
39. Lu, M., *et al.*, Physical interaction between aldolase and vacuolar H^+-ATPase is essential for the assembly and activity of the proton pump. *J Biol. Chem.*, 2007, 282(34): pp. 24495–24503.
40. Lu, M., *et al.*, The glycolytic enzyme aldolase mediates assembly, expression, and activity of vacuolar H^+-ATPase. *J. Biol. Chem.*, 2004, 279(10): pp. 8732–8739.
41. Lu, M., *et al.*, Interaction between aldolase and vacuolar H^+-ATPase: evidence for direct coupling of glycolysis to the ATP-hydrolyzing proton pump. *J Biol. Chem.*, 2001, 276(32): pp. 30407–30413.
42. Merkulova, M., *et al.*, Aldolase directly interacts with ARNO and modulates cell morphology and acidic vesicle distribution. *Am. J. Physiol. Cell Physiol.*, 2011, 300(6): pp. C1442–C1455.
43. Smardon, A.M., M. Tarsio, and P.M. Kane, The RAVE complex is essential for stable assembly of the yeast V-ATPase. *J. Biol. Chem.*, 2002, 277(16): pp. 13831–13839.

44. Kane, P.M., The long physiological reach of the yeast vacuolar H^+-ATPase. *J. Bioenerg. Biomembr.*, 2007, 39(5/6): pp. 415–421.
45. Smardon, A.M. and P.M. Kane, RAVE is essential for the efficient assembly of the C subunit with the vacuolar H(+)-ATPase. *J. Biol. Chem.*, 2007, 282(36): pp. 26185–26194.
46. Kresnowati, M.T., *et al.*, When transcriptome meets metabolome: fast cellular responses of yeast to sudden relief of glucose limitation. *Mol. Syst. Biol.*, 2006, 2: p. 49.
47. Walther, T., *et al.*, Control of ATP homeostasis during the respiro-fermentative transition in yeast. *Mol. Syst. Biol.*, 2010, 6: p. 344.
48. Dechant, R. and M. Peter, The *N*-terminal domain of the V-ATPase subunit 'a' is regulated by pH in vitro and in vivo. *Channels (Austin)*, 2011, 5(1): pp. 4–8.
49. Marshansky, V., The V-ATPase a2-subunit as a putative endosomal pH-sensor. *Biochem. Soc. Trans.*, 2007, 35(Pt. 5): pp. 1092–1099.
50. Maranda, B., *et al.*, Intra-endosomal pH-sensitive recruitment of the Arf-nucleotide exchange factor ARNO and Arf6 from cytoplasm to proximal tubule endosomes. *J. Biol. Chem.*, 2001, 276(21): pp. 18540–18550.
51. Portillo, F., Regulation of plasma membrane H(+)-ATPase in fungi and plants. *Biochim. Biophys. Acta*, 2000, 1469(1): pp. 31–42.
52. Diakov, T.T. and P.M. Kane, Regulation of vacuolar proton-translocating ATPase activity and assembly by extracellular pH. *J. Biol. Chem.*, 2010, 285(31): pp. 23771–23778.
53. Qi, J. and M. Forgac, Cellular environment is important in controlling V-ATPase dissociation and its dependence on activity. *J. Biol. Chem.*, 2007, 282(34): pp. 24743–24751.
54. Jasti, J., *et al.*, Structure of acid-sensing ion channel 1 at 1.9 A resolution and low pH. *Nature*, 2007, 449(7160): pp. 316–323.
55. Wilkens, S., Z. Zhang, and Y. Zheng, A structural model of the vacuolar ATPase from transmission electron microscopy. *Micron*, 2005, 36(2): pp. 109–126.
56. Wilkens, S. and M. Forgac, Three-dimensional structure of the vacuolar ATPase proton channel by electron microscopy. *J. Biol. Chem.*, 2001, 276(47): pp. 44064–44068.
57. Peters, C., *et al.*, Trans-complex formation by proteolipid channels in the terminal phase of membrane fusion. *Nature*, 2001, 409(6820): pp. 581–588.
58. Clare, D.K., *et al.*, An expanded and flexible form of the vacuolar ATPase membrane sector. *Structure*, 2006, 14(7): pp. 1149–1156.

59. Srinivasan, S., *et al.*, Crystal structure of the cytoplasmic *N*-terminal domain of subunit I, a homolog of subunit a, of V-ATPase. *J. Mol. Biol.*, 2011, 412(1): pp. 14–21.
60. Inoue, T. and M. Forgac, Cysteine-mediated cross-linking indicates that subunit C of the V-ATPase is in close proximity to subunits E and G of the V1 domain and subunit a of the V0 domain. *J. Biol. Chem.*, 2005, 280(30): pp. 27896–27903.
61. Young, B.P., *et al.*, Phosphatidic acid is a pH biosensor that links membrane biogenesis to metabolism. *Science*, 2010, 329(5995): pp. 1085–1088.
62. Martinez-Munoz, G.A. and P.M. Kane, Vacuolar and plasma membrane proton pumps collaborate to achieve cytosolic pH homeostasis in yeast. *J. Biol. Chem.*, 2008.
63. Serrano R. Transport across yeast vacuolar and plasma membranes. In: Broach J.R., Jones E.W., Pringle J.R., editors. The Molecular and Cellular Biology of the Yeast *Saccharomyces*: Genome Dynamics, Protein Synthesis, and Energetics. Cold Spring Harbor, N.Y.: Cold Spring Harbor Laboratory Press; 1991, pp. 523–585.
64. Forster, C. and P.M. Kane, Cytosolic Ca^{2+} homeostasis is a constitutive function of the V-ATPase in *Saccharomyces cerevisiae*. *J. Biol. Chem.*, 2000, 275(49): pp. 38245–38253.
65. Kraft, C., F. Reggiori, and M. Peter, Selective types of autophagy in yeast. Biochim. Biophys. Acta, 2009, 1793(9): pp. 1404–1412.
66. Cebollero, E. and F. Reggiori, Regulation of autophagy in yeast *Saccharomyces cerevisiae*. *Biochim. Biophys. Acta*, 2009, 1793(9): pp. 1413–1421.
67. Hyslop, P.A., *et al.*, Mechanisms of oxidant-mediated cell injury. The glycolytic and mitochondrial pathways of ADP phosphorylation are major intracellular targets inactivated by hydrogen peroxide. *J. Biol. Chem.*, 1988, 263(4): pp. 1665–1675.
68. Orij, R., *et al.*, In vivo measurement of cytosolic and mitochondrial pH using a pH-sensitive GFP derivative in Saccharomyces cerevisiae reveals a relation between intracellular pH and growth. *Microbiology*, 2009, 155(Pt. 1): pp. 268–278.
69. Busa, W.B. and R. Nuccitelli, Metabolic regulation via intracellular pH. *Am. J. Physiol.*, 1984, 246(4, Pt. 2): pp. R409–R438.
70. Casey, J.R., S. Grinstein, and J. Orlowski, Sensors and regulators of intracellular pH. *Nat. Rev. Mol. Cell Biol.*, 2009, 11(1): pp. 50–61.

71. Kurkdjian, A. and J. Guern, Intracellular pH: Measurement and Importance in Cell Activity. *Annu. Rev. Plant Physiol. Plant Mol. Biol.*, 1989, 40(1): pp. 271–303.

72. Perona, R. and R. Serrano, Increased pH and tumorigenicity of fibroblasts expressing a yeast proton pump. *Nature*, 1988, 334(6181): pp. 438–340.

73. Perona, R., *et al.*, Transformation and pH homeostasis of fibroblasts expressing yeast H(+)-ATPase containing site-directed mutations. *Mol. Cell. Biol.*, 1990, 10(8): pp. 4110–4115.

74. Molina, J.R., *et al.*, PTEN, NHERF1 and PHLPP form a tumor suppressor network that is disabled in glioblastoma. *Oncogene*, 2012, 31(10): pp. 1264–1274.

75. Stiernet, P., *et al.*, Glucose-induced cytosolic pH changes in beta-cells and insulin secretion are not causally related: studies in islets lacking the Na^+/H^+ exchangeR NHE1. *J. Biol. Chem.*, 2007, 282(34): pp. 24538–24546.

Chapter 6

Vacuolar H^+-ATPase (V-ATPase) Activated by Glucose: A Possible Link to Diabetes

Suguru Nakamura
Department of Biological Sciences, Murray State University, 2112D Biological Science Building, Murray, KY 42071, USA
snakamura@murraystate.edu

6.1 Introduction

Glucose is a ubiquitous fuel in biology. It is used as an energy source in most organisms, from bacteria to humans. Use of glucose as an energy source in cells is via aerobic or anaerobic respiration. Both of these start with the early steps of the glycolysis metabolic pathway.

Metabolic control of ion transport in renal tubular epithelium provides not only a means to reduce cellular ATP consumption during limited metabolic substrate or oxygen availability but also has an important role in regulating transport during substrate availability. Studies in turtle urinary bladder [1, 2], mammalian proximal tubules [3], and mammalian collecting ducts [4] have demonstrated the importance of metabolic substrates in the control

Handbook of H^+-ATPases
Edited by Suguru Nakamura

ISBN 978-981-4411-91-2 (Hardcover), 978-981-4411-92-9 (eBook)
www.panstanford.com

of epithelial H^+ transport and have shown that glucose is a preferred substrate

Vacuolar H^+-ATPase (V-ATPase)-mediated H^+ transport is coupled with glycolysis in urinary epithelia. Steinmetz *et al.* [5] demonstrated close coupling of proton secretion with anaerobic lactate production in the turtle urinary bladder. Kurtz [3] found that H^+-ATPase activity in the rabbit S3 proximal tubule was inhibited by the sulfhydryl reagent iodoacetate, an inhibitor of glyceraldehyde-3-phosphate dehydrogenase (GAPDH), suggesting dependence of the H^+-ATPase on glycolysis. In the genetic disorder hereditary fructose intolerance, caused by deficiency in the glycolytic enzyme aldolase B, fructose induces rapid, severe proximal renal tubular acidosis [6], which is thought to be caused by inhibition of residual aldolase activity from fructose metabolites [7].

V-ATPase binds directly to the glycolytic enzyme aldolase and colocalizes with V-ATPase in two proton-transporting cell types: renal proximal tubule cells and osteoclasts [8]. The functional importance of the interaction was demonstrated in Saccharomyces, in which deletion of the aldolase gene resulted in disassembly of V-ATPase [8], similar to the disassembly observed after glucose removal [9], providing further evidence for direct coupling of V-ATPase with the glycolytic pathway. Our study shows that glucose activates V-ATPase activity in renal epithelial cells through a pathway requiring aerobic glycolysi; coupling between V-ATPase activity and glycolysis. Glucose is a potent and rapid activator of V-ATPase-mediated proton secretion in both LLC-PK1 cells and isolated renal proximal tubules. We also examined the activity of V-ATPase in the individual intercalated cells (ICs) of outer medullary collecting ducts (OMCDs) in normal and diabetic animals. These findings provide preliminary evidence suggestive of an association between V-ATPase and diabetes.

6.2 Glucose Activates V-ATPase Activity in LLC-PK1 Cells

The brewer's yeast *Saccharomyces cerevisiae* has V-ATPase on the vacuolar membrane that is similar in structure and properties to mammalian V-ATPases [10–14]. Yeast use glucose preferentially as a substrate for anaerobic glycolysis in ethanol production [15]. Glucose removal induced rapid disassembly of yeast V-ATPase. Rapid reassembly occurred in a concentration-dependent manner with glucose retreatment [9, 16] and required metabolism of glucose beyond the formation of glucose 6-phosphate [16]. Other previous studies have shown that V-ATPase binds directly to aldolase and that V-ATPase disassembly occurs in yeast strains that are deficient in aldolase and other glycolytic enzymes [8]. Collectively, these experiments demonstrate that glycolysis is essential for assembly and function of the *Saccharomyces* V-ATPase.

We examined the activities of V-ATPase in pig kidney epithelial cell line, LLC-PK_1 cells [17]. Acute removal of glucose from medium markedly reduced V-ATPase-dependent pH_i recovery activity. Readdition of glucose induced concentration-dependent reactivation of V-ATPase pH_i recovery activity within 2 min (Fig. 6.1). The glucose concentration dependence of V-ATPase activation was determined by stimulating cells with varying (0–30 mM) concentrations of glucose after acute glucose removal (Fig. 6.2). The data demonstrate that LLC-PK_1 cells contain glucose-activated V-ATPase, the activity of which is dependent on the continual presence of glucose and the concentration of glucose.

Amlal *et al.* [18] reported H^+-ATPase activity in LLC-PK_1 cells activated by hypotonicity. To determine whether glucose-induced Vacuolar H^+-ATPase activation was a result of changing the solution osmolarity in our study, we used mannitol as a substitute for glucose. As shown in Fig. 6.3, mannitol failed to activate V-ATPase activity, showing that glucose-induced V-ATPase activation is not due to an increase in solution osmolarity [17].

6.3 Glucose Activates Vacuolar H^+-ATPase Activity in Rat Proximal Tubule

LLC-PK_1 cells have several morphological and physiological properties resembling those of proximal tubule cells [19–22], including Na^+-glucose cotransport [21], but they do not possess a complete proximal tubule phenotype. To determine whether glucose activation of Vacuolar H^+-ATPase occurs in renal tubular epithelial

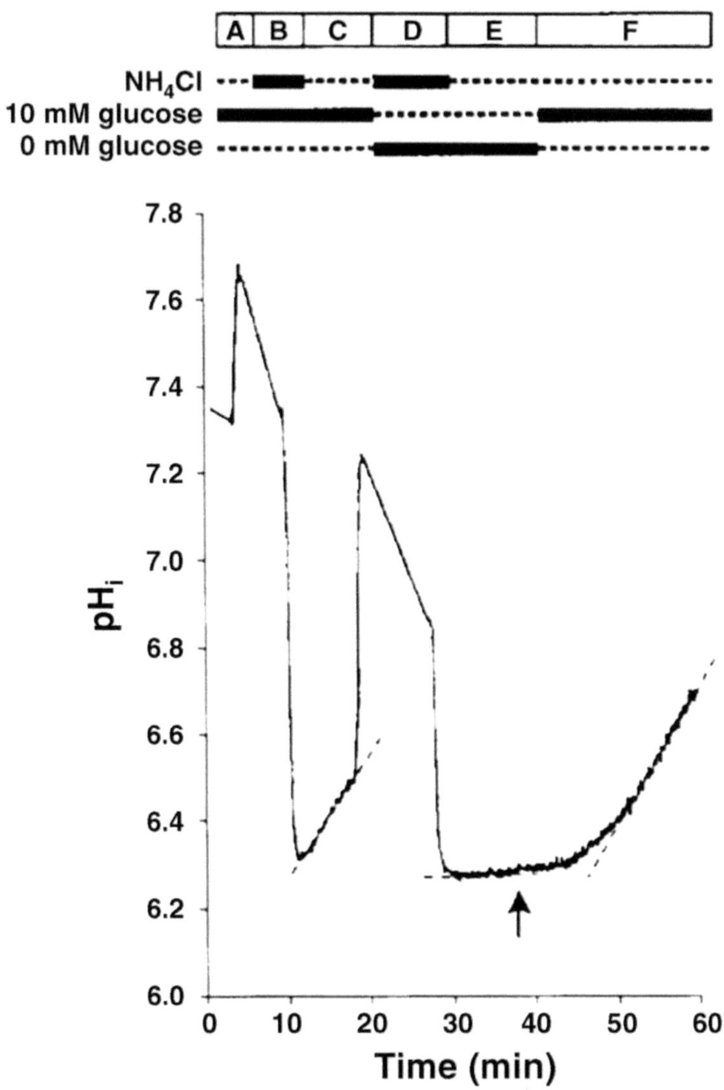

Figure 6.1 Glucose activates V-ATPase activity in LLC-P_{K1} cells. pHi measurement and intracellular acidification (periods A and B) were performed. V-ATPase activity was present in 10 mM glucose (period C). After reacidification (period D), V-ATPase activity was nearly abolished in absence of glucose (period E) but recovered rapidly to initial rate after addition of 10 mM glucose (period F).

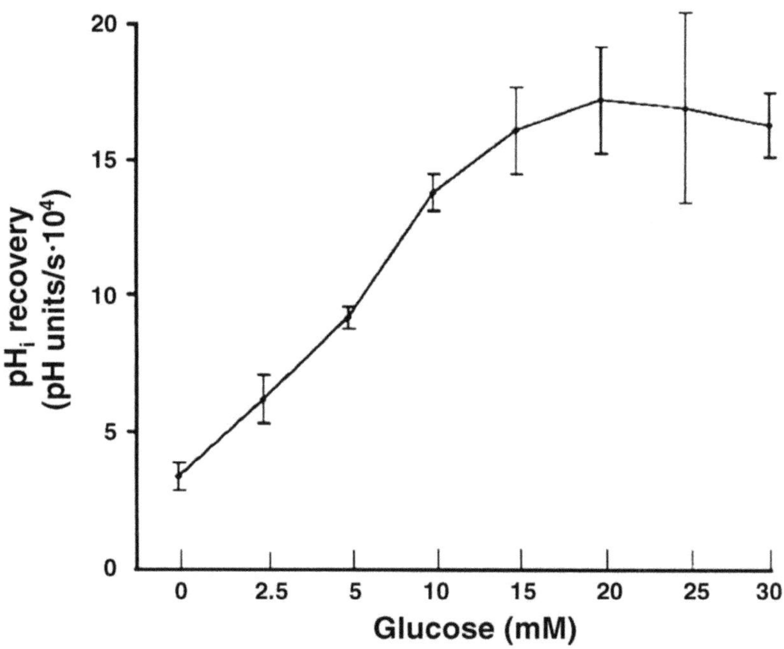

Figure 6.2 Glucose concentration dependence of V-ATPase activation. pH_i recovery rates were measured at varying concentrations as indicated. Values are means ± SE for pH_i recovery (dpH_i/dt in pH units/s).

cells from intact proximal tubules, the effect of glucose on the V-ATPase activity was examined in isolated perfused rat PT S3 segments. As shown in Figs. 6.4A and B, changing the glucose concentration from 2.5 to 10.5 mM produced a significant increase in pH_i recovery rates. The results suggest that intact rat renal proximal tubule cells have glucose-activated V-ATPase activity similar to that observed in LLC-PK1 cells [21].

6.4 Glucose Activation of V-ATPase Requires Glycolysis

In principle, glucose could activate V-ATPase through metabolism, through signaling pathways, or both. To determine whether activation of V-ATPase activity by glucose requires metabolism through

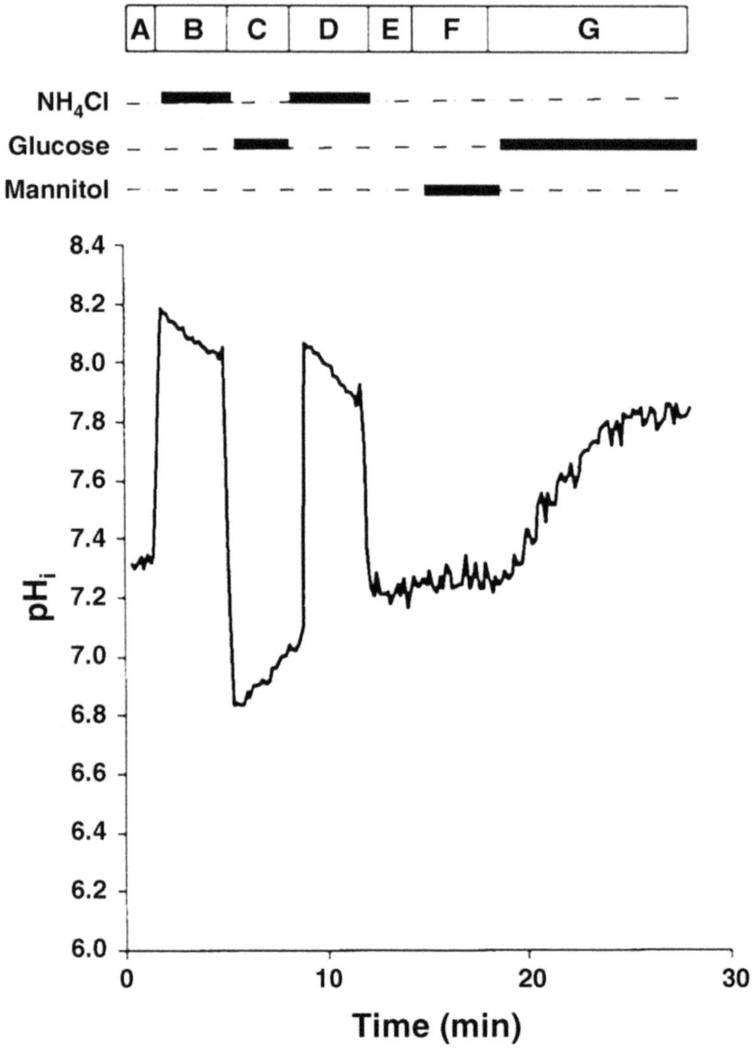

Figure 6.3 Glucose-induced V-ATPase activation is not due to change in medium osmolarity. Procedures used were identical to those in Fig. 6.1, except that 10 mM mannitol replaced glucose (period F), which produced no pH_i recovery. Recovery of V-ATPase activity was observed after readdition of 10 mM glucose (period G).

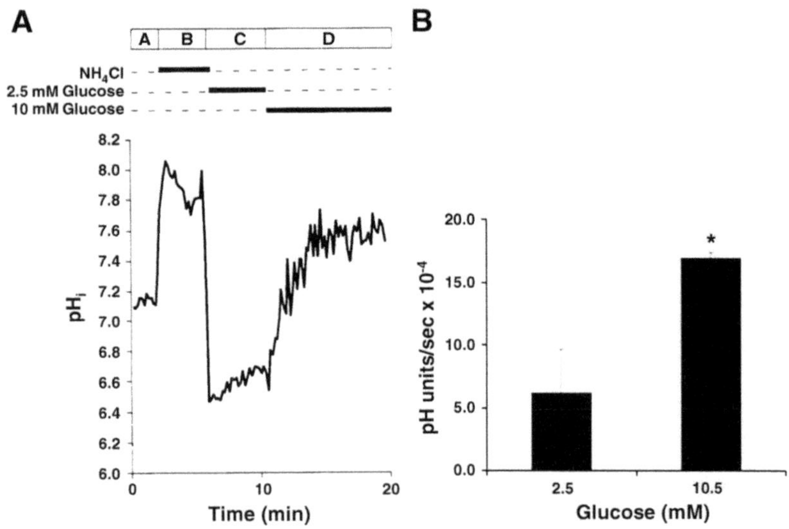

Figure 6.4 Glucose activates V-ATPase in isolated rat renal proximal tubule. A: intracellular pH measurements with BCECF were performed on isolated microperfused rat proximal tubules in Na^+-free conditions. After NH_4Cl induced pH_i acidification (periods A and B), pH_i recovery rates were measured in 2.5 mM glucose (period C) and 10 mM glucose (period D). B: pH_i recovery rates in 2.5 and 10 mM glucose. Values are means ± SE; $n = 14$. $P < 0.05$.

glycolysis [17], the effect of 20 mM of 2-DG, a glycolytic inhibitor, was examined. As shown in Fig. 6.5, pretreatment with 2-DG eliminated glucose-activated V-ATPase activity, suggesting that the response requires metabolism of glucose through glycolysis. Treatment with the mitochondrial electron transport inhibitor antimycin A (20 μM) partially inhibited activation of V-ATPase by glucose (Fig. 6.6), suggesting that activation requires aerobic rather than anaerobic glycolysis. In support of this interpretation, it was found that 10 mM of pyruvate, the mitochondrial substrate for aerobic glycolysis, induced concanamycin-sensitive pH recovery, indicating activation of H^+-ATPase activity (Fig. 6.7). In contrast, sodium acetate produced cytosolic alkalinization that was unaffected by concanamycin A (Fig. 6.8), probably by metabolism to bicarbonate in mitochondria. Alanine (10 mM) also was unable to activate V-ATPase activity (Fig. 6.9)

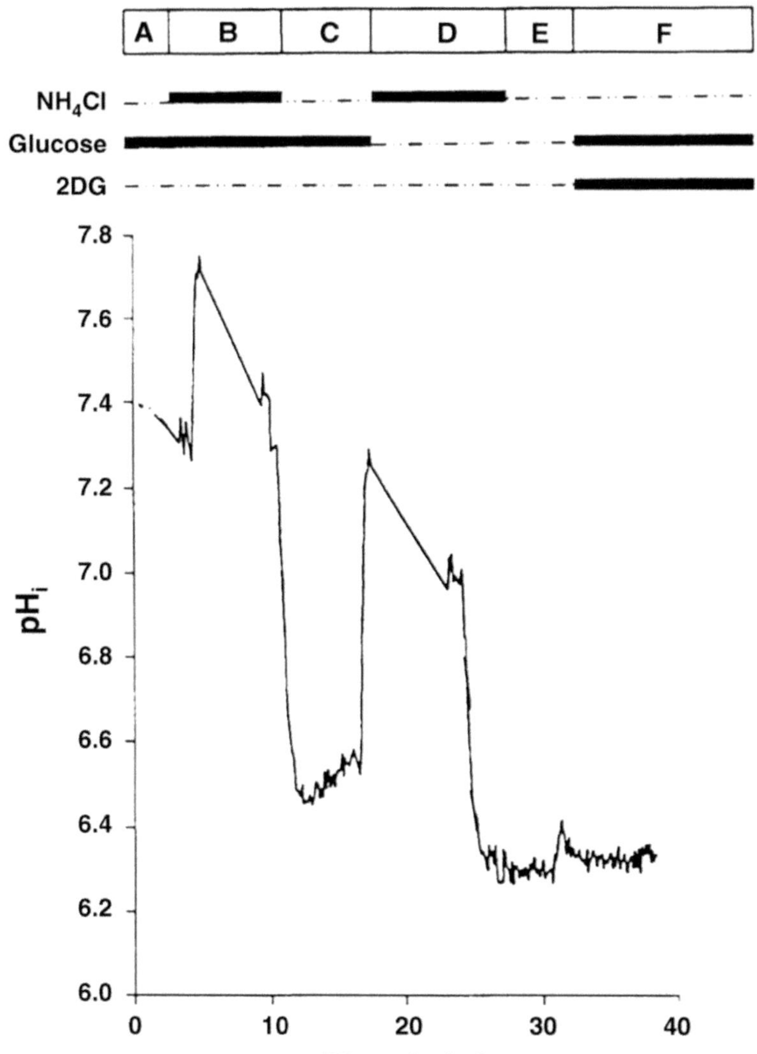

Figure 6.5 Glycolysis is required for glucose-induced V-ATPase activation. Procedures used were identical to those in Fig. 6.1, except that glycolytic inhibitor 2-deoxy-D-glucose (2-DG; 20 mM) was added during second glucose treatment (*period F*), which completely inhibited V-ATPase activation.

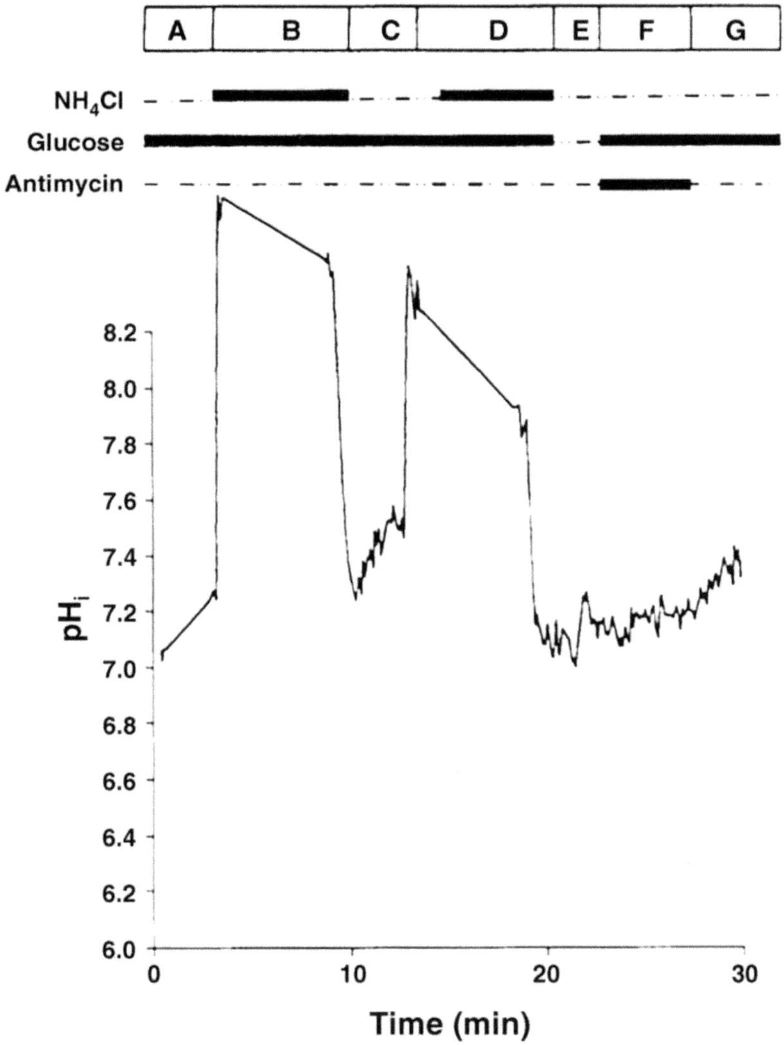

Figure 6.6 Effect of antimycin A on glucose-induced V-ATPase activation. Procedures used were identical to those in Fig. 6.1, except that mitochondrial electron transport inhibitor antimycin A (20 μM) was added during second glucose treatment (*period F*), which partially inhibited V-ATPase activation. Incomplete recovery of V-ATPase activity was observed after antimycin A removal (*period G*).

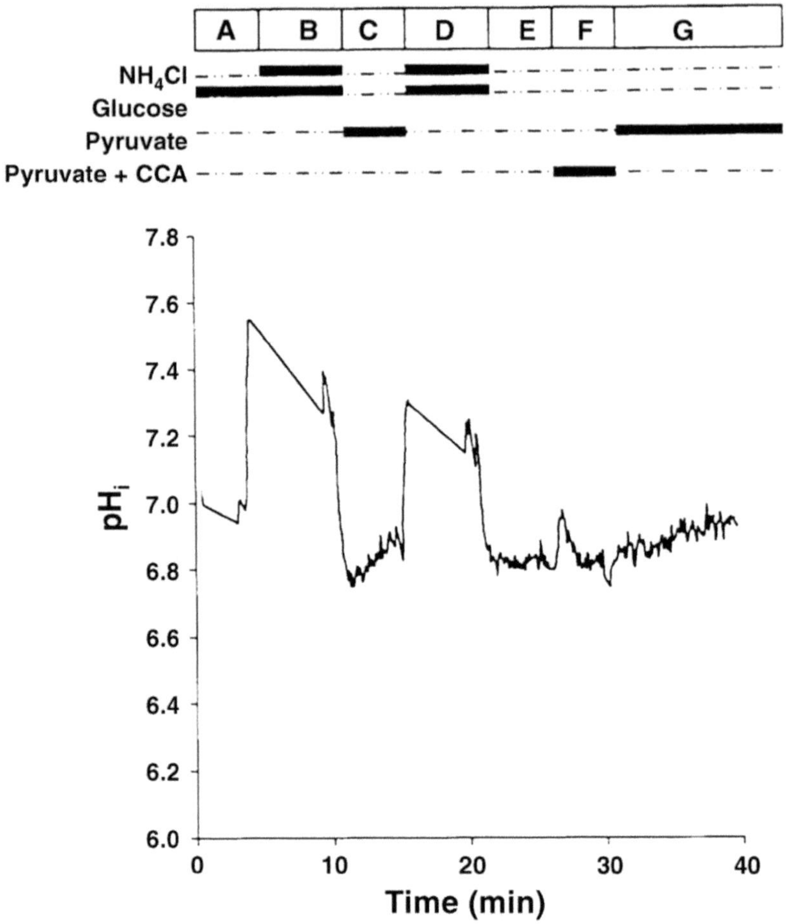

Figure 6.7 Pyruvate activates V-ATPase in absence of glucose. Cells were treated with NH_4Cl in presence of 10 mM glucose to induce pH_i acidification (*periods A* and *B*) followed by removal of glucose and treatment with 10 mM pyruvate (*period C*), which produced significant pH recovery. After repeat pH_i acidification in presence of glucose (*period D*), cells were treated with glucose-free solution (*period E*), which produced no pH recovery, and then with 10 mM pyruvate containing 100 nM CCA (*period F*), which also produced no recovery. Partial pH_i recovery was observed after removal of CCA (*period G*).

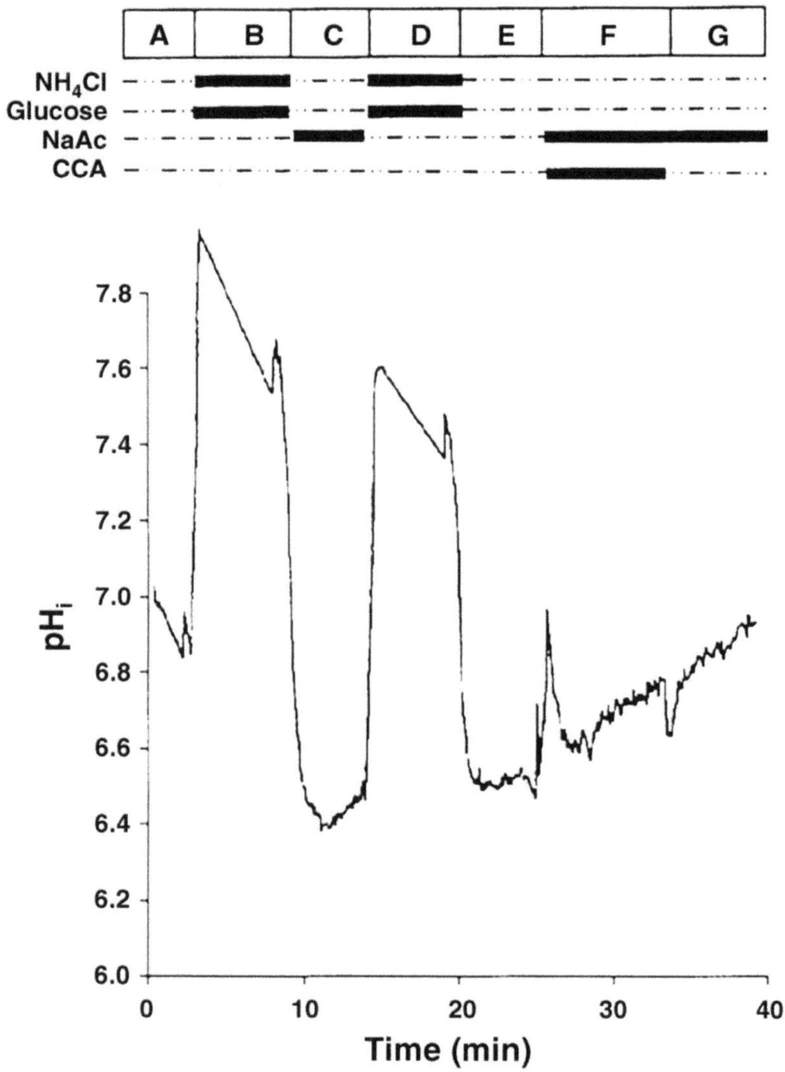

Figure 6.8 Sodium acetate (NaAc) alkalinizes intracellular pH but does not activate V-ATPase. Procedures used were identical to those in Fig. 6.7, except that 10 mM sodium acetate replaced pyruvate. Acetate produced alkalinization of pH_i (*period C*) that was not observed after acetate removal (*period E*) and was not affected by V-ATPase inhibitor CCA (*period F*).

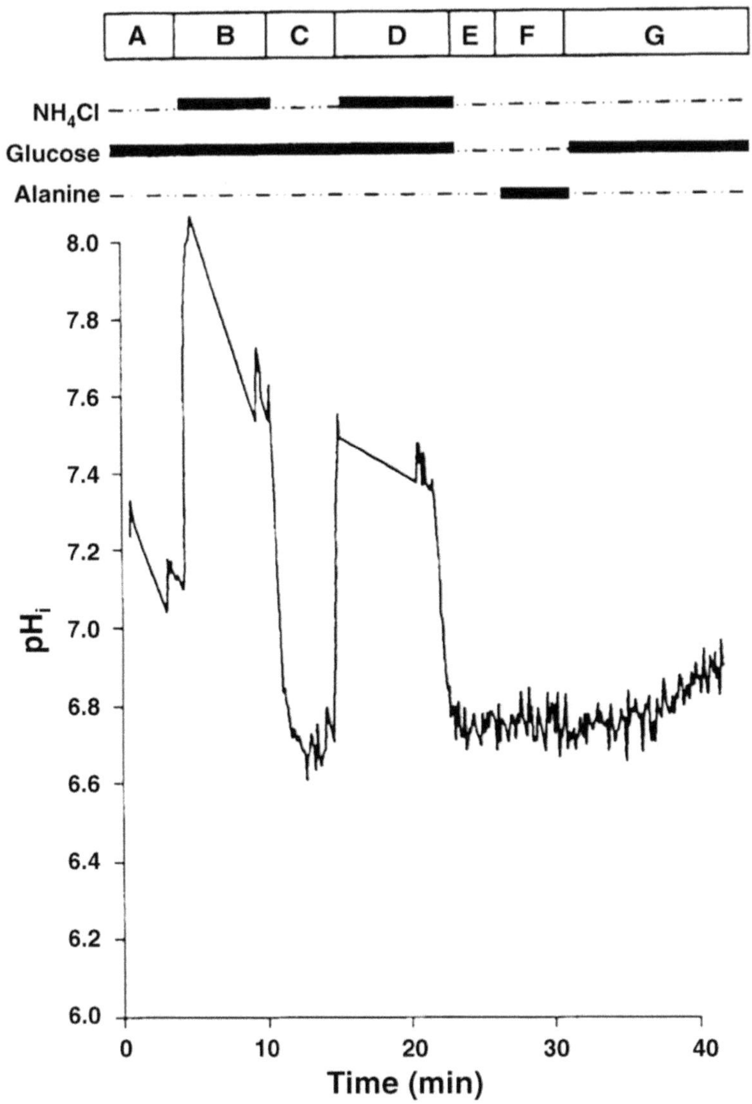

Figure 6.9 Alanine does not activate V-ATPase. Procedures used were similar to those in Fig. 6.3, except that 10 mM alanine replaced mannitol. No V-ATPase activity was observed with alanine in the absence of glucose (*period F*), but activity resumed after replacement of glucose (*period G*).

6.5 Glucose Activation of V-ATPase Requires PI3K Activity

In insulin-responsive tissues [23–25], lymphocytes [26], and other cells [23], signaling pathways involving phosphatidylinositol 3-kinase (PI3K) are involved in controlling glucose entry [25] and metabolism [24]. To determine whether a PI3K-dependent signaling pathway contributes to the effect of glucose on V-ATPase activity, the effect of 500 nM wortmannin, a PI3K inhibitor, was studied. As shown in Fig. 6.10, V-ATPase activity was inhibited completely by wortmannin, and activity was restored after wortmannin removal.

Our studies demonstrate that glucose is a potent and rapid activator of V-ATPase-mediated proton secretion in both LLC-PK1 cells and isolated renal proximal tubules [17]. The glucose effect was concentration dependent, with a half-maximal effect at 5.1 mM of glucose and a maximal effect at 20 mM. Glucose-induced V-ATPase activation was rapidly reversible. V-ATPase activity was lost after 5–10 min of glucose removal and was restored rapidly by the readdition of glucose, with mean timing of 2–3 min to initial recovery and 5.5–7 min to half-maximal recovery. The effect was not due to changes in extracellular fluid osmolarity, because mannitol produced no significant effect. Glucose-induced V-ATPase activation required metabolism of glucose through the glycolytic pathway because it was inhibited entirely by 2-DG. Metabolism likely occurs through aerobic glycolysis because the mitochondrial complex III inhibitor antimycin partially inhibited V-ATPase activation. Pyruvate, an end product of the glycolytic pathway, activated V-ATPase activity in the absence of glucose, but alanine and acetate did not activate V-ATPase activity. These results extend those of the recent studies of the interaction of the V-ATPase with aldolase [8] and provide further evidence for coupling between V-ATPase activity and glycolysis.

Several studies have examined metabolic pathways supporting proton transport by V-ATPases in urinary epithelia [1, 3]. In the turtle urinary bladder, a model epithelium resembling the mammalian kidney cortical collecting duct [27, 28] with an electrogenic plasma membrane V-ATPase [29, 30], glucose stimulated electrogenic V-ATPase-mediated proton transport [1, 5, 31, 32]. Under standard

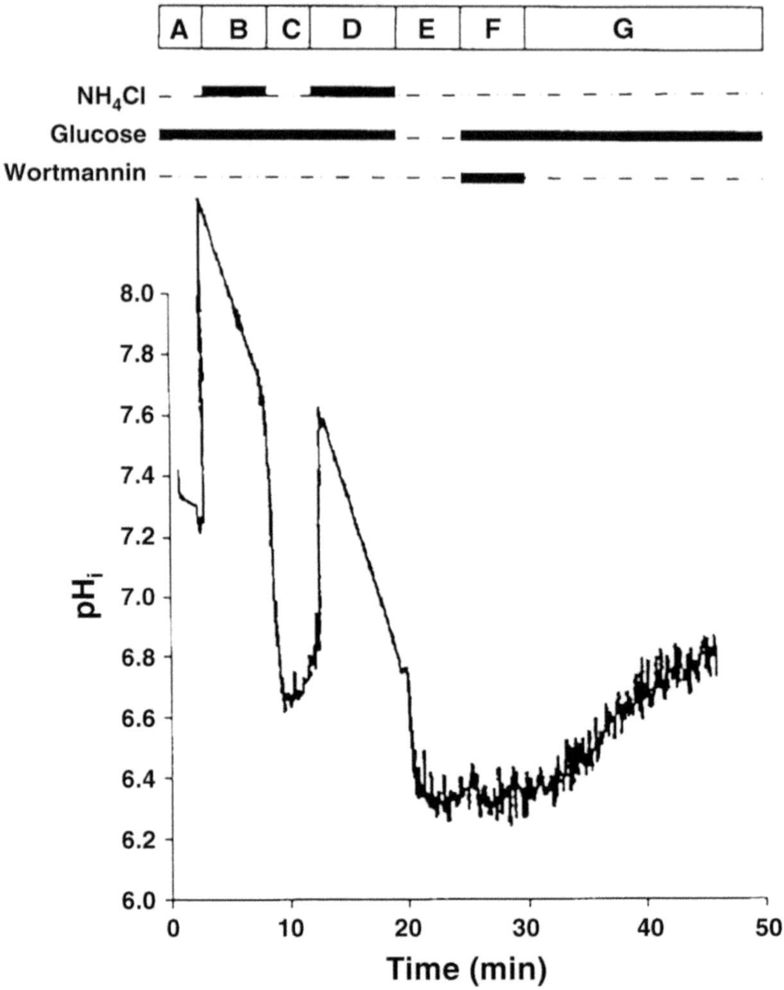

Figure 6.10 Glucose-induced V-ATPase activation requires PI3K activity. Procedures used were similar to those in Fig. 6.9, except that PI3K inhibitor wortmannin (500 nM) replaced alanine. Glucose-induced V-ATPase activity was abolished by wortmannin *(period* F) and partially recovered after removal of wortmannin (*period G*).

conditions, transport was inhibited by 2-DG and deoxygenation [21] and was not coupled with lactate production [32]. In bladders treated with aldosterone to activate H^+ transport, active proton secretion showed a greater response to glucose addition and was stimulated by pyruvate [33]. These studies suggest coupling of H^+ transport with aerobic glycolysis under these conditions. Under anaerobic conditions, H^+ transport was tightly coupled with lactate production through anaerobic glycolysis [5].

H^+-ATPase activity was identified in rabbit renal proximal tubules that were inhibited by the glycolytic inhibitor iodoacetate, suggesting coupling of activity with glycolysis [3]. Proximal tubules have a low rate of lactate production [34], indicating that glucose metabolism in this segment occurs primarily by aerobic glycolysis. The present study shows that H^+-ATPase activity in rat proximal tubules is stimulated by increasing the extracellular glucose concentration, a response similar to the glucose-induced V-ATPase activation observed in LLC-PK1 cells.

Glucose does not stimulate V-ATPase activity by changing cell ATP content. In principle, low glucose concentrations could suppress V-ATPase activity by reducing cell ATP levels. Cell ATP concentrations in kidney [35] and cultured cells [36] are estimated to be at least 2–3 mM. Thus, cell ATP levels would need to fall dramatically to attribute the effects of glucose to altered cell ATP levels [17].

Nakamura *et al.* [37] previously examined the effect of glucose on the levels of ATP, ADP, and ATP-to-ADP ratio in serum- and/or glucose-starved LLCPK cells. Their study showed that incubation in glucose-free medium for 16 h reduced cell ATP content by 37% and increased ADP content slightly, changing the ATP-to-ADP ratio significantly from 2.53 to 1.22. However, stimulation with 10 mM glucose for periods ranging from 2.5 to 30 min produced no significant changes in ATP or ADP content. These results indicate that stimulation of V-ATPase activity by glucose, at least within short time intervals, likely does not occur by increasing ATP availability

Recently, other studies have shown that ATP level remained practically unchanged when yeast cells were grown in either 2% or 0.025% glucose [38]. Krauss *et al.* [39] demonstrated that under physiological conditions, hyperglycemia-induced mitochondrial superoxide production activates uncoupling protein 2, which decreases the ATP-to-ADP ratio.

Other previous studies have shown that glucose stimulates glycolysis and increases the activity of several glycolytic enzymes in LLC-PK1 cells [40]. The present study revealed that the glycolytic inhibitor 2-DG prevented glucose-induced V-ATPase activation in LLC-PK1 cells. Taken together, these results suggest that glucose activates V-ATPase by stimulating glycolysis.

The present study shows that pyruvate also activated V-ATPase activity in the absence of glucose. It is possible that pyruvate could be converted to glucose through the gluconeogenic pathway and subsequently metabolized by glycolysis [410]. In the presence of the V-ATPase inhibitor concanamycin A, it was observed that pyruvate-induced V-ATPase activity was inhibited and that V-ATPase activity was restored by pyruvate after CCA removal. As shown in Fig. 6.8, however, the V-ATPase inhibitor (i.e., CCA) had no effect on acetate-induced pH recovery, likely due to a mechanism other than V-ATPase activity involvement for the regulation of intracellular pH. The alkalinization observed with acetate might be due to metabolic generation of alkali. Ishikawa *et al.* [42] showed that acetate induced cytosolic alkalinization that was not affected by H^+-ATPase inhibitors and probably was a result of mitochondrial metabolism of acetate to bicarbonate. Studies in the kidney [41, 43] have demonstrated the importance of such "futile" cycles of glycolysis and gluconeogenesis, which provide the capacity for rapid changes in glycolytic flux [44, 45]. Net ATP consumption in futile cycles may be reduced by separation and compartmentalization of the glycolytic and gluconeogenic pathways [44].

Both pyruvate and acetate are metabolized by the citric acid cycle. The metabolism of acetate through acetyl coenzyme A-synthetase consumes one ATP and generates one less NADH and CO2 [42] than the metabolism of pyruvate through pyruvate dehydrogenase. Acetate also inhibits some of the pathways for pyruvate metabolism [46, 47]. The observed differences between these two agents on V-ATPase activation strongly implicate mitochondrial metabolism in the activation pathway, consistent with the partial inhibition of activation observed with antimycin.

It is also significant that alanine was unable to activate V-ATPase. Although alanine can be converted to pyruvate by transamination [48, 49] and is a potential substrate for gluconeogenesis [48, 50], studies in both isolated renal proximal tubules [51, 52] and human volunteers [53, 54] have shown that alanine is a poor substrate for renal gluconeogenesis compared with lactate. These studies support the interpretation that substrate flux through the glycolytic pathway is required for V-ATPase activation. The results observed are not specific for V-ATPase, however, because Hering-Smith and Hamm [4] found that alanine was not an effective metabolic substrate for collecting duct Na^+ transport in the absence of glucose.

The results of our study indicate that PI3K activity is required for glucose activation of V-ATPase (Fig. 6.10). PI3K activity is required for several signaling pathways involved in glucose control, including glucose entry and glycogen metabolism [23–25]. Glucose activates V-ATPase activity in renal epithelial cells through a pathway requiring aerobic glycolysis and PI3K activity, providing evidence for coupling between V-ATPase activity and glycolysis [17].

6.6 V-ATPase Activity in the Perfused Collecting Duct in Diabetic Animals

To examine the effect of glucose on Vacuolar H^+-ATPase-mediated net acid secretion in the diabetic kidney, we dissected the rat OMCD after three days of steptozotocin treatment. Diabetes was induced by intraperitoneal injection (40–45 mg/kg) of streptozotocin in mature 19-week-old male Sprague–Dawley rats. Hyperglycemia occurs normally in 24 h after the steptozotocin treatment. Plasma glucose was 7.4 $\pm$ 0.4 mM in normal rats (n = 21), and increased significantly to 30.9 $\pm$ 0.7 mM in diabetic rats (n = 30) (Fig. 6.11).

In vitro microperfusion was performed as shown in Figs. 6.12A and B. Intracellular pH was measured in individual ICs of OMCD tubule cells [55, 56]. Fig. 6.12C shows heterogeneity of ICs and principal cells (PCs) uptake of BCECF; fluorescence intensity is remarkably higher in ICs.

As shown in Figs. 6.13A and B, following NH_3/NH_4^+ prepluse, pH_i recovery rates were 10.8 $\pm$ 0.8$\times$ 10^{-4} pH units/sec in

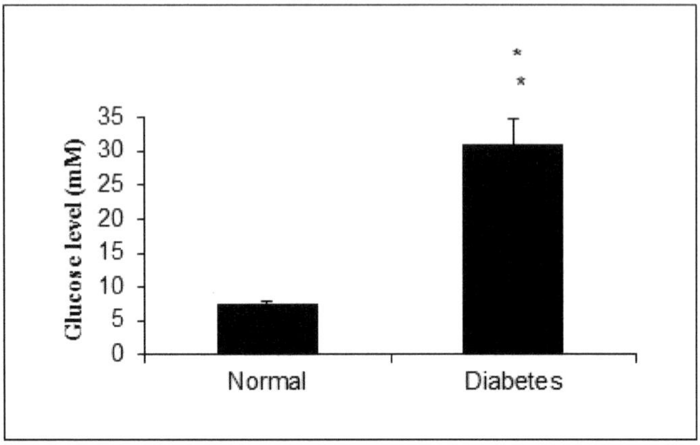

Figure 6.11 Streptozotocin-induced diabetic rats were examined in 24 hours. Plasma glucose was 7.4 ± 0.4 mM in normal rats (n = 21), and increased significantly to 30.9 ± 0.7 mM in diabetic rats (n = 30).

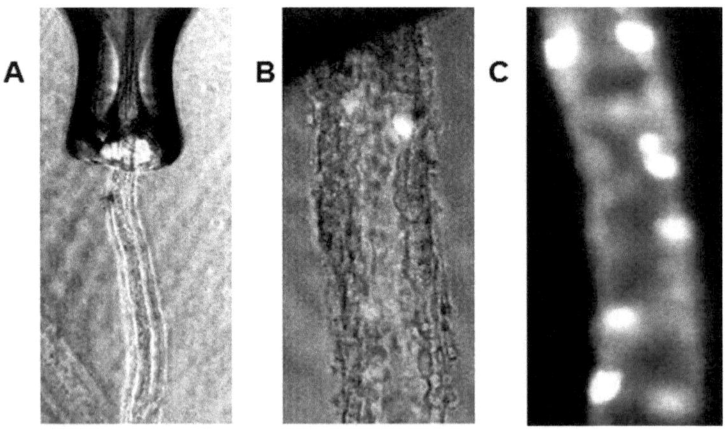

Figure 6.12 *In vitro* microperfusion was performed and Intracellular pH was measured in individual intercalated cells (ICs) of OMCD tubule cells. A: perfused OMCD tubule. B: zoomed OMCD tubule. C: heterogeneity of intercalated cells (ICs) and principal cells (PCs) uptake of BCEC; fluorescence intensity is remarkably higher in ICs of OMCD tubule cells.

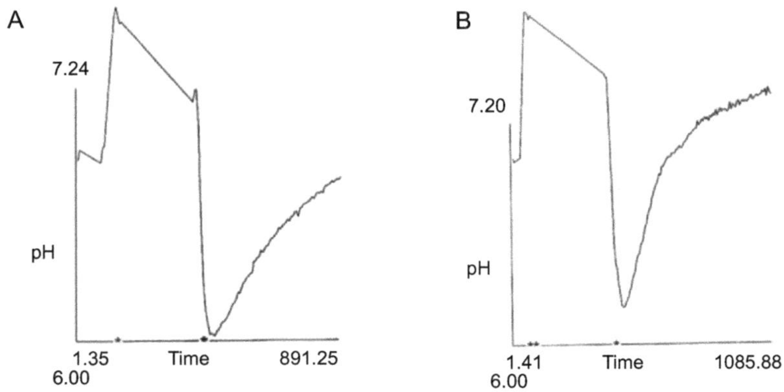

Figure 6.13 Intracellular pH was measured in the diabetic and non-diabetic rats. pH_i recovery rates were $10.8 \pm 0.8 \times 10^{-4}$ pH units/sec in non-diabetes (A, $n = 4$) and increased significantly to $15.6 \pm 0.3 \times 10^{-4}$ pH units/sec in diabetes (B, $p < 0.0$; $n = 4$).

non-diabetes (Fig. 6.13A) ($n = 4$ tubules) and significantly increased to $15.6 \pm 0.3 \times 10^{-4}$ pH units/sec in diabetes (Fig. 6.13B) ($p < 0.01$; $n = 4$). The microperfusion rates were 1–13 nl/min and the solutions used were glucose-free. The activity of V-ATPase, monitored by the pH_i recovery after NH_3/NH_4^+ removal, is comparable in the diabetic and non-diabetic groups. Our results suggest that glucose is an important physiologic regulator of V-ATPase activity in the diabetic OMCD tubules.

It has been reported that gluconeogenesis in diabetic rabbit kidney tubules is increased by about 20–30% [57]. Chloroquine is a potent inhibitor of glutamate dehydrogenase and inhibits the rate of glucose formation [15, 58–60]. In patients with non-insulin-dependent diabetes mellitus, a study has shown that chloroquine improves glucose tolerance, increases peripheral glucose disposal, and decreases metabolic clearance rate of insulin [61, 62].

To identify the gluconeogenic steps responsible for the activation of V-ATPase in the IC of OMCD, we used 200 μM chloroquine as an inhibitor of glucose formation through gluconeogenesis. As shown in Fig. 6.14, in the presence of chloroquine, the pH_i recovery was blocked (Fig. 6.14, marked period) and the pH_i was restored following chloroquine removal. The result suggests that

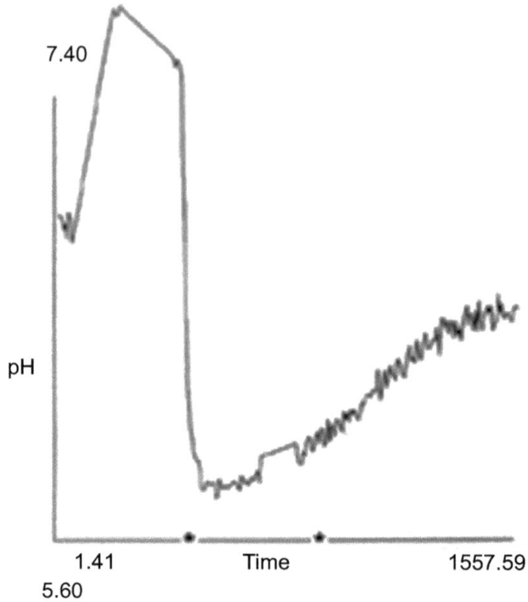

Figure 6.14 Gluconeogenic inhibitor, chloroquine inhibits V-ATPase activity in the IC of OMCD tubule. V-ATPase activity was abolished by 200 μM chloroquine (marked period) and recovered after removal of chloroquine.

the inhibitory effect of chloroquine on the V-ATPase activity could possibly link it to a pathological mechanism by gluconeogenesis in diabetic disease.

6.7 Immunocytochemical Studies in the Collecting Duct in Diabetic Animals

As shown in Fig. 6.15, the diagram illustrates the mechanism of the apical Vacuolar H^+-ATPase activity in the IC of OMCD [63].

The immunocytochemical localization and polar distribution of Vacuolar H^+-ATPase in ICs have been well studied in the intact kidney tissue [63–65]. The tubular immuno-staining is important to our microperfusion study in the isolated tubules. To localize the V-ATPase pump, we used a murine monoclonal antibody (E11) against

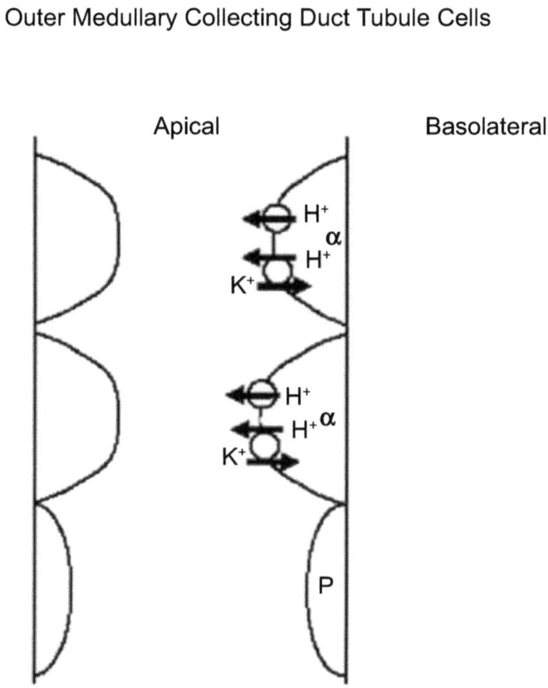

Figure 6.15 Illustration of H^+ secretion into the lumen due to Vacuolar H^+-ATPase and H^+-K+ -ATPase activities in the apical membrane of the intercalated cell of OMCD tubule. The renal outer medullary collecting duct (OMCD) is a heterogenous epithelium consisting of intercalated cells (ICs) and principal cells (PCs). "α" indicates alpha-type intercalated cells (ICs) and "p" indicate principal cells (PCs).

the 31 kD subunit of the V-ATPase in the isolated OMCD in normal (Fig. 6.16) and diabetic rats (Fig. 6.17).

In the normal rat OMCD, the ICs demonstrate V-ATPase staining, which is polarized to either apical or basolateral pole or distributed diffusely throughout the cell (Fig. 6.16). In the diabetic OMCD, there is prominent degree of V-ATPase staining, which is sharply delineated along the apical cytoplasmic vesicular staining (Fig. 6.17). These results indicate the apical V-ATPase activity in the IC of OMCD is the crucial membrane transporter for the luminal proton secretion. This is consistent with results from our previous microperfusion studies [56, 66].

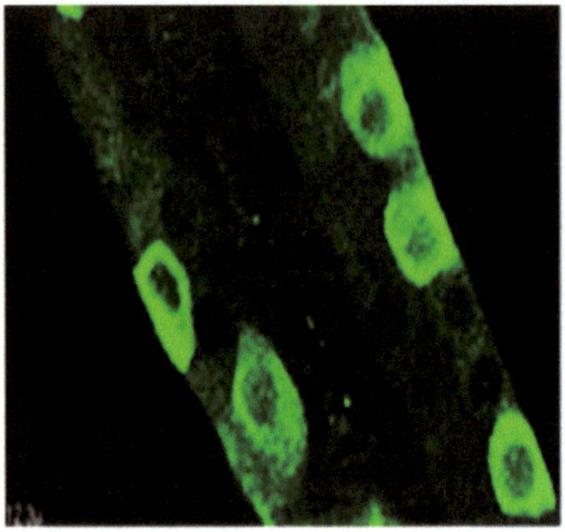

Figure 6.16 Vacuolar H^+-ATPase distribution was examined by immunostaining with an antibody of V-ATPase E-11 in isolated OMCD in normal rats. As the results show, V-ATPase staining is diffusely distributed throughout the intercalated cells (ICs) of the OMCD tubule.

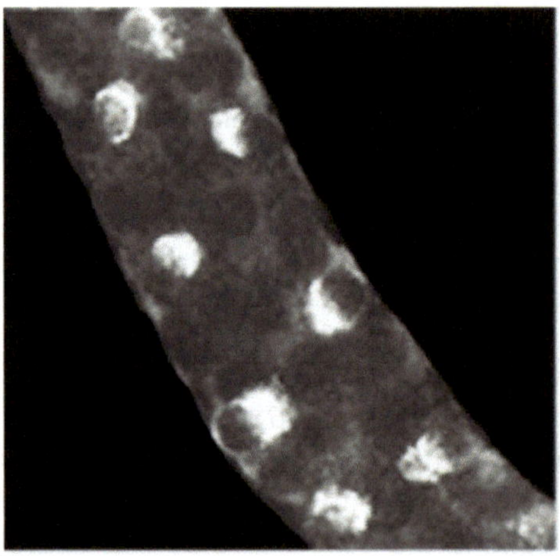

Figure 6.17 In the diabetic rats, V-ATPase staining was polarized to the apical plasma membrane of the ICs of the OMCD tubule. The degree of V-ATPase polarization at the apical membrane could indicate V-ATPase activation through V-ATPase reassembling process. V-ATPase activity under these circumstances was shown to be via enhanced insertion of this pump into the luminal membrane of α-type ICs.

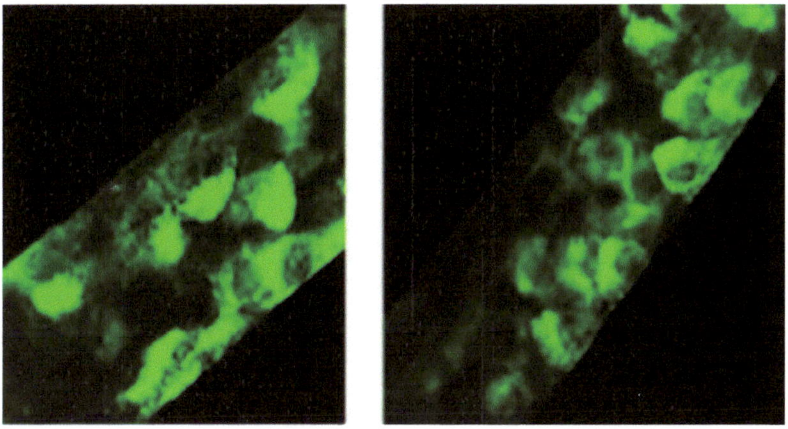

Figure 6.18 In the non-diabetic rats, the effect of glucose on V-ATPase distribution was examined by adding high glucose (50 mM glucose) to the staining media. The intercalated cells show greatly in brightness of Vacuolar H^+-ATPase staining with well polarized from the basal to the apical poles.

We have also confirmed the effect of glucose on the distribution of V-ATPase in the non-diabetic OMCD by adding high glucose (50 mM glucose) to the staining media. As shown in Fig. 6.18, the ICs show greatly in brightness of Vacuolar H^+-ATPase staining with well polarized from the basal to the apical poles. These studies demonstrate that glucose is an important physiologic regulator of Vacuolar H^+-ATPase activities in the individual ICs of OMCDs of normal and diabetic animals.

References

1. Al-Awqati Q, Mueller A, and Steinmetz PR. Transport of H^+ against electrochemical gradients in turtle urinary bladder. *Am J Physiol Renal Fluid Electrolyte Physiol* 233: F502–F508, 1977.
2. Schwartz JH and Steinmetz PR. Metabolic energy and PCO2 as determinants of H^+ secretion by turtle urinary bladder. *Am J Physiol Renal Fluid Electrolyte Physiol* 233: F145–F149, 1977.
3. Kurtz I. Apical Na^+/H^+ antiporter and glycolysis-dependent H^+-ATPase regulate intracellular pH in the rabbit S3 proximal tubule. *J Clin Invest* 80: 928–935, 1987.

4. Hering-Smith KS and Hamm LL. Metabolic support of collecting duct transport. *Kidney Int* 53: 408–415, 1998.
5. Steinmetz PR, Husted RF, Mueller A, and Beauwens R. Coupling between H^+ transport and anaerobic glycolysis in turtle urinary bladder: effect of inhibitors of H^+-ATPase. *J Membr Biol* 59: 27–34, 1981.
6. Morris RC Jr. An experimental renal acidification defect in patients with hereditary fructose intolerance. II. Its distinction from classic renal tubular acidosis; its resemblance to the renal acidification defect associated with the Fanconi syndrome of children with cystinosis. *J Clin Invest* 47: 1648–1663, 1968.
7. Ali M, Rellos P, and Cox TM. Hereditary fructose intolerance. *J Med Genet* 35: 353–365, 1998.
8. Lu M, Holliday LS, Zhang L, Dunn WA Jr, and Gluck SL. Interaction between aldolase and vacuolar H^+-ATPase: evidence for direct coupling of glycolysis to the ATP-hydrolyzing proton pump. *J Biol Chem* 276: 30407–30413, 2001.
9. Kane PM. Disassembly and reassembly of the yeast vacuolar H^+-ATPase in vivo. *J Biol Chem* 270: 17025–17032, 1995.
10. Forgac M. Structure and properties of the vacuolar (H^+)-ATPases. *J Biol Chem* 274: 12951–12954, 1999.
11. Gluck SL, Underhill DM, Iyori M, Holliday LS, Kostrominova TY, and Lee BS. Physiology and biochemistry of the kidney vacuolar H^+-ATPase. *Annu Rev Physiol* 58: 427–445, 1996.
12. Nelson N and Harvey WR. Vacuolar and plasma membrane proton-adenosinetriphosphatases. *Physiol Rev* 79: 361–385, 1999.
13. Stevens TH and Forgac M. Structure, function and regulation of the vacuolar (H^+)-ATPase. *Annu Rev Cell Dev Biol* 13: 779–808, 1997.
14. Wieczorek H, Grüber G, Harvey WR, Huss M, Merzendorfer H, and Zeiske W. Structure and regulation of insect plasma membrane H^+-V-ATPase. *J Exp Biol* 203: 127–135, 2000.
15. Johnston M. Feasting, fasting and fermenting. Glucose sensing in yeast and other cells. *Trends Genet* 15: 29–33, 1999.
16. Parra KJ and Kane PM. Reversible association between the V1 and V0 domains of yeast vacuolar H^+-ATPase is an unconventional glucose induced effect. *Mol Cell Biol* 18: 7064–7074, 1998.
17. Nakamura S. Glucose activates H^+-ATPase in kidney epithelial cells. *Am J Physiol Cell Physiol.* 287(1): C97–105, 2004.

18. Amlal H, Goel A, and Soleimani M. Activation of H^+-ATPase by hypotonicity: a novel regulatory mechanism for H^+ secretion in IMCD cells. *Am J Physiol Renal Physiol* 275: F487–F501, 1998.
19. Cole LA, Scheid JM, and Tannen RL. Induction of mitochondrial metabolism and pH-modulated ammoniagenesis by rocking LLC-PK1cells. *Am J Physiol Cell Physiol* 251: C293–C298, 1986.
20. Nielsen R, Birn H, Moestrup SK, Nielsen M, Verroust P, and Christensen EI. Characterization of a kidney proximal tubule cell line, LLCPK1, expressing endocytotic active megalin. *J Am Soc Nephrol* 9: 1767–1776, 1998.
21. Rabito CA and Ausiello DA. Na^+-dependent sugar transport in a cultured epithelial cell line from pig kidney. *J Membr Biol* 54: 31–38, 1980.
22. Temm-Grove C, Helbing D, Wiegand C, Honer B, and Jockusch BM. The upright position of brush border-type microvilli depends on myosin filaments. *J Cell Sci* 101: 599–610, 1992.
23. Cantley LC. The phosphoinositide 3-kinase pathway. *Science* 296: 1655–1657, 2002.
24. Hall RK and Granner DK. Insulin regulates expression of metabolic genes through divergent signaling pathways. *J Basic Clin Physiol Pharmacol* 10: 119–133, 1999.
25. Tengholm A and Meyer T. A PI3-kinase signaling code for insulin triggered insertion of glucose transporters into the plasma membrane. *Curr Biol* 12: 1871–1876, 2002.
26. Plas DR and Thompson CB. Cell metabolism in the regulation of programmed cell death. *Trends Endocrinol Metab* 13: 75–78, 2002.
27. Al-Awqati Q. H^+ transport in urinary epithelia. *Am J Physiol Renal Fluid Electrolyte Physiol* 235: F77–F88, 1978.
28. Steinmetz PR. Cellular organization of urinary acidification. *Am J Physiol Renal Fluid Electrolyte Physiol* 251: F173–F187, 1986.
29. Gluck S, Kelly S, and Al-Awqati Q. The proton translocating ATPase responsible for urinary acidification. *J Biol Chem* 257: 9230–9233, 1982.
30. Youmans SJ and Barry CR. Bafilomycin A1 at nanomolar concentrations saturably inhibits a portion of turtle bladder acidification current. *J Exp Biol* 204: 2911–2919, 2001.
31. Al-Awqati Q. Effect of aldosterone on the coupling between H^+ transport and glucose oxidation. *J Clin Invest* 60: 1240–1247, 1977.
32. Kelly S, Dixon TE, and Al-Awqati Q. Metabolic pathways coupled to H^+ transport in turtle urinary bladder. *J Membr Biol* 54: 237–243, 1980.

33. Al-Awqati Q, Norby LH, Mueller A, and Steinmetz PR Characteristics of stimulation of H^+ transport by aldosterone in turtle urinary bladder. *J Clin Invest* 58: 351–358, 1976.
34. Bagnasco S, Good D, Balaban R, and Burg M. Lactate production in isolated segments of the rat nephron. *Am J Physiol Renal Fluid Electrolyte Physiol* 248: F522–F526, 1985.
35. Li HY, Dai LJ, Quamme GA. Effect of chemical hypoxia on intracellular ATP and cytosolic Mg2+ levels. *J Lab Clin Med.* 122(3): 260–72, 1993.
36. Snowdowne KW, Freudenrich CC, Borle AB. The effects of anoxia on cytosolic free calcium, calcium fluxes, and cellular ATP levels in cultured kidney cells. *J Biol Chem.* 260(21): 11619–26, 1985.
37. Nakamura S, Lundberg AG, Zhang L, Lu M, Sautin Y, and Gluck SL. Glucose activates H^+-ATPase in kidney epithelial cells (Abstract). *J Am Soc Nephrol* 13: 267A–268A, 2002.
38. Osorio H, Moradas-Ferreira P, Gunther Sillero MA, and Sillero A. In *Saccharomyces cerevisiae*, the effect of H_2O_2 on ATP, but not on glyceraldehyde-3-phosphate dehydrogenase, depends on the glucose concentration. *Arch Microbiol* 181: 231–236, 2004.
39. Krauss S, Zhang CY, Scorrano L, Dalgaard LT, St-Pierre J, Grey ST, and Lowell BB. Superoxide-mediated activation of uncoupling protein 2 causes pancreatic beta cell dysfunction. *J Clin Invest* 112: 1831–1842, 2003.
40. Gstraunthaler G, Seppi T, and Pfaller W. Impact of culture conditions, culture media volumes, and glucose content on metabolic properties of renal epithelial cell cultures. Are renal cells in tissue culture hypoxic? *Cell Physiol Biochem* 9: 150–172, 1999.
41. Schering B, Reinacher M, and Schoner W. Localization and role of pyruvate kinase isoenzymes in the regulation of carbohydrate metabolism and pyruvate recycling in rat kidney cortex. *Biochim Biophys Acta* 881: 62–71, 1986.
42. Ishikawa M, Fujino T, Sakashita H, Morikawa K, and Yamamoto T. Kinetic properties and structural characterization of highly purified acetyl-CoA synthetase from bovine heart and tissue distribution of the enzyme in rat tissues. *Tohoku J Exp Med* 175: 55–67, 1995.
43. Katz J and Rognstad R. Futile cycles in the metabolism of glucose. *Curr Top Cell Regul* 10: 237–289, 1976.
44. Jones ME, Berry MN, and Phillips JW. Futile cycles revisited: a Markov chain model of simultaneous glycolysis and gluconeogenesis. *J Theor Biol* 217: 509–523, 2002.

45. Pilkis SJ and Granner DK. Molecular physiology of the regulation of hepatic gluconeogenesis and glycolysis. *Annu Rev Physiol* 54: 885–909, 1992.

46. Chauvin MF, Megnin-Chanet F, Martin G, Mispelter J, and Baverel G. The rabbit kidney tubule simultaneously degrades and synthesizes glutamate. A ^{13}C NMR study. *J Biol Chem* 272: 4705–4716, 1997.

47. Conjard A, Dugelay S, Chauvin MF, Durozard D, Baverel G, Martin G. The anaplerotic substrate alanine stimulates acetate incorporation into glutamate and glutamine in rabbit kidney tubules. A (13)C NMR study. *J Biol Chem* 277: 29444–29454, 2002

48. Voet D and Voet JG. Biochemistry. New York: Wiley, 1995.

49. Yang RZ, Blaileanu G, Hansen BC, Shuldiner AR, and Gong DW. cDNA cloning, genomic structure, chromosomal mapping, and functional expression of a novel human alanine aminotransferase. *Genomics* 79: 445–450, 2002.

50. Meyer C, Stumvoll M, Dostou J, Welle S, Haymond M, and Gerich J. Renal substrate exchange and gluconeogenesis in normal postabsorptive humans. *Am J Physiol Endocrinol Metab* 282: E428–E434, 2002.

51. Lietz T, Rybka J, and Bryla J. Fatty acids and glycerol or lactate are required to induce gluconeogenesis from alanine in isolated rabbit renal cortical tubules. *Amino Acids* 16: 41–58, 1999.

52. Pimenta WP and Silva Veiga JA. Increased glucose synthesis in renal tubule fragments from hyperthyroid rats. *Can J Physiol Pharmacol* 77: 143–146, 1999.

53. Cersosimo E, Garlick P, and Ferretti J. Renal substrate metabolism and gluconeogenesis during hypoglycemia in humans. *Diabetes* 49: 1186–1193, 2000.

54. Stumvoll M, Meyer C, Perriello G, Kreider M, Welle S, and Gerich J. Human kidney and liver gluconeogenesis: evidence for organ substrate selectivity. *Am J Physiol Endocrinol Metab* 274: E817–E826, 1998.

55. Nakamura S, Amlal H, Galla JH, and Soleimani M. NH4+ secretion in inner medullary collecting duct in potassium deprivation: role of colonic H^+-K+-ATPase. *Kidney Int* 56: 2160–2167, 1999.

56. Nakamura S, Amlal H, Schultheis PJ, Galla JH, Shull GE, and Soleimani M. HCO_3^- reabsorption in renal collecting duct of NHE-3-deficient mouse: a compensatory response. *Am J Physiol Renal Physiol* 276: F914–F921, 1999.

57. Kiersztan A, Modzelewska A, Jarzyna R, Jagielska E and Bryla J. Inhibition of gluconeogenesis by vanadium and metformin in kidney-

cortex tubules isolated from control and diabetic rabbits. *Biochem Pharmacol* 63(7): 1371–82, 2002.

58. Crabb DW, Jersild RA Jr, McCune SA, Swartzentruber MS and Harris RA. Inhibition of hepatocyte proteolysis and lactate gluconeogenesis by chloroquine. *Arch Biochem Biophys.* 203(1): 49–57, 1980.
59. Jarzyna R, Kiersztan A, Lisowa O and Bryla J. The inhibition of gluconeogenesis by chloroquine contributes to its hypoglycaemic action. *Eur J Pharmacol* 428(3): 381–8, 2001.
60. Stumvoll M, Perriello G, Meyer C and Gerich J. Role of glutamine in human carbohydrate metabolism in kidney and other tissues. *Kidney Int* 55(3): 778–92, 1999.
61. Powrie JK, Smith GD, Shojaee-Moradie F, Sonksen PH and Jones RH. Mode of action of chloroquine in patients with non-insulin-dependent diabetes mellitus. *Am J Physiol* 260(6 Pt 1): E897–904, 1991.
62. Smith GD, Amos TA, Mahler R and Peters TJ. Effect of chloroquine on insulin and glucose homoeostasis in normal subjects and patients with non-insulin-dependent diabetes mellitus. *Br Med J (Clin Res Ed)* 294(6570): 465–7, 1987.
63. Nakamura S. H^+-ATPase activity in selective disruption of H^+-K+-ATPase alpha 1 gene of mice under normal and K-depleted conditions. *J Lab Clin Med* 147(1): 45–51, 2006.
64. Bastani B. Colocalization of H-ATPase and H,K-ATPase immunoreactivity in the rat kidney. *J Am Soc Nephrol* 5(7): 1476–82, 1995.
65. Verlander JW, Madsen KM, Low PS, Allen DP, Tisher CC. Immunocytochemical localization of band 3 protein in the rat collecting duct. *Am J Physiol* 255(1 Pt 2): F115–25, 1988.
66. Nakamura S, Amlal H, Soleimani M, and Galla JH. Pathways for HCO_3^- reabsorption in mouse medullary collecting duct segments. *J Lab Clin Med* 136: 218–223, 2000.

Chapter 7

Vacuolar Proton Pump (V-ATPase) and Insulin Secretion

Ge-Hong Sun-Wada[a] **and Yoh Wada**[b]

[a] *Faculty of Pharmaceutical Sciences, Doshisha Women's College of Liberal Arts, Kohdo, Kyotanabe, Kyoto 610-0395, Japan*

[b] *Division of Biological Science, Institute of Scientific and Industrial Research, Osaka University, 8-1 Mihogaoka, Ibaraki, Osaka 567-0047, Japan*

kwada@dwc.doshisha.ac.jp

7.1 Introduction

The vacuolar-type H^+-ATPases (V-ATPase) are a family of multi-subunit ATP-dependent proton pumps involved in a wide variety of physiological processes. They are present in endomembrane organelles such as vacuoles, lysosomes, endosomes, the Golgi apparatus, chromaffin granules, and coated vesicles, and acidify the luminal pH of these intracellular compartments. They also pump protons across the plasma membranes of specialized cells, including osteoclasts and epithelial cells, in kidneys and male genital tracts.

The insulin system constitutes one of the most important endocrine circuits for maintaining physiological integrity. Its action is not restricted to energy metabolism/catabolism, but is also

Handbook of H^+-ATPases
Edited by Suguru Nakamura

ISBN 978-981-4411-91-2 (Hardcover), 978-981-4411-92-9 (eBook)
www.panstanford.com

involved in the growth and differentiation of various tissues and cell types. In mammals, insulin production is limited to a small population of cells in the pancreatic Langerhans islets called the β-cells. Despite localized production of insulin, it has versatile roles and systemic effects.

In this chapter, we will focus first on the structure and regulation of V-ATPase, and then, on recently obtained knowledge concerning its role in insulin secretion.

7.2 Structure and Regulation of V-ATPase

7.2.1 *The V-ATPase Complex*

V-ATPases are highly conserved multi-subunit enzymes that use the energy of ATP hydrolysis to transport protons across membranes (for review, see [1]). V-ATPase consists of two major functional sectors known as V_1 and V_O (Fig. 7.1). The V_1 sector comprises eight different subunits: *A, B, C, D, E, F, G,* and *H*. This sector contains three catalytic sites for ATP hydrolysis formed from the *A* and *B* subunits ($A3B3$). The V_O sector, which is responsible for proton translocation across the membranes, contains up to 6 subunits: subunit *a*; subunit *d*; the proteolipids c, c', and c''; and subunit *e* [2]. The proteolipids are highly conserved small transmembrane proteins that span the membrane four times, with both termini residing in the organelle lumen, and form a ring-like structure [3]. The number of proteolipid subunits in the ring has been reported to be six, based on chemical analysis and cryoelectron microscopy of the bovine-coated vesicle complex from electron microscopy analysis of the *c* ring from *Nephrops norvegicus* [1].

Mammalian genes and their corresponding yeast genes are summarized in Table 7.1. V-ATPase is structurally and mechanistically related to F-ATPase (ATP synthase), which is responsible for ATP synthesis in mitochondria, chloroplasts, and bacteria; however, the physiological roles of these two enzymes are very different [4]. In addition to the subunits listed above, V-ATPase also contains several accessory proteins that are essential for its function [5, 6]. One of the accessory proteins, Ac 8–9 (Fig. 7.1), has been

Table 7.1 Characteristics of V-ATPase subunits

	Domain	Subunit	Yeast gene	Mouse isoforms (expression)	Mouse gene
V_1	Catalytic hexamer (A3B3)	A	*VMA1*		*Atp6A1*
		B	*VMA2*	*B*1 (renal, epididymis, otic)	*Atp6V1B1*
				*B*2 (ubiquitous)	*Atp6V1B2*
		C	*VMA5*	*C*1 (ubiquitous)	*Atp6V1C1*
				*C*2–a (lung)	*Atp6V1C2a*
				*C*2–b (lung, kidney)	*AtpV1C2b*
		D	*VMA8*	–	*Atp6V1D1*
		E	*VMA4*	*E*1 (testis)	*Atp6V1E1*
				*E*2 (ubiquitous)	*Atp6V1E2*
		F	*VMA7*		*Atp6V1F1*
	Stalks	G	*VMA10*	*G*1 (ubiquitous)	*Atp6V1G1*
				*G*2 (neural)	*Atp6V1G2*
				*G*3 (renal, epididymis)	*Atp6V1G3*
		H	*VMA13*	*H*1 (two alternatively spliced isoforms found in human)	*Atp6V1H1*
	Proton pathway	d	*VMA6*	*d*1 (ubiquitous)	*Atp6Vod1*
				*d*2 (renal, epididymis)	*Atp6Vod2*
V_O		e	*VMA9*		*Atp6v0e*
		a	*STV1 VPH1*	*a*1(ubiquitous, synaptic	*Atp6Voa1*
				vesicle localization)	*Atp6Voa2*
				*a*2 (ubiquitous, Golgi localization)	*Atp6Voa3*
				*a*3 (ubiquitous, lysosomal localization) *Atp6Voa4*	
				*a*4 (renal, epididymis, optic)	
		c	*VMA3*		*Atlp6Voc*
		c'	*VMA11*	No mammalian gene	
		c"	*VMA16*		*Atp6Vof*
		Ac45	No yeast gene		*ATP6AP1*
		Ac8–9	No yeast gene (pro)renin receptor	Identical to	*ATP6AP2*

shown to be a truncated form of the (pro)renin receptor, which is encoded by *ATP6ap2* and plays a key role in the activation of the local renin–angiotensin system. Ablation of *Atp6ap2* selectively suppresses expression of the V_O subunit of V-ATPase, resulting in the deacidification of intracellular vesicles [5]. In addition, Niehrs

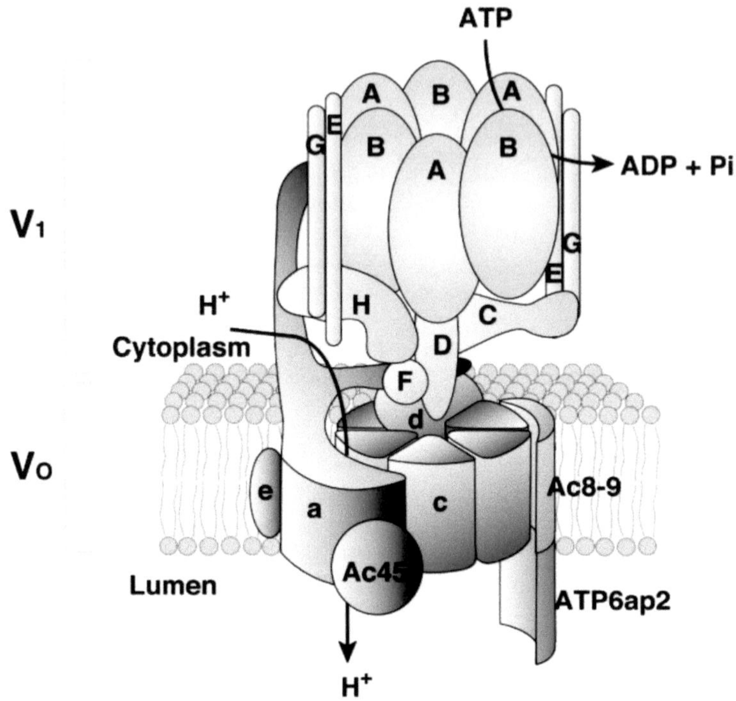

Figure 7.1 Structure of the V-ATPase. The V-ATPase complex is composed of a peripheral domain (V_1), which is responsible for ATP hydrolysis, and an integral domain (V_O), which is involved in proton translocation across the membrane. The core of the V_1 domain is composed of a hexameric arrangement of alternating A and B subunits, which participate in ATP binding and hydrolysis. The V_O domain includes a ring of proteolipid subunits (*c*, *c'*, and *c"*) that are adjacent to subunits *a* and *e*. The V_1 and V_O domains are connected by a central stalk, composed of subunits *D* and *F* of V_1 and subunit *d* of V_O, and multiple peripheral stalks, composed of subunits *C*, *E*, *G*, *H* and the *N*-terminal domain of subunit *a*. V-ATPase contains accessory proteins Ac45 and Ac8–9, both of which are essential for the function of the enzyme.

and colleagues reported that the (pro)renin receptor, along with V-ATPase-mediated acidification, is required for Wnt signaling [7]. It will be interesting to examine the relationship between the function of V-ATPase and pathological function of the (pro)renin receptor.

7.2.2 *The Rotational Catalysis of V-ATPase*

It has been established that F-ATPase can synthesize ATP when coupled with an electrochemical proton gradient, or hydrolyze ATP forming the gradient, whereas it is believed that V-ATPases can only perform the latter function under physiological conditions. The catalytic mechanism of F-ATPase has been studied extensively. It has been shown that the peripheral ATP hydrolysis domain F_1 and the integral proton-translocation domain F_O operate by a rotary mechanism (for review see [4]). V-ATPases carry out similar rotational catalysis. It has been shown that the V_1 subunits, other than A and B, form two types of stalks—peripheral and central—that connect the V_1 and V_O domains. These stalks have distinct functions in the rotary mechanism by which V-ATPases couple ATP hydrolysis to proton transport. The central stalk, consisting of subunits F and D, serves as a rotor that couples the energy released from the hydrolysis of ATP to rotation of a ring of proteolipid subunits in V_O. The peripheral stalks, consisting of subunits C, E, G, H and the N-terminal domain of subunit a, serve to prevent rotation of the $A3B3$ head during ATP hydrolysis, and therefore, serve a stator function [1] (Fig. 7.1).

We studied the rotational activity of V-ATPase by attaching an actin filament to the G subunit of a whole yeast V_1V_O complex that was immobilized on a glass surface via the rotor domain V_O. Upon the addition of ATP, we observed continuous counter–clockwise rotation of the actin filament [8]. This rotation was inhibited by concanamycin, a specific V-ATPase inhibitor, but not by azide, an inhibitor of F-ATPase. Since bafilomycin A, an antibiotic similar to concanamycin, has been shown to bind to Vo [9], possibly to its a subunit [10], the rotation was likely blocked by tightly bound concanamycin.

7.2.3 *Regulation of the Activity of V-ATPase*

The activity of multi-subunit V-ATPases is regulated through a number of mechanisms, including control of their cellular localization, changes in the efficiency of coupling proton transport with

ATP hydrolysis, reversible dissociation of V_1V_O complexes, and regulation of other related anion channels.

7.2.3.1 Function of isoforms of subunit a in enzyme targeting

Distinct isoforms of V-ATPase are preferentially localized to different compartments within a single cell or to different cell types in a multicellular system. Subunit a, the largest integral V_O subunit at 100 kDa, plays critical roles in this diverse cellular distribution.

The genome of the yeast *Saccharomyces cerevisiae* encodes two *a* subunit isoforms (Vph1p and Stv1p) for V-ATPase. They are different in energy-coupling efficiency (H^+/ATP ratio), subcellular localization (in the vacuole and the late Golgi compartment, respectively), and metabolic regulation for subunit association under cellular conditions, for example, glucose deprivation (see below at 7.2.3.2) [11–13]. The Vph1p-containing V-ATPase is localized to the vacuolar membrane and required for acidification of the central vacuoles, whereas the Stv1p-containing enzyme is needed for acidification of the lumen of Golgi apparatus and prevacuolar compartments [11–13].

The *Caenorhabditis elegans* genome encodes four *a* subunits that differ in cell-type localization and expression during developmental [14]. Four *a* subunit isoforms have been identified in mouse, of which *a*1, *a*2 and *a*3 are expressed ubiquitously, and *a*4 is predominantly expressed in the kidney [15–18]. Four *a* isoforms have also been found in humans [19]. Since they are related to specific cell functions and diseases, many aliases have been used. The nomenclature for the *a* subunit genes, along with other subunits, has been revised recently [19].

Immunostaining with *a* isoform-specific antibodies revealed that human *a*3 was colocalized with lamp2 (lysosome-associated membrane protein 2), a marker for late endosomes and lysosomes [20]. In macrophages and professional secretory cells, *a*3 was localized in phagosomes and secretory granules, respectively. Staining for GM130, a *cis*-Golgi marker, overlapped with both *a*1 and *a*2, suggesting that these isoforms are localized to the Golgi complex in NIH3T3 cells and in RAW264.7 cells, a lymphocyte derived cell line [20]. Furthermore, *a*1 was also localized to organelles other than lysosomes/late endosomes or the Golgi apparatus [20].

7.2.3.2 Regulation of V-ATPase enzyme activity via reversible dissociation

Other regulatory features include reversible dissociation/reassembly of the V-ATPase complex, changes in the coupling efficiency between ATP hydrolysis and proton transport, and low–molecular mass activator and inhibitor proteins (reviewed in [1]). In yeast, a drop in extracellular nutrients, initiated by a brief glucose deprivation, resulted in disassembly of fully assembled and active V-ATPase complexes into cytosolic V_1 complexes and membrane bound V_0 complexes. Remarkably, this disassembly was reversible, and does not require new protein synthesis, nor does it involve many of the signal-transduction pathways that are activated by nutrient depletion [21]. It is also not due to direct effects of glucose on the V-ATPase complex. Further, dissociation and reassembly appear to be independently controlled processes because dissociation requires an intact microtubular network, whereas reassembly requires a novel protein complex called RAVE (regulator of the ATPase of vacuolar and endosomal membranes) [1].

It has been suggested that the glycolytic enzyme aldolase functions as the glucose sensor that signals for the dissociation of the V-ATPase complex. In yeast, the physical association between aldolase and V-ATPase is essential for V-ATPase assembly and activity. The interaction of V-ATPase with aldolase was lost upon removal of glucose from growth media, and loss of association caused by mutations in the aldolase gene leads to dissociation of the V-ATPase complex [22].

7.2.3.3 Regulation of luminal acidification via other ion channels

Although V-ATPase translocates protons into the lumen of endomembrane organelles, the luminal pH is regulated by a variety

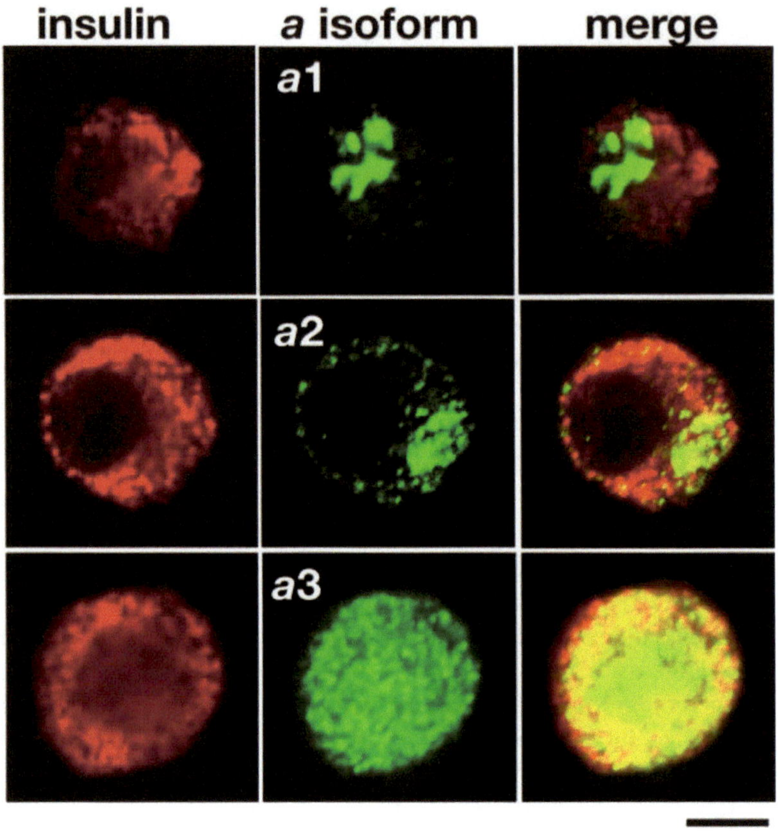

Figure 7.2 Expressions and localizations of *a* isoforms in mouse pancreatic β-cells. Pancreatic β-cells prepared from wild-type mice were stained with antibodies against the *a* isoforms (green) and insulin (red). Merged images are also shown (Bar, 5 μm). Republished with permission of Company of Biologists Ltd., from Ref. 54.

of factors. Continuous proton influx into an organelle would result in the generation of a net inside-positive membrane potential that would inhibit further proton translocation. It is believed that inwardly directed Cl^- channels, as well as cation channels, play an essential role in regulating the pH of an organelle [23, 24]. The Cl^- channels, in turn, may be regulated by protein kinase A-dependent phosphorylation [25].

7.3 Secretory Granules and V-ATPase

7.3.1 *Secretory Hormones and Luminal Acidic pH*

Maintenance of an acidic luminal environment in the exocytic and endocytic pathways is required at various steps for proper processing and sorting of vesicle contents, which include a variety of signaling molecules, secretory and surface proteins, and ligand–receptor complexes [26, 27]. Secretory granules have a luminal pH of 5.0–5.5 [28–30], and their acidification plays important roles in the condensation and maturation of their contents [31]. In addition, the electrochemical proton gradient across the membrane provides the energy for accumulation of neurotransmitters in synaptic vesicles [32, 33].

Prohormone processing to peptide hormones occurs in the regulated secretory pathway via endoproteolytic cleavage by prohormone convertase 1/3 (PC1) and prohormone convertase 2 (PC2) (for review see [34]). The enzymatic activities of PC1 and PC2 are highly dependent on the local intraorganellar pH [35]. In case of proinsulin, it has been shown that the conversion is inhibited by ionophores, which disrupt the intracellular proton gradient [29], demonstrating that an acidic environment is necessary for the enzyme reaction *in vivo*. The primary site of this conversion, revealed by immunoelectron microscopy, is clathrin-coated vesicles, which are less acidic than mature secretory granules and more proximal to the Golgi apparatus [29, 36].

7.3.2 *Function of V-ATPase in Glucose Signaling*

In pancreatic β-cells, glucose activates the cAMP-dependent protein kinase A (PKA) pathway, which is required to stimulate insulin secretion (for review, see [37]). Recently, Dechant *et al.* showed that the activity of V-ATPase is regulated by glucose, via effects on cytosolic pH, and contributes to PKA activation and insulin secretion [38]. Genetic analysis has also revealed that the assembly and disassembly of the V-ATPase complex is regulated by cytosolic pH in yeast. The activity of yeast V-ATPase is required for complete activation of PKA in response to glucose, and this glucose-sensing pathway is conserved in mammalian pancreatic β-cells. Addition of concanamycin A, a V-ATPase-specific inhibitor, before glucose stimulation of starved β-cells abolishes the glucose-dependent luminal acidification of organelles. This V-ATPase inhibitor has also been shown to reduce the glucose-dependent phosphorylation of CREB1, and thus reduce insulin secretion.

7.3.3 *Acidification of Secretory Granules by V-ATPase with a3 Subunit Isoforms*

As described above, the diverse intracellular localization of V-ATPase is achieved in part by expressing distinct isoforms of the membrane-integral *a* subunit [13, 20].

We found that one of the *a* isoforms, *a*3, was localized to the membranes of insulin-containing secretory granules in pancreatic β-cells, whereas the other subunit isoforms, *a*1 and *a*2, were localized to compartments without insulin (Fig. 7.2). Therefore, V-ATPase with *a*3 was targeted to the insulin-containing secretory granules in β-cells. In undifferentiated cells, the *a*3 isoform localized to the late endosomes and lysosomes [20], and participated in the acidification of these lytic compartments. In osteoclasts and macrophages, two other specifically differentiated cells, the *a*3 isoform together with the lysosomal protein lamp2 are targeted to plasma and phagosomal membranes, respectively. In β-cells, V-ATPase containing the *a*3 isoform is highly expressed in pancreatic islets, and localized to the membranes of insulin-containing secretory granules. However, *a*3 and lamp2 are not precisely colocalized within the intracellular compartments. Therefore, the sorting mechanisms for lamp2 and *a*3 are not the same in all cellular contexts. Similarly, in neuron cells, the *a*1 isoform localizes to synaptic vesicles and the presynaptic plasma membrane. In renal proximal tubule cells, the *a*2 isoform is present only in apical endosomes. However, in undifferentiated fibroblast cells, both isoforms are colocalized with GM130 [20]. Overall, the intracellular sorting events for the different V-ATPase isoforms of *a* subunit seem to be differentiation-dependent [20].

Neither the *a*3 isoform, nor *a*3-mediated acidification, is an absolute requirement for insulin processing. From immunoblotting analysis and observing morphology of the dense-core granules by electron microscopy, we found that the maturation of insulin was normal in the islets of *oc/oc* (*oc* stands for osteosclerosis) mutant mice, which lack the $a3$ isoform. As the primary site of insulin conversion is clathrin-coated vesicle where $a1$ isoform is localized, it is possible that other a subunit isoforms are responsible for this acidification.

However, the mutant *oc/oc* mice (without the $a3$ isoform) exhibited a significantly lower level of plasma insulin than the wild-type, despite producing and holding substantial amounts of insulin in the pancreas. In addition, insulin secretion in response to glucose or depolarizing stimulation was significantly impaired in *oc/oc* mutant islets. Inhibition of granule acidification with bafilomycin A in βHC9, a stabilized cell line of pancreatic β-cell, showed that acidification itself is not directly involved in the secretion of existing secretory granules. Together, these observations suggest that the $a3$ mutants are defective in exocytosis of secretory granules, and insulin secretion may be dependent on the physical presence of functional V-ATPase a subunit.

In addition to pancreatic cells, we also examined the expression and distribution of a subunit isoforms in endocrine tissues—including adrenal, thyroid, parathyroid, and pituitary gland—and found that the $a3$ isoform was highly expressed in all endocrine cells [39]. These results suggest that functions of the $a3$ isoform involved in exocytosis in regulated secretion are shared across multiple tissues.

7.4 V-ATPase in Vesicle Fusion

It has been suggested for decades that proton gradients may be involved in the fusion of secretory vesicles to target membranes [40, 41]. Barg *et al.* showed that acidic pH may regulate priming of the granules for secretion, a process involving pairing of SNARE (soluble *N*-ethylmaleimide-sensitive factor attachment protein receptor) proteins on the vesicles and target membranes to establish fusion competence [42].

Recently, as the case of the *a*3 isoform in mouse pancreatic cells [20], the V_O subunits of V-ATPase have been implicated in membrane fusion or secretion independent of intracompartmental acidification. In yeast vacuolar fusion, this membrane intrinsic V_O sector has been demonstrated to be an essential component [43]. The current model proposes that fusion pores are assembled from V_O domains in opposing membranes that pair in *trans*, and the V_O–V_O *trans*-complex on the paired membranes provides the initial point of membrane continuity between the two separate membranes [43–45]. V_O has been shown to promote fusion in multiple steps in the late exocytic and endocytic pathways, including secretion of neurotransmitters, phagosome-lysosome fusion, and secretion of exosomes [46–49]. It has also been shown that V_O interacts with SNAREs on synaptic vesicles in rat, fish, and fly [49–52]. These results are highly indicative that the V-ATPase is directly or indirectly involved in dynamic membrane interaction.

The precise molecular mechanisms underlying the behavior or regulation of the V_O components are largely unknown. Hiesinger and colleagues showed that a mutant *a* subunit in *Drosophila melanogaster*, $v100^{R755A}$, which lacks the proton translocation function, rescues synaptic vesicles fusion through interaction with syntaxin1A and early endosomal fusion through interaction with syntaxin7. In contrast, the $v100^{R755A}$ mutant does not rescue the maturation of endosomes to degradative compartments that requires the pump activity of V_1V_O-ATPase holo-enzyme [52]. Most recently, Mayer and colleagues have screened for yeast proteolipid alleles that interfere with fusion, but support proton translocation. They found fusion-deficient point mutations in all of the proteolipids, and these substitutions support ATP-dependent proton translocation. Several mutants reduced the efficiency of SNARE complex formation, whereas others showed reduced fusion activity with normal SNARE activation and *trans*-SNARE pairing. These observations support separate roles for proteolipids in SNARE pairing, lipid mixing, and proton translocation. Interestingly, several fusion-deficient proteolipid mutants stabilized the V_1V_O holoenzyme, suggesting the possibility that the fusion deficiency results from a depletion of free V_O [53].

It has been shown that the large cytosolic *N*-terminal region of the fly V-ATPase *a* subunit, *V*100, contains a calmodulin-binding site. The *a* subunit forms a tight complex with calmodulin in a Ca^{2+}-dependent manner. Mutations in the calmodulin-binding site in the *a* subunit lead to a loss of calmodulin recruitment to synapses, suggesting a V-ATPase V_O-dependent regulation of calmodulin in synaptic fusion.

Interestingly, three loci linked with the development of diabetes map close to genes encoding V-ATPase (Online Mendelian Inheritance in Man, OMIM™), consistent with a potential genetic contribution of alterations of V-ATPase to the development of the disease.

References

1. Forgac, M. (2007). Vacuolar ATPases: rotary proton pumps in physiology and pathophysiology. *Nat. Rev. Mol. Cell. Biol.* 8, 917–929.
2. Anraku, Y. (1996) Structure and function of the yeast vacuolar membrane H^+-ATPase. In: Konings, W. N., Kaback, H. R., and Lolkema, J. S. (eds). *Handbook of Biological Physics*, Elsevier Science, Amsterdam.
3. Powell, B. W., Graham, L. A., and Stevens, T. H. (2000). Molecular characterization of the yeast vacuolar H^+-ATPase proton pore. *J. Biol. Chem.* 275(31), 23654–23660.
4. Futai, M., Omote, H., Sambongi, Y., and Wada, Y. (2000). ATP synthase (H^+-ATPase): coupling between catalysis, mechanical work, and proton translocation. *Biochim. Biophys. Acta* 1458(2–3), 276–288.
5. Kinouchi, K., Ichihara, A., Sano, M., Sun-Wada, G. H., Wada, Y., Kurauchi-Mito, A., Bokuda, K., Narita, T., Oshima, Y., Sakoda, M., Tamai, Y., Sato, H., Fukuda, K., and Itoh, H. (2010). The (pro)renin receptor/ATP6AP2 is essential for vacuolar H^+-ATPase assembly in murine cardiomyocytes. *Circ. Res.* 107, 30–34.
6. Jansen, E. J., Hafmans, T. G., and Martens, G. J. (2010). V-ATPase–mediated granular acidification is regulated by the V-ATPase accessory subunit Ac45 in POMC-producing cells. *Mol. Biol. Cell* 21(19), 3330–3339.
7. Cruciat, C. M., Ohkawara, B., Acebron, S. P., Karaulanov, E., Reinhard, C., Ingelfinger, D., Boutros, M., and Niehrs, C. (2010). Requirement of

prorenin receptor and vacuolar H^+-ATPase-mediated acidification for Wnt signaling. *Science* 327(5964), 459–463.

8. Hirata, T., Iwamoto-Kihara, A., Sun-Wada, G. H., Okajima, T., Wada, Y., and Futai, M. (2003). Subunit rotation of vacuolar-type proton pumping ATPase: relative rotation of the *G* as to *c* subunit. *J. Biol. Chem.* 278(26), 23714–23719.
9. Hanada, H., Moriyama, Y., Maeda, M., and Futai, M. (1990). Kinetic studies of chromaffin granule proton ATPase and effects of bafilomycin A-1. *Biochem. Biophys. Res. Commun.* 170(2), 873–878.
10. Zhang, J., Feng, Y., and Forgac, M. (1994). Proton conduction and bafilomycin binding by the V_0-domain of the coated vesicle V-ATPase. *J. Biol. Chem.* 269(38), 23518–23523.
11. Manolson, M. F., Proteau, D., Preston, R. A., Stenbit, A., Roberts, B. T., Hoyt, M. A., Preuss, D., Mulholland, J., Botstein, D., and Jones, E. W. (1992). The *VPH1* gene encodes a 95-kDa integral membrane polypeptide required for in vivo assembly and activity of the yeast vacuolar H^+-ATPase. *J. Biol. Chem.* 267(20), 14294–14303.
12. Manolson, M. F., Wo, B., Proteau, D., Taillon, B. E., Roberts, B. T., Hoyt, M. A., and Jones, E. W. (1994). *STV1* gene encodes functional homologue of 95-kDa yeast vacuolar H^+-ATPase subunit Vph1p. *J. Biol. Chem.* 269(19), 14064–14074.
13. Kawasaki-Nishi, S., Nishi, T., and Forgac, M. (2001). Yeast V-ATPase complexes containing different isoforms of the 100-kDa a-subunit differ in coupling efficiency and in vivo dissociation. *J. Biol. Chem.* 276(21), 17941–17948.
14. Oka, T., Toyomura, T., Honjo, K., Wada, Y., and Futai, M. (2001). Four subunit *a* isoforms of *Caenorhabditis elegans* vacuolar H^+-ATPase cell–specific expression during development. *J. Biol. Chem.* 276, 33079–33085.
15. Toyomura, T., Oka, T., Yamaguchi, C., Wada, Y., and Futai, M. (2000). Three subunit *a* isoforms of mouse vacuolar H^+-ATPase. Preferential expression of the *a*3 isoform during osteoclast differentiation. *J. Biol. Chem.* 275(12), 8760–8765.
16. Nishi, T., and Forgac, M. (2000). Molecular cloning and expression of three isoforms of the 100-kDa a subunit of the mouse vacuolar proton-translocating ATPase. *J. Biol. Chem.* 275, 6824–6830.
17. Oka, T., Murata, Y., Namba, M., Yoshimizu, T., Toyomura, T., Yamamoto, A., Sun-Wada, G. H., Hamasaki, N., Wada, Y., and Futai, M. (2001). *a*4, a unique kidney-specific isoform of mouse vacuolar H^+-ATPase subunit *a*. *J. Biol. Chem.* 276, 40050–40054.

18. Smith, A. N., Finberg, K. E., Wagner, C. A., Lifton, R. P., Devonald, M. A., Su, Y., and Karet, F. E. (2001). Molecular cloning and characterization of *Atp6n1b*: a novel fourth murine vacuolar H^+-ATPase a-subunit gene. *J. Biol. Chem.* 276(45), 42382–42388.

19. Smith, A. N., Lovering, R. C., Futai, M., Takeda, J., Brown, D., and Karet, F. E. (2003). Revised nomenclature for mammalian vacuolar-type H^+-ATPase subunit genes. *Mol. Cell* 12(4), 801–803.

20. Toyomura, T., Murata, Y., Yamamoto, A., Oka, T., Sun-Wada, G. H., Wada, Y., and Futai, M. (2003). From lysosomes to plasma membrane: Localization of vacuolar type H^+-ATPase with the *a*3 isoform during osteoclast differentiation. *J. Biol. Chem.* 278(24), 22023–22030.

21. Kane, P. M. (2006). The where, when, and how of organelle acidification by the yeast vacuolar H^+-ATPase. *Microbiol. Mol. Biol. Rev.* 70(1), 177–191.

22. Lu, M., Ammar, D., Ives, H., Albrecht, F., and Gluck, S. L. (2007). Physical interaction between aldolase and vacuolar H^+-ATPase is essential for the assembly and activity of the proton pump. *J. Biol. Chem.* 282(34), 24495–24503.

23. Futai, M., Oka, T., Sun-Wada, G.-H., Moriyama, Y., Kanazawa, H., and Wada, Y. (2000). Luminal acidification of diverse organelles by V-ATPase in animal cells. *J. Exp. Biol.* 203(Pt 1), 107–116.

24. Sun-Wada, G.-H., Wada, Y., and Futai, M. (2003). Vacuolar H^+ pumping ATPases in luminal acidic organelles and extracellular compartments: common rotational mechanism and diverse physiological roles. *J. Bioenerg. Biomembr.* 35, 347–358.

25. Forgac, M. (1998). Structure, function and regulation of the vacuolar H^+-ATPases. *FEBS Lett.* 440(3), 258–263.

26. Futai, M., Oka, T., Sun-Wada, G., Moriyama, Y., Kanazawa, H., and Wada, Y. (2000). Luminal acidification of diverse organelles by V-ATPase in animal cells. *J. Exp. Biol.* 203 Pt 1, 107–116.

27. Weisz, O. A. (2003). Acidification and protein traffic. *Int. Rev. Cytol.* 226, 259–319.

28. Arvan, P., and Castle, J. D. (1986). Isolated secretion granules from parotid glands of chronically stimulated rats possess an alkaline internal pH and inward-directed H^+ pump activity. *J. Cell Biol.* 103, 1257–1267.

29. Orci, L., Ravazzola, M., Amherdt, M., Madsen, O., Perrelet, A., Vassalli, J.-D., and Anderson, R. G. W. (1986). Conversion of proinsulin to insulin occurs coordinately with acidification of maturing secretory vesicles. *J. Cell Biol.* 103, 2273–2281.

30. Urbe, S., Dittie, A. S., and Tooze, S. A. (1997). pH-dependent processing of secretogranin II by the endopeptidase PC2 in isolated immature secretory granules. *Biochem. J.* 321(Pt 1), 65–74.
31. Hutton, J. C. (1994). Insulin secretory granule biogenesis and the proinsulin-processing endopeptidases. *Diabetologia* 37(Suppl 2), S48–S56.
32. Moriyama, Y., Maeda, M., and Futai, M. (1990). Energy coupling of L-glutamate transport and vacuolar proton-ATPase in brain synaptic vesicles. *J. Biochemistry* 108(4), 689–693.
33. Nelson, N. (1993). Presynaptic events involved in neurotransmission. *J. Physiol. Paris. 1993;* 87(3), 171–178.
34. Seidah, N. G. (2011). The proprotein convertases, 20 years later. *Methods Mol. Biol.* 768, 23–57.
35. Anderson, R. G., and Orci, L. (1988). A view of acidic intracellular compartments. *J. Cell Biol.* 106(3), 539–543.
36. Orci, L., Ravazzola, M., Storch, M. J., Anderson, R. G., Vassalli, J. D., and Perrelet, A. (1987). Proteolytic maturation of insulin is a post-Golgi event which occurs in acidifying clathrin-coated secretory vesicles. *Cell* 49(6), 865–868.
37. Kasai, H., Hatakeyama, H., Ohno, M., and Takahashi, N. (2010). Exocytosis in islet β-cells. *Adv. Exp. Med. Biol.* 654, 305–338.
38. Dechant, R., Binda, M., Lee, S. S., Pelet, S., Winderickx, J., and Peter, M. (2010). Cytosolic pH is a second messenger for glucose and regulates the PKA pathway through V-ATPase. *EMBO J.* 29(15), 2515–2526.
39. Sun-Wada, G. H., Tabata, H., Kawamura, N., Futai, M., and Wada, Y. (2007). Differential expression of *a* subunit isoforms of the vacuolar-type proton pump ATPase in mouse endocrine tissues. *Cell Tissue Res.* 329(2), 239–248.
40. Al-Awqati, Q., Gluck, S., Reeves, W., and Cannon, C. (1983). Regulation of proton transport in urinary epithelia. *J. Exp. Biol.* 106, 135–141.
41. Burgess, T. L., and Kelly, R. B. (1987). Constitutive and regulated secretion of proteins. *Annu. Rev. Cell Biol.* 3, 243–293.
42. Barg, S., Huang, P., Eliasson, L., Nelson, D. J., Obermuller, S., Rorsman, P., Thevenod, F., and Renstrom, E. (2001). Priming of insulin granules for exocytosis by granular Cl^- uptake and acidification. *J. Cell Sci.* 114(Pt 11), 2145–2154.
43. Peters, C., Bayer, M. J., Buhler, S., Andersen, J. S., Mann, M., and Mayer, A. (2001). Trans-complex formation by proteolipid channels in the terminal phase of membrane fusion. *Nature* 409(6820), 581–588.

44. Bayer, M. J., Reese, C., Buhler, S., Peters, C., and Mayer, A. (2003). Vacuole membrane fusion: V_0 functions after trans-SNARE pairing and is coupled to the Ca^{2+}-releasing channel. *J. Cell Biol.* 162(2), 211–222.

45. Muller, O., Bayer, M. J., Peters, C., Andersen, J. S., Mann, M., and Mayer, A. (2002). The Vtc proteins in vacuole fusion: coupling NSF activity to V(0) trans-complex formation. *EMBO J.* 21(3), 259–269.

46. Hiesinger, P. R., Fayyazuddin, A., Mehta, S. Q., Rosenmund, T., Schulze, K. L., Zhai, R. G., Verstreken, P., Cao, Y., Zhou, Y., Kunz, J., and Bellen, H. J. (2005). The V-ATPase Vo subunit *a*1 is required for a late step in synaptic vesicle exocytosis in *Drosophila*. *Cell* 121(4), 607–620.

47. Dunant, Y., Corderiro, J. M., and Goncalves, P. P. (2009). Exocytosis, mediatophore, and vesicular Ca^{2+}/H^+ antiport in rapid neurotransmission. *Mech. Exocytosis* 1152, 100–112.

48. Peri, F., and Nusslein-Volhard, C. (2008). Live imaging of neuronal degradation by microglia reveals a role for v0-ATPase a1 in phagosomal fusion in vivo. *Cell* 133(5), 916–927.

49. Di Giovanni, J., Boudkkazi, S., Mochida, S., Bialowas, A., Samari, N., Leveque, C., Youssouf, F., Brechet, A., Iborra, C., Maulet, Y., Moutot, N., Debanne, D., Seagar, M., and El Far, O. (2010). V-ATPase membrane sector associates with synaptobrevin to modulate neurotransmitter release. *Neuron* 67(2), 268–279.

50. Galli, T., McPherson, P. S., and De Camilli, P. (1996). The V_0 sector of the V-ATPase, synaptobrevin, and synaptophysin are associated on synaptic vesicles in a Triton X-100-resistant, freeze-thawing sensitive, complex. *J. Biol. Chem.* 271(4), 2193–2198.

51. Morel, N. (2003). Neurotransmitter release: the dark side of the vacuolar-H^+-ATPase. *Biol. Cell* 95(7), 453–457.

52. Williamson, W. R., Wang, D., Haberman, A. S., and Hiesinger, P. R. (2010). A dual function of V_0-ATPase a1 provides an endolysosomal degradation mechanism in *Drosophila melanogaster* photoreceptors. *J. Cell Biol.* 189(5), 885–899.

53. Strasser, B., Iwaszkiewicz, J., Michielin, O., and Mayer, A. (2011). The V-ATPase proteolipid cylinder promotes the lipid-mixing stage of SNARE–dependent fusion of yeast vacuoles. *EMBO J.* 30(20), 4126–4141.

54. Sun-Wada, G.-H., Toyomura, T., Murata, Y., Yamamoto, A., Futai, M., and Wada, Y. (2006). The a3 isoform of V-ATPase regulates insulin secretion from pancreatic beta-cells. *J. Cell Sci.* 119, 4531–4540.

Chapter 8

Role of V-ATPase, Cytohesin-2/Arf6, and Aldolase in Regulation of Endocytosis: Implications for Diabetic Nephropathy

Vladimir Marshansky
Center for Systems Biology, Program in Membrane Biology and Renal Division, Simches Research Center, Massachusetts General Hospital and Harvard Medical School, Boston, MA 02114, USA
Marshansky.Vladimir@mgh.harvard.edu, Vladimir_Marshansky@hms.harvard.edu

Vacuolar-type H^+-ATPase (V-ATPase), cytohesins, and Arf-family GTP-binding proteins (Arfs) are essential for vesicular trafficking of receptors and their signaling along endocytic pathway of eukaryotic cells. Deficient function of V-ATPase, cytohesins, and Arfs in endocytic trafficking and signaling of receptors have been recently recognized as important mechanisms in a variety of human diseases including diabetes and its complications. In the past few years, significant progress has been made in our understanding of function and regulation of V-ATPase and their cell biological role in crosstalk with cytohesins and Arfs. Here, we review these studies on novel roles of V-ATPase, cytohesins, and Arfs in the regulation of vesicular trafficking and signaling events within endocytic pathway. We also discuss novel role of aldolase in regulation of cross-talk between V-ATPase and cytohesins/Arfs. Finally, we give special emphasis on the

Handbook of H^+-ATPases
Edited by Suguru Nakamura

ISBN 978-981-4411-91-2 (Hardcover), 978-981-4411-92-9 (eBook)
www.panstanford.com

potential role of these proteins in diabetic complication of kidney proximal tubule function and their use as emerging therapeutic targets.

8.1 Endocytic Pathways: Uptake, Signaling, Recycling, or Degradation of Receptors/Ligand Complexes

Endocytosis is a fundamental cellular process that is used by eukaryotic cells to communicate with external milieu and internalize an enormous variety of macromolecules and microorganisms, including nutrients, hormones, DNA, and viruses. It is also crucial for the retrieval, signaling and recycling of a cellular receptors localized at the plasma membrane [1]. The best studied is the clathrin-dependent endocytosis (CDE) pathway, which mediates internalization of transferrin, EGFR/ErbB, megalin-cubilin receptors among many others. However during the last decade various clathrin-independent endocytosis (CIE) pathways have been identified and include: (i) Arf6-dependent endocytosis and (ii) macropinocytosis pathways [1, 2]. In HeLa and COS cells the Arf6-dependent CIE pathway operates constitutively and intersects in early endosomes with the CDE pathway [3, 4]. In the "resting" state of these cells the Arf6-dependent pathway is involved in constitutive trafficking from plasma membrane of a variety of proteins including MHC1, β-integrins, and GPI-anchored proteins. Macropinocytosis has initially been viewed as a non-regulated bulk uptake of soluble macromolecules. However, it is now generally accepted that it is a tightly regulated form of endocytosis which co-exist and crosstalk with other endocytic pathways. It is noteworthy that function of this pathway is cell specific and its regulation could be dependent upon particular cellular requirements. For example, while macrophages and dendritic cells have got the highly active constitutive macropinocytosis, in other cell types it remains inactive but could be transiently activated by protein kinase C or by growth factors [5, 6]. While at low concentration of epidermal growth factor (EGF) the EGFR receptors were internalized via CDE pathway, at high concentrations of EGF the EGFR receptors could be internalized via robust macropinocytosis pathway [7]. Thus, these studies uncover the ligand-dependent activation of macropinocytosis and crosstalk between CDE and CIE pathways for endocytosis of endogenous EGFR/ErbB receptors in HeLa cells [7–9]. Finally, effects in vesicular trafficking within the endocytic pathways have been recently recognized as an important cell biological mechanism in human diseases including bacterial and viral infection, bone and neurodegenerative disorders, cancer and its metastasis [10, 11]. In particular, an important role of kidney proximal tubules endosomal/lysosomal protein degradation pathway in pathogenesis of early stages of diabetic nephropathy will be later discussed in this review.

8.2 Regulation of Endocytosis by V-ATPase and the Luminal pH of Endosomal/Lysosomal Compartments

Endocytic compartments of eukaryotic cells maintain different luminal pH due to function of V-ATPase. The pH becomes more acidic in lysosomes as endosomal pathway reaches its destination. Function of this pathway is absolutely requires the V-ATPase-driven endosomal/lysosomal acidification for receptors/ligands signaling, recycling and degradation. The regulation of organellar acidification is achieved by either: (i) specific targeting and trafficking of V-ATPase to organelles or by (ii) regulation of V-ATPase activity. Thus, next we will discuss structure, targeting, and function of V-ATPase followed by recent findings on the direct role of V-ATPase as well as V-ATPase-dependent acidification in function of various receptors along endocytic pathway.

8.2.1 *Structure, Targeting, and Function of the V-ATPase Proton-Pumping Rotary Nano-Motor*

The V-ATPase is a multi-subunit complex that functions as a proton pumping rotary nano-motor and is composed of two sectors (V_1V_O). The cytoplasmic V_1-sector of the V-ATPase is composed of eight different subunits with following stoichiometry (A_3:B_3:C:D:E_3:F:G_3:H) and is responsible for ATP hydrolysis. The transmembrane V_O-sector is composed of five different subunits (a:c_5:c'':d:e) and is responsible for proton translocation across membrane (Fig. 8.1a). The catalytic hexamer A_3B_3 is connected to the proton pathway by stalks. The central stalk (formed by F- and D-subunits) is attached to a ring of hydrophobic subunits (c, c″) which is operating as a "rotor". The cytosolic *N*-terminal domain of a-subunit (aN) together with C and H-subunits form a "collar", while three heterodimers of G- and E-subunits form tree peripheral stalks which are attached to the A_3B_3 hexamer and form a "stator". The transmembrane C-terminal domain of a-subunit (aC) together with transmembrane c-ring forms a proton channel. Thus, one of the mechanisms of differential acidification of endosomal compartments is achieved by regulation of the coupling between ATP hydrolysis and proton pumping during function of this rotary nano-motor.

Another mechanism of differential acidifications includes the compartment specific targeting of V-ATPase subunits isoforms. In human and mice following subunit isoforms are found: (i) two isoforms for the B, E, H and d-subunits; (ii) three isoforms for the C and G-subunits; and (iii) four isoforms for a-subunit (a1, a2, a3, and a4). The expression and targeting of all these isoforms is cell and organelle specific. Importantly, targeting and localization of V-ATPase in different compartments along both endocytic and exocytic pathways as well as plasma membrane depends upon a-subunits isoforms [12]. In yeast targeting of V-ATPase to vacuole or Golgi is achieved by two a-subunit isoforms Vph1p and Stv1p, respectively. In mammalian cells targeting and compartment specific localization of V-ATPase is also depends upon a-subunit isoforms. In particular, the a1-subunit isoform is highly expressed in neuronal tissue. In nerve terminals, *the a1-isoform* is specifically targeted to synaptic vesicles in which V-ATPase is involved in uptake of various neurotransmitters and their release upon stimulation [13, 14]. However, in neuronal cells a1-isoform could also be targeted to compartments of endocytic pathways. For example, in presenilin-1 (PS1) null blastocysts, neurons from mice hypomorphic for or conditionally depleted of PS1, the a1-isoform is involved in targeting of V-ATPase to lysosomes [15]. Importantly, this study identified the cell biological mechanism of a1-isoform targeting to lysosomes, in which physical interaction of unglycosylated a1-isoform with PS1 is required for its *N*-glycosylation and efficient delivery from ER to lysosomes. In our previous studies, *the a2-isoform* was found to be targeting V-ATPase to the early endosomes of megalin/cubilin receptor–mediated protein degradative pathway in kidney proximal tubules both *in situ* and in MTC cultured cells [16–18]. It is noteworthy that in MTC cells a2-isoform does not targeted to Golgi, while in cultured osteoclast cells [19] and B16 cell [20] both endogenous a2- and a1-isoforms are targeted to the Golgi complex. In addition, in neuroendocrine PC12 cells the stable expression of fluorescence-tagged a2-isoform (a2-EmGFP) also revealed targeting of V-ATPase with this subunit to Golgi apparatus [14]. In initial studies, *the a3-isoform* was identified as osteoclast lysosomal specific which could be relocated to plasma membrane during osteoclast differentiation [19–21]. Also recent studies on macrophages convincingly demonstrated that during bacterial infection the nascent phagosomes progressively acquiring a3-subunit containing V-ATPase from lysosomes [21]. However, in neuroendocrine PC12 cells the stable expressed a3-EmGFP was targeting to early endosomes of endocytic pathway, while this endogenous a3-isoform was specifically targeted to insulin containing secretory granules of exocytic pathway in pancreatic β-cells [12]. In contrast to tissue ubiquitous a1-, a2- and a3-isoforms, the expression of *a4-isoform* is highly specific for kidney, cochlea and epididymis. Moreover, in these tissues a4-isoform is specifically targeted to the apical plasma membrane of collecting duct intercalated and epididymal clear cells [22]. In conclusion, depending of the cell type the endosomal and lysosomal V-ATPase is specifically targeted by the a1-, a2-, and a3-isoforms, which could give rise to differential function and acidification in these compartments. However, the mechanisms of their specific targeting are largely unknown and only recently have started emerging [15]. It is noteworthy, that recent studies on knockdown of specific a-isoforms revealed functional compensation by a-subunit isoforms which may also take place in function of endocytic pathway under physiological conditions [14]. Previously we proposed the V-ATPase itself is functioning as a pH-sensing receptor which via interaction of a2-isoform with cytohesin-2 and Arf6 modulate vesicular trafficking within endosomal/lysosomal pathway (Fig. 8.1a) [12]. Recently, the interactions of V-ATPase with cytohesin-2 were also extended

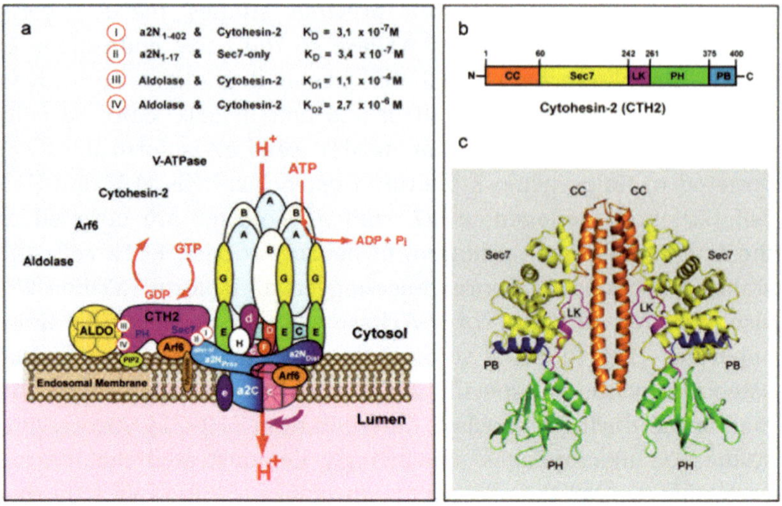

Figure 8.1 Structure, subunit, and domain composition of V-ATPase and cytohesin-2. (a) The diagram shows the complex subunit structure of the V-ATPase. ATP hydrolysis drives clockwise rotation of the central stalk and ring of proteolipid subunits. This rotation leads to the translocation of protons from the cytosol to acidify lumen of endomembrane organelles. It also illustrates the binding sites for cytohesin-2 (CTH2) with a2-subunit (a2N); cytohesin-2 with aldolase (ALDO) as well as Arf6 with c-subunit of V-ATPase. Roman numbers indicate interfaces and affinities of interaction: (I) CTH2 with $a2N_{1-402}$ and (II) Sec7 domain with $a2N_{1-17}$ and (III and IV) CTH2 with aldolase (ALDO). (b) Schematic representation of domains and regulatory elements of cytohesin-2. These structures are indicated as follows: (i) CC-domain (1–60aa) in orange (ii) Sec7-domain (61–242aa) in yellow (iii) Sec-7-PH-linker (LK) (243–261aa) in magenta (iv) PH-domain (262–375aa) in green, and (v) PB-domain (376–400aa) in blue. Boundaries of the domains are indicated as amino acid numbers. (c) Ribbon representation of a molecular model of full-length cytohesin-2(wt) in its dimeric state. Structural domains and elements of cytohesin2 are indicated as in (b). Adapted from Refs. 12 and 97.

to other a1-, a3-, and a4-isoforms, which may provide potential mechanism of functional compensation by a-isoforms in regulation of acidification capacity of endosomal compartments. Finally, cross-talk between V-ATPase and cytohesin-2/Arf6 pointed out the possibility that these proteins might function as "molecular on/off switches" in disassembly/assembly of V-ATPase and thus regulation of its function and acidification capacity endosomal/lysosomal compartments [12, 23]. Implications of these findings to diabetes will be discussed in more details later in this review.

8.2.2 *Role of V-ATPase in Trafficking and Function of Receptors and Their Regulatory Proteins along Endocytic Pathway*

Mounting evidence indicates that establishing an acidic environment of intracellular organelles is not the only function of V-ATPase. This multi-subunit complex is also involved in variety of direct protein–proteins interactions and regulation of cell biological events along both exocytotic and endocytotic pathways as was previously reviewed [12]. Here we will briefly analyze exciting recent discoveries with emphasis on the role of V-ATPase in trafficking and function of various receptors and their regulatory proteins along endocytic pathway. *Following novel roles of V-ATPase in function of early endosomes has been recently uncovered. The signaling by the Notch receptor* between cells is crucial during development and its pathological activation is hallmark of many cancers [24]. Following binding of ligands, the Notch receptor is activated by its cleavage via γ-secretase-mediated intra-membrane proteolysis giving rise to transcriptional activation of important target genes. Recent study demonstrated that in signal-receiving cells V-ATPase via acidification of early endosomes promotes γ-secretase-dependent S3 cleavage and therefore activation and signaling of Notch receptors [25]. *The signaling by the Wnt/β-catenin pathway* is another fundamental mechanism that control cell proliferation, polarity, embryonic tissue development and also implicated in pathogenesis of metastatic cancers [26]. Recent study on the Wnt signal transmission uncovered that activation of low-density receptor-related protein 6 (LRP6) is taking place in early endosomes and requires direct interaction with V-ATPase via its accessory subunit M8-9, which is also known to be functioning as prorenin receptor (PRR). Thus, these results uncovered an unsuspected direct role for V-ATPase as well as V-ATPase-dependent acidification of early endosomes during Wnt/β-catenin signaling [27]. *The signaling by the insulin-like growth factor (IGF-I) receptor* is inducing expression of heme-binding protein (HRG-1) which is targeted specifically to early endosomes but not to lysosomes [28]. In this study it was also found that HRG-1 promotes activation of V-ATPase via direct interaction with c-subunit of the proton pump. Moreover, levels of HGR-1 expression in HeLa cells correlates with endosomal acidification and endocytic trafficking of Alexa-488 labeled transferrin receptors. Thus, this study demonstrated the novels role of HRG-1 in modulation of V-ATPase activity and regulation of endocytic receptors trafficking which may facilitate tumor growth and cancer progression [28]. *Other novel roles of V-ATPase in function of late endosomes/lysosomes have been also recently revealed. The mTOR complex 1 (mTORC1)* is multi-component protein kinase complex that is involved in growth regulation and is activated by amino acids during nutrient response. It was found that V-ATPase is involved in amino acid sensing, recruitment and activation of the mTORC1 on lysosomal membranes via its interaction with Rag GTPases. Authors proposed, that V-ATPase is essential component of the mTOR-pathways which is involved in amino acids sensing in lysosomal lumen [29]. *Finally, novel role of PS1* for targeting of V-ATPase to lysosomes and proper lysosomal acidification during autophagy was already discussed above. Thus, these recent studies are underlying the crucial role of V-ATPase-dependent lysosomal acidification and protein degradation in pathogenesis of neurogenerative diseases and metabolic disorders including diabetes.

8.3 Regulation of Endocytic Trafficking, Receptors Signaling and Gene Expression by Cytohesin-2 and Arf6 GTP–binding Proteins

8.3.1 *Role of Cytohesin-2 and Arf6 in Regulation of Endocytosis*

The Ras-superfamily small GTPases, which includes Arf-family GTP-binding proteins, regulate an extraordinary variety of cell functions [30]. They are functioning as "molecular switches" in which transition between "on" and "off" states of this molecular device is mediated by a GDP/GTP cycle. Six members of Arf-family (Arf1-Arf6) have been identified in mammalian cells, as being important in the regulation of both exocytic and endocytic pathways [31, 32]. In particular, Arf6 has been implicated in the regulation of endocytosis, organelle biogenesis and actin cytoskeleton remodeling [33–35]. Activation of Arfs is mediated by **g**uanine-nucleotide **e**xchange **f**actors (GEFs). All known Arf-GEFs contain a conserved Sec7 catalytic domain, which is responsible for GEF activity. Cytohesin-2 (CTH2) (also known as ARNO, **A**DP-**r**ibosylation factor **n**ucleotide site **o**pener) is a member of the cytohesin subfamily of Arf-GEFs, which includes another three members: cytohesin-1, cytohesin-3 (also called ARNO3 or GRP1) and cytohesin-4. Their structure is divided into the following four regions: (i) an *N*-terminal coiled-coil (CC), (ii) a central Sec7 domain, (iii) a pleckstrin homology (PH) domain, and (iv) a C-terminal polybasic (PB) motif region (Fig. 8.1b,c). Regulation of targeting and activation of Arfs to different intracellular membranes takes place via specific GEFs that are associated with target membranes [36, 37]. In summary, depending on cell type and endocytic pathway function, the Arf6 was found to localized to CCP and CCV [38], early endosomes [16, 39], recycling endosomes [40], late endosomes [41] and micropinosomes [3, 4]. Recently, a novel role of Arf6 was also uncovered in regulation of fast endocytic recycling of Tfr receptors via classical CDE pathway [38]. This study revealed that while activated Arf6-GFP is localized at the plasma membrane in CCPs it does not participate in recruiting AP-2 adaptors. On the other hand, the activated Arf6-GFP was also found on endocytic vesicles interacting with JIP proteins and selectively involved in microtubule-dependent endocytic recycling of tranferrin receptors. Thus this study uncovered a novel mechanism of separation of Arf6 activation from its effector function in regulation of fast recycling pathway [38]. In conclusion, depending on the differential targeting of Arf6 and cytohesin-2, they may be involved in regulation of both degradation endosomal/lysosomal CDE pathway and micropinocytosis recycling CIE pathways.

8.3.2 *Emerging Role of Cytohesins in Receptors Signaling and Gene Expression*

During the last few years cytohesin family Arf-GEFs has emerged as a crucial molecules which are involved in regulation of signaling and early trafficking events of various important receptors. This cell biological role of cytohesins is exerted either at the plasma membrane or in early endosomes. Mammalian cytohesin-1 was originally identified as cytoplasmic protein interacting with LFA-1 β2-chain of CD18 and required for β-integrin-mediated adhesion in Jukart T cells [42, 43]. Interestingly, that the direct interaction between cytohesin-1 and β2-tail it taking place via its Sec7 domain but does not required its GEF activity. Thus, these early events of cytosolic protein–protein interactions are regulating the cytohesin-dependent inside-out signaling resulting in an increased avidity of β2-integrin-substrate binding. Interestingly, cytohesin-2/3 but not cytohesin-1 were recently identified as a crucial down-stream effectors for the insulin-receptor signaling cascade [44–46]. In particular, these studies demonstrated that the direct interaction of these cytohesins with phosphorylated insulin receptor was involved in its coupling to the PI3K signaling pathway. Thus, in *Drosophila* cytohesin function was shown to be essential for insulin signaling and cell growth, while in mouse the defective cytohesin signaling give rise to insulin resistance and consequent development of metabolic syndrome during diabetes [46]. In cancer cells cytohesins have been also recently identified as crucial cytoplasmic activators of EGFR/ErbB receptors that are involved in phosphorylation-dependent dimerization, and cancerogenic signaling of these receptors [47].

Importantly, in addition to their role as Arf-GEFs or receptors-signaling molecules the cytohesins have recently emerged as a transcriptional regulators. It was shown that cytohesin-1 is involved in activating the IL-2 gene promoter in T cells [48]. In particular, the scaffolding protein specifically interacting with cytohesin-1 in T cells called **c**ytohesin **b**inder and **r**egulator (Cybr) is involved in MAPK-mediated and T cell antigen receptor-dependent gene transcription [49]. On the other hand, cytohesin-2 was shown is involved in regulation of gene expression via the MAPK signaling pathway during serum-mediated transcriptional activation in non-immune cells [50]. These studies indicate that transcriptional regulation by cytohesins may be a widespread phenomenon among the small GEFs and particular members of the cytohesin family may be regulated by cell-type specific gene expression.

In conclusion, the downstream pathways which are regulated by cytohesins and Arfs include: (i) recruiting coat components during early events of endocytosis, (ii) modifying phospholipids at plasma membrane and endosomal compartments, (iii) dimerization and signaling of receptors at plasma membrane, (iv) sorting and trafficking of receptors at early stages of endocytosis, (v) modulating JIPs/kinesin-1 or JIPs/dynein/dynactin microtubule-dependent vesicular movement, and (vi) remodeling the actin cytoskeleton [31, 32, 37, 51] including via interaction of cytohesin-2 with aldolase, as recently described in our laboratory [52]. Below we will address in more details the roles of V-ATPase, cytohesin-2/Arf6 and aldolase in regulation of endocytic protein degradation pathway in kidney proximal tubules (PT) and its potential role in development of diabetic nephropathy.

8.4 Regulation of Endocytic Protein Degradation Pathway in Kidney Proximal Tubules: Implications for Development of Diabetic Nephropathy

8.4.1 *Kidney Proximal Tubule Megalin/Cubilin-Mediated Endosomal/Lysosomal Proteins Degradation Pathway*

Reabsorption of filtered proteins is of central importance for function of kidney proximal tubule epithelial cells (Fig. 8.2a). Trafficking and degradation of these proteins via endosomal/lysosomal degradation pathway is important for normal protein homeostasis and for maintenance of physiological levels of hormones, vitamins and cytokines in blood. Many low molecular weight filtered plasma proteins include: (i) albumin; (ii) transferrin; (iii) vitamin-binding proteins; (iv) cytokines; and (iv) hormones such as insulin, as well as drugs, such as gentamycin and amicacin, that are reabsorbed in PT cells through receptor-mediated endocytosis [53–55]. This process involves three apically located multiligand-binding receptors, megalin, cubilin and amnionless [53, 54, 56, 57]. Megalin is a transmembrane glycoprotein that belongs to the low-density lipoprotein receptor gene family and functions as a low-selectivity, high-capacity scavenger receptor. The cubilin molecule together with the protein amnionless forms the cubam receptor. However, cubilin also interacts with megalin and may function as a dual-receptor complex (Fig. 8.2a). While the function of megalin and cubilin receptors as well as their ligand specificity is well studied, the role of amnionless in function of endocytic pathway is just emerging [54, 56, 57]. It is also noteworthy that the existence of the alternative endocytic pathways including transcytosis or macropinocytosis in kidney PT epithelial cells is currently unknown. Thus, in kidney PT cells *in situ* as well as in cultured MTC and OK cells, megalin/cubilin receptors and their ligands enter cells exclusively via the constitutively operated CDE endosomal/lysosomal pathway [54, 55, 58]. In this pathway the ligands are delivered via clathrin-coated vesicle (CCV), to early endosomes (Fig. 8.2a–c), late endosomes (Fig. 8.2a,d,e), and then to lysosomes for degradation (Fig. 8.2a), while megalin/cubilin-receptors return to the plasma membrane via recycling endosomes (Fig. 8.2a). Recently, the differential role of megalin and cubilin in protein reabsorption has been examined using conditional Cre-LoxP knockout technology in mice [59]. This important study revealed a critical role of cubilin in albumin handling, while megalin plays an important auxiliary role in the internalization of cubilin/albumin complexes. Importantly, using this new genetic model [59, 60], authors convincingly demonstrated that normally small amount of filtrated albumin is uniquely reabsorbed via megalin/cubilin-mediated endosomal/lysosomal pathway and delivered to lysosomes for degradation. It is noteworthy, that while the notion of small glomerular filtration of albumin was recently challenged, these studies have received an intensive critique and assessment as artifactual observations [56, 57, 59, 61–64]. Thus, the study with genetically modified mouse model definitely resolved this recent controversy by demonstrating: (i) limited amounts of albumin filtration and (ii) its exclusive handling by megalin/cublin-mediated endosomal/lysosomal pathway, supporting previously reported

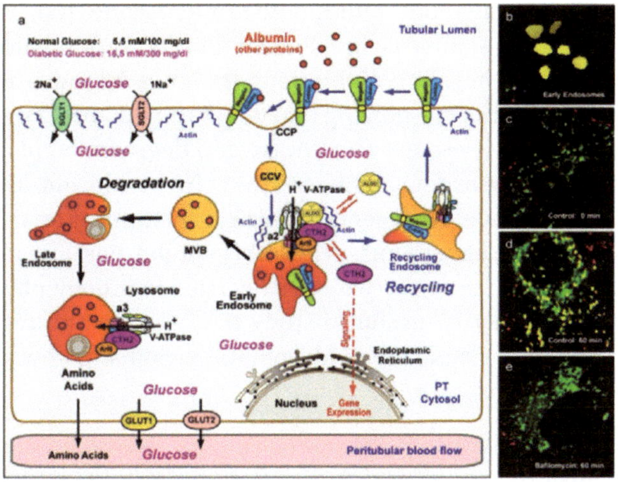

Figure 8.2 Reabsorption of filtered proteins and glucose by kidney proximal tubule cells. (a) Diagram showing endocytic uptake and trafficking of filtrated albumin and other proteins as well as reabsorption and handling of glucose by proximal tubules. The compartments of endocytic pathway are shown in yellow/red. Vesicular trafficking steps for degradation branch are indicated in black arrows and for recycling branch in blue arrows. Differential targeting of V-ATPase is cell-specific and compartment-specific. Particularly in proximal tubules V-ATPase with a2-isoform is targeted to early endosomes, while in some cells V-ATPase with a3-isoform is targeted to lysosomes. In early endosomes a2-isoform functions as pH-sensor by recruiting cytohesin-2/Arf6 in acidification-dependent manner and involved in the formation of endosomal carrier vesicles also known as **m**ulti**v**esicular **b**odies (MVB). These vesicular intermediates are involved in the trafficking between early and late endosomes. Aldolase (ALDO) is also found in early endosomes where it could interact with cytohesin-2 and V-ATPase. Depending on the segment of proximal tubules filtered glucose is reabsorbed at the apical pole by either SGLT1 or SGLT2 Na^{+}/glucose-cotransporters. Glucose is diffused via its gradient in PT cytosol and returned to the blood stream at the basolateral pole by either GLUT1 or GLUT2 transporters. (b) Confocal image of *in situ* proximal tubule early endosomes double-labeled with antibodies against a2-subunit V-ATPase and EEA1 protein. (c–e) Inhibition of acidification by bafilomycin A1 prevents the delivery of internalized albumin-Alexa594 from early to late endosomes. Proximal tubule cells in culture were transfected with a vector expressing Rab7-EGFP, a marker of late endosomes. They were then pulsed for 5 min with albumin-Alexa594 and chased for about 75 min. These images are single frames taken from real-time movies of the chase period imaged by spinning disk confocal microscopy. Adapted from Ref. 16.

revised data using both multiphoton microscopy [65, 66] and micropuncture [64] approaches. Finally, another recent study using this conditional megalin/cubilin-deficient mice has also resolved another controversy by showing that generation of urinary albumin fragments does not require PTs uptake and degradation of albumin via endosomal/lysosomal pathway [67]. In conclusion, this new genetic mouse model represents powerful tool to study this pathway in health and disease. Thus, while this model has already been essential in order to solve two controversies in kidney physiology, it will undoubtedly be useful to study the function and regulation of megalin/cubilin-mediated endosomal/lysosomal pathway under diabetic conditions.

8.4.2 *Diabetes, Glucose Handling by Kidney Proximal Tubules and Diabetic Nephropathy*

In healthy individuals glucose homeostasis is controlled via the tight hormonal regulation of glucose consumption by the central nervous and peripheral tissues, with its production by the liver and in some smaller extent by the kidney. The kidney has also got an unique role in maintenance of glucose blood level via its glomerular filtration and proximal tubular reabsorption (Fig. 8.2a). Two glucose/sodium co-transporters SGLT1 and SGLT2 are involved in reabsorption of glucose on the apical pole of proximal tubule cells followed by its passive cytosolic diffusion and delivery of glucose across basolateral membrane of PTs via GLUT1 and GLUT2 transporter (Fig. 8.2a) [68, 69]. This function makes renal PT cells especially sensitive to variations of blood glucose levels and adaptive to its reabsorptive handling. Under normoglycemic conditions (5,5 mM/100 mg/dl) about 180 g is daily filtered and reabsorbed by proximal tubules practically without loss of the glucose in urine. However, under hyperglycemic conditions (16,5 mM/300 mg/dl) during diabetes the filtered glucose saturated the transport capacity of SGLT1 and SGLT2 transporters giving rise to “glucose toxicity” and accompany detrimental renal effect in diabetes (Fig. 8.2a) [70]. Diabetes is one of the leading causes of a slow deterioration of the kidneys (nephropathy) leading to end-stage renal disease (ESRD) and finally to kidney failure. In the Unites States, approximately 35% of patients who develop chronic renal failure have diabetes. About half of these patients have insulin-dependent diabetes mellitus (IDDM, or type 1 diabetes) and the other half non-insulin-dependent diabetes mellitus (NIDDM, or type 2 diabetes).

The pathogenesis of late stages of diabetic complications leading to diabetic nephropathy and ESRD are relatively well studied and include glomerular defects linked with progressive deterioration of proximal tubular function and tubulointerstitial inflammation and fibrosis. One of the features of late stages diabetic nephropathy is diminished proteins reabsorption by proximal tubules via megalin/cubilin-mediated endosomal/lysosomal protein degradative pathway leading to proteinuria. Patients with IDDM are more likely to develop proteinuria (mainly albuminuria) than those with NIDDM [71]. Microalbuminuria (albumin excretion >30 mg/d) can predict the development of diabetic nephropathy in both IDDM and NIDDM, but persistent albuminuria (>300 mg/d) is the hallmark of diabetic nephropathy and is used for clinical diagnosis. Importantly, albuminuria is not only a consequence of diabetes, but is also a causative factor in the progressive renal failure associated with late stages of diabetes [72]. In particular, elevated urinary albumin has been implicated in the induction of proinflammatory chemokines and growth factors, followed by infiltration of macrophages, initiation of inflammation and development of fibrosis in diabetic kidney [70].

However, the pathogenesis of important early events that link poorly controlled diabetes with repeated events of high glucose levels remains largely unexplored. In particular, the role of proximal tubule in the early stages of pathogenesis of diabetic nephropathy was underestimated and its importance has been appreciated only recently [70]. Following cell biological events have been recognized as a crucial in early hyperglycemia during diabetes: (i) enhancement of the glucose reabsorption by proximal tubule cells, (ii) unique growth and early hyperplasia of the proximal tubules, and (iii) primary hyper-reabsorption by the proximal tubules. Importantly, it was established that the enhanced glucose reabsorption and tubular growth contribute to primary proximal tubular hyper-reabsorption, while primary proximal tubular hyper-reabsorption also contributes to glomerular hyper-filtration [70]. On the other hand, recent *in vivo* studies demonstrated that acute hyper-glycemia gives rise to rapid and significant increase in primary glomerular permeability to macromolecules, thus, could inversely promote proximal tubule hyper-reabsorption and increase in function of megalin/cubilin endosomal/lysosomal protein degradative pathway. Below we will discuss potential mechanism of the phenomenon and the emerging crucial roles of V-ATPase, cytohesin-2/Arf6 and aldolase in regulation of endocytic pathway and protein reabsorption during early stages of diabetes.

8.4.3 *Aldolase: Novel Roles of the "Old" Enzyme in Regulation of Endocytic Trafficking and Actin Cytoskeleton Rearrangement*

Fructose bisphosphate aldolase (ALDO) (EC 4.1.2.13) catalyses the reversible cleavage of fructose 1,6-bisphosphate into glyceraldehyde 3-phosphate and dihydroxyacetone phosphate. The role of aldolase is well established as a crucial regulatory enzyme of glycolysis. Aldolase is very well characterized protein due to its high abundance and relatively easy purification from natural sources. Crystal structures of various aldolase isoforms were solved shedding light on its molecular mode of action in glucose metabolism [73–75]. However, various alternative so called "moonlighting" functions for this enzyme were recently identified. It was demonstrated that aldolase could interact with variety of proteins unrelated to glucose metabolism such as: (i) phospholipase D_2, (ii) endocytotic sorting protein nexin-9 and cytoskeletal proteins, (iii) F-actin, (iv) actin nucleation promoting factor WASP, (v) α-tubulin, and (vi) light chain 8 of dynein [76–81]. These studies suggested that aldolase also functions as a scaffolder protein in the coordination of: (i) cytoskeleton rearrangement and cell motility, (ii) trafficking and recycling of membrane proteins, and (iii) signal transduction during endocytosis.

In addition the direct interaction between aldolase and the V-ATPase was also reported [82]. Three different subunits of the V-ATPase: the transmembrane a-subunit of the V_0 sector and the soluble E- and B-subunits of the V_1-sector are interacting with aldolase [83]. In *Saccharomyces cerevisiae* the interaction between aldolase and the V-ATPase is modulated by glucose [84, 85]. It was suggested that aldolase is acting as a "glucose sensor" which via its physical association with V-ATPase is involved in reversible assembly/disassembly of V_1V_O sectors of V-ATPase [86]. This important regulatory mechanism of V-ATPase function is also found in response to ceased feeding in *Manduca sexta* [87] and in response to glucose depletion in kidney proximal tubule epithelial cells [88]. It is noteworthy that, while in yeast glucose-dependent assembly/disassembly of V-ATPase and its interaction with aldolase is modulated by the Ras/cAMP/PKA pathway [89], in kidney proximal tubule cells the effect of glucose on V-ATPase is mediated by phosphatidylinositol 3-kinase (PI3K)-dependent signaling pathway [88].

8.4.4 *Role of V-ATPase, Cytohesin-2/Arf6 and Aldolase in Regulation of Megalin/Cubilin-Mediated Endosomal/Lysosomal Protein Degradation Pathway: Implications for Diabetic Nephropathy*

In proximal tubule cells similarly to other eukaryotic cells, acidification of endocytic compartments becomes increasingly acidic from early endosomes to late endosomes and finally to lysosomes [10]. Intravesicular acidification of these endocytic compartments is also driven by proton pumping V-ATPase and could be inhibited by bafilomycin (Fig. 8.2c–e). It is functioning in conjunction with a parallel chloride conductance facilitated by CLC-5 and CLC-4 electrogenic Cl^-/H^+-exchangers as previously reviewed [10, 12]. In Dent disease and Fanconi syndrome reduction of chloride-driven electrical shunt inhibits endosomal V-ATPase-dependent acidification giving rise to diminished function of megalin/cubilin-mediated endocytic pathway and urinary wasting of albumin [10]. However, the biochemical mechanism which coupled deficient endosomal acidification with impaired endocytic trafficking remained unclear. Recent data from our laboratory has uncovered this crucial link between V-ATPase-driven endosomal acidification and Arf GTP-binding proteins in regulation of the megalin/cubilin-mediated endosomal/lysosomal protein degradation pathway [16, 17]. This involves a direct, acidification-dependent interaction of

V-ATPase with cytosolic Arfs and their regulatory proteins [16]. We uncovered, that a2-subunit isoform of V-ATPase is targeted to early endosomes in the PT and directly interacts with the cytohesin-2 (CTH2) (Figs. 8.1a, 8.2a, and 8.3b), while the Arf6 is binding to the c-subunit of the V-ATPase (Figs. 8.1a and 8.2a). Thus, the a2-isoform containing endosomal V-ATPase was discovered as novel pH-sensor which interaction with cytohesin-2/Arf6 regulates endosomal protein trafficking [16–18]. It is important to underline that our data showing the interaction of cytohesin-2 with the "endosomal" V-ATPase a2-subunit were extended to the other a-isoforms (a1, a3 and a4) of the V-ATPase (Figs 8.3a) [52]. This new finding indicates that regulatory interaction of V-ATPases with cytohesin-2 and signaling between these proteins is a general cell biological phenomenon. We hypothesized that cross-talk between V-ATPase and cytohesin-2/Arf6 may also exist on lysosomes, where it could regulate function of V-ATPase (Fig. 8.2a). In the line with this hypothesis, it is noteworthy that regulation of V-ATPase activity by assembly of V_1V_0 sectors onto the lysosomal membrane was previously shown in mammalian dendritic and liver cells [90, 91]. In dendritic cells this mechanism is critical for lysosomal acidification, activation of proteases, protein degradation and antigen presentation during maturdation [90]. In liver cells, however, the assembly of V_1V_0 sectors is triggered by EGFR/ErbB-receptors signaling and activation of Erk/Akt pathway. This, in turn, give rise to an increased lysosomal V-ATPase driven acidification, acceleration of protein degradation, amino acids production, their release into cytosol and activation of Rheb/mTORC1 complex [91]. It is important to underline that, one member of the EGFR/ErbB-receptor family, ErbB4 is also expressed in kidney proximal tubule epithelial cells and involved in their polarization and proliferation [92, 93]. Accordingly, the recent Genetics of Nephropathy and International Effort (GENIE) consortium GWAS study identified this ErbB4 gene as strongly associated with development of diabetic nephropathy (DN) caused by type 1 diabetes [94, 95]. Finally, recent a multi-ethnic GWAS meta-analysis has also identified a2-subunit isoform gene (*ATP6V0A2*) of V-ATPase as a common genetic variant associated with serum adiponectin levels and risk of type 2 diabetes in humans [96].

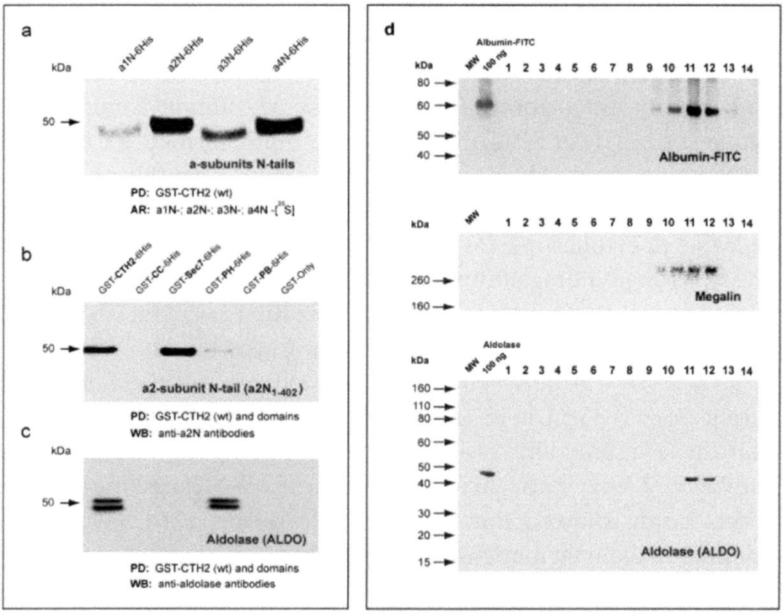

Figure 8.3 Direct interaction of cytohesin-2 with all four a-subunit isoforms of V-ATPase as well as with a2N and aldolase localized in early endosomes. (a) Direct interaction of cytohesin-2 with all four a-subunit isoforms demonstrated in pull-down experiments using recombinant proteins. Cytosolic tails of a1-, a2-, a3-, and a4-subunit isoforms were *in-vitro* translated and metabolically labeled with [^{35}S]-methionine and used in pull-down assays with GST-CTH2(wt). Detection of a1N, a2N, a3N, and a4N was performed by autoradiography. (b) Cytohesin-2 interacts with a2N$_{1-402}$ mainly via its Sec7-domain. Recombinant GST-CTH2(wt) and its domains were immobilized on glutathione-beads, incubated with *in vitro* translated a2N$_{1-402}$, and eluted complexes were analyzed by western blot analysis using an anti-a2N antibody. (c) Cytohesin-2 interacts with aldolase via its PH-domain. Recombinant GST-CTH2(wt) and its domains were immobilized on glutathione-beads, incubated with *in vitro* translated aldolase-B, and eluted complexes were analyzed by western blot analysis using an anti-aldolase antibody. (d) Aldolase is found on early endosomes of proximal tubules cells. Cell fractionation and localization of endogenous aldolase in early endosomal fractions containing albumin-FITC ligand and megalin receptor as a markers. All proteins were detected by western blot analysis. Adapted from Refs. 52 and 97.

Our recent studies were also focused on molecular mechanism of interaction between a2-subunit V-ATPase and cytohesin-2 [97]. We identified two structural elements involved in specific and high affinity association of the V-ATPase a2-subunit isoform with cytohesin-2: (i) an N-terminal binding motif formed by the first seventeen amino acids of the a2N called $a2N_{1-17}$ peptide and (ii) an interaction pocket formed by the catalytic Sec7 and regulatory PB-domains of cytohesin-2 (Fig. 8.1b,c) [97]. Biacore analysis revealed a very strong binding affinity between this $a2N_{1-17}$ peptide and the Sec7-domain of cytohesin-2, with dissociation constant $K_D = 3.44 \times 10^{-7}$M (Fig. 8.1a, interface II), similar to the binding affinity $K_D = 3.13 \times 10^{-7}$M between wild-type a2N(wt) and the full length cytohesin-2 (wt) protein (Fig. 8.1a, interface I). We hypothesized that these interactions are crucial for signaling and regulation of cytohesin-2 enzymatic Arf-GEF activity by V-ATPase. Indeed our recent study showed that peptide of V-ATPase $a2N_{1-17}$ potently modulates the enzymatic GDP/GTP-exchange activity of cytohesin-2 via direct binding with its Sec7 domain (Fig. 8.1a–c, 8.3b) [98, 99]. In live PT cells, this V-ATPase-derived, cell-penetrating and biologically active $a2N_{1-17}$ peptide strongly activates the macropinocytosis pathway [100, 101]. Thus, we have uncovered a novel function of the V-ATPase as a signaling receptor that modulates activity of the Arf-GEF cytohesin-2 and cognate Arf GTP-binding proteins. Signaling between the V-ATPase and Arfs may regulate crosstalk between distinct endocytic receptor-mediated and macropinocytosis pathways [98–101].

Although this previous work uncovered a functional cross-talk between V-ATPase, cytohesin-2 and Arfs, other downstream effectors and related cell biological events have not been unraveled. Thus, we suggested that since V-ATPase interacts with both cytohesin-2 and aldolase, these two proteins could in turn interact with each other and coordinate endocytic vesicle trafficking and cytoskeleton rearrangements. Indeed, using a combination of protein-protein interaction techniques we demonstrated that in MTC cells cytohesin-2 specifically interacts with aldolase [52]. Importantly, using purified rat kidney PTs we performed additional pull-down experiments followed by mass-spectrometry analysis and confirmed interaction of cytohesin-2 with aldolase *in situ* as well. The direct interaction of aldolase with cytohesin-2 via its PH-domain was further studied in pull-down (Fig. 8.3c) and Biacore experiments. This approach revealed a two step interaction between these two proteins with $K_{D1} = 1.1 \times 10^{-4}$ M and $K_{D2} = 2.7 \times 10^{-6}$ M clearly indicating a potential regulatory mechanism of this interaction (Fig. 8.1a, interfaces III and IV). Moreover, using an iodixanol gradient-based cell fractionation approach we also demonstrated for the first time the association of aldolase, which is generally considered as a predominantly cytosolic protein, with early endosomes containing the megalin/albumin-FITC receptor/ligand complexes (Fig. 8.3d) and where it could directly interact with cytohesin-2 and V-ATPase (Fig. 8.1a and 8.2b) [52]. Thus, our data revealed a novel specific interaction between cytohesin-2 and aldolase in the V-ATPase/Arf6/cytohesin-2/aldolase complex on early endosomes (Fig. 8.1a), which may regulate the activities of its components and function of the protein degradative pathway under physiologic and pathologic conditions (Fig. 8.2a) [10, 12, 53].

In search for functional significance of interactions within V-ATPase/Arf6/cytohesin-2/aldolase complex we next studied the effect of aldolase knock-down on regulation of gene expression, the actin cytoskeleton–dependent cell morphology and distribution of V-ATPase containing endosomal/lysosomal compartments. These studies demonstrated that the direct interaction between aldolase and cytohesin-2 are important in: (i) gelsolin gene expression, (ii) actin cytoskeletal rearrangement, and (iii) redistribution of endosomal vesicles [52]. Thus, our findings suggest a potential important role of aldolase as "glucose sensor" during early stages of diabetes in modulation of cytohesin-2/Arf6 signaling in gene expression, V-ATPase function, actin cytoskeleton rearrangement, and thus, regulation of proximal tubule endocytic pathway (Fig. 8.2a). Indeed, using **n**on-**o**bese **d**iabetic (NOD) mice as a model of early stages of human type 1 diabetes we demonstrated an increase of expression and apical redistribution of both V-ATPase (Fig. 8.4a,b,e) and megalin (Fig. 8.4c–e) at early stages (2 weeks) of diabetes in NOD mice. These results were also confirmed by quantitative RT-PCR and western-blot analysis. Thus, these data support our hypothesis that the protein degradation pathway in PT (Fig. 8.2a), represented by these critical components, is upregulated

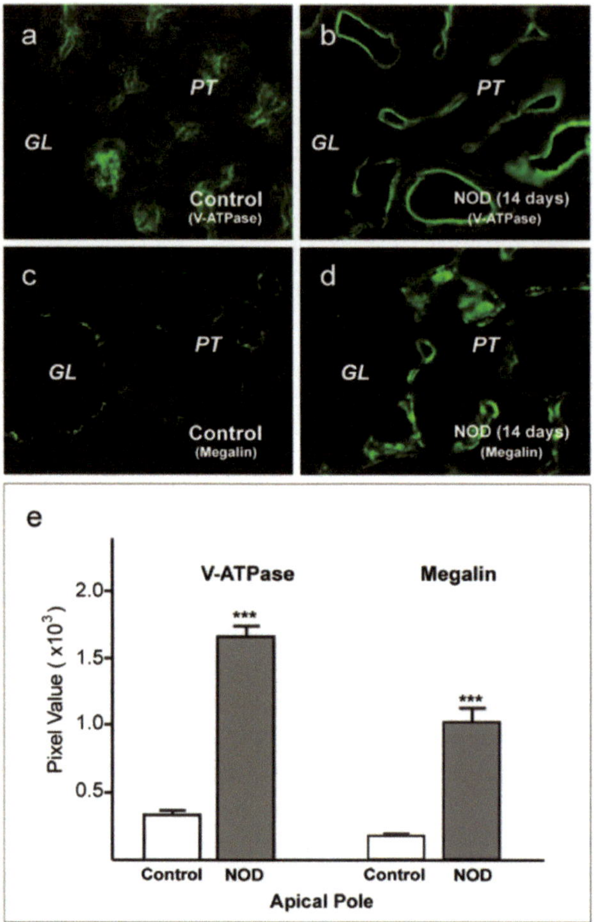

Figure 8.4 Function of V-ATPase-dependent megalin/cubilin-mediated endosomal/lysosomal pathway under normal and diabetic conditions in NOD mice as a model of early stages of human type 1 diabetes. (a, b) Immunocytochemistry detection of increased expression and apical redistribution of V-ATPase in control (a) and in early stages (2 weeks) of diabetes NOD (b) mice. Specific monoclonal antibodies against E-subunit of V-ATPase were used in these experiments. (c, d) Immunocytochemistry detection of increased expression and apical redistribution of megalin in control (c) and in early stages (2 weeks) of diabetes NOD (d) mice. Specific monoclonal antibodies against megalin were used in these experiments. (e) Semi-quantitative estimation of the levels of expression and targeting of V-ATPase and megalin in apical pole of control and NOD mice.

by high levels of glucose which would affect its function during early stages of diabetes.

In conclusion, recent studies from our laboratory provided new insights into regulation of megalin/cubilin-mediated endosomal/lysosomal protein degradation pathway under physiological and pathological conditions such as early stages of renal diabetic complication. We uncovered a novel role of endosomal V-ATPase as pH-sensing receptor which is involved in acidification-dependent transmembrane signaling and regulation of activity of cytohesin-2/Arf6 GTP-binding proteins. We also uncovered aldolase as a novel modulator of V-ATPase/Arf6/cytohesin-2/aldolase super complex which may be involved in glucose-dependent regulation of gene expression, modulation of gelsolin/actin cytoskeleton and function of V-ATPase-driven acidic endocytic compartments (Figs. 8.1a and 8.2a). However, the exact pathophysiological role of "glucose toxicity" during early stages of diabetes in modulation of interplay between gene expression, endosomal signaling and trafficking within the megalin/cubilin-mediated endosomal/lysosomal pathway remains to be uncovered. Future research of these pathways in diabetes should also be focused on studies of the molecular mechanisms and interactions interfaces between V-ATPase, cytohesin-2, and aldolase. These studies will allow to develop pharmaceutical (peptides) or perform the screening or rational design of small molecules (drugs) that will be targeting novel V-ATPase/Arf6/cytohesin-2/aldolase complex (Figs. 8.1a and 8.2a). Thus, the ultimate goal of the development of these therapies is to normalize the function of megalin/cubilin-mediated endosomal/lysosomal protein degradation pathway during "glucose toxicity" in diabetes in order to prevent and treat diabetic nephropathy.

Acknowledgments

This work was supported by a National Institutes of Health (NIH) grant DK038452 and by Boston Area Diabetes Endocrinology Research Center (BADERC) grant DK057521. The Microscopy Core Facility of the Program in Membrane Biology received additional support from the Boston Area Diabetes and Endocrinology Research

Center (NIH DK57521) and from the Center for the Study of Inflammatory Bowel Disease (NIH DK43351).

References

1. Doherty, G.J. & McMahon, H.T. Mechanisms of endocytosis. *Annu Rev Biochem* 78, 857–902 (2009).
2. Mayor, S. & Pagano, R.E. Pathways of clathrin-independent endocytosis. *Nat Rev Mol Cell Biol* 8, 603–612 (2007).
3. Donaldson, J.G., Porat-Shliom, N. & Cohen, L.A. Clathrin-independent endocytosis: a unique platform for cell signaling and PM remodeling. *Cell Signal* 21, 1–6 (2009).
4. Grant, B.D. & Donaldson, J.G. Pathways and mechanisms of endocytic recycling. *Nat Rev Mol Cell Biol* 10, 597–608 (2009).
5. Gong, Q., Huntsman, C. & Ma, D. Clathrin–independent internalization and recycling. *J Cell Mol Med* 12, 126–144 (2008).
6. Lim, J.P. & Gleeson, P.A. Macropinocytosis: an endocytic pathway for internalising large gulps. *Immunol Cell Biol* 89, 836–843 (2011).
7. Sigismund, S. *et al.* Clathrin-independent endocytosis of ubiquitinated cargos. *Proc Natl Acad Sci USA* 102, 2760–2765 (2005).
8. Orth, J.D., Krueger, E.W., Weller, S.G. & McNiven, M.A. A novel endocytic mechanism of epidermal growth factor receptor sequestration and internalization. *Cancer Res* 66, 3603–3610 (2006).
9. Orth, J.D. & McNiven, M.A. Get off my back! Rapid receptor internalization through circular dorsal ruffles. *Cancer Res* 66, 11094–11096 (2006).
10. Marshansky, V., Ausiello, D.A. & Brown, D. Physiological importance of endosomal acidification: potential role in proximal tubulopathies. *Curr Opin Nephrol Hypertens* 11, 527–537 (2002).
11. Brown, D., Marshansky, V. Renal V-ATPase: Physiology and pathophysiology, in *Handbook of ATPases* (ed. Futai, Y.W.M. and Kaplan, J.H.) 414–442 (Wiley-VCH Verlag GmbH & Co. KGaA, Weinheim (2004).
12. Marshansky, V. & Futai, M. The V-type H(+)-ATPase in vesicular trafficking: targeting, regulation and function. *Curr Opin Cell Biol* 20, 415–426 (2008).
13. Morel, N., Dedieu, J.C. & Philippe, J.M. Specific sorting of the a1 isoform of the V-H+ATPase a subunit to nerve terminals where it associates

with both synaptic vesicles and the presynaptic plasma membrane. *J Cell Sci* 116, 4751–4762 (2003).

14. Saw, N.M. *et al.* Vacuolar H(+)-ATPase subunits Voa1 and Voa2 cooperatively regulate secretory vesicle acidification, transmitter uptake, and storage. *Mol Biol Cell* 22, 3394–3409 (2011).
15. Lee, J.H. *et al.* Lysosomal proteolysis and autophagy require presenilin 1 and are disrupted by Alzheimer-related PS1 mutations. *Cell* 141, 1146–1158 (2010).
16. Hurtado-Lorenzo, A. *et al.* V-ATPase interacts with ARNO and Arf6 in early endosomes and regulates the protein degradative pathway. *Nat Cell Biol* 8, 124–136 (2006).
17. Recchi, C. & Chavrier, P. V-ATPase: a potential pH sensor. *Nat Cell Biol* 8, 107–109 (2006).
18. Marshansky, V. The V-ATPase a2-subunit as a putative endosomal pH-sensor. *Biochem Soc Trans* 35, 1092–1099 (2007).
19. Toyomura, T. *et al.* From lysosomes to the plasma membrane: localization of vacuolar-type H^+-ATPase with the a3 isoform during osteoclast differentiation. *J Biol Chem* 278, 22023–22030 (2003).
20. Sun-Wada, G.H. & Wada, Y. Vacuolar-type proton pump ATPases: roles of subunit isoforms in physiology and pathology. *Histol Histopathol* 25, 1611–1620 (2010).
21. Sun-Wada, G.H., Tabata, H., Kawamura, N., Aoyama, M. & Wada, Y. Direct recruitment of H^+-ATPase from lysosomes for phagosomal acidification. *J Cell Sci* 122, 2504–2513 (2009).
22. Pietrement, C. *et al.* Distinct expression patterns of different subunit isoforms of the V-ATPase in the rat epididymis. *Biol Reprod* 74, 185–194 (2006).
23. Merkulova, M., Hosokawa, H., Bakulina, A., Dip, P. V., Young, R. T., Khatri, A., Ausiello, D. A., Grüber, G., and Marshansky, V. Structural modal of a2-subunit *N*-terminus and its binding interface for cytohesin-2: Implication for regulation of V-ATPase function. "*Experimental Biology Meetting*" (Boston, MA), Abstract of the Poster Presentation, p. 333 (2013).
24. Kopan, R. & Ilagan, M.X. The canonical Notch signaling pathway: unfolding the activation mechanism. *Cell* 137, 216–233 (2009).
25. Vaccari, T., Duchi, S., Cortese, K., Tacchetti, C. & Bilder, D. The vacuolar ATPase is required for physiological as well as pathological activation of the Notch receptor. *Development* 137, 1825–1832 (2010).

26. MacDonald, B.T., Tamai, K. & He, X. Wnt/beta-catenin signaling: components, mechanisms, and diseases. *Dev Cell* 17, 9–26 (2009).
27. Cruciat, C.M. *et al.* Requirement of prorenin receptor and vacuolar H^+-ATPase-mediated acidification for Wnt signaling. *Science* 327, 459–463 (2010).
28. O'Callaghan, K.M. *et al.* Heme-binding protein HRG-1 is induced by insulin-like growth factor I and associates with the vacuolar H^+-ATPase to control endosomal pH and receptor trafficking. *J Biol Chem* 285, 381–391 (2010).
29. Zoncu, R. *et al.* mTORC1 senses lysosomal amino acids through an inside-out mechanism that requires the vacuolar H-ATPase. *Science* 334, 678–683 (2011).
30. Bourne, H.R., Sanders, D.A. & McCormick, F. The GTPase superfamily: a conserved switch for diverse cell functions. *Nature* 348, 125–132 (1990).
31. Donaldson, J.G. & Jackson, C.L. Regulators and effectors of the ARF GTPases. *Curr Opin Cell Biol* 12, 475–482 (2000).
32. D'Souza-Schorey, C. & Chavrier, P. ARF proteins: roles in membrane traffic and beyond. *Nat Rev Mol Cell Biol* 7, 347–358 (2006).
33. Schafer, D.A., D'Souza-Schorey, C. & Cooper, J.A. Actin assembly at membranes controlled by ARF6. *Traffic* 1, 892–903 (2000).
34. Balasubramanian, N., Scott, D.W., Castle, J.D., Casanova, J.E. & Schwartz, M.A. Arf6 and microtubules in adhesion-dependent trafficking of lipid rafts. *Nat Cell Biol* 9, 1381–1391 (2007).
35. Myers, K.R. & Casanova, J.E. Regulation of actin cytoskeleton dynamics by Arf-family GTPases. *Trends Cell Biol* 18, 184–192 (2008).
36. Frank, S., Upender, S., Hansen, S.H. & Casanova, J.E. ARNO is a guanine nucleotide exchange factor for ADP-ribosylation factor 6. *J Biol Chem* 273, 23–27 (1998).
37. Casanova, J.E. Regulation of Arf activation: the Sec7 family of guanine nucleotide exchange factors. *Traffic* 8, 1476–1485 (2007).
38. Montagnac, G. *et al.* Decoupling of activation and effector binding underlies ARF6 priming of fast endocytic recycling. *Curr Biol* 21, 574–579 (2011).
39. van der Goot, F.G. Separation of early steps in endocytic membrane transport. *Electrophoresis* 18, 2689–2693 (1997).
40. Peters, P.J. *et al.* Characterization of coated vesicles that participate in endocytic recycling. *Traffic* 2, 885–895 (2001).

41. Lafourcade, C., Sobo, K., Kieffer-Jaquinod, S., Garin, J. & van der Goot, F.G. Regulation of the V-ATPase along the endocytic pathway occurs through reversible subunit association and membrane localization. *PLoS One* 3, e2758 (2008).
42. Kolanus, W. *et al.* Alpha L beta 2 integrin/LFA-1 binding to ICAM-1 induced by cytohesin-1, a cytoplasmic regulatory molecule. *Cell* 86, 233–242 (1996).
43. Kolanus, W. Guanine nucleotide exchange factors of the cytohesin family and their roles in signal transduction. *Immunol Rev* 218, 102–113 (2007).
44. Hafner, M. *et al.* Inhibition of cytohesins by SecinH3 leads to hepatic insulin resistance. *Nature* 444, 941–944 (2006).
45. Fuss, B., Becker, T., Zinke, I. & Hoch, M. The cytohesin Steppke is essential for insulin signalling in Drosophila. *Nature* 444, 945–948 (2006).
46. Jackson, C. Diabetes: kicking off the insulin cascade. *Nature* 444, 833–834 (2006).
47. Bill, A. *et al.* Cytohesins are cytoplasmic ErbB receptor activators. *Cell* 143, 201–211 (2010).
48. Perez, O.D. *et al.* Leukocyte functional antigen 1 lowers T cell activation thresholds and signaling through cytohesin-1 and Jun-activating binding protein 1. *Nat Immunol* 4, 1083–1092 (2003).
49. Chen, Q., Coffey, A., Bourgoin, S.G. & Gadina, M. Cytohesin binder and regulator augments T cell receptor-induced nuclear factor of activated T Cells. AP-1 activation through regulation of the JNK pathway. *J Biol Chem* 281, 19985–19994 (2006).
50. Theis, M.G. *et al.* Discriminatory aptamer reveals serum response element transcription regulated by cytohesin-2. *Proc Natl Acad Sci USA* 101, 11221–11226 (2004).
51. Randazzo, P.A., Nie, Z., Miura, K. & Hsu, V.W. Molecular aspects of the cellular activities of ADP-ribosylation factors. *Sci STKE* 2000, re1 (2000).
52. Merkulova, M. *et al.* Aldolase directly interacts with ARNO and modulates cell morphology and acidic vesicle distribution. *Am J Physiol Cell Physiol* 300, C1442–1455 (2011).
53. Marshansky, V. *et al.* Receptor-mediated endocytosis in kidney proximal tubules: recent advances and hypothesis. *Electrophoresis* 18, 2661–2676 (1997).

54. Christensen, E.I. & Birn, H. Megalin and cubilin: multifunctional endocytic receptors. *Nat Rev Mol Cell Biol* 3, 256–266 (2002).
55. Nielsen, R. *et al.* Endocytosis provides a major alternative pathway for lysosomal biogenesis in kidney proximal tubular cells. *Proc Natl Acad Sci USA* 104, 5407–5412 (2007).
56. Christensen, E.I., Verroust, P.J. & Nielsen, R. Receptor-mediated endocytosis in renal proximal tubule. *Pflugers Arch* 458, 1039–1048 (2009).
57. Nielsen, R. & Christensen, E.I. Proteinuria and events beyond the slit. *Pediatr Nephrol* 25, 813–822 (2010).
58. Gekle, M., Mildenberger, S., Freudinger, R., Schwerdt, G. & Silbernagl, S. Albumin endocytosis in OK cells: dependence on actin and microtubules and regulation by protein kinases. *Am J Physiol* 272, F668–F677 (1997).
59. Amsellem, S. *et al.* Cubilin is essential for albumin reabsorption in the renal proximal tubule. *J Am Soc Nephrol* 21, 1859–1867 (2010).
60. Weyer, K. *et al.* Mouse model of proximal tubule endocytic dysfunction. *Nephrol Dial Transplant* 26, 3446–3451 (2011).
61. Christensen, E.I., Birn, H., Rippe, B. & Maunsbach, A.B. Controversies in nephrology: renal albumin handling, facts, and artifacts! *Kidney Int* 72, 1192–1194 (2007).
62. Gekle, M. Renal albumin handling: a look at the dark side of the filter. *Kidney Int* 71, 479–481 (2007).
63. de Borst, M.H. On the origin of albuminuria. *Kidney Int* 72, 1409; author reply 1409–1410 (2007).
64. Remuzzi, A., Sangalli, F., Fassi, A. & Remuzzi, G. Albumin concentration in the Bowman's capsule: multiphoton microscopy vs. micropuncture technique. *Kidney Int* 72, 1410–1411; author reply 1411 (2007).
65. Peti-Peterdi, J. Independent two-photon measurements of albumin GSC give low values. *Am J Physiol Renal Physiol* 296, F1255–F1257 (2009).
66. Tanner, G.A. Glomerular sieving coefficient of serum albumin in the rat: a two-photon microscopy study. *Am J Physiol Renal Physiol* 296, F1258–1265 (2009).
67. Weyer, K., Nielsen, R., Christensen, E.I. & Birn, H. Generation of urinary albumin fragments does not require proximal tubular uptake. *J Am Soc Nephrol* 23, 591–596 (2012).
68. Chao, E.C. & Henry, R.R. SGLT2 inhibition—a novel strategy for diabetes treatment. *Nat Rev Drug Discov* 9, 551–559 (2010).

69. Vallon, V. Molecular determinants of renal glucose reabsorption. Focus on "Glucose transport by human renal Na^+/D-glucose cotransporters SGLT1 and SGLT2". *Am J Physiol Cell Physiol* 300, C6–C8 (2011).
70. Vallon, V. The proximal tubule in the pathophysiology of the diabetic kidney. *Am J Physiol Regul Integr Comp Physiol* 300, R1009–R1022 (2011).
71. Hosteter, T.H. *CECIL Textbook of Medicine* (ed. Bennet J.C.), 599–602 (W.B. Saunders Company, Philadelphia) (1996).
72. Parving, H.H. & Osterby, R. *The Kidney* (ed. Brenner, B.M.), 1731–1773 (W.B. Saunders Company, Philadelphia) (2000).
73. Dalby, A.R., Tolan, D.R. & Littlechild, J.A. The structure of human liver fructose-1,6-bisphosphate aldolase. *Acta Crystallogr D Biol Crystallogr* 57, 1526–1533 (2001).
74. Arakaki, T.L. *et al.* Structure of human brain fructose 1,6-(bis)phosphate aldolase: linking isozyme structure with function. *Protein Sci* 13, 3077–3084 (2004).
75. Malay, A.D., Allen, K.N. & Tolan, D.R. Structure of the thermolabile mutant aldolase B, A149P: molecular basis of hereditary fructose intolerance. *J Mol Biol* 347, 135–144 (2005).
76. Wang, J., Morris, A.J., Tolan, D.R. & Pagliaro, L. The molecular nature of the F-actin binding activity of aldolase revealed with site-directed mutants. *J Biol Chem* 271, 6861–6865 (1996).
77. Wang, J., Tolan, D.R. & Pagliaro, L. Metabolic compartmentation in living cells: structural association of aldolase. *Exp Cell Res* 237, 445–451 (1997).
78. Volker, K.W. & Knull, H. A glycolytic enzyme binding domain on tubulin. *Arch Biochem Biophys* 338, 237–243 (1997).
79. Buscaglia, C.A., Penesetti, D., Tao, M. & Nussenzweig, V. Characterization of an aldolase-binding site in the Wiskott-Aldrich syndrome protein. *J Biol Chem* 281, 1324–1331 (2006).
80. Rangarajan, E.S., Park, H., Fortin, E., Sygusch, J. & Izard, T. Mechanism of aldolase control of sorting nexin 9 function in endocytosis. *J Biol Chem* 285, 11983–11990 (2010).
81. Kim, J.H. *et al.* Phospholipase D2 directly interacts with aldolase via its pH domain. *Biochemistry* 41, 3414–3421 (2002).
82. Lu, M., Holliday, L.S., Zhang, L., Dunn, W.A., Jr. & Gluck, S.L. Interaction between aldolase and vacuolar H^+-ATPase: evidence for direct coupling of glycolysis to the ATP-hydrolyzing proton pump. *J Biol Chem* 276, 30407–30413 (2001).

83. Lu, M., Sautin, Y.Y., Holliday, L.S. & Gluck, S.L. The glycolytic enzyme aldolase mediates assembly, expression, and activity of vacuolar H^+-ATPase. *J Biol Chem* 279, 8732–8739 (2004).
84. Kane, P.M. Regulation of V-ATPases by reversible disassembly. *FEBS Lett* 469, 137–141 (2000).
85. Kane, P.M. The where, when, and how of organelle acidification by the yeast vacuolar H^+-ATPase. *Microbiol Mol Biol Rev* 70, 177–191 (2006).
86. Lu, M., Ammar, D., Ives, H., Albrecht, F. & Gluck, S.L. Physical interaction between aldolase and vacuolar H^+-ATPase is essential for the assembly and activity of the proton pump. *J Biol Chem* 282, 24495–24503 (2007).
87. Beyenbach, K.W. & Wieczorek, H. The V-type H+ATPase: molecular structure and function, physiological roles and regulation. *J Exp Biol* 209, 577–589 (2006).
88. Sautin, Y.Y., Lu, M., Gaugler, A., Zhang, L. & Gluck, S.L. Phosphatidylinositol 3-kinase-mediated effects of glucose on vacuolar H^+-ATPase assembly, translocation, and acidification of intracellular compartments in renal epithelial cells. *Mol Cell Biol* 25, 575–589 (2005).
89. Bond, S. & Forgac, M. The Ras/cAMP/protein kinase A pathway regulates glucose-dependent assembly of the vacuolar (H+)-ATPase in yeast. *J Biol Chem* 283, 36513–36521 (2008).
90. Trombetta, E.S., Ebersold, M., Garrett, W., Pypaert, M. & Mellman, I. Activation of lysosomal function during dendritic cell maturation. *Science* 299, 1400–1403 (2003).
91. Xu, Y., Parmar, A., Roux, E., Balbis, A., Dumas, V., Chevalier, S., & Posner, B. I. Epidermal growth factor-induced vacuolar (H+)-ATPase assembly: a role in signaling via mTORC1 actvation. *J Biol Chem* 287, 26409–26422 (2012).
92. Veikkolainen, V., et al. ErbB4 modulates tubular cell polarity and lumen diameter during kidney development. *J Am Soc Nephorl* 23, 112–122 (2012).
93. Zeng, F., Zhang, M. Z., Singh, A. B., Zent, R., and Harris, R.C. ErbB4 isoforms selectivity regilate growth factor induced Madin-Darby canine kidney cell tubulogenesis. *Mol Biol Cell* 18, 4446–4456 (2007).
94. Sandholm, N., et al. New susceptibility Loci associated with kidney disease in type 1 diabetes. *PLoS Genet* 8, e1002921 (2012).
95. Boger, C. A., and Sedor, J. R. GWAS of Diabetic Nephropathy: Is the GENIE out of the Bottle? *PLoS Genet* 8, e1002989 (2012).

96. Dastani, Z., et al. Novel loci for adiponectin levels and their influence on type 2 diabetes and metabolic traits: A multi-ethnic meta-analysis of 45,981 indifiduals. *PLoS Genet* 8, e1002607 (2012).
97. Merkulova, M., Bakulina, A., Thaker, Y.R., Grüber, G. & Marshansky, V. Specific motifs of the V-ATPase a2-subunit isoform interact with catalytic and regulatory domains of ARNO. *Biochim Biophys Acta* 1797, 1398–1409 (2010).
98. Hosokawa, H., et al. The N termini of a-subunit isoforms are involved in signaling between vacuolar H^+-ATPase (V-ATPase) and cytohesin-2. *J Biol Chem* 288, 5896–5913 (2013).
99. Marshansky, V., et al. V-ATPase is a novel evolutionarily conserved cytohesin-singaling receptor, "*Gordon Research Conference: Bioenergetics*" (Proctor Academy, Andover, NH) Abstract of the Poster Presentation and Invited Talk (2013).
100. Marshansky, V., Hosokawa, H., Zhuang, Z., Randazzo, P. A., Grüber, G., Ausiello, D. A. novel role of cytohesin-2 in regulation of macropinocytosis pathway and proliferation. "*Experimental Biology Meeting*" (Boston, MA) Abstract of Poster Presentation and Invited Talk, p. 54 (2013).
101. Marshansky, V., Rubinstein, J. L., Grüber, G. Eukaryotic V-ATPase: Novel structure findings and functional insights. *Biochim Biophys Acta - Bioenergetics*, In Press (2014).

Chapter 9

Kidney Vacuolar H^+-ATPase Regulation

Patricia Valles[a] **and Daniel Batlle**[b]

[a]*Area de Fisiopatología, Departamento de Patología, Facultad de Ciencias Médicas, Universidad Nacional de Cuyo, Mendoza, Argentina*
[b]*Division of Nephrology/Hypertension, Northwestern University Feinberg School of Medicine, Chicago, Illinois, USA*
d-batlle@northwestern.edu

Vacuolar H^+-ATPases are ubiquitous multisubunit complexes mediating the ATP-dependent transport of protons. V-ATPases are highly expressed in the plasma membranes of some epithelial cells in the kidney. They are also located on the limiting membranes of intracellular organelles in the degradative and secretory pathways of all cells.

V-ATPases are composed of a peripheral V_1 domain containing eight different subunits that are responsible for ATP hydrolysis. They are also composed of an integral V_0 domain containing six different subunits that translocate protons. In mammalian cells, most of the V-ATPase subunits exist in multiple isoforms which are often expressed in a tissue-specific manner. Isoforms of one of the V_0 subunits (subunit a) have been shown to possess information that targets the V-ATPase to distinct cellular destinations.

Handbook of H^+-ATPases
Edited by Suguru Nakamura

ISBN 978-981-4411-91-2 (Hardcover), 978-981-4411-92-9 (eBook)
www.panstanford.com

This chapter focuses on the function and regulation of renal vacuolar H^+-ATPases. The localization of vacuolar H^+-ATPases in the kidney, and their role in intracellular pH (pHi) regulation, transepithelial proton transport, and acid–base homeostasis will be discussed.

9.1 Introduction

Vacuolar H^+-ATPases are ubiquitous multisubunit complexes mediating ATP-driven vectorial transport of protons across membranes [1–5]. They are expressed in virtually all eukaryotic cells in intracellular membranes or in the plasma membrane of specialized cells. V-ATPases couple hydrolysis of cytosolic ATP to proton transport out of the cytosol, resulting in acidification of intracellular compartments such as the vacuole/lysosome, the endosome, and the Golgi apparatus and energization of the plasma membrane in a variety of cells. V-ATPases have adapted to a remarkably wide variety of physiological roles ranging from the control of cytosolic pH homeostasis to driving secondary transport of ions and nutrients to regulating protein sorting [1–3].

Vacuolar acidification is essential for a variety of cellular processes [4, 5] including ligand-receptor dissociation and receptor recycling following receptor-mediated endocytosis, intracellular targeting of newly synthesized lysosomal enzymes, protein processing and degradation, and coupled transport of small molecules, such as neurotransmitters. Acidification of endosomal compartments is also necessary for budding of transport vesicles [6, 7] and infection by certain envelope viruses [8]. However, the V-ATPase is also expressed at very high density in the plasma membranes of several specialized cells that are involved in extracellular pH regulation, *via* proton transport, in different organs including the kidney, the inner ear, the epididymis, and bone [6, 9, 10].

Acidification of the cellular environment is intricately linked to specialized cell function as exemplified in osteoclasts where protons generated by the vacuolar H^+-ATPases are used to dissolve bone matrix, or, macrophages where acidic pH is involved in killing and digesting neighboring cells or pathogens [11]. In other cells, vacuolar H^+-ATPases mediate the regulation of extracellular pH of closed compartments such as in the inner ear and endolymph fluid [12]. Perhaps the most evident function of vacuolar H^+-ATPases is in the kidney epithelia, where their role in acid–base transport and transepithelial transport is crucial for many physiologic processes.

This review will address the function, regulation, and role of vacuolar H^+-ATPases on renal physiology. Monogenic defects in two subunits (ATP6V0A4, ATP6V1B1) of the vacuolar H^+-ATPase causing distal renal tubular acidosis in humans and the potential involvement of this pump in other disease states will be discussed elsewhere in this book.

9.2 Structure and Molecular Organization of Vacuolar H^+-ATPases

Vacuolar H^+-ATPases belong to the large super family of ATPases which is subdivided into three major subclasses: (1) P-type ATPases such as Na^+/K^+-ATPases, Ca^{2+}-ATPases, and H^+/K^+-ATPases; (2) mitochondrial F_1F_0-ATPases; and (3) V-type (vacuolar) H^+-ATPases [13].

Mitochondrial F_1F_0-ATPases and vacuolar H^+-ATPases share many structural features in their subunit composition such as amino acid sequences and subunit arrangements [14]. Functionally, however, they are distinguished by the fact that F_1F_0-ATPases use a proton gradient for ATP-synthesis, whereas vacuolar H^+-ATPases use ATP-hydrolysis to generate a proton gradient [10].

The vacuolar H^+-ATPases are hetero-oligomeric complexes composed of 13 polypeptide types. These can be fractionated into a soluble cytoplasmically disposed V_1 domain of 640 kDa, and a membrane-associated V_0 domain of 240 kDa, together forming a protein complex of approximately 900 kDa (Fig. 9.1). Both domains are connected through a stalk-like structure which belongs to the V_1 domain [2]. The peripheral domain (V_1) hydrolyzes ATP and the integral domain (V_0) conducts protons. Dissociation of V_1 and V_0 is an important *in vivo* mechanism of controlling vacuolar ATPase activity as described in detail later in this chapter.

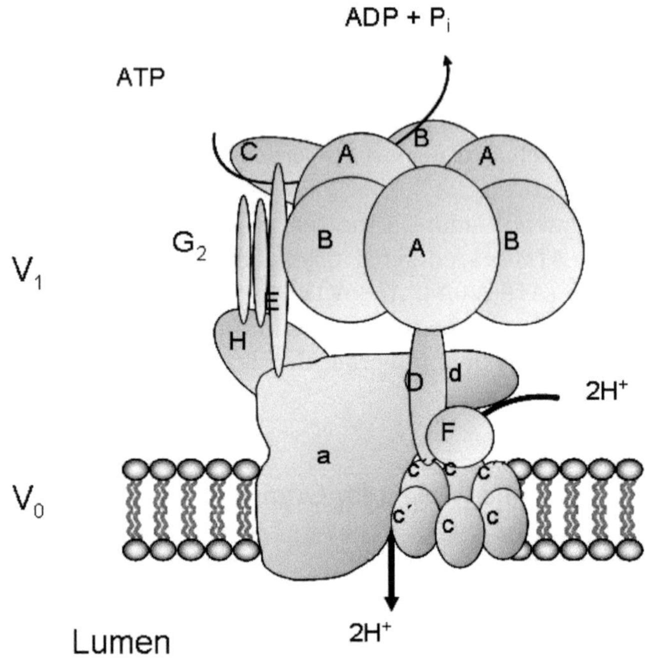

Figure 9.1 Structural model of Vacuolar H^+-ATPase. Vacuolar H^+-ATPase is a multisubunit complex composed of a peripherical cytosolic domain V_1 and an integral domain V_0. The cytosolic V_1 domain consists of eight subunits termed A–H. The membrane-bound V_0 domain is composed of the subunits a–d with several isoforms of the c subunits. Reprinted from *FEBS Lett*, 545, Kawasaki-Nishi, S., Nishi, T., Forgac, M., Proton translocation driven by ATP hydrolysis in V-ATPases, pp. 76–85, 2003, with permission from Elsevier.

The names of the respective subunits are given according to the recently revised HUGO (Human Genome Organization) nomenclature [15]. The subunit names consist of the family name (ATP6) followed by the domain name (V_1 or V_0) and then the specific polypeptide designation (e.g., *ATP6V1B1*). Thus, ATP6 designates all subunits of the vacuolar H^+-ATPase as a subfamily of the super family of ATPases. V_1 delineates subunits belonging to the peripheral, catalytic V_1 domain, whereas V_0 subunits belong to the membrane bound V_0 domain. B1 designates the specific polypeptide subunit.

9.2.1 *The Cytosolic V_1 Domain*

The cytosolic V_1 domain of the vacuolar H^+-ATPase consists of eight different subunits termed A–H. The major bulk of the V_1 domain is composed of three copies of subunit A that alternate with copies of subunit B to form the ATPase catalytic site. This hexameric structure can be easily seen in conventional electron micrographs in the form of dense, stud-like projections that emanate from the cytosolic side of the plasma membrane of proton-secreting cells [16]. ATP hydrolysis occurs at catalytic sites located at the interface of the A- and B-subunits [17, 18], which are each present in three copies per complex and are arranged in alternating fashion in a ring. Most of the catalytic site residues are contributed by the A-subunit [19]. A second set of nucleotide binding sites is located at the other A/B-subunit interface (termed "noncatalytic" sites), which are composed primarily of B-subunit residues and may function to regulate activity [20, 21]. The A- and B-subunits exhibit a high homology with the α- and β-subunits of the F_1F_0-ATPase. Only one isoform of the A-subunit has been identified. Two highly homologous B-subunits (ATP6V1B1 and ATP6V1B2) with a tissue- and cell-specific expression pattern exist in many species [5, 21–25]. The sequence comparisons demonstrate that the vacuolar H^+-ATPase B-subunits share approximately 20–25% amino acid identity not only with the vacuolar H^+-ATPase A-subunit, but also with the α- and β-subunits of the mitochondrial F_1F_0-ATPase, suggesting that all four of these proteins evolved from a common ancestral nucleotide-binding protein. Mutations of conserved residues in these sequences in the A-subunit lead to significant losses of activity [18, 19], as previously observed for the β-subunit of the F_1F_0-ATPase. Inhibition of vacuolar H^+-ATPase activity also occurred after modification of a single noncatalytic site of the β-subunit by a photo-activated, nonhydrolysable nucleotide analog [26]. This raises the possibility that the nucleotide-binding site of the B-subunit may participate in the regulation of H^+-ATPase activity. Comparison of the B1- and B2-subunit isoform amino acid sequences reveals that the central 469 amino acids are highly conserved, while the 20–25 amino acids at both the amino- and carboxy-termini have diverged greatly, thus raising the possibility that these terminal regions provide specialized isoform-specific functions of the B-subunit [27]. The B1-subunit of the vacuolar H^+-ATPase has a C-terminal DTAL amino acid motif typical of PDZ-binding proteins that associate with the PDZ protein, NHE-RF (Na^+/H^+ exchanger regulatory factor). The PDZ domain family of plasma membrane proteins and associated PDZ-binding proteins were named after the initial three members of the family (PSD-95, *Drosophila* discs large protein, and ZO-1). PDZ proteins contain 80–100 amino acid stretches that allow them to interact with other proteins that have a four amino acid PDZ-binding cassette, usually at the extreme C terminus of the cytoplasmic domain [28]. This B1 isoform is amplified in renal intercalated cells [29]. In contrast, proximal tubules express the B2 isoform that lacks the C-terminal PDZ-binding motif. Alterations in the PDZ-binding motif could potentially affect insertion and function of the vacuolar H^+-ATPase, although none of the *ATP6V1B1* missense mutations identified to date in distal renal tubular acidosis (dRTA) kindreds are located in the carboxy-terminal region of the protein [30].

The remaining V_1 subunits are distributed between one of two types of stalks—peripheral and central—that connect the V_1 and V_0 domains. These stalks have distinct functions in the rotary mechanism by which the V-ATPases couple ATP hydrolysis to proton transport. The central stalk serves as a rotor that couples the energy that is released from the hydrolysis of ATP to the rotation of a ring of proteolipid subunits in V_0. The peripheral stalks serve to prevent rotation of the A_3B_3 head during ATP hydrolysis and therefore serve a stator function [5].

9.2.2 *The Membrane-Associated V_0 Domain*

The V_0 domain is a 260 kDa integral complex composed of five different subunits (subunits a, d, c, c′, and c″ with molecular masses of 100–117 kDa) that function in proton translocation [23]. All the V_0 subunits are integral membrane proteins except for subunit d, which is a peripheral membrane protein and is tightly associated with the V_0 subcomplex [25].

The V_0 subunits appear to be present in a stoichiometry of four or five copies of subunit c and single copies of the remaining subunits. The proteolipid subunits (c, c′, and c″) are highly hydrophobic proteins that are arranged to form a ring (31). Each proteolipid subunit contains a single buried glutamic acid (Glu) residue located in transmembrane helix-4 (TM4) of subunits c and c′ and TM3 of subunit c″ (31). These Glu residues undergo reversible protonation during proton transport, and also crucial for proton transport is subunit a, a 100-kDa protein that has a hydrophilic *N*-terminal domain that is oriented toward the cytoplasmic surface and a C-terminal domain that contains eight or nine transmembrane helices [6]. Mutagenesis studies suggest that, like the a subunit of the F-ATPases [32], the V-ATPase a-subunit contains several buried charged residues that participate in proton translocation [32, 33]. Although not absolutely required for proton transport for either the V- [33] or F-ATPases [34], these residues appear to be in a position to influence proton movement, as reflected in the pH dependence of activity [33]. Thus, the a-subunit may form water-filled channels that allow protons to gain access to the buried carboxyl group of subunit c and to exit this site to the opposite aqueous compartment, as has been suggested for the F-ATPase a-subunit [32]. Evidence has also been obtained suggesting that the 100-kDa subunit possesses the binding site for the highly specific V-ATPase inhibitor bafilomycin [35].

The a-subunit has recently received special attention due to the fact that at least two human diseases are caused by mutations in two isoforms of this subunit (*ATP6V0A3* and *ATP6V0A4*). In humans, the four a-subunit isoforms display 47–61% identity at the amino acid level [36, 37]. Mutations in the *ATP6V0A3* gene result in one type of infantile malignant autosomal recessive osteoporosis [38]. The a3 isoform has also been localized to insulin-containing secretory vesicles in pancreatic islet cells, where it likely contributes to the acidic environment required for proteolytic processing of insulin. Interestingly, mice with genetic ablation of the a3 gene are still able to process insulin but show a defect in insulin secretion [39]. Immunolocalization studies of islet cells from these mice suggest that the a2 isoform may compensate for the loss of a3 in acidifying the insulin-containing compartment, but a2 appears unable to replace a3 in its role in insulin secretion. The a3 isoform is also expressed in the adrenal, parathyroid, thyroid, and pituitary glands [6], and in immature melanosomes, where it functions to keep this organelle acidic [40].

The a4 isoform is expressed almost exclusively in the apical membrane of alpha-intercalated cells of the late distal tubule and collecting duct [10]. Loss-of-function mutations in the *ATP6V0A4* gene cause autosomal recessive distal renal tubular acidosis [36]. *ATP6V0A4* (a4) is only expressed in the kidney and epididymis [36, 37]. Mutations in a4 cause the defective urinary acid secretion associated with renal tubular acidosis [36]. Interestingly, a4 is also involved in targeting of V-ATPases to the apical membrane of epididymal clear cells, which are responsible for acidification of the epididymal lumen [41].

The a1 isoform has been shown to localize to both the presynaptic membrane and synaptic vesicles present at nerve terminals [42]. V-ATPase complexes containing a1 likely function to acidify synaptic vesicles, but V_0 complexes containing a1 present in both the synaptic vesicle and the synaptic plasma membrane have been suggested to play a more direct role infusion between these membranes during neurotransmitter release [43]. The a1-subunit has also been implicated in fusion between phagosomes and lysosomes during phagocytosis in the brain [44]. By applying live pH indicators, it has been shown that the a1-subunit mediates fusion between phagosomes and lysosomes during phagocytosis, a function that is independent of its proton pump activity [45].

The a2 isoform has been identified in apical endosomes of proximal tubule cells of the kidney, where it functions to provide the low pH necessary to release endocytosed peptides from the receptors megalin and cubilin that are involved in their absorptive uptake from the tubular fluid [6, 46]. The transmembrane topology of the a-subunits of the vacuolar H^+-ATPase is, however, still a controversial issue and additional structural/crystallography studies are needed to clarify its transmembrane arrangement [22, 47].

9.2.3 *The Stalk-Subunit Arrangement in the Stalk Regions*

The A- and B-subunits of the V_1 domain are connected to the V_0 domain through several protein subunits termed the "stalk" which also belongs to the V_1 domain. The subunits C, E, G, H, and the *N*-terminal domain of subunit a all form part of a peripheral stalk, whereas subunits D and F form the central stalk [48]. The subunit

F might have a regulatory role in controlling the coupling of ATP hydrolysis and proton transport, which might be facilitated by the two conformational states. Subunits C and H both appear to be located near the interface of the V_1 and V_0 domains, although functional studies indicate that both proteins are important in modulating the interaction between V_1 and V_0 [4, 6]. High-resolution crystal structures of subunits C and H reveal that each are composed of two domains [49]. Domain analysis of subunit H reveals that the *N*-terminal domain is required for the activation of ATP hydrolysis, whereas the C-terminal domain is involved in the coupling of proton transport and ATP hydrolysis domain, analysis of subunit E has demonstrated that the *N*-terminal domain of this subunit can interact with both subunit H and subunit C [6]. Analysis of the crystal structure of the C subunit provides further evidence that it is part of the stalk and may help to attach the central stalk to the V_0 domain [49]. This subunit C is of particular interest because of its putative role in regulating dissociation of the V-ATPase complex *in vivo* [50, 51]. Dissociation of V_1 and V_0 represents an important mechanism of controlling vacuolar H^+-ATPase activity in cells [51, 52]. By interacting with key subunits of the peripheral stalk (including subunits E, G, and a), subunit C may be able to disrupt interactions between these subunits that would otherwise lead to a stably assembled peripheral stalk [53]. In addition to true vacuolar H^+-ATPase subunits, associated proteins have been identified. Some of these proteins are expressed in a highly tissue-specific manner, suggesting that these proteins may be involved in tissue-specific functions of the vacuolar H^+-ATPase or may be involved in the targeting of the pump to specific intracellular structures [10, 54].

9.3 Proton Transport by the V-ATPases

Similar to the homologous family of F_1F_0 ATP synthases, the V-ATPases operate by a rotary mechanism [55]. In V_1V_0, hydrolysis of ATP in the V_1 domain causes rotation of the entire rotary assembly, which includes subunits D, F, d, and the ring of proteolipid subunits (c, c′, c″). Rotation of the proteolipid ring relative to subunit a causes active transport of protons [56]. Subunit a provides for entry and exit of protons to the protonatable carboxyl groups of Glu residues of the proteolipid ring by aqueous hemi-channels, as previously proposed for the ATP synthase. The rotation of the protonatable sites past the cytoplasmic hemi-channel forces their protonation as they are subsequently placed in contact with the hydrophobic environment. On reaching the luminal hemi-channel, deprotonation of the buried carboxyl groups is induced by their interaction with the crucial Arg residue of subunit a that stabilizes the carboxyl group in its charged form, thus releasing the proton into the luminal hemi-channel [57]. Evidence for close interaction between TM7 of subunit a containing the critical arginine residue and TM4 of subunit c′ containing the essential glutamic acid residue has come from zero-length cross-linking studies between engineered cysteine residues in each protein [58]. Both TM3 and TM5 of subunit c″ contain buried glutamic acid residues, but only that in TM3 is essential for proton translocation. Cross-linking of subunit a to both subunits c′ and c″ shows evidence for helical swiveling (that is, rotation about an axis through the center of the helix) in both subunit a and the proteolipid subunits [58, 59]. That is, zero-length cross-linking can be observed between residues that would require considerable radial mobility in both helices. It is tempting to speculate that such helical swiveling in subunit a may help to control proton accessibility through the postulated hemi-channels that participate in proton translocation. The specific and highly potent inhibitors bafilomycin and concanamycin have been shown to bind at helical interfaces within the proteolipid ring [60] and it is possible that they inhibit activity by preventing helical swiveling within the ring. Subunit a may also participate to a lesser degree in conferring sensitivity to these inhibitors [61]. As there are 3 catalytic nucleotide-binding sites in V_1 and 6–10 protonatable sites on the proteolipid ring of V_0, the predicted H^+/ATP stoichiometry for the V-ATPases is 2–3.3. This is in reasonable agreement with the measured value [57].

9.4 Kidney-Specific Subunits of the Vacuolar H^+-ATPase

The ubiquitous and intercalated cell-specific isoforms of the vacuolar H^+-ATPase subunits are differentially expressed in mature kidney and during mouse nephrogenesis. Ubiquitous A, B2, E1, G1, and C1 show an early and stable expression throughout nephrogenesis, followed by a slight increase around birth. This expression pattern contrasts with the later and progressive expression of the intercalated cell-specific isoforms B1, G3, C2, and a4. Subcellular fractionation studies indicated a preferential location of both ubiquitous and intercalated cell-specific subunits to endosomes in both developing and mature kidneys [62].

Little is known about the expression patterns of these intercalated cell-specific isoforms in the kidney with the exception of the B1 isoform (*ATP6V1B1*) and a4-subunit (*ATP6V0A4*). Mutations in *ATP6V1B1* and *ATP6V0A4* genes, encoding intercalated cell-specific B1- and a4-subunits, respectively, have been associated with early-onset cases of distal renal tubular acidosis (dRTA) [30, 63, 64].

9.5 Distribution and Role of the Vacuolar H^+-ATPase in the Kidney

The kidney plays a major role in the regulation of acid–base homeostasis. This is achieved by two basic processes: (a) the reabsorption of bicarbonate along the nephron and (b) the secretion of H^+ and the excretion of ammonium and phosphate and other titratable acids [65]. The vacuolar H^+-ATPase is the main regulating protein involved in this process [66].

9.5.1 *Proximal Tubule*

A bafilomycin and *N*-ethylmaleimide (NEM)-sensitive vacuolar H^+-ATPase is located in the apical membrane and in intracellular organelles of proximal tubule epithelial cells [67]. Evidence from this includes numerous functional studies using intact tubules and isolated brush border membrane vesicles as well as immunocytochemical analysis using specific antibodies raised against various subunits of the vacuolar H^+-ATPase. Specifically, vacuolar H^+-ATPase is present in apical invaginations at the base of the microvilli, distinct from clathrin-coated regions [68]. The presence of the 31 kDa E-subunit of the vacuolar H^+-ATPase along with the "brain" or B2 (*ATP6V1B2*) isoform of the 56 kDa subunit has been demonstrated in rat proximal tubule segments. The B2-subunit of the vacuolar H^+-ATPase is expressed in the kidney proximal convoluted tubule as well as in the thick ascending limb (TAL), distal convoluted tubule (DCT), connecting segment and in both α- and β-type intercalated cells of the collecting ducts [69, 70]. Under baseline conditions, the B2 isoform was found predominantly on intracellular vesicles. However, α-intercalated cells show high levels of B2-subunit expression in the apical membrane, where it co-localizes with E and B1 isoforms. These findings indicate that in addition to its role in the acidification of intracellular organelles, the B2 isoform could contribute to transepithelial proton secretion [70]. The B2-subunit isoform of the vacuolar H-ATPase has a truncated carboxy-terminal domain that lacks the PDZ-binding domain found in the B1 (intercalated cell) isoform [29]. This subunit does not interact with the PDZ protein, NHERF1 in proximal tubules, and the localization of these proteins in the apical region of proximal tubule epithelial cells is distinct [28]. While NHERF1 is expressed mainly in the brush-border microvilli (where it co-localizes with NHE-3), the vacuolar H^+-ATPase is found in the submicrovillar region. At the light microscope level, the proximal tubule vacuolar H^+-ATPase localizes at the base of the brush border, which is the membrane domain showing a high level of clathrin-mediated endocytotic activity [71].

High-resolution immunogold labeling revealed that the proton pumps are not concentrated in the clathrin-coated domains of the apical membrane but are located at the neck of the apical invaginations, between clathrin-coated domains and the microvilli [69]. However, endosomes isolated from renal cortex, and which are derived mainly from proximal tubules, show ATP-dependent acidification *in vitro*. This acidification is inhibited by bafilomycin and confirms the role of the vacuolar H^+-ATPase in acidifying proximal tubule endosomes.

Recent work from Hurtado-Lorenzo *et al.* has identified a novel role for transmembrane V-ATPase subunits as endosomal pH sensors that are involved in the recruitment of cytosolic downstream trafficking proteins to acidified vesicles, ensuring their appropriate and correct progression along the degradative pathway from early endosomes to lysosomes [46]. Thus, the V-ATPase not only generates intravesicular acidification but also senses and responds to the acidic pH by mobilizing vesicle coat components, which are critical for vesicle trafficking, from the cytosol to the endosomal membrane.

The proximal tubule reabsorbs 70–80% of the filtered bicarbonate which occurs mainly in the initial segments. Two main transport systems mediate H^+-secretion on the apical side, the first step in bicarbonate reabsorption, Na^+/H^+ exchange, and H^+ secretion via the vacuolar H^+ ATPase (Fig. 9.2A). Several isoforms of Na^+/H^+-exchangers, localized in the brush border membrane of the proximal tubule, including NHE-2, NHE-3, and NHE-8 are involved in proton secretion. About 50% of the overall apical NHE activity may be mediated by NHE-3, the remainder by another isoform(s) [72]. Up to 40% of proximal tubule bicarbonate reabsorption is Na^+ independent. It is sensitive to the vacuolar H^+-ATPase inhibitor, bafilomycin, and it is mediated by vacuolar H^+-ATPases expressed in the brush border membrane.

Secreted H^+ ions combine with filtered HCO_3^- to produce H_2O and CO_2, a process catalyzed by the membrane-bound carbonic anhydrase IV. CO_2 then diffuses into the proximal tubule cells where it reacts with H_2O (catalyzed by the cytosolic carbonic anhydrase II isoform) to form H^+ and HCO_3^-. The generated HCO_3^- is exported into blood via the basolateral electrogenic Na^+/HCO_3^- (kNBC-1, SLC4A4), whereas the proton is recycled back into the tubule lumen by the Na^+/H^+-exchangers and vacuolar H^+-ATPase where it can react with another HCO_3^- ion [10] (Fig. 9.2A).

9.5.2 *Loop of Henle*

The TAL reabsorbs 15–20% of filtered bicarbonate. Proton secretion and bicarbonate transport in the TAL of Henle's loop are mediated by both the vacuolar H^+-ATPase and other transporters including the Na^+/H^+-exchanger, NHE-3 [73]. As in the proximal tubule, the apical mechanism responsible for this transport is predominantly via the Na^+-H^+-exchanger. Both the electroneutral Na^+-HCO_3^- co-transporter, NBC(N)1, and electrogenic Na^+-HCO_3^- co-transporter, NBC4, are expressed in the TAL. The Cl^-/HCO_3^- exchanger, AE2,

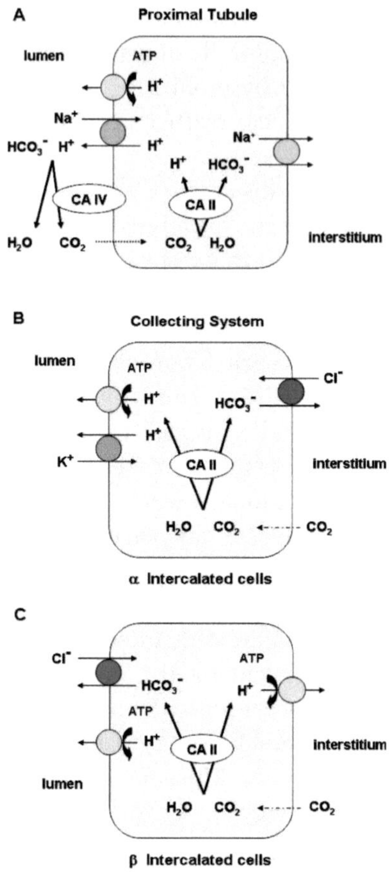

Figure 9.2 Schematic model of transport processes involved in renal bicarbonate absorption and proton secretion. (A) In the proximal tubule, protons are secreted via apical Na^+/H^+ exchangers and vacuolar H^+-ATPases. The secreted H^+ combine with filtered HCO_3^- under the influence of a membrane-bound carbonic anhydrase (CA IV) to form H_2O and CO_2. (B) Type A intercalated cells are characterized by the expression of the basolateral Cl^-/HCO_3^- exchanger AE-1 and the presence of an apical vacuolar H^+-ATPase. H^+ and HCO_3^- are formed by a cytosolic CA II and secreted into the lumen and the interstitium, respectively. Type-A intercalated cells may also express an apical H^+-K^+-ATPase. (C) Type-B intercalated cells are characterized by the absence of the AE-1 Cl^-/HCO_3^- exchanger on the basolateral side and the presence of the vacuolar H^+-ATPase which can be found on both sides of the cell. In addition, type-B intercalated cells express an apical Cl^-/HCO_3^- exchanger, pendrin. Adapted from Ref. [10].

has also been localized in the basolateral membrane of the TAL in rat and mouse kidney [74] and provides a route for bicarbonate reabsorption in this membrane domain. Vacuolar H^+-ATPase in this segment is located in numerous cytoplasmic vesicles, which are concentrated at the apical pole of the cell. Immunogold electron microscopy has clearly shown that a part of the vacuolar H^+-ATPase is located in the apical plasma membrane, and another part of it is associated with subapical vesicles [74]. In the thin limbs of Henle, plasma membrane vacuolar H^+-ATPase was detected in the initial portion of the thin descending limb, immediately after the S3 segment of the proximal tubule. Both apical and basolateral membranes were labeled. A bafilomycin-sensitive vacuolar H^+-ATPase has also been detected by functional measurements in this initial segment [75]. No other segments of long or short thin limbs have been reported to contain the vacuolar H^+-ATPase.

9.5.3 *Cortical Collecting Duct*

Intercalated cells of the collecting duct express the highest levels of vacuolar H^+-ATPases among all acid–base transporting cells in the kidney. The subcellular localization of the vacuolar H^+- ATPase within an intercalated cell subtype is dependent on whether the cell secretes or absorbs net H^+ equivalents. α-type intercalated cells mediate secretion of net H^+ equivalents through an apical H^+-ATPase, which functions in series with the basolateral Cl^-/HCO_3^- exchanger, AE1 (Fig. 9.3A, B). B-type intercalated cells (B-IC), however, are identified as a population of V-ATPase-rich IC in the cortical collecting duct. These cells do not express basolateral AE1 [10, 76] but express the V-ATPase in various locations, including on the basolateral plasma membrane while expressing anion exchanger, pendrin, in the apical membrane [77]. In β-type intercalated cells, the apical Cl^-/HCO_3^- exchanger, pendrin, functions in series with a basolateral H^+-ATPase to mediate secretion of OH^- equivalents, particularly during metabolic alkalosis [78]. B-IC but not A-IC also expresses the PDZ-binding protein NHERF1 (sodium/hydrogen exchanger regulatory factor), which interacts directly with the C-terminus of the B1 isoform of the V-ATPase B-subunit [28]. In addition, while many B-IC have either a basolateral, diffuse

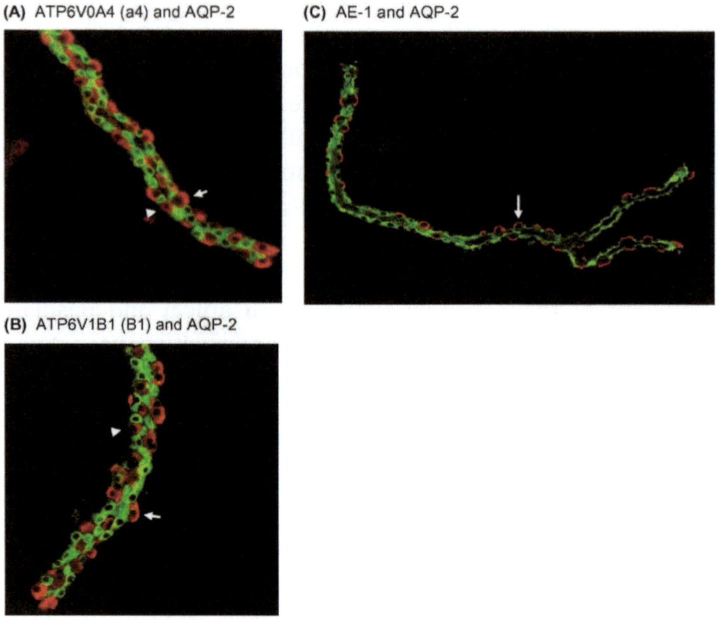

Figure 9.3 Localization of acid–base transporting proteins in cortical and medullary collecting ducts. Isolated mouse cortical and medullary collecting ducts were labeled with antibodies against the principal cell-specific water channel aquaporin-2 (AQP-2, *green*), and then co-labeled with antibodies against the *ATP6V0A4 (a4)* or *ATP6V1B1 (B1)* subunits of the vacuolar H^+-ATPase, or against anion exchanger-1 (*AE-1*) or against pendrin, and viewed with confocal microscopy. (A, B) Expression of both vacuolar H^+-ATPase subunits in cortical collecting duct fragments was restricted to cells negative for AQP-2, thus representing intercalated cells. In many cells expression of the vacuolar H^+-ATPase subunits was basolateral (*arrow*) or diffuse. Some cells also showed apical staining of the vacuolar H^+-ATPase (*arrow head*). (C) Localization of the Cl^-/HCO_3^- exchanger, AE1, in cortical and medullary collecting duct. The expression of AE1 is restricted to the basolateral side of intercalated cells, thus marking these cells as α-intercalated cells. Reprinted from *Semin Nephrol*, 26(5), Valles P., Lapointe M. S., Wysocki J., Batlle D., Kidney vacuolar H^+-ATPase: physiology and regulation, 361–374, Copyright (2006), with permission from Elsevier.

cytoplasmic or even bipolar distribution of the V-ATPase, some B-IC have predominantly apical V-ATPase, giving rise to a population of IC that look like A-IC with respect to the V-ATPase localization but have no basolateral AE1 [76, 77]. These cells, which can have apical pendrin, are similar to the so-called non-A and non-B cells [78]. All intercalated cells contain a large amount of cytoplasmic carbonic anhydrase (CAII), which is also a key protein in supporting the proton transporting capacity of these cells by providing membrane-associated pumps and transporters with a supply of protons and bicarbonate. The 56 kDa B1-subunit of the vacuolar H^+-ATPase, which contains a C-terminal PDZ-binding motif (DTAL) that interacts with NHERF1sodium/hydrogen exchanger regulatory factor, is expressed in B intercalated cells but not in A intercalated cells. The specific role of NHERF1 in B-IC remains unclear [28]. Whether A-IC and B-IC are variants of the same cell type but that change polarity and expression of some membrane proteins or whether they represent two distinct cell types has long been discussed [79, 80]. The fact that the apical and basolateral anion exchange proteins are not the same (pendrin *vs.* AE1, respectively) clearly indicates that A-IC and B-IC are not simple mirror images of each other that are produced by transcytosis of V-ATPase and an anion exchanger into opposite poles of the cell [81, 82].

In the cortical collecting duct, all subtypes of intercalated cells are detectable. In the outer stripe of the outer medulla, α cells predominate but a few residual β cells can be found. In the inner stripe of the outer medulla, only α-intercalated cells are present, which represent about 40% of the epithelial cell population of the collecting duct. Principal cells make up the remaining 60% of the tubule epithelium. In the inner medulla, the epithelium initially contains between 5% and 10% α-intercalated cells. These cells disappear from the epithelium in the middle and terminal portions of the inner medullary collecting duct (IMCD) [69].

The late distal tubule, connecting segment, and cortical collecting duct, together reabsorb 5% of the filtered bicarbonate. In addition to the reabsorption of filtered bicarbonate, the kidney must also generate new bicarbonate to buffer approximately 70 mEq of acid produced by daily metabolism. In the H^+ secreting α-type intercalated cells, net excretion of protons occurs through apically localized vacuolar H^+-ATPases, where the H^+ is produced by the cytosolic carbonic anhydrase II (Fig. 9.2B). The HCO_3^- generated by the intracellular hydration of CO_2 exits the cell across the basolateral membrane by the kidney-specific isoform of the Cl^-/HCO_3^- exchanger AE-1 (SLC4A1) [83]. On the contrary, under conditions of bicarbonate secretion such as in metabolic alkalosis, βp-type intercalated cells are activated [84]. These are only found in the connecting segment and cortical collecting duct. β-type intercalated cells secrete bicarbonate into the tubular fluid via an apically located Cl^-/HCO_3^- exchanger, whereas basolaterally expressed vacuolar H^+-ATPases extrude protons into the interstitium (Fig. 9.2C). This apical Cl^-/HCO_3^- exchanger is insensitive to 4,4′-diisothiocyanostilbene-2,2′-disulfonic acid (DIDS) in the presence of Cl^- but becomes DIDS sensitive in its absence [85]. Pendrin localizes in the apical membrane of all non-α-intercalated cells and is regulated by acid–base status [86]. Genetic loss of pendrin function (PDS knock-out mice) leads to a reduced capacity of bicarbonate secretion in isolated perfused cortical collecting ducts [78]. Pendrin seems to have some sensitivity to DIDS [87].

9.5.4 *Medullary Collecting Duct*

The medullary collecting duct also has a high abundance of H^+ secreting cells but lacks principal cells. There are no HCO_3^- secreting cells in the medullary collecting duct. However, in addition to α-intercalated cells [76], the outer medullary collecting duct (OMCD) contains principal cells which are involved in water and Na^+ and K^+ transport. *In vitro* perfusion studies of isolated OMCDs demonstrated that proton secretion is mediated by vacuolar H^+-ATPases and bicarbonate absorption is Cl^--dependent from the bath side due to the presence of the basolateral Cl^-/HCO_3^- exchanger, AE-1 (band 3 protein) (Fig. 9.3C). The activity of the vacuolar H^+-ATPase in this segment is under the control of several hormones. α-intercalated cells likely have other physiological roles in addition to regulating acid–base balance. For example, intercalated cells secrete Cl^-, which may participate in the regulation of fluid and electrolyte balance [79]. In α-intercalated cells of the rat OMCD, basolateral Cl^- uptake occurs through the Cl^-/HCO_3^- exchanger AE-1, as well as the $Na^+/K^+/2Cl^-$ exchanger, NKCC1 [88]. Moreover, the Cl^-/HCO_3^- exchanger, Slc26a7, is expressed on the basolateral membrane of α-intercalated cells of OMCD and is greatly upregulated with water restriction *in vivo* [89].

An important feature of the medullary collecting duct is the high capacity for H^+ secretion that leads to the titration of phosphate, and ammonia which leads to the excretion of titratable acids and ammonium, respectively [90]. The buffering of protons is necessary to maintain a favorable proton gradient across the apical membrane in order to facilitate vacuolar H^+-ATPase activity. Without titratable acids, the proton gradient would exceed more than 3 pH units and reduce the efficiency of urinary acidification. Excretion of ammonium is a major regulatable component of net acid secretion and is thought to occur through secretion of H^+ in parallel with nonionic diffusion of NH_3 [91]. NH_3 and H^+ form NH_4^+, a nonpermeable ion, which is trapped in the collecting duct lumen. In addition to vacuolar H^+-ATPase, a H^+/K^+-ATPase is expressed in the OMCD [92].

The contribution of H^+/K^+-ATPases to proton secretion in this segment and thus to bicarbonate reabsorption is relatively minor compared to the vacuolar H^+-ATPase. H^+/K^+-ATPases may rather play an important role during systemic potassium depletion [93].

The IMCD is composed of two separate segments: the initial portion and the terminal region, which comprises approximately the last two-thirds of the IMCD [94]. From the initial third of the IMCD, intercalated cells gradually disappear, and only cells called "IMCD" are found in the terminal portion of this tubule segment. These cells express AQP-2 and are involved in the final concentration of urine [94].

9.6 Function of the Vacuolar H^+-ATPase: Chloride Dependence

Vacuolar H^+-ATPase is an electrogenic pump that contributes to the translocation of H^+-ions across the cell membrane, thus rendering the cell interior negative with respect to the exterior, and therefore leads to the generation of a transmembrane electrical gradient. The presence of a parallel Cl^- conductance at the cell plasma membranes provides an electric shunt that compensates for the positive charge transferred by the pump [95]. This, therefore, dissipates the electrical gradient and permits the continued pumping of H^+. The presence of an anion conductance was inferred from the higher rate of acidification of the organelles in the presence of chloride, or the generation of a higher transmembrane potential in its absence [95].

More recently, direct measurements of intraorganellar chloride during acidification alone or in parallel with monitoring intraorganellar acidification confirmed the direct relationship between both processes. An increase in endosomal chloride concentration was observed in parallel with endosomal pH acidification by near 1.5 pH units. Both processes were blocked by inhibition of vacuolar H^+-ATPase activity with bafilomycin [96]. In the kidney, vacuolar H^+-ATPase-dependent acidification is Cl^- dependent in endosomal fractions, and in brush border membrane vesicle [10]. In endocytic vesicles from rabbit proximal tubule, the chloride conductance was activated by phosphorylation through a cyclic AMP (cAMP)-dependent protein kinase [97].

Angiotensin II stimulates H^+-ATPase-dependent proton extrusion, via a process involving a Cl^--dependent insertion of vesicles into the brush border membrane in proximal tubule cells [98]. Moreover, in isolated rabbit S3 proximal tubules, the apical insertion of V-ATPase-containing vesicles was also shown to be dependent on chloride, being delayed in the absence of these anion [99].

In vivo microperfusion experiments in the late distal tubule showed a reduction of electrogenic H^+ secretion by the Cl^- channel blocker, 5-nitro-2-(3-phenylpropylamino)-benzoate (NPPB), indicating coupling between H^+-ATPase and Cl^- transport [100].

The mechanisms for the apparent differences in the requirement for chloride in the different nephron segments have remained elusive, and there is some controversy as to the molecular identity of the associated Cl^- conducting proteins for each segment, including ClC-5 and the cystic fibrosis transmembrane conductance regulator and CFTR [101]. In proximal tubule cells, the co-localization of proton pumps with ClC-5 chloride channels in apical endosomes is related to an X-chromosome-linked disease, Dent's disease characterized by low-molecular weight proteinuria, hypercalciuria, and kidney stones [102, 103]. This disease is caused by mutations of the *CLCN5* gene, predominantly expressed in kidney. This channel conducts outwardly rectifying chloride currents that represent a counter-ion flux to the electrogenic influx of protons into the endosomes, suggesting that ClC-5 is essential for renal endocytosis [103]. These endosomes form part of the receptor-mediated endocytic pathway that transports proteins, and therefore the lack of ClC-5 might impair endosomal acidification, compromising the reabsorption of filtered proteins and causing the proteinuria observed in Dent's disease [102].

In the collecting duct, the expression of ClC-5 co-localized with H^+-ATPase in intracellular vesicles and in the plasma membrane has been described in both β-intercalated cells and in acid-secreting α-intercalated cells [103]. ClC-5 is involved in the plasma membrane insertion and recycling of these vesicles. Mutations of ClC-5 may affect the expression of apical H^+-ATPase in α-intercalated cells and impair urinary acidification [104]. Thus, these data also suggest a role for ClC-5 in the proton secretion mechanism in these cells [104].

Another channel proposed to have a role in the regulation of V-ATPase is CFTR. This channel is expressed in renal endosomes, membranes of proximal and distal tubules, and intercalated cells of collecting ducts, especially in type β cells [105]. Defective acidification of intracellular organelles in cystic fibrosis suggest possible role for CFTR in organelle acidification dependent on V-type H^+-ATPase [106]. Recently, Tararthuch *et al.* demonstrated that cAMP stimulates the extrusion of H^+ in MDCK-C11 cells in the presence of Na^+-free solution [107]. This finding suggests a participation of chloride currents through CFTR channels stimulated by cAMP, increasing the activity of H^+-ATPase. In contrast, no changes in the acidification rates of several intracellular organelles were demonstrable after activation of chloride currents stimulated by cAMP [108]. Thus, the role of CFTR in the activity of H^+-ATPase has been controversial.

9.7 Endocytosis and Acidification of Intracellular Vesicles

Regulation of vacuolar H^+-ATPase function may occur at several levels, such as assembly or disassembly of V_1 and V_0 domains, trafficking and sorting to and from the membrane, interactions with other (activator or inhibitory) proteins, alteration of cells expressing vacuolar H^+-ATPases (polarity of intercalated cells), or by increased transcription/ translation of vacuolar H^+-ATPases.

There is ample *in vivo* and *in vitro* evidence for regulation of H^+-ATPase activity by trafficking/exocytosis [109, 110]. Likewise, H^+-ATPase-dependent acidification is required for exocytosis to occur. The secretory and endocytotic pathways in membrane-bound compartments of epithelial and nonepithelial cells are acidified via the action of a vacuolar H^+-ATPase [111].

Both immunocytochemical and functional studies have demonstrated the presence of proton pumps on endosomal membranes and parts of the Golgi/TGN [112, 113] as well as on lysosomal membranes [114]. In receptor-mediated endocytosis, a low pH within early endosomes triggers the dissociation of internalized ligand–receptor complexes (Fig. 9.4). This allows for recycling of receptors to the plasma membrane, which provides a mechanism for controlling both the rate of ligand uptake and the density of cell surface receptors [56]. For hormone and growth-factor receptors, receptor density can, in turn, control the sensitivity of cells to the ligand. A similar strategy is used for the targeting of newly synthesized lysosomal enzymes from the Golgi to lysosomes (Fig. 9.4). Acidification of late endosomes causes the release of lysosomal enzymes from the mannose 6-phosphate receptor and permits recycling of the receptor to the *trans*-Golgi network [5]. Neutralization of vesicle luminal pH may inhibit some steps of the intracellular trafficking pathway by preventing the recruitment of coat proteins that are required for vesicle formation and budding [115]. Many of the acidification-dependent steps involve vesicle fission and fusion via specialized COP (coat protein)-coated transport vesicles [116].

Kidney proximal tubule epithelial cells have an extensive apical endocytotic apparatus that is involved in the reabsorption of low-molecular weight proteins which traverse the glomerular filtration barrier. They are also involved in the extensive recycling of functionally important apical plasma membrane transporters. The Arf-nucleotide exchange factor, ARNO (ADP-ribosylation factor nucleotide site opener), Arf6, and Arf1, small GTPases, are part of

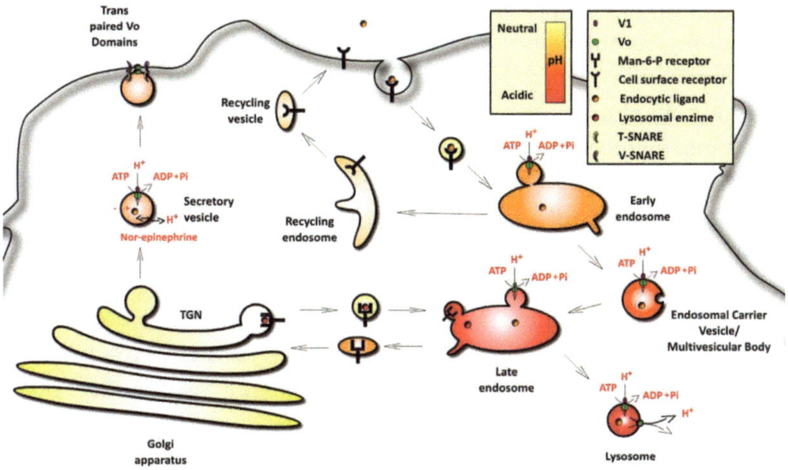

Figure 9.4 Role of intracellular V-ATPases in normal and disease processes. Role of intracellular V-ATPase in membrane trafficking, endocytosis, and secretion. Extracellular ligands are internalized by receptor mediated endocytosis and trafficked to the sorting endosome. Acidification of the endosome by the V-ATPase allows for release of the ligand and the recycling of the receptor back to the membrane. Budding of endosomal carrier vesicles and multivesicular bodies are also dependent on the acidic environment. Reprinted from Ref. [56], Copyright 2008, with permission from Elsevier.

the kidney proximal tubule receptor-mediated endocytosis pathway [117, 118], Arf6 and ARNO, but not Arf1, are recruited from the cytosol to proximal tubule endosomal membranes in response to a decrease in pH of the endosomal lumen [117].

9.8 Interaction of the Vacuolar H^+-ATPase with SNARE Proteins

Soluble *N*-ethylmaleimide-sensitive factor attachment protein receptor (SNARE) complexes are a class of vesicle-associated membrane proteins. SNARE complexes bridge opposing membrane bilayers that appear to mediate specific membrane fusion in the endomembrane system [119]. Although SNARE complexes appear to represent a universal membrane fusion machine, their role in determining the specificity of intracellular membrane fusion is still being established. Each SNARE complex characterized to date appears to consist of a thermostable parallel helix bundle composed of four heptad repeat-containing SNARE motifs [120]. These proteins, involved in the trafficking and fusion of synaptic vesicles, have been recently renamed as R-SNAREs (arginine-containing SNAREs) or Q-SNAREs (glutamine-containing SNAREs) [121]. More than 30 mammalian SNARE proteins have been identified and classified into three distinct groups, the syntaxin, VAMP, or SNAP-25 families, based on their sequence homology and domain structure [119]. Certain subunits of the vacuolar H^+-ATPase have the capacity to bind to SNARE proteins, suggesting that the SNARE machinery may participate in vacuolar H^+-ATPase trafficking. Clostridial toxins, which are specific SNARE proteases, inhibit acid-induced vacuolar H^+-ATPase exocytosis in cultured IMCD cells [122]. Subunit E of the vacuolar H^+-ATPase co-immunoprecipitates with α-SNAP, SNAP-23, syntaxin, and VAMP2. Interaction between syntaxin-1A and the vacuolar H^+-ATPase is also involved in the exocytosis of the pump to the apical membrane of IMCD cells [123]. While SNARE proteins have clearly been implicated in the attachment of vesicles to their target membranes, the subsequent process of membrane fusion still remains largely uncharacterized. The V_0 domain of the vacuolar H^+-ATPase, particularly subunit c, was proposed to be involved in the lipid bilayer mixing that occurs during membrane fusion [124].

In addition to SNARE proteins, the vacuolar H^+-ATPase has the capacity to interact with various other proteins indicating its role in a multitude of regulatory functions. The B1 (*ATP6V1B1*) subunit of the vacuolar H^+-ATPase possesses a COOH-terminal "DTAL" motif typical of a PDZ-interacting domain [29]. The association of the B1-subunit with the PDZ protein NHERF-1 in rat kidney β intercalated cells has been demonstrated. NHERF-1 co-localizes with the vacuolar H^+-ATPase in either the apical or the basolateral pole. NHERF-1 is not detectable in α-intercalated cells suggesting that interaction of the vacuolar H^+-ATPase with NHERF-1 might play a role in modulating the variable vacuolar H^+-ATPase polarity that characterizes the β cell phenotype [28]. While NHERF-1 is located in the apical brush border in the proximal tubule, the vacuolar H^+-ATPase is located in a distinct membrane domain at the base of the brush border consistent with the expression of the truncated B2-subunit isoform, lacking the DTAL motif, in this tubule segment [28, 81]. Association between the B1-subunit and another PDZ binding protein, the Na^+/bicarbonate transporter, NBC3, has also been demonstrated in kidney lysates [125]. Thus, PDZ binding may prove a mechanism for the functional binding of the vacuolar H^+-ATPase with other transport proteins in the kidney. Indirect interaction between the vacuolar H^+-ATPase and the actin cytoskeleton occurs via NHERF binding. In addition, the V_1 complex can directly bind actin. The amino termini of both the B1- and B2-subunits of the vacuolar H^+-ATPase contain high affinity F-actin binding sites [126]. A direct interaction of the V-ATPase subunits B1, B2, and C with actin has been demonstrated but the role played by actin in V-ATPase recycling remains unclear [81, 127].

9.9 Mechanisms of Regulation of Kidney Vacuolar H^+-ATPase Activity

The distribution and phenotypic expression of intercalated cells varies in the different regions of the connecting segment and collecting duct. Although all intercalated cells can modulate their cell surface expression of vacuolar H^+-ATPase by vesicle trafficking, the situation is especially complex in the cortical collecting duct and the connecting segment of the distal tubule. Intercalated cells of the collecting duct express high levels of the vacuolar H^+-ATPase in intracellular vesicles and/or on their plasma membrane [128]. The cortical collecting duct has two distinct functional subtypes of intercalated cells. The H^+-secreting or α type has an apical H^+-ATPase and basolateral Cl^-/HCO_3^- exchanger and exhibits vigorous apical endocytosis [82]. The HCO_3^- secreting or β type has its polarization reversed compared to α cells. That is, it has an apical Cl^-/HCO_3^- exchanger (pendrin) and a basolateral H^+-ATPase [128]. Also, α- and β-intercalated cells have been distinguished by a differential expression of other proteins, including NHERF-1 [28].

During metabolic acidosis, the number of β-intercalated cells is reduced, while that of α-intercalated cells increases without a change in the total number of intercalated cells. This suggests the conversion of one cell type to another, a process termed *plasticity of functional epithelial polarity* [129]. In the past few years, Al-Awqati *et al.* have addressed the question of A-IC and B-IC interconversion and provided evidence that this process involves a cell differentiation event that is mediated by a protein called hensin [130]. Hensin (later called the Mouse Genome Project, DMBT1) is a large multidomain protein expressed in most epithelial cells, but in the kidney it is found only in collecting ducts where it is expressed by both principal and intercalated cells [131]. This protein is secreted into the extracellular environment under certain conditions (e.g., acidosis), where it polymerizes and binds to cell surface integrins that, in turn, induce terminal differentiation signals resulting in transformation of B-IC into A-IC [130, 131]. In fact, bicarbonate-secreting intercalated cells in culture could be converted to proton-secreting cells by growing them on a matrix which contains hensin [132]. In this model, the HCO_3^- secreting β-intercalated cells removed Cl^-/HCO_3^- exchangers from the apical membrane and began to insert Cl^-/HCO_3^- exchangers on the basolateral membrane. This can occur within a few hours and requires protein synthesis. Cyclophilin, via its cis/trans prolyl isomerase activity, is required for this polymerization, and galectin-3 is needed to aggregate the protein and activation of integrins for the development of the hensin fiber [133]. An intact cytoskeletal network of actin and tubulin, as well as degradation and internalization of the apically located peanut lectin-binding protein, is also required in this process. By using collecting ducts incubated *in vitro*, it has been shown that anti-hensin antibodies applied to the basolateral bathing medium can inhibit the induction of acid secretion and bicarbonate reabsorption that normally occurs following incubation of the tubules in acidic medium [133]. Clearly, the change in phenotype between α- and β-intercalated cells involves more than a simple change in the polarity of membrane transporters. Rather it requires restructuring of the entire sorting machinery [134]. Recently, a conditional deletion of hensin in the intercalated cells using a Cre driven by the promoter of the B1 V-ATPase subunit was obtained [135]. The results of the conditional deletion of hensin demonstrated that all IC cells in the cortex were of the pendrin-expressing β form, and there were no α-IC in either cortex or medulla. In the medulla, the intercalated cells were present, but on ultra structure they resembled the cortical β-IC. However, unlike the cortical β-IC, they did not express pendrin nor was their cytoplasm as dense, and they had some vesicles underneath the apical membrane [136]. The conditional knockout studies demonstrated that there is a path of differentiation in which the β cell is a progenitor of the α type. However, this simplified scheme seems to apply only to the cortex, suggesting that medullary α-IC cells are derived from a different type of cell. Al-Awqati *et al.* proposed that the new cell type that bears many similarities to the β-IC but does not express pendrin is a progenitor of the medullary α-intercalated cell [82, 137].

Both α- and β-intercalated cells are present in inner medulla of neonatal rats, and the number of β cells is greater in pups from alkalotic mothers [138]. In addition, in the cortex, the relative numbers of α *versus* β cells are constantly changing in response to variations in acid–base status. Thus, the model of one plastic intercalated cell phenotype would require that the hensin-induced terminal differentiation of β-cells into α-intercalated cells be a reversible phenomenon so that the appropriate response to alkalosis could occur (i.e., the production of more β-type intercalated cells by α-cell "dedifferentiation"). The apical anion exchanger in β-cells was not detectable using antibodies against AE-1. In cultured intercalated cells, the apical anion exchanger was identified as AE-1, whereas in the kidney collecting duct, pendrin has now been identified as the major Cl^-/HCO_3^- exchanger responsible for apical anion exchange in β-intercalated cells [76]. Furthermore, apical pendrin is relocated to the cytosol of β-intercalated cells in acid-loaded mouse kidney, consistent with previous reports of acid-induced apical anion exchanger internalization in rabbit collecting ducts [133]. In the medullary collecting tubule, fully differentiated α-intercalated cells actively proliferate in response to acidosis, demonstrating another difference between cortical and medullary acid-secreting cells [139].

9.10 Metabolic Regulation of the Vacuolar H^+-ATPase Activity

9.10.1 *Metabolic Acidosis*

The vacuolar H^+-ATPase plays an important role in the adaptive response of the kidney to altered acid–base or electrolyte status [139]. In metabolic acidosis and in respiratory acidosis (hypercapnia), upregulation of H^+ secretion occurs in the CCD, OMCD, and initial IMCD [140, 141]. In experimental models, this acid/base disturbance, induced by the addition of NH_4Cl to the drinking water, increases expression of vacuolar H^+-ATPase subunits in the luminal membrane of α-type intercalated cells in the late distal tubule, the connecting segment, and the collecting duct [142]. Either acidosis or high CO_2 leads to redistribution of H^+-ATPases to the membrane involving microtubular trafficking from intracellular stores [143]. In isolated perfused rabbit proximal tubules and collecting ducts a rise in CO_2 leads to the exocytotic insertion of vacuolar H^+-ATPases, thus increasing its activity in the plasma membrane. This translocation of vacuolar H^+-ATPase into the membrane is paralleled by increased activity of NEM, bafilomycin, or concanamycin-sensitive ATPase activity as well as proton extrusion [144]. Acidosis was also reported to result in the removal by endocytosis of the apical C^-/HCO_3^- transporter in B-IC, prior to its molecular identification as pendrin [139]. Metabolic acidosis associated with low urinary Cl^- excretion downregulates pendrin expression or at least prevent its stimulation. The sensing mechanism for metabolic acidosis must be located either intracellularly or at the basolateral side of the cells as only metabolic acidosis but not urinary acidification *per se* influence pendrin expression [145].

9.10.2 *Metabolic Alkalosis*

Transport processes during metabolic alkalosis are regulated in a mirror-like manner to metabolic acidosis. H^+ secretion through the H^+-ATPase is downregulated along the collecting duct by internalization of proton pumps. This is characterized by shifts of vacuolar H^+-ATPase immunoreactivity from the apical pole to subapical compartments in α-type intercalated cells and a more pronounced basolateral staining in β-type intercalated cells [146]. The activation of HCO_3^- secretion in β-type intercalated cells by metabolic alkalosis involves the apical pendrin protein [78]. Interestingly, alkalosis induces the endocytotic retrieval of V-ATPase from the apical membrane of all IC but specifically stimulates the accumulation of V-ATPase in the basolateral membrane only of B-IC. Thus, during alkalosis, not only is apical proton secretion by A-IC and B-IC reduced but apical bicarbonate secretion (*via* pendrin) and basolateral proton extrusion (*via* the V-ATPase) are also increased likely in an attempt to correct the alkalosis. These phenotypic changes can be detected over relatively short time frames (less than 6 hours) in the kidney [147] and in perfused tubules [139]. While the molecular mechanisms underlying V-ATPase redistribution remain poorly understood, exo- and endocytosis of acid–base transporters is certainly involved. A-IC in particular has a very high rate of apical endocytosis that can be detected by introducing various probes, including FITC-dextran or horseradish peroxidase into the collecting duct lumen [76]. Metabolic alkalosis due to hypokalemia is caused in part by the shift of intracellular K^+ to the extracellular space resulting in the uptake of protons into cells. During hypokalemia caused by low K^+ in the diet or by the use of loop diuretics stimulates the Na^+/HCO_3^- co-transporter NBC-1 in the proximal tubule [148]. During hypokalemia, intercalated cells found in the distal tubule and cortical collecting duct exhibit hypertrophy [149]. This is accompanied by a more pronounced apical location of the vacuolar H^+-ATPase and its increased activity [150]. These transport mechanisms contribute to the development of metabolic alkalosis under these conditions. In contradistinction to metabolic alkalosis, acute and chronic hypocapnia (respiratory alkalosis) has no effect on NEM-sensitive ATPase activity along the nephron. A key unresolved issue specific to renal physiology is how extracellular acid–base status is sensed by renal epithelial cells to initiate their homeostatic response to these stimuli. The factors that have been suggested include, not surprisingly, pH, CO_2, and bicarbonate as well as a number of potential hormonal stimuli. Early studies showed that basolateral CO_2 elevation, together with an initial increase in calcium, stimulate proton secretion by proximal tubules and collecting duct IC [151, 152]. This happens at least in part by inducing the apical insertion of V-ATPase in these cells. Some candidate protein, as family of G protein-coupled receptors, can generate cAMP or inositol 1,4,5-trisphosphate/calcium signals in response to acidic extracellular pH in some cell types [153].

A study by Pastor *et al.* on proton-secreting cells in the epididymis revealed that apical V-ATPase accumulation is stimulated by luminal bicarbonate via a soluble adenylate cyclase (sAC) mediated increase in intracellular cAMP [154]. sAC is highly expressed in IC and, moreover, that its localization partially overlaps with that of the V-ATPase [155]. Thus, in A-IC, sAC is mainly apical, whereas in B-IC, sAC is bipolar. sAC-regulated cAMP signaling may, therefore, constitute a general sensing mechanism for regulating V-ATPase-mediated proton transport. Thus, the presence of sAC could allow IC to respond to an elevation of apical (luminal) HCO_3^-. IC respond to cAMP treatment by accumulating V-ATPase in their apical membrane and by developing their apical microvilli A-IC respond to cAMP treatment by accumulating V-ATPase in their apical membrane and by developing their apical microvilli [156]. Subapical vesicles in A-IC are enriched in V-ATPase compared with the apical membrane of nonstimulated cells, suggesting that exocytosis of these vesicles would be sufficient to explain the increase in apical membrane V-ATPase density and activity upon cAMP. However, no evidence that cAMP induces a significant change in V-ATPase distribution in the B-IC population after acute cAMP exposure was found [156].

9.11 Regulation of the Vaculoar H$^+$-ATPase by the Renin-Angiotensin-Aldosterone System

Stimulation of vacuolar H$^+$-ATPase activity by both angiotensin II and aldosterone has been reported for various nephron segments and cell types. Angiotensin II is a potent vasoconstrictor which, in addition to its effects on Na$^+$ homeostasis and blood pressure regulation, also plays a major role in the regulation of acid–base balance. Intrarenal angiotensin II is regulated by several complex processes, involving formation of both systemically delivered and intrarenally formed substrate, as well as receptor-mediated internalization. Angiotensin II signals mainly through two receptor subtypes: AT_1 and AT_2 [157]. Although both receptors are expressed in the kidney, AT_1 is the predominant subtype found in blood vessels and almost ubiquitously along the nephron [158].

Angiotensin II is a potent activator of bicarbonate reabsorption and proton secretion along the nephron. However, different results regarding the effect of angiotensin II on bicarbonate reabsorption and proton secretion have been reported at the functional level, depending on the angiotensin II concentration and tubule segment studied [159]. The target transport mechanisms include the Na^+/HCO_3^- co-transporter and Na^+/H^+ exchanger (NHE-3) in the proximal tubule, and the vacuolar H^+-ATPase in the proximal tubule, DCT, and intercalated cells of the cortical collecting tubule. Luminal perfusion of angiotensin II stimulates Na^+ $/H^+$ exchange in both early and late distal tubule, and the vacuolar H^+-ATPase in the late distal tubule [159]. In the late cortical collecting tubule and outer medullary collecting tubule, angiotensin II enhances proton secretion by type A intercalated cells [160, 161].

Recent study by Pech *et al.* has demonstrated that angiotensin II stimulates chloride absorption in isolated mouse cortical collecting ducts [162]. Chloride absorption was dependent in the presence of pendrin and blocked by a H^+-ATPase inhibitor suggesting that H^+-ATPases in non-type A intercalated cells may also be stimulated by angiotensin II [162]. Along the same lines, angiotensin II enhances bicarbonate secretion in rabbit early cortical collecting ducts [163]. Taken together, these observations suggest that angiotensin II may have a direct stimulatory effect on H^+-ATPases in type B-intercalated cells and may thereby drive bicarbonate secretion and chloride absorption in the presence of pendrin. Moreover, stimulation of pendrin activity may also indirectly enhance electroneutral NaCl reabsorption via the recently described Na^+-driven $Cl^-/HCO_3^-/$ exchanger (NDCBE/SLC4A8) [164].

Angiotensin II in the physiological range stimulates the activity of H^+-ATPases in intercalated cells in isolated mouse CNT/CCDs [165]. Angiotensin II most likely acts via AT_1 receptors since the AT_1 receptor antagonist saralasin completely abolished the effect [165]. In type A-intercalated cells, angiotensin II stimulates H^+-

ATPases by a process that requires an intact microtubular network sensitive to colchicine [161] and leads to an accumulation of H^+-ATPases at the luminal membrane [160]. The stimulated H^+-ATPase activity would then energize enhanced chloride absorption or luminal alkalinization via pendrin [162, 163]. Wagner *et al.* have demonstrated that angiotensin II stimulates H^+-ATPase activity in isolated mouse CNT/CCD intercalated cells involving AT_1 receptors and leading to enhanced membrane abundance of H^+-ATPases [165]. Increased H^+-ATPase activity is a prerequisite for enhancing pendrin activity by altering driving forces for bicarbonate secretion and chloride absorption [165]. In addition to acute effects of angiotensin II on vacuolar H^+-ATPase activity, effects of chronic AT_1 blockade have also been reported. In *in vivo* microperfused distal tubules from remaining nephrons in 2/3 nephrectomized rats, chronic AT_1 blockade reduced bicarbonate reabsorption in close association with the reduction of synthesis and insertion of apical H^+-ATPase [166]. AT_1 receptors couple intracellularly mainly to phospholipase C, Ca^{2+}, and the PKC pathway [167].

Aldosterone also induces apical V-ATPase accumulation in intercalated cells from the OMCD in a PKC-dependent manner [168]. Thus, the HCO_3/sAC system is but one of the possible sensors linked to signaling pathways that could induce a rise in intracellular cAMP and V-ATPase trafficking in these cells. The signaling pathway involved in activation of vacuolar H^+-ATPase activity by angiotensin II, however, needs to be elucidated.

Regulation of H^+-ATPase activity includes mechanisms such as trafficking from submembraneous pools, regulation by cytosolic activator and inhibitor proteins, or gene expression and protein expression [169]. Colchicine, an agent that disrupts the microtubular network, prevents the angiotensin II induced stimulation of vacuolar H^+-ATPase in intercalated cells of isolated mouse cortical collecting ducts. This suggested that vacuolar H^+-ATPases, some of its subunits, or other regulatory proteins must traffic to the membrane. Studies to date indicate that H^+-ATPase is regulated to only a minor extent by transcriptional and translational mechanisms [10].

Little is known about the *in vivo* regulation of the various subunits of the H^+-ATPase by angiotensin II *in vivo*. To circumvent the potential effect of aldosterone, we examined the expression of H^+-ATPase subunits, B1 and a4, in an animal model of selective aldosterone deficiency (adrenalectomized [ADX] rats maintained on glucocorticoid replacement) [170]. Our data showed an increase in the relative protein abundance of B1-subunit isoform of the H^+-ATPase after chronic infusion of angiotensin II in the renal medulla of ADX rats. This increase in H^+-ATPase expression by angiotensin II was limited to intercalated cells in the medullary collecting ducts. The expression of the a4-subunit, by contrast, was not altered by either ADX or exogenous angiotensin II [170].

Aldosterone has both indirect and direct effects on proton secretion. In the connecting segment and cortical collecting duct, aldosterone stimulates the electrogenic vacuolar H^+-ATPase through increased reabsorption of Na^+ in principal cells. This renders the tubule lumen more negative and thus facilitates proton diffusion into the lumen [171]. In the OMCD, aldosterone also stimulates proton secretion, but the effect persists in the absence of Na^+ reabsorption, suggesting direct activation of vacuolar H^+-ATPase activity [172].

Although aldosterone stimulates H^+-ATPase–dependent bicarbonate reabsorption in all collecting duct segments, different sensitivities of vacuolar H^+-ATPase enzymatic activity in distinct collecting duct segments have been described [173]. Aldosterone, like other steroid hormones, works mainly through increasing gene transcription. Thus, activation of the vacuolar H^+-ATPase is likely due to increased transcription of one or more subunits. The regulation of other ion-transport processes by aldosterone involves genomic changes in transcription and synthesis of transport proteins, such as subunits of the epithelial Na^+ channel ENaC or the Na^+/K^+-ATPase [174].

Our findings in ADX rats suggested that the removal of aldosterone results in a decrease in the relative abundance and expression of B1 H^+-ATPase in the medullary, but not the cortical collecting duct [170]. Immunofluorescence with confocal microscopy studies revealed that aldosterone promotes trafficking of the a4-subunit toward plasma membrane in alpha-intercalated cells [168]. A dual effect of aldosterone (i.e., to promote trafficking of the a4-subunit to the cell membrane and to increase a4 protein synthesis) is likely required to facilitate assembly of the H^+-ATPase proton pump and thus increased hydrogen ion secretion. In microdissected cortical and medullary collecting duct, segments from ADX rats, a pharmacological dose of aldosterone, increased enzyme H^+-ATPase activity irrespective of potassium levels [175]. Winter *et al.* have demonstrated that nongenomic stimulatory effect of aldosterone on H(+)- ATPase activity in acid secretory intercalated cells of isolated mouse (OMCDs [168]. They investigated the intracellular signaling cascade mediating this stimulating effect. Aldosterone stimulated H^+-ATPase activity in isolated mouse and human OMCDs. This effect was blocked by suramin, a general G protein inhibitor, and GP-2A, a specific G(αq) inhibitor, whereas pertussis toxin was without effect. Inhibition of phospholipase C with U-73122, chelation of intracellular Ca^{2+} with BAPTA, and blockade of protein kinase C prevented the stimulation of H^+-ATPases. Stimulation of PKC by DOG mimicked the effect of aldosterone on H^+-ATPase activity. Similarly, aldosterone and DOG induced a rapid translocation of H^+-ATPases to the luminal side of OMCD cells *in vivo*. In addition, PD098059, an inhibitor of ERK1/2 activation, blocked the aldosterone and DOG effects. Inhibition of protein kinase A (PKA) with H89 or KT2750 prevented and incubation with 8-bromoadenosine-cAMP mildly increased H^+-ATPase activity. Thus, the nongenomic modulation of H^+-ATPase activity in OMCD-intercalated cells by aldosterone involves several intracellular pathways and may be mediated by a G(αq) protein-coupled receptor and PKC. PKA and cAMP appear to have a modulatory effect. This rapid nongenomic action of aldosterone may participate in the regulation of H^+-ATPase activity and contribute to final urinary acidification [176].

Alzamora *et al.* have demonstrated that Ser-175A is a major PKA phosphorylation site in the A-subunit in the cytoplasmic V(1) sector of the V-ATPase in a polarized rabbit cell line of collecting duct A-type intercalated cell. Mutation in S-175a had shown the decrease in V-ATPase activity that was unresponsive to PKA activation [177]. Their results suggest a novel mechanism for the regulation of V-ATPase localization and activity in kidney cells via direct PKA-dependent phosphorylation of the A-subunit at Ser-175 [177].

This nongenomic stimulatory effect of aldosterone required PKC and was associated with a rapid transient increase in intracellular Ca^{2+}.

A level of regulation involving changes in protein levels, however, may be important as well. Our results in medullary collecting tubule segments suggest that both aldosterone and angiotensin II provide an additional level of regulation by altering *in vivo* subunit protein abundances [170]. This may occur either by changing protein synthesis or by altering protein degradation. Taken together, these studies suggest that angiotensin II and aldosterone affect vacuolar H^+-ATPase activity through different mechanisms, which may be specific to both the segment and the cell type.

9.12 Functional Characterization of (pro)Renin Receptor in Association with V-ATPase

A receptor for renin and for the inactive proenzyme form of renin prorenin was cloned in 2002 and was called PRR [(pro)renin receptor] [178]. The (pro)renin receptor [(P)RR] is a unique molecule that binds prorenin and renin in tissues, not only leading to their activation, but also inducing intracellular signaling involving ERK (extracellular-signal-regulated kinase) in an Ang II-independent manner.

The PRR gene is named *ATP6AP2* (ATPase 6 accessory protein 2) because PRR was initially found as a truncated form co-purifying with V-ATPase (vacuolar H^+-ATPase) [179]. ATP6ap1 and ATP6ap2 are also named Ac45 and PRR, respectively, and both are sometimes called accessory "sub-units" of the V-ATPase and can be found associated to the V_0 domain of the proton pump. Recent studies have shown that Ac45 interacts with the V(0)-sector of the V-ATPase complex, thereby regulating the intragranular pH and Ca^{2+}-dependent exocytotic membrane fusion [180]. Thus, Ac45 can be considered as a V-ATPase regulator in the neuroendocrine secretory pathway.

ATP6AP2 has recently been found to be identical to the (pro)renin receptor and has a dual role: (i) it regulates V-ATPase activity and (ii) it acts as an adapter by binding to both the V-ATPase and the Wnt receptor complex, thereby recruiting the receptor complex into an acidic microenvironment [180]. The human ATP6AP2 gene located on the X chromosome at locus p11.4, codes for a protein with 350-amino acids. PRR is a small protein of an apparent molecular mass of 35–39 kDa organized into a large EC (extracellular) domain, a soluble PRR (sPRR) found in plasma and urine [181]. The truncated form of PRR initially described to be associated with the V-ATPase in bovine chromaffin granules is composed of the TM–IC domains and a short portion of the EC domain, cleaved at a close vicinity of a putative furin-cleavage site (RKTR) [182].

(P)RR activation may play a possible role in the development of cardiac fibrosis, proteinuria, hypertension, and diabetes. The binding of prorenin to the receptor has two important consequences. First, prorenin undergoes nonproteolytic activation, displays full enzymatic activity, and initiates Ang II-dependent effects. However, unexpectedly, different studies using various RAS inhibitors show that (P)RR/ATP6ap2 activation triggers intracellular pathways in an Ang II-independent manner [183]. Indeed, the mitogen-activated protein kinase (MAPK) ERK1/2 is activated resulting in the upregulation of pro-fibrotic genes. In kidney tissues, (P)RR/ATP6ap2 upregulates via ERK1/2 stimulation inflammatory mediators such as cyclooxygenase-2 (COX2), interleukin-1 (IL-1b), and tumor necrosis factor alpha. [184].

The subcellular localization of (P)RR/ATP6ap2 is unusual for a receptor. Only a minor amount of the protein was found on the plasma membrane, the majority is located on intracellular vesicles [185]. In addition, a furin cleavage site on the extracellular part of (P)RR/ATP6ap248 allows the release of a 28 kDa fragment during passage through the Golgi apparatus, which could be detected in the supernatant of cells expressing the protein [186].

Because PRR was identified as a component of the RAS (renin–angiotensin system) and because of its potential pro-fibrotic effects, it has been primarily studied in cardiovascular and renal diseases. It has been also recently identified in the glomerular epithelial cells (podocyte) [186]. The PRR is also highly expressed in neurons and on plasma membranes, where it binds renin and mediates ERK1/2 phosphorylation [187, 188], but it is also present in synaptic vesicles, where V-ATPase is essential for neurotransmitter concentration and maturation, suggesting again a link between PRR and V-ATPase.

A functional link between PRR and V-ATPase has been suggested for the first time in renal cells by Advani *et al.* [189]. Using a combination of *in situ* hybridization, immunohistochemistry, and electron microscopy, it was shown that the (P)RR is expressed predominantly by collecting duct intercalated acid-transporting A cells, where it co-localizes with the vacuolar H^+-ATPase. Consistent with this local function, *in vitro* studies show that intracellular signaling through the (P)RR is critically dependent on the activity of the vacuolar H^+-ATPase. Advani *et al.* have shown that blocking V-ATPase function with bafilomycin in MDCK (Madin–Darby canine kidney) cells of a collecting duct/distal lineage also inhibited PRR activation and ERK phosphorylation induced by (pro)renin, thus establishing a link between PRR and V-ATPase [189]. This finding and other recent observations indicate a primarily "tubular" pattern of expression [190]. Since the initial description of the (P)RR receptor in 2002 by Nguyen *et al.* [191], the field in this area is rapidly evolving.

References

1. Stevens, T.H., and Forgac, M. (1997). Structure, function and regulation of the vacuolar (H+)-ATPase. *Annu Rev Cell Dev Biol*, 13, pp. 779–808.
2. Nelson, N., and Harvey, W.R. (1999). Vacuolar and plasma membrane proton-adenosinetriphosphatases. *Physiol Rev*, 79, pp. 361–385.
3. Wieczorek, H., Brown, D, Grinstein, S, Ehrenfeld, J, and Harvey, W.R (1999). Animal plasma membrane energization by proton-motive V-ATPases. *Bioessays* 21, pp. 637–648.
4. Forgac, M. (1999). Structure and properties of the vacuolar (H+)-ATPases. *J Biol Chem*, 274, pp. 12951–12954.
5. Nishi, T., and Forgac, M. (2002). The vacuolar (H+)-ATPases-nature's most versatile proton pumps. *Nat Rev Mol Cell Biol*, 3, pp. 94–103.
6. Forgac, M. (2007). Vacuolar ATPases: rotary proton pumps in physiology and pathophysiology. *Nat Rev Mol Cell Biol*, 8, pp. 917–929.
7. Gruenberg, J. (2001). The endocytic pathway: a mosaic of domains. *Nat Rev Mol Cell Biol*, 2, pp. 721–730.

8. Stegmann, T. (2000). Membrane fusion mechanisms: the influenza hemagglutinin paradigm and its implications for intracellular fusion. *Traffic*, 1, pp. 598–604.
9. Breton, S., Smith, P.J., Lui, B., and Brown, D. (1996). Acidification of the male reproductive tract by a proton pumping (H+)-ATPase. *Nat Med*, 2(4), pp. 470–472.
10. Wagner, C.A., Finberg, K.E., Breton, S., Marshansky, V., Brown, D., and Geibel, J.P. (2004). Renal vacuolar H$^+$-ATPase. *Physiol Rev*, 84, pp. 1263–1314.
11. Lee, B.S., Holliday, L.S., Ojikutu, B., Krits, I., and Gluck, S.L. (1996). Osteoclasts express the B2 isoform of vacuolar H_-ATPase intracellularly and on their plasma membranes. *Am J Physiol Cell Physiol*, 270, pp. C382–C388.
12. Couloigner, V., Teixeira, M., Hulin, P., Sterkers, O., Bichara, M., Escoubet, B., Planelles, G., and Ferrary, E. (2000). Effect of locally applied drugs on the pH of luminal fluid in the endolymphatic sac of guinea pig. *Am J Physiol Regul Integr Comp Physiol*, 279, pp. R1695–R1700.
13. Muller, V., and Gruber, G. (2003). ATP synthases: structure, function and evolution of unique energy converters. *Cell Mol Life Sci*, 60, pp. 474–494.
14. Gruber, G., Wieczorek, H., Harvey, W.R., and Muller, V. (2001). Structure-function relationships of A-, F- and V-ATPases. *J Exp Biol*, 204, pp. 2597–2605.
15. Smith, A.N., Lovering, R.C., Futai, M., Takeda, J., Brown, D., and Karet, F.E. (2003). Revised nomenclature for mammalian vacuolar-type H$^+$-ATPase subunit genes. *Mol Cell*, 12, pp. 801–803.
16. Brown, D., Gluck, S., and Hartwig, J. (1987). Structure of the novel membrane-coating material in proton-secreting epithelial cells and identification as an H$^+$-ATPase. *J Cell Biol*, 105, pp. 1637–1648.
17. Liu, Q., Kane, P.M., Newman, P.R., and Forgac, M. (1996). Site-directed mutagenesis of the yeast V-ATPase B subunit (Vma2p). *J Biol Chem*, 271, pp. 2018–2022.
18. Liu, Q., Leng, X.H., Newman, P.R., Vasilyeva, E., Kane, P.M., and Forgac, M. (1997). Site-directed mutagenesis of the yeast V-ATPase A subunit. *J Biol Chem*, 272, pp. 11750–11756.
19. MacLeod, K.J., Vasilyeva, E., Baleja, J.D., and Forgac, M. (1998). Mutational analysis of the nucleotide binding sites of the yeast vacuolar proton-translocating ATPase. *J Biol Chem*, 273, pp. 150–156.

20. Vasilyeva, E., Liu, Q., MacLeod, K.J., Baleja, J.D., and Forgac, M. (2000). Cysteine scanning mutagenesis of the noncatalytic nucleotide binding site of the yeast V-ATPase. *J Biol Chem*, 275, pp. 255–260.
21. Niederstatter, H., and Pelster, B. (2000). Expression of two vacuolar-type ATPase B subunit isoforms in swim bladder gas gland cells of the European eel: nucleotide sequences and deduced amino acid sequences. *Biochim Biophys Acta*, 1491, pp. 133–142.
22. Nishi, T., and Forgac, M. (2000). Molecular cloning and expression of three isoforms of the 100-kDa a subunit of the mouse vacuolar proton-translocating ATPase. *J Biol Chem*, 275, pp. 6824–6830.
23. Oka, T., Yamamoto, R., and Futai, M. (1998). Multiple genes for vacuolar-type ATPase proteolipids in *Caenorhabditis elegans*. A new gene, vha-3, has a distinct cell-specific distribution. *J Biol Chem*, 273, pp. 22570–22576.
24. Pannabecker, T.L., Brokl, O.H., Kim, Y.K., Abbott, D.E., and Dantzler, W.H. (2002). Regulation of intracellular pH in rat renal inner medullary thin limbs of Henle's loop. *Pflugers Arch*, 443, pp. 446–457.
25. Parra, K.J., and Kane, P.M. (1998). Reversible association between the V1 and V0 domains of yeast vacuolar H^+-ATPase is an unconventional glucose-induced effect. *Mol Cell Biol*, 18, pp. 7064–7074.
26. Vasilyeva, E., and Forgac, M. (1996). 3′-O-(4-Benzoyl)benzoyladenosine 5′-triphosphate inhibits activity of the vacuolar (H+)-ATPase from bovine brain clathrin-coated vesicles by modification of a rapidly exchangeable, noncatalytic nucleotide binding site on the B subunit. *J Biol Chem*, 271, pp. 12775–12782.
27. van Hille, B., Richener, H., Schmid, P., Puettner, I., Green, J.R., and Bilbe, G. (1994). Heterogeneity of vacuolar H(+)-ATPase: differential expression of two human subunit B isoforms. *Biochem J*, 303(Pt. 1), pp. 191–198.
28. Breton, S., Wiederhold, T., Marshansky, V., Nsumu, N.N., Ramesh, V., and Brown, D. (2000). The B1 subunit of the H^+-ATPase is a PDZ domain-binding protein. Colocalization with NHE-RF in renal B-intercalated cells. *J Biol Chem*, 275, pp. 18219–18224.
29. Nelson, R.D., Guo, X.L., Masood, K., Brown, D., Kalkbrenner, M., and Gluck, S. (1992). Selectively amplified expression of an isoform of the vacuolar H(+)-ATPase 56-kilodalton subunit in renal intercalated cells. *Proc Natl Acad Sci USA*, 89, pp. 3541–3545.
30. Karet, F.E., Finberg, K.E., Nelson, R.D., Nayir, A., Mocan, H., Sanjad, S.A., Rodriguez Soriano, J., Santos, F., Cremers, C.W., Di Pietro, A., Hoffbrand,

B.I., Winiarski, J., Bakkaloglu, A., Ozen, S., Dusunsel, R., Goodyer, P., Hulton, S.A., Wu, D.K., Skvorak, A.B., Morton, C.C., Cunningham, M. J., Jha, V., and Lifton, R.P. (1999). Mutations in the geneen coding B1 subunit of H-ATPase cause renal tubular acidosis with sensorineural deafness. *Nat Genet*, 21, pp 84–90.

31. Wang, Y., Cipriano, D.J., and Forgac, M. (2007). Arrangement of subunits in the proteolipid ring of the V-ATPase. *J Biol Chem*, 282, pp. 34058–34065.
32. Vik, S.B., and Antonio, B.J. (1994). A mechanism of proton translocation by F1F0 ATP synthases suggested by double mutants of the a subunit. *J Biol Chem*, 269, pp. 30364–30369.
33. Leng, X.H., Manolson, M.F., and Forgac, M. (1998). Function of the COOH-terminal domain of Vph1p in activity and assembly of the yeast V-ATPase. *J Biol Chem*, 273, pp. 6717–6723.
34. Leng, X.H., Manolson, M.F., Liu, Q., and Forgac, M. (1996). Site-directed mutagenesis of the 100-kDa subunit (Vph1p) of the yeast vacuolar (H+)-ATPase. *J Biol Chem*, 271, pp. 22487–22493.
35. Zhang, J., Feng, Y., and Forgac, M. (1994). Proton conduction and bafilomycin binding by the V0 domain of the coated vesicle V-ATPase. *J Biol Chem*, 269, pp. 23518–23523.
36. Smith, A.N., Skaug, J., Choate, K.A., Nayir, A., Bakkaloglu, A., Ozen, S., Hulton, S.A., Sanjad, S.A., Al-Sabban, E.A., Lifton, R.P., *et al.* (2000). Mutations in ATP6N1B, encoding a new kidney vacuolar proton pump 116-kD subunit, cause recessive distal renal tubular acidosis with preserved hearing. *Nat Genet*, 26, pp. 71–75.
37. Smith, A.N., Finberg, K.E., Wagner, C.A., Lifton, R.P., Devonald, M.A., Su, Y., and Karet, F.E. (2001). Molecular cloning and characterization of Atp6n1b: a novel fourth murine vacuolar H^+-ATPase a-subunit gene. *J Biol Chem*, 276, pp. 42382–42388.
38. Frattini, A., Orchard, P. J., Sobacchi, C., Giliani, S., Abinun, M., Mattsson, J. P., Keeling, D.J., Andersson, A. K., Wallbrandt, P., Zecca, L., Notarangelo, L. D., Vezzoni, P., and Villa, A. (2000). Defects in TCIRG1 subunit of the vacuolar proton pump are responsible for a subset of human autosomal recessive osteopetrosis, *Nat Genet*, 25, pp. 343–346.
39. Sun-Wada, G.H., Toyomura, T., Murata, Y., Yamamoto, A., Futai, M., and Wada, Y. (2006). The a3 isoform of V-ATPase regulates insulin secretion from pancreatic beta-cells. *J Cell Sci*, 119, pp. 4531–4540.

40. Tabata, H., Kawamura, N., Sun-Wada, G.H., and Wada, Y. (2008). Vacuolar-type H(+)-ATPase with the a3 isoform is the proton pump on premature melanosomes. *Cell Tissue Res*, 332, pp. 447–460.

41. Pietrement, C., Sun-Wada, G.H., Silva, N.D., McKee, M., Marshansky, V., Brown, D., Futai, M., and Breton, S. (2006). Distinct expression patterns of different subunit isoforms of the V-ATPase in the rat epididymis. *Biol Reprod*, 74, pp. 185–194.

42. Morel, N., Dedieu, J.C., and Philippe, J.M. (2003). Specific sorting of the a1 isoform of the V-H^+-ATPase a subunit to nerve terminals where it associates with both synaptic vesicles and the presynaptic plasma membrane. *J Cell Sci*, 116, pp. 4751–4762.

43. Hiesinger, P.R., Fayyazuddin, A., Mehta, S.Q., Rosenmund, T., Schulze, K.L., Zhai, R.G., Verstreken, P., Cao, Y., Zhou, Y., Kunz, J., *et al.* (2005). The V-ATPase V0 subunit a1 is required for a late step in synaptic vesicle exocytosis in Drosophila. *Cell*, 121, pp. 607–620.

44. Peri, F., and Nusslein-Volhard, C. (2008). Live imaging of neuronal degradation by microglia reveals a role for v0-ATPase a1 in phagosomal fusion in vivo. *Cell*, 133, pp. 916–927.

45. Peri, F., and Nüsslein-Volhard, C. (2008). Live imaging of neuronal degradation by microglia reveals a role for v0-ATPase a1 in phagosomal fusion in vivo. *Cell*, 133(5), pp. 916–927.

46. Hurtado-Lorenzo, A., Skinner, M., El Annan, J., Futai, M., Sun-Wada, G.H., Bourgoin, S., Casanova, J., Wildeman, A., Bechoua, S., Ausiello, D.A., Brown, D., and Marshansky, V. (2006). V-ATPase interacts with ARNO and Arf6 in early endosomes and regulates the protein degradative pathway. *Nat Cell Biol*, 8, pp. 124–136.

47. Leng, X.H., Nishi, T., and Forgac, M. (1999). Transmembrane topography of the 100-kDa a subunit (Vph1p) of the yeast vacuolar proton-translocating ATPase. *J Biol Chem*, 274, pp. 14655–14661.

48. Wilkens, S., and Capaldi, R. A. (1998). ATP synthase's second stalk comes into focus. *Nature*, 393, pp. 29–30.

49. Iwata, M., Imamura, H., Stambouli, E., Ikeda, C., Tamakoshi, M., Nagata, K., Makyio, H., Hankamer, B., Barber, J., Yoshida, M., Yokoyama, K. and Iwata, S. (2004). Crystal structure of a central stalk subunit C and reversible association/dissociation of vacuole-type ATPase. *Proc Natl Acad Sci USA*, 101, pp. 59–64.

50. Curtis, K.K., Francis, S.A., Oluwatosin, Y., and Kane, P.M. (2002). Mutational analysis of the subunit C (Vma5p) of the yeast vacuolar H^+-ATPase. *J Biol Chem*, 277, pp. 8979–8988.

51. Kane, P.M. (1995). Disassembly and reassembly of the yeast vacuolar H(+)-ATPase in vivo. *J Biol Chem*, 270, pp. 17025–17032.
52. Sumner, J.P., Dow, J.A., Earley, F.G., Klein, U., Jager, D., and Wieczorek, H. (1995). Regulation of plasma membrane V-ATPase activity by dissociation of peripheral subunits. *J Biol Chem*, 270, pp. 5649–5653.
53. Inoue, T., and Forgac, M. (2005). Cysteine-mediated cross-linking indicates that subunit C of the V-ATPase is in close proximity to subunits E and G of the V1 domain and subunit a of the V0 domain. *J Biol Chem*, 280, pp. 27896–27903.
54. Holthuis, J.C., Jansen, E.J., Schoonderwoert, V.T., Burbach, J.P., and Martens, G.J. (1999). Biosynthesis of the vacuolar H^+-ATPase accessory subunit Ac45 in Xenopus pituitary. *Eur J Biochem*, 262, pp. 484–491.
55. Hirata, T., Iwamoto-Kihara, A., Sun-Wada, G.H., Okajima, T., Wada, Y., and Futai, M. (2003). Subunit rotation of vacuolar-type proton pumping ATPase: relative rotation of the G and C subunits. *J Biol Chem*, 278(26), pp. 23714–23719.
56. Jefferies, K.C., Cipriano, D.J., and Forgac, M. (2008). Function, structure and regulation of the vacuolar (H+)-ATPases. *Arch Biochem Biophys*, 476, pp. 33–42.
57. Cipriano, D.J., Wang, Y., Bond, S., Hinton, A., Jefferies, K.C., Qi, J., and Forgac, M. (2008). Structure and regulation of the vacuolar ATPases. *Biochim Biophys Acta*, 1777, pp. 599–604
58. Kawasaki-Nishi, S., Nishi, T., and Forgac, M. (2003). Interacting helical surfaces of the transmembrane segments of subunits a and c′ of the yeast V-ATPase defined by disulfide-mediated cross-linking. *J Biol Chem*, 278, pp. 41908–41913.
59. Wang, Y., Inoue, T., and Forgac, M. (2004). TM2 but not TM4 of subunit c′ interacts with TM7 of subunit a of the yeast V-ATPase as defined by disulfide-mediated cross-linking. *J Biol Chem*, 279, pp. 44628–44638.
60. Bowman, B.J., McCall, M.E., Baertsch, R., and Bowman, E.J. (2006). A model for the proteolipid ring and bafilomycin/concanamycin-binding site in the vacuolar ATPase of *Neurospora crassa*. *J Biol Chem*, 281, pp. 31885–31893.
61. Wang, Y., Inoue, T., and Forgac, M. (2005). Subunit a of the yeast V-ATPase participates in binding of bafilomycin. *J Biol Chem*, 280, pp. 40481–40488.
62. Jouret, F., Auzanneau, C., Debaix, H., Wada, G.H., Pretto, C., Marbaix, E., Karet, F.E., Courtoy, P.J., and Devuyst, O. (2005). Ubiquitous and kidney-specific subunits of vacuolar H^+-ATPase are differentially expressed during nephrogenesis. *J Am Soc Nephrol*, 16, pp. 3235–3246.

63. Karet, F.E., Finberg, K.E., Nayir, A., Bakkaloglu, A., Ozen, S., Hulton, S.A., Sanjad, S.A, Al-Sabban, E.A., Medina, J.F. and Lifton, R.P. (1999). Localization of a gene for autosomal recessive distal renal tubular acidosis with normal hearing (rdRTA) to 7q33–34. *Am J Hum Genet*, 65(6), pp. 1656–1665.

64. Stover, E.H., Borthwick, K.J., Bavalia, C., Eady, N., Fritz, D.M., Rungroj, N, Morton, C.C., Axon, P.R., Akil, I., Al-Sabban, E.A., Baguley, D.M., Bianca, S., Bakkaloglu, A., Bircan, Z., Chauveau, D., Clermont, M.J., Guala, A., Hulton, S.A., Kroes, H., Li Volti, G., Mir, S., Mocan, H., Nayir, A., Ozen, S., Rodriguez Soriano, J., Sanjad, S.A., Tasic, V., Taylor, C.M., Topaloglu, R., Smith, A.N., Karet, F.E. (2002). Novel ATP6V1B1 and ATP6V0A4 mutations in autosomal recessive distal renal tubular acidosis with new evidence for hearing loss. *J Med Genet* 39(11), pp. 796–803.

65. Batlle, D., and Shah, M. (2007). Physiologic principles in the clinical evaluation of electrolyte, water, and acid–base, disorders. In: *The Kidney: Physiology and Pathophysiology* (Alpern, R.J., and Hebert, S.E., eds.). New York: Elsevier, pp. 2113–2141.

66. Hamm, L.L., and Alpern, R.J. (2000). Cellular mechanisms of renal tubular acidification. In: *The Kidney: Physiology and Pathophysiology* (3rd ed.), edited by Seldin, D.W. and Giebisch, G. Philadelphia, PA: Lippincott Williams & Wilkins, pp.1935–1979.

67. Jehmlich, K., Sablotni, J., Simon, B.J., and Burckhardt, G. (1991). Biochemical aspects of H(+)-ATPase in renal proximal tubules: inhibition by N,N'-dicyclohexylcarbodiimide, N-ethylmaleimide, and bafilomycin. *Kidney Int Suppl*, 33, pp. S64–S70.

68. Sabolic, I., and Burckhardt, G. (1986). Characteristics of the proton pump in rat renal cortical endocytotic vesicles. *Am J Physiol*, 250, pp. F817–F826.

69. Brown, D., Hirsch, S., and Gluck, S. (1988). Localization of a proton-pumping ATPase in rat kidney. *J Clin Invest*, 82, pp. 2114–2126.

70. Paunescu, T.G., Da Silva, N., Marshansky, V., McKee, M., Breton, S., and Brown, D. (2004). Expression of the 56-kDa B2 subunit isoform of the vacuolar H(+)-ATPase in proton-secreting cells of the kidney and epididymis. *Am J Physiol Cell Physiol*, 287, pp. C149–C162.

71. Rodman, J.S., Kerjaschki, D., Merisko, E., and Farquhar, M.G. (1984). Presence of an extensive clathrin coat on the apical plasmalemma of the rat kidney proximal tubule cell. *J Cell Biol*, 98, pp. 1630–1636.

72. Choi, J.Y., Shah, M., Lee, M.G., Schultheis, P.J., Shull, G.E., Muallem, S., and Baum, M. (2000). Novel amiloride-sensitive sodium-dependent proton

secretion in the mouse proximal convoluted tubule. *J Clin Invest*, 105, pp. 1141–1146.

73. Capasso, G., Rizzo, M., Pica, A., Di Maio, F.S., Moe, O.W., Alpern, R.J., and De Santo, N.G. (2002). Bicarbonate reabsorption and NHE-3 expression: abundance and activity are increased in Henle's loop of remnant rats. *Kidney Int*, 62, pp. 2126–2135.
74. Alper, S.L., Stuart-Tilley, A.K., Biemesderfer, D., Shmukler, B.E., and Brown, D. (1997). Immunolocalization of AE2 anion exchanger in rat kidney. *Am J Physiol*, 273, pp. F601–F614.
75. Pannabecker, T.L., Brokl, O.H., Kim, Y.K., Abbott, D.E., and Dantzler, W.H. (2002). Regulation of intracellular pH in rat renal inner medullary thin limbs of Henle's loop. *Pflugers Arch*, 443, pp. 446–457.
76. Alper, S.L., Natale, J., Gluck, S., Lodish, H.F., and Brown, D. (1989). Subtypes of intercalated cells in rat kidney collecting duct defined by antibodies against erythroid band 3 and renal vacuolar H^+-ATPase. *Proc Natl Acad Sci USA*, 86, pp. 5429–5433.
77. Brown, D., Hirsch, S., and Gluck, S. (1988). An H^+-ATPase in opposite plasma membrane domains in kidney epithelial cell subpopulations. *Nature*, 331, pp. 622–624.
78. Royaux, I.E., Wall, S.M., Karniski, L.P., Everett, L.A., Suzuki, K., Knepper, M.A., and Green, E.D. (2001). Pendrin, encoded by the Pendred syndrome gene, resides in the apical region of renal intercalated cells and mediates bicarbonate secretion. *Proc Natl Acad Sci USA*, 98, pp. 4221–4226.
79. Wagner, C.A., Mohebbi, N., Uhlig, U., Giebisch, G.H., Breton, S., Brown, D., and Geibel, J.P. (2011). Angiotensin II stimulates H-ATPase activity in intercalated cells from isolated mouse connecting tubules and cortical collecting ducts. *Cell Physiol Biochem*, 28, pp. 513–520.
80. Al-Awqati, Q. (2011). Terminal differentiation in epithelia: the role of integrins in hensin polymerization. *Annu Rev Physiol.*, 73, pp. 401–412.
81. Brown, D., Paunescu, T.G., Breton, S., and Marshansky, V. (2009). Regulation of the V-ATPase in kidney epithelial cells: dual role in acid–base homeostasis and vesicle trafficking. *J Exp Biol*, 212, pp. 1762–1772.
82. Al-Awqati, Q, and Gao, X.B. (2011). Differentiation of intercalated cells in the kidney. *Physiology (Bethesda)*, 26(4), pp. 266–272.
83. Alper, S.L. (2002). Genetic diseases of acid–base transporters. *Annu Rev Physiol*, 64, pp. 899–923.

84. Frische, S., Kwon, T.H., Frokiaer, J., Madsen, K.M., and Nielsen, S. (2003). Regulated expression of pendrin in rat kidney in response to chronic NH4Cl or $NaHCO_3$ loading. *Am J Physiol Renal Physiol*, 284, pp. F584–F593.

85. Emmons, C. (1999). Transport characteristics of the apical anion exchanger of rabbit cortical collecting duct beta-cells. *Am J Physiol*, 276, pp. F635–F643.

86. Wagner, C.A., Finberg, K.E., Stehberger, P.A., Lifton, R.P., Giebisch, G.H., Aronson, P.S., and Geibel, J.P. (2002). Regulation of the expression of the Cl-/anion exchanger pendrin in mouse kidney by acid–base status. *Kidney Int*, 62, pp. 2109–2117.

87. Rillema, J.A., and Hill, M.A. (2003). Prolactin regulation of the pendrin-iodide transporter in the mammary gland. *Am J Physiol Endocrinol Metab*, 284, pp. E25–E28.

88. Wall, S.M. (2005). Recent advances in our understanding of intercalated cells. *Curr Opin Nephrol Hypertens*, 14, pp. 480–484.

89. Barone, S., Amlal, H., Xu, J., Kujala, M., Kere, J., Petrovic, S., and Soleimani, M. (2004). Differential regulation of basolateral Cl-/HCO_3^- exchangers SLC26A7 and AE1 in kidney outer medullary collecting duct. *J Am Soc Nephrol*, 15, pp. 2002–2011.

90. Hamm, L.L., and Simon, E.E. (1987). Roles and mechanisms of urinary buffer excretion. *Am J Physiol*, 253, pp. F595–F605.

91. Yip, K.P., and Kurtz, I. (1995). NH3 permeability of principal cells and intercalated cells measured by confocal fluorescence imaging. *Am J Physiol*, 269, pp. F545–F550.

92. Wingo, C.S., Madsen, K.M., Smolka, A., and Tisher, C.C. (1990). H-K-ATPase immunoreactivity in cortical and outer medullary collecting duct. *Kidney Int*, 38, pp. 985–990.

93. Armitage, F.E., and Wingo, C.S. (1994). Luminal acidification in K-replete OMCDi: contributions of H-K-ATPase and bafilomycin-A1-sensitive H-ATPase. *Am J Physiol*, 267, pp. F450–F458.

94. Clapp, W.L., Madsen, K.M., Verlander, J.W., and Tisher, C.C. (1989). Morphologic heterogeneity along the rat inner medullary collecting duct. *Lab Invest*, 60, pp. 219–230.

95. Marshansky, V., and Vinay, P. (1996). Proton gradient formation in early endosomes from proximal tubules. *Biochim Biophys Acta*, 1284, pp. 171–180.

96. Sonawane, N.D., Thiagarajah, J.R., and Verkman, A.S. (2002). Chloride concentration in endosomes measured using a ratio able fluorescent

Cl^- indicator: evidence for chloride accumulation during acidification. *J Biol Chem*, 277, pp. 5506–5513.

97. Bae, H.R., and Verkman, A.S. (1990). Protein kinase A regulates chloride conductance in endocytic vesicles from proximal tubule. *Nature*, 348, pp. 637–639.
98. Wagner, C.A, Giebisch, G., Lang, F., and Geibel, J. (1998). Angiotensin II stimulates vesicular H^+-ATPase in rat proximal tubular cells. *Proc Natl Acad Sci USA*, 95, pp 9665–9668.
99. Malnic, G., and Geibel, J.P. (2000). Cell pH and H(+) secretion by S3 segment of mammalian kidney: role of H(+)-ATPase and Cl(-). *J Membr Biol*, 178, pp. 115–125.
100. Carraro-Lacroix, L.R., Lessa, L.M., Fernandez, R., and Malnic, G. (2009). Physiological implications of the regulation of vacuolar H^+-ATPase by chloride ions. *Braz J Med Biol Res*, 42, pp. 155–163.
101. Fernandez, R., Bosqueiro, J.R., Cassola, A.C., and Malnic, G. (1997). Role of Cl^- in electrogenic H^+ secretion by cortical distal tubule. *J Membr Biol*, 157, pp. 193–201.
102. Jentsch, T.J. (2007). Chloride and the endosomal-lysosomal pathway: emerging roles of CLC chloride transporters. *J Physiol*, 578, pp. 633–640.
103. Gunther, W., Luchow, A., Cluzeaud, F., Vandewalle, A., and Jentsch, T.J. (1998). ClC-5, the chloride channel mutated in Dent's disease, colocalizes with the proton pump in endocytotically active kidney cells. *Proc Natl Acad Sci USA*, 95, pp. 8075–8080.
104. Han, J.S., Kim, G.H., Kim, J., Jeon, U.S., Joo, K.W., Na, K.Y., Ahn, C., Kim, S., Lee, S.E., and Lee, J.S. (2002). Secretory-defect distal renal tubular acidosis is associated with transporter defect in H(+)-ATPase and anion exchanger-1. *J Am Soc Nephrol*, 13, pp. 1425–1432.
105. Morales, M.M., Carroll, T.P., Morita, T., Schwiebert, E.M., Devuyst, O., Wilson, P.D., Lopes, A.G., Stanton, B.A., Dietz, H.C., Cutting, G.R., and Guggino, WB. (1996). Both the wild type and a functional isoform of CFTR are expressed in kidney. *Am J Physiol*, 270, pp. F1038–F1048.
106. Barasch, J., Kiss, B., Prince, A., Saiman, L., Gruenert, D., and al-Awqati, Q. (1991). Defective acidification of intracellular organelles in cystic fibrosis. *Nature*, 352, pp. 70–73.
107. Tararthuch, A.L., Fernandez, R., and Malnic, G. (2007). Cl^- and regulation of pH by MDCK-C11 cells. *Braz J Med Biol Res*, 40, pp. 687–696.

108. Biwersi, J., Emans, N., and Verkman, A.S. (1996). Cystic fibrosis transmembrane conductance regulator activation stimulates endosome fusion in vivo. *Proc Natl Acad Sci USA*, 93, pp. 12484–12489.

109. Marshansky, V., Ausiello, D.A., and Brown, D. (2002). Physiological importance of endosomal acidification: potential role in proximal tubulopathies. *Curr Opin Nephrol Hypertens*, 11, pp- 527–537.

110. Brown, D., and Stow, J.L. (1996). Protein trafficking and polarity in kidney epithelium: from cell biology to physiology. *Physiol Rev*, 76, pp. 245–297.

111. Mellman, I. (1992). The importance of being acid: the role of acidification in intracellular membrane traffic. *J Exp Biol*, 172, pp 39–45.

112. Hilden, S.A., Johns, C.A., and Madias, N.E. (1988). Cl-dependent ATP-driven H^+-transport in rabbit renal cortical endosomes. *Am J Physiol*, 255, pp. F885–F897.

113. Moriyama, Y., and Nelson, N. (1989). H^+-translocating ATPase in Golgi apparatus. Characterization as vacuolar H^+-ATPase and its subunit structures. *J Biol Chem*, 264, pp. 18445–18450.

114. Moriyama, Y., and Nelson, N. (1989). Lysosomal H+-translocating ATPase has a similar subunit structure to chromaffin granule H^+-ATPase complex. *Biochim Biophys Acta*, 980, pp. 241–247.

115. Aniento, F., Gu, F., Parton, R.G., and Gruenberg, J. (1996). An endosomal beta COP is involved in the pH-dependent formation of transport vesicles destined for late endosomes. *J Cell Biol*, 133, pp. 29–41.

116. Bremser, M., Nickel, W., Schweikert, M., Ravazzola, M., Amherdt, M., Hughes, C.A., Sollner, T.H., Rothman, J.E., and Wieland, F.T. (1999). Coupling of coat assembly and vesicle budding to packaging of putative cargo receptors. *Cell*, 96, pp. 495–506.

117. Gu, F., and Gruenberg, J. (2000). ARF1 regulates pH-dependent COP functions in the early endocytic pathway. *J Biol Chem*, 275, pp. 8154–8160.

118. Maranda, B., Brown, D., Bourgoin, S., Casanova, J.E., Vinay, P., Ausiello, D.A., and Marshansky, V. (2001). Intra-endosomal pH-sensitive recruitment of the Arf-nucleotide exchange factor ARNO and Arf6 from cytoplasm to proximal tubule endosomes. *J Biol Chem*, 276, pp. 18540–18550.

119. Chen, Y.A., and Scheller, R.H. (2001). SNARE-mediated membrane fusion. *Nat Rev Mol Cell Biol*, 2, pp. 98–106.

120. Hay, J.C. (2001). SNARE complex structure and function. *Exp Cell Res*, 271, pp. 10–21.

121. Fasshauer, D., Sutton, R.B., Brunger, A.T., and Jahn, R. (1998). Conserved structural features of the synaptic fusion complex: SNARE proteins reclassified as Q- and R-SNAREs. *Proc Natl Acad Sci USA*, 95, pp. 15781–15786.

122. Banerjee, A., Shih, T., Alexander, E.A., and Schwartz, J.H. (1999). SNARE proteins regulate H(+)-ATPase redistribution to the apical membrane in rat renal inner medullary collecting duct cells. *J Biol Chem*, 274, pp. 26518–26522.

123. Li, G., Alexander, E.A., and Schwartz, J.H. (2003). Syntaxin isoform specificity in the regulation of renal H^+-ATPase exocytosis. *J Biol Chem*, 278, pp. 19791–19797.

124. Bayer, M.J., Reese, C., Buhler, S., Peters, C., and Mayer, A. (2003). Vacuole membrane fusion: V0 functions after trans-SNARE pairing and is coupled to the Ca^{2+}-releasing channel. *J Cell Biol*, 162, pp. 211–222.

125. Pushkin, A., Abuladze, N., Newman, D., Muronets, V., Sassani, P., Tatishchev, S., and Kurtz, I. (2003). The COOH termini of NBC3 and the 56-kDa H^+-ATPase subunit are PDZ motifs involved in their interaction. *Am J Physiol Cell Physiol*, 284, pp. C667–C673.

126. Vitavska, O., Wieczorek, H., and Merzendorfer, H. (2003). A novel role for subunit C in mediating binding of the H^+-V-ATPase to the actin cytoskeleton. *J Biol Chem*, 278, pp. 18499–18505.

127. Vitavska, O., Merzendorfer, H., and Wieczorek, H. (2005). The V-ATPase subunit C binds to polymeric F-actin as well as to monomeric G-actin and induces cross-linking of actin filaments. *J Biol Chem*, 280, pp. 1070–1076.

128. Wagner, C.A., Devuyst, O., Bourgeois, S., and Mohebbi, N. (2009). Regulated acid–base transport in the collecting duct. *Pflugers Arch—Eur J Physiol*, 458, pp. 137–156.

129. Kim, Y.H., Kwon, T.H., Frische, S., Kim, J., Tisher, C.C., Madsen, K.M., and Nielsen, S. (2002). Immunocytochemical localization of pendrin inintercalated cell subtypes in rat and mouse kidney. *Am J Physiol Renal Physiol*, 283, pp. F744–F754.

130. Schwartz, G.J., Barasch, J., and Al-Awqati, Q. (1985). Plasticity of functional epithelial polarity. *Nature*, 318, pp. 368–371.

131. Al-Awqati, Q., Vijayakumar, S., Hikita, C., Chen, J., and Takito, J. (1998). Phenotypic plasticity in the intercalated cell: the hensin pathway. *Am J Physiol*, 275, pp. F183–F190.

132. Takito, J., Hikita, C., and Al-Awqati, Q. (1996). Hensin, a new collecting duct protein involved in the in vitro plasticity of intercalated cell polarity. *J Clin Invest*, 98, pp. 2324–2331.

133. Schwartz, G.J., and Al-Awqati, Q. (2005). Role of hensin in mediating the adaptation of the cortical collecting duct to metabolic acidosis. *Curr Opin Nephrol Hypertens*, 14, pp. 383–388.

134. Schwartz, G.J., Tsuruoka, S., Vijayakumar, S., Petrovic, S., Mian, A., and Al-Awqati, Q. (2002). Acid incubation reverses the polarity of intercalated cell transporters, an effect mediated by hensin. *J Clin Invest*, 109, pp. 89–99.

135. Vijayakumar, S., Erdjument-Bromage, H., Tempst, P., and Al-Awqati, Q. (2008). Role of integrins in the assembly and function of hensinin intercalated cells. *J Am Soc Nephrol*, 19, pp. 1079–1091.

136. Miller, R.L., Lucero, O.M., Riemondy, K.A., Baumgartner, B.K., Brown, D., Breton, S., and Nelson, R.D. (2009). The V-ATPase B1-subunit promoter drives expression of Cre recombinase in intercalated cells of the kidney. *Kidney Int*, 75, pp. 435–439.

137. Gao, X., Eladari, D., Leviel, F., Tew, BY., Mirò-Julia, C., Cheema, F., Miller, L., Nelson, R., Paunescu, T.G., Mc-Kee, M., Brown, D., and Al-Awqati, Q. (2010). Deletion of hensin/DMBT1 blocks conversion of β- to α-intercalated cells and induces distal renal tubular acidosis. *Proc Natl Acad Sci USA*, 107, pp. 21872–21877.

138. Bonnici, B., and Wagner, C.A. (2004). Postnatal expression of transport proteins involved in acid–base transport in mouse kidney. *Pflugers Arch*, 448, pp. 16–28.

139. Van Huyen, J.P., Cheval, L., Bloch-Faure, M., Belair, M.F., Heudes, D., Bruneval, P., Doucet A. (2008). GDF15 triggers homeostatic proliferation of acid-secreting collecting duct cells. *J Am Soc Nephrol*, 19, pp. 1965–1974.

140. Satlin, L.M., and Schwartz, G.J. (1989). Cellular remodeling of HCO_3^- secreting cells in rabbit renal collecting duct in response to an acidic environment. *J Cell Biol*, 109, pp. 1279–1288.

141. Eiam-ong, S., Laski, M.E., Kurtzman, N.A., and Sabatini, S. (1994). Effect of respiratory acidosis and respiratory alkalosis on renal transport enzymes. *Am J Physiol*, 267, pp. F390–F399.

142. Tsuruoka, S., Kittelberger, A.M., and Schwartz, G.J. (1998). Carbonic anhydrase II and IV mRNA in rabbit nephron segments: stimulation during metabolic acidosis. *Am J Physiol*, 274, pp. F259–F267.

143. Fejes-Toth, G., and Naray-Fejes-Toth, A. (1995). Effect of acid/base balance on H-ATPase 31 kD subunit mRNA levels in collecting duct cells. *Kidney Int*, 48, pp. 1420–1426.

144. Brown, D., Sabolic, I., and Gluck, S. (1991). Colchicine-induced redistribution of proton pumps in kidney epithelial cells. *Kidney Int*, Suppl. 33, pp. S79–S83.

145. Khadouri, C., Marsy, S., Barlet-Bas, C., Cheval, L., and Doucet, A. (1992). Effect of metabolic acidosis and alkalosis on NEM-sensitive ATPase in rat nephron segments. *Am J Physiol*, 262, pp. F583–F590.

146. Hafner, P., Grimaldi, R., Capuano, P., Capasso, G., and Wagner, C.A. (2008). Pendrin in the mouse kidney is primarily regulated by C^{l}-excretion but also by systemic metabolic acidosis. *Am J Physiol Cell Physiol*, 295, pp. C1658–C1667.

147. Bastani, B., Purcell, H., Hemken, P., Trigg, D., and Gluck, S. (1991). Expression and distribution of renal vacuolar proton-translocating adenosine triphosphatase in response to chronic acid and alkali loads in the rat. *J Clin Invest*, 88, pp. 126–136.

148. Sabolic, I., Brown, D., Gluck, S.L., and Alper, S.L. (1997). Regulation of AE1 anion exchanger and H(+)-ATPase in rat cortex by acute metabolic acidosis and alkalosis. *Kidney Int*, 51, pp. 125–137.

149. Amlal, H., Habo, K., and Soleimani, M. (2000). Potassium deprivation upregulates expression of renal basolateral Na(+)-HCO(3)(-) cotransporter (NBC-1). *Am J Physiol Renal Physiol*, 279, pp. F532–F543.

150. Elger, M., Bankir, L., and Kriz, W. (1992). Morphometric analysis of kidney hypertrophy in rats after chronic potassium depletion. *Am J Physiol*, 262, pp. F656–F667.

151. Bailey, M., Capasso, G., Agulian, S., Giebisch, G., and Unwin, R. (1999). The relationship between distal tubular proton secretion and dietary potassium depletion: evidence for up-regulation of H^+-ATPase. *Nephrol Dial Transplant*, 14, pp. 1435–1440.

152. Schwartz, G.J., and Al-Awqati, Q. (1985). Carbon dioxide causes exocytosis of vesicles containing H+ pumps in isolated perfused proximal and collecting tubules. *J Clin Invest*, 75, pp. 1638–1644.

153. van Adelsberg, J., and Al-Awqati, Q. (1986). Regulation of cell pH by Ca2+-mediated exocytotic insertion of H^+-ATPases. *J Cell Biol*, 102, pp. 1638–1645.

154. Ludwig, M.G., Vanek, M., Guerini, D., Gasser, J.A, Jones, C.E, Junker, U., Hofstetter, H., Wolf, R.M., and Seuwen, K. (2003). Proton-sensing G-protein-coupled receptors. *Nature*, 425, pp. 93–98.

155. Pastor-Soler, N., Beaulieu, V., Litvin, T.N., Da Silva, N., Chen, Y., Brown, D., Buck, J., Levin, L.R., and Breton, S. (2003). Bicarbonate-regulated adenylyl cyclase (sAC) is a sensor that regulates pH-dependent V-ATPase recycling. *J Biol Chem*, 278, pp. 49523–49529.

156. Paunescu, T.G., Da Silva, N., Russo, L.M., McKee, M., Lu, H.A., Breton, S., and Brown, D. (2008). Association of soluble adenylyl cyclase with the V-ATPase in renal epithelial cells. *Am J Physiol Renal Physiol*, 294, pp. F130–F138.

157. Paunescu, T.G., Ljubojevic, M., Russo, L.M., Winter, C., McLaughlin, M.M., Wagner, C.A., Breton, S., and Brown, D. (2010). cAMP stimulates apical V-ATPase accumulation, microvillar elongation, and proton extrusion in kidney collecting duct A-intercalated cells. *Am J Physiol Renal Physiol*, 298, pp. F643–F654.

158. Allen, A.M., Zhuo, J., and Mendelsohn, F.A. (2000). Localization and function of angiotensin AT1 receptors. *Am J Hypertens*, 13, pp. 31S–38S.

159. Burns, K.D., Regnier, L., Roczniak, A., and Hebert, R.L. (1996). Immortalized rabbit cortical collecting duct cells express AT1 angiotensin II receptors. *Am J Physiol*, 271, pp. F1147–F1157.

160. Navar, L.G., Harrison-Bernard, L.M., Wang, C.T., Cervenka, L., and Mitchell, K.D. (1999). Concentrations and actions of intraluminal angiotensin II. *J Am Soc Nephrol*, 10(Suppl. 11), pp. S189–S195.

161. Pech, V., Zheng, W., Pham, T.D., Verlander, J.W., and Wall, S.M. (2008). Angiotensin II activates H^+-ATPase in type A intercalated cells. *J Am Soc Nephrol*, 19, pp. 84–91.

162. Rothenberger, F., Velic, A., Stehberger, P.A., Kovacikova, J., and Wagner, C.A. (2007). Angiotensin II stimulates vacuolar H^+-ATPase activity in renal acid-secretory intercalated cells from the outer medullary collecting duct. *J Am Soc Nephrol*, 18, pp. 2085–2093.

163. Pech, V., Kim, Y.H., Weinstein, A.M., Everett, L.A., Pham, T.D., and Wall, S.M. (2007). Angiotensin II increases chloride absorption in the cortical collecting duct in mice through a pendrin-dependent mechanism. *Am J Physiol Renal Physiol*, 292, pp. F914–F920.

164. Weiner, I.D., New, A.R, Milton, A.E., and Tisher, C.C. (1995). Regulation of luminal alkalinization and acidification in the cortical collecting duct by angiotensin II. *Am J Physiol*, 269, pp. F730–F738.

165. Leviel, F., Hubner, C.A., Houillier, P., Morla, L., El Moghrabi, S., Brideau, G., Hassan, H., Parker, M.D., Kurth, I., Kougioumtzes, A., *et al.* (2010). The Na^+-dependent chloride-bicarbonate exchanger SLC4A8 mediates

an electroneutral Na^+ reabsorption process in the renal cortical collecting ducts of mice. *J Clin Invest*, 120, pp. 1627–1635.

166. Wagner, C.A., Giebisch, G., Lang, F., and Geibel, J.P. (1998). Angiotensin II stimulates vesicular H^+-ATPase in rat proximal tubular cells. *Proc Natl Acad Sci USA*, 95, pp. 9665–9668.
167. Levine, D.Z., Iacovitti, M., Luck, B., Hincke, M.T., Burns, K.D., and Fryer, J.N. (2000). Surviving rat distal tubule bicarbonate reabsorption: effects of chronic AT(1) blockade. *Am J Physiol Renal Physiol*, 278, pp. F476–F483.
168. Bouby, N., Hus-Citharel, A., Marchetti, J., Bankir, L., Corvol, P., and Llorens-Cortes, C. (1997). Expression of type 1 angiotensin II receptor subtypes and angiotensin II-induced calcium mobilization along the rat nephron. *J Am Soc Nephrol*, 8, pp. 1658–1667.
169. Winter, C., Schulz, N., Giebisch, G., Geibel, J.P., and Wagner, C.A. (2004). Nongenomic stimulation of vacuolar H^+-ATPases in intercalated renal tubule cells by aldosterone. *Proc Natl Acad Sci USA*, 101, pp. 2636–2641.
170. Wagner, C.A., and Geibel, J.P. (2002). Acid-base transport in the collecting duct. *J Nephrol*, 15 (Suppl. 5), pp. S112–S127.
171. Vallés, P., Wysocki, J., Salabat, M.R., Cokic, I., Ye, M., LaPointe, M.S., and Batlle, D. (2005). Angiotensin II increases H^+-ATPase B1 subunit expression in medullary collecting ducts. *Hypertension*, 45, pp. 818–823.
172. Brown, D., and Breton, S. (2000). *The Kidney: Physiology and Pathophysiology*, eds. Seldin, D.W., and Giebisch, G. Philadelphia: Lippincott Williams & Wilkins, pp. 171–191.
173. Stone, D.K., Seldin, D.W., Kokko, J.P., and Jacobson, H.R. (1983). Mineralocorticoid modulation of rabbit medullary collecting duct acidification. A sodium-independent effect. *J Clin Invest*, 72, pp. 77–83.
174. Garg, L.C., and Narang, N. (1988). Effects of aldosterone on NEM-sensitive ATPase in rabbit nephron segments. *Kidney Int*, 34, pp. 13–17.
175. Verrey, F. (1999). Early aldosterone action: toward filling the gap between transcription and transport. *Am J Physiol*, 277, pp. F319–F327.
176. Eiam-Ong, S., Kurtzman, N.A., and Sabatini, S. (1993). Regulation of collecting tubule adenosine triphosphatases by aldosterone and potassium. *J Clin Invest*, 91, pp. 2385–2392.

177. Winter, C., Kampik, N.B., Vedovelli, L., Rothenberger, F., Paunescu, T.G., Stehberger, P.A., Brown, D., John, H., and Wagner, C.A. Aldosterone stimulates vacuolar H(+)-ATPase activity in renal acid-secretory intercalated cells mainly via a protein kinase C-dependent pathway. *Am J Physiol Cell Physiol*, 301, pp. C1251–C1261.

178. Alzamora, R., Thali, R.F., Gong, F., Smolak, C., Li, H., Baty, C.J., Bertrand, C.A., Auchli, Y., Brunisholz, R.A., Neumann, D., Hallows, K.R., Pastor-Soler, N.M. (2010). PKA regulates vacuolar H^+-ATPase localization and activity via direct phosphorylation of the a subunit in kidney cells. *J Biol Chem*, 285, pp. 24676–24685.

179. Nguyen, G., Delarue, F., Burckle, C., Bouzhir, L., Giller, T., and Sraer, J.D. (2002). Pivotal role of the renin/prorenin receptor in angiotensin II production and cellular responses to renin. *J Clin Invest*, 109, pp. 1417–1427.

180. L'Huillier, N., Sharp, M.G.F., Dunbar, D.R., and Mullins, J.J. (2006). On the relationship between the renin receptor and the vacuolar proton ATPase membrane sector associated protein (M8–M9). In *The Local Cardiac Renin Angiotensin-Aldosterone System* (Frolich, E.D., and Richard, N.R.E., eds.) Heidelberg: Springer, pp. 17–34.

181. Nguyen, G. (2011). Renin, (pro)renin and receptor: an update. *Clin Sci (Lond)*, 120, pp. 169–178.

182. Cousin, C., Bracquart, D., Contrepas, A., Corvol, P., Muller, L., and Nguyen, G. (2009). Soluble form of the (pro)renin receptor generated by intracellular cleavage by furin is secreted in plasma. *Hypertension*, 53, pp. 1077–1082.

183. Ludwig, J., Kerscher, S., Brandt, U., Pfeiffer, K., Getlawi, F., Apps, D. K. and Schägger, H. (1998). Identification and characterization of a novel 9.2-kDa membrane sector-associated protein of vacuolar proton-ATPase from chromaffin granules. *J Biol Chem*, 273, pp. 10939–10947.

184. Schefe, J.H., Menk, M., Reinemund, J., Effertz, K., Hobbs, R.M., Pandolfi, P.P., Ruiz, P., Unger, T., and Funke-Kaiser, H. (2006). A novel signal transduction cascade involving direct physical interaction of the renin/prorenin receptor with the transcription factor promyelocytic zinc finger protein. *Circ Res*, 99, pp. 1355–1366.

185. Sihn, G., Rousselle, A., Vilianovitch, L., Burckle, C., and Bader, M. (2010). Physiology of the (pro)renin receptor: Wnt of change? *Kidney Int*, 78, pp. 246–256.

186. Saris, J.J., t Hoen, P.A., Garrelds, I.M., Dekkers, D.H., den Dunnen, J.T., Lamers, J.M., and Jan Danser, A.H. (2006). Prorenin induces

intracellular signaling in cardiomyocytes independently of angiotensin II. *Hypertension*, 48, pp. 564–571.

187. Sakoda, M., Ichihara, A., Kurauchi-Mito, A., Narita, T., Kinouchi, K., Murohashi-Bokuda, K., Saleem, M.A., Nishiyama, A., Suzuki, F., and Itoh, H. (2010). Aliskiren inhibits intracellular angiotensin II levels without affecting (pro)renin receptor signals in human podocytes. *Am J Hypertens*, 23, pp. 575–580.
188. Contrepas, A., Walker, J., Koulakoff, A., Franek, K.J., Qadri, F., Giaume, C., Corvol, P., Schwartz, C.E., and Nguyen, G. (2009). A role of the (pro)renin receptor in neuronal cell differentiation. *Am J Physiol Regul Integr Comp Physiol*, 297, pp. R250–R257.
189. Advani, A., Kelly, D.J., Cox, A.J., White, K.E., Advani, S.L., Thai, K., Connelly, K.A., Yuen, D., Trogadis, J., Herzenberg, A.M., *et al.* (2009). The (Pro)renin receptor: site-specific and functional linkage to the vacuolar H^+-ATPase in the kidney. *Hypertension*, 54, pp. 261–269.
190. Feldman, D.L., Jin, L., Xuan, H., Contrepas, A., Zhou, Y., Webb, R.L., Mueller, D.N., Feldt, S., Cumin, F., Maniara, W., *et al.* (2008). Effects of aliskiren on blood pressure, albuminuria, and (pro)renin receptor expression in diabetic TG(mRen-2)27 rats. *Hypertension*, 52, pp. 130–136.
191. Nguyen, G. (2006). Increased cyclooxygenase-2, hyperfiltration, glomerulosclerosis, and diabetic nephropathy: put the blame on the (pro)renin receptor? *Kidney Int*, 70, pp. 618–620.

Chapter 10

Long-Term Regulation of Vacuolar H^+-ATPase by Angiotensin II in Proximal Tubule Cells

Luciene Regina Carraro-Lacroix,[a] Thaíssa Dantas Pessoa,[b] and Gerhard Malnic[b]

[a] *Federal University of Paraná, Brazil*

[b] *Department of Physiology and Biophysics, Institute of Biomedical Sciences, University of São Paulo, Av. Prof. Lineu Prestes 1524, São Paulo, SP 05508-000, São Paulo, Brazil*

gemalnic@usp.br

10.1 General Considerations

The function of the kidney is to keep constant the volume and composition of the body fluids, by filtering the entire plasma volume every 30 minutes and reabsorbing approximately two-thirds of the filtrate in the proximal tubule (PT) [1, 2]. The glomerular filtrate enters the lumen of the PT, and the majority of filtered sodium and HCO_3^- are reclaimed along the early (S1) and convoluted (S2) portions of the PT in a manner coupled to the secretion of H^+ into the lumen [2]. Furthermore, large amounts of chloride and water are also reabsorbed in this nephron segment. Considering the role

Handbook of H^+-ATPases
Edited by Suguru Nakamura

ISBN 978-981-4411-91-2 (Hardcover), 978-981-4411-92-9 (eBook)
www.panstanford.com

of PTs and other nephron segments, less than 1% of water, sodium, chloride, and variable amounts of other solutes are excreted in the urine.

Fine control of sodium and water excretion, necessary for blood pressure and osmolality homeostasis has been traditionally attributed to regulation of transport in the distal nephron and collecting ducts. However, increasing evidence has been indicating that the PT transport machinery, responsible for the massive reabsorption of sodium and water, is acutely and chronically regulated by blood pressure, extracellular fluid volume, the renin-angiotensin system (RAS), and the sympathetic nervous system [3].

10.2 Angiotensin II Action on H^+-ATPase

The RAS plays an important role in the regulation of arterial pressure and body fluid and electrolyte homeostasis through its major active effector peptide Angiotensin II (Ang II) [4, 5]. The hormone elicits its biological actions through a combination of intra- and extra-renal effects that regulate renal salt and water excretion as well as fluid intake [6]. In addition to its role as a potent vasoconstrictor involved in renal blood flow and glomerular filtration regulation, Ang II plays an important role in the direct regulation of solute transport by renal tubules [7]. In other words, the hormone is extremely important for Na^+, water, and Cl^- reabsorption by the kidney, in part *via* stimulation of argipressin or antidiuretic hormone (ADH) (vasopressin) secretion. Moreover, Ang II exerts important stimulatory effects on aldosterone secretion and much of the sodium-conserving actions are also mediated by the renal effects of aldosterone [5].

As shown in Fig. 10.1, a number of transport mechanisms have extensively been reported as targets of Ang II action in kidney, including the Na^+–HCO_3^- co-transporter [8–10], Na^+/H^+ exchanger [11–17], H^+-ATPase [7, 18–22], Cl^-/HCO_3^- exchanger [23], Na^+/K^+-ATPase [24–26], Ca^{+2}-ATPase [27], Na–Pi co-transporter [28], epithelial Na^+ channel [29–31], Na–Glucose co-transporter [32], K^+ channels [33, 34], Na^+–K^+–$2Cl^-$ co-transporter and chloride channels [35, 36].

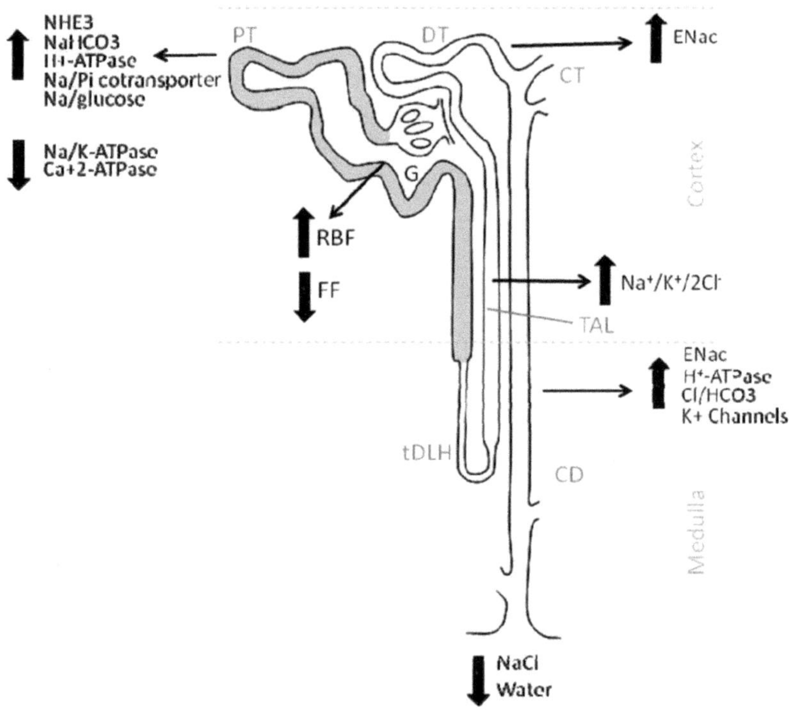

Figure 10.1 Schematic representation of the nephron depicting the most important transport mechanisms modulated by Angiotensin II in kidney. (PT, proximal tubules; DT, distal tubules; CT, collecting tubule; TAL, thick ascending limb; tDLH, thin descending limb of Henle's loop; CD, collecting duct; G, glomerulus; RBF, renal blood flow; FF, filtration fraction).

Since the Na^+/H^+ exchanger is the predominant mechanism for reabsorption of Na^+ and secretion of H^+ across the apical membrane of PT cells [37], it constitutes the main target for the regulating factors in PT. However, up to 40% of PT bicarbonate reabsorption is Na^+-independent and believed to be mediated by the H^+-ATPase expressed in the brush-border membrane [38]. In this segment, the vacuolar H^+-ATPase has been shown to be modulated by Ang II in either animal or cell culture models [18, 19, 39]. In this system, Ang II (10^{-9} M) was seen to stimulate the vacuolar H^+-ATPase activity after both acute [39] and chronic treatment [19].

In general, Ang II binds to one of two major receptor subtypes to exert its biological actions, AT_1 and AT_2, the first being the responsible for most of the known physiological actions of the hormone [40]. In the kidney, both receptors are expressed, but AT_1 is predominant [41]. Immunohistochemical studies have demonstrated an abundance of AT_1 receptors on the luminal surface of proximal and distal tubule cells as well as on vascular smooth muscle cells of afferent and efferent arterioles and on glomerular mesangial cells [42]. In general, binding of Ang II to AT_1 receptor causes stimulation of mechanisms involved with bicarbonate reabsorption and acid secretion, with consequent increase in overall bicarbonate reabsorption [38].

Several studies have demonstrated a stimulatory effect of Ang II on the H^+-ATPase activity and/or expression [18, 21, 22, 39, 43], but inhibition of the ATPase by the hormone has also been reported [7]. In all cases where the pharmacological profile was characterized with specific antagonists, the hormonal action occurred through AT_1 receptors [38].

The regulation of H^+-ATPase by Ang II depends also on the concentration of the hormone. Inhibition of H^+-ATPase by Ang II in cortical collecting duct was observed at 10^{-10} to 10^{-5} M, with maximum inhibition at 10^{-8} M [7]. On the other hand, in PTs, 10^{-9} M Ang II provoked a stimulation of the H^+-ATPase activity [18, 39] in a similar way as was observed in collecting duct cells (Madin Darby Canine kidney cells (MDCK cells)), where the maximum stimulation was at 10^{-7} M Ang II [22]. In addition, Wagner *et al.* found that 10^{-9} M Ang II activates H^+-ATPase in intercalated cells *via* AT_1 receptors, with consequent stimuli of the trafficking of the ATPase into the membrane [18]. Even low concentrations of the hormone (10^{-12} M) were able to stimulate vacuolar H^+-ATPase in late distal segments [44].

It has been well documented that Ang II can exert its effect on the Na^+/H^+ exchanger located in the brush-border membrane *via* multiple signaling pathways [12, 15, 16, 45]. Low concentrations stimulate, whereas high concentrations inhibit Na^+ transport in rats [16, 17, 46]. Low concentrations of Ang II inhibit adenylate cyclase, which leads to the stimulation of the Na^+/H^+ exchanger [45]. In addition, in most tissues, the physiological effects of Ang II are transduced by phospholipase C (PLC) activation, which leads to the hydrolysis of phosphatidylinositol 4,5-biphosphate to inositol triphosphate (IP_3) and diacylglycerol, and consequent activation of protein kinase C (PKC), which also stimulates the Na^+/H^+ exchanger [47].

In contrast, the inhibitory effect of high concentrations of hormone on sodium transport is believed to be, at least in part, dependent of an increase in intracellular calcium [12]. The signaling pathways involved with the effect of Ang II on the vacuolar H^+-ATPase activity, however, are poorly understood. One single study reported the signaling pathways in PT cells, where it was observed that Ang II regulates H^+-ATPase through a PKA-independent mechanism and the final effect of the hormone results from a complex balance between PKC and calcium [39]. Furthermore, in collecting duct cells, Oliveira-Souza *et al.* observed that the stimulatory effect of Ang II on H^+-ATPase is dependent on intracellular increases of Ca^{+2} [22]. Beyond PKC and calcium, activation of tyrosine kinase by AT_1 [48] was also observed and the authors speculated that this agent could be the responsible for the activation of H^+-ATPase by Ang II.

In isolated rat PTs and in intercalated cells, stimulation of H^+-ATPase by Ang II has been shown to be prevented by colchicine, an inhibitor of microtubule-dependent vesicular trafficking, indicating that vesicular insertion of H^+-ATPase-containing vesicles is involved in this stimulation [18]. In fact, there is strong evidence for the regulation of H^+-ATPase activity by trafficking and exocytosis. Holliday *et al.* reported that the amino-terminal domains of both isoforms of the B subunit, B1 and B2 of H^+-ATPase contain binding sites to F-actin and are likely responsible for the interaction between V-ATPase and actin filaments *in vivo* [49]. In the same way, Vitavska *et al.* have found a binding site of subunit C to F-actin [50], which also could allow for the trafficking. In addition, the interaction of the B1 subunit of the proton pump with sodium–hydrogen exchanger regulatory factor (NHERF) in B-intercalated cells could be responsible for the anchoring and/or targeting of membrane-associated H^+-ATPase molecules in this cell type [51], possibly providing a target for regulation by hormones.

Furthermore, in collecting duct cells, evidence has been shown that regulation of H^+-ATPase exocytosis is dependent on soluble

N-ethylmaleimide sensitive factor attachment receptor (SNARE), important for vesicle insertion into the membrane during exocytosis processes [52–54].

In our aforementioned work [39], we have shown that acute addition of Ang II (10^{-9} M) to immortalized rat proximal tubule cell (IRPTC) cells led to a significant increase in H^+ extrusion *via* H^+-ATPase. This acute effect may be related to insertion of pre-formed H^+-ATPase complexes and/or its subunits into the cell membrane, since the observed effect was visible within a period of minutes. Besides this acute effect, we next observed a chronic increase in H^+-ATPase activity by Ang II when IRPTC cells were followed for a period of up to 24 h. Our results have shown a significant increase in H^+-ATPase activity in 10 h Ang II-treated cells. This stimulatory effect was shown to be *via* AT_1 receptor, since the treatment with losartan totally prevented the effect of the hormone on the pH_i recovery mediated by vacuolar H^+-ATPase. A long-term effect of Ang II was also observed by other authors, however on NHE3-mediated transport in the PT cell line opossum kidney proximal (OKP). In this cellular model, the chronic effects of the hormone were shown to be dependent, at least in part, on transcriptional modulation of the exchanger [14].

Levine *et al.* have shown that chronic AT_1 inhibition of *in vivo* micro-perfused distal tubules from surviving nephrons of nephrectomized rats reduced bicarbonate reabsorption in close association with the reduction of synthesis and insertion of apical H^+-ATPase [55]. Nevertheless, another *in vivo* study has shown an increase in the relative protein abundance of the B1 isoform of the H^+-ATPase after chronic infusion of Ang II into the renal medulla of adrenalectomized rats [21].

The 56 kDa B subunit of H^+-ATPase is expressed as two highly homologous isoforms encoded, respectively by two different genes, B1 (ATP6V1B1), encoded by Atp6v1b1, and B2 (ATP6V1B2), encoded by Atp6v1b2 [38]. In intercalated cells isolated from mouse outer medullary collecting duct, Ang II was shown to stimulate the vacuolar H^+-ATPase, where the B1 subunit was required for the stimulation [56]. In addition, increases in B1 subunit expression were observed in medullary collecting ducts of Ang II-treated rats and it was speculated that this effect can up-regulate H^+ secretion in this tubule segment [21].

Unlike the B1 subunit, PT cells normally express apical membrane-associated V-ATPases containing the B2 isoform [57, 58]. Despite its predominantly cytoplasmic distribution in most cells under baseline conditions, where it contributes to the acidification of intracellular organelles, the B2-containing V-ATPase has also a role in epithelia under some conditions (Paunescu *et al.* 2004), possibly providing a target for regulation by hormones [58]. In fact, Mohebbi *et al.* have found a downregulation of the B2 subunit in the PT by mild hypothyroidism [59].

Besides studying the long-term effect of Ang II on H^+-ATPase [19], we evaluated the molecular basis of Ang II-induced stimulation of H^+-ATPase activity initially by analyzing the relative mRNA and protein abundance of the B2 subunit isoform of H^+-ATPase after chronic incubation with the hormone. In our model, we did not find significant differences in both B2 subunit mRNA and protein abundance in Ang II-treated cells in comparison to the controls. This shows that the chronic stimulatory effect of Ang II on vacuolar H^+-ATPase in PT cells is modulated by a mechanism(s) other than RNA and protein expression of the B2 subunit of the ATPase. Nevertheless, although we did not observe transcriptional–translational modulation of the B2 subunit by Ang II, we could not exclude a genomic regulation of other H^+-ATPase subunits by the hormone, since the utilization of actinomycin and cycloheximide (inhibitors of transcription and translation, respectively) partially, but not significantly, prevented the stimulatory effect of the long-term treatment with Ang II on the H^+-ATPase activity.

Although the determination of the H^+-ATPase subunit(s) genomically modulated by the long-term treatment with Ang II cannot be established on the basis of the results of this study, our observations suggest that Ang II modulates the synthesis of one or more subunits of vacuolar H^+-ATPase. In fact, Valles *et al.* have found an effect of Ang II in increasing the H^+-ATPase B1 subunit expression in medullary collecting ducts, unlike the membrane-sector a4 subunit [21]. On the other hand, Paunesco *et al.* observed, in renal medullary intercalated cells of B1-deficient mice, a compensatory membrane expression of the B2 subunit, probably due to a redistribution of B2-containing V-ATPases from the cytoplasm to the apical membrane domain and not by modulation of either mRNA or protein levels of this subunit [57].

Beyond control of V-ATPase subunit expression in specialized cells, regulation of H^+-ATPase activity is also accomplished through a number of other mechanisms, including reversible assembly and disassembly of V_0 and V_1, including domain changes leading to the efficiency of coupling of proton transport with ATP hydrolysis and control of their cellular localization by trafficking [60–63]. The V-ATPase translocation by vesicular trafficking between intracellular compartments is an essential mechanism for V-ATPase recruitment in response to a number of stimuli [64].

Our data [19] show that the stimulatory effect of Ang II on the H^+-ATPase activity in chronically treated cells is accompanied by an increase in cell surface expression of the V-ATPase, indicating that the hormone induces the trafficking of the ATPase from intracellular to plasma membranes. Furthermore, as the presence of V_1 subunit B2 among the biotinylated proteins is an indication that the plasma membrane contains assembled vacuolar H^+-ATPase, it could be speculated that Ang II ultimately leads to an increase in functionally active H^+-ATPase in the plasma membrane.

In collecting duct cells, it has been shown that regulation of H^+-ATPase exocytosis is dependent on SNARE proteins [52–54, 65]. In addition, the interaction of the B1 subunit of the proton pump with NHE-RF in B-intercalated cells could be responsible for the anchoring and/or targeting of membrane-associated H^+-ATPase molecules in this cell type [51], possibly providing a target for regulation by hormones (Fig. 10.2).

Furthermore, Holliday *et al.* reported that the amino-terminal domains of both isoforms of the B subunit, B1 and B2, contain binding sites to F-actin which may be responsible for the interaction between V-ATPase and actin filaments *in vivo* [49] and could allow for the trafficking. The interaction between F-actin and subunit B was also shown to be dependent on phosphatidylinositol 3-kinase (PI3K) [66]. On the basis of this interaction, Sautin *et al.* have demonstrated an involvement of PI3K-dependent signaling in the control of V-ATPase trafficking, assembly, and function by glucose in renal epithelial cells [67], suggesting that the glucose effects on H^+-ATPase trafficking can be mediated by changes in actin microfilaments promoted by PI3K. Our findings now suggest that Ang II effects on vacuolar H^+-ATPase activity are also dependent on PI3K, since the stimulatory effect of the hormone on the H^+-ATPase activity was completely reverted by wortmannin, a specific inhibitor of PI3K. Activation of the tyrosine kinase receptor-dependent and -independent pathways could be involved in the upstream mechanism leading to PI3K. In fact, our results show an involvement of tyrosine kinase phosphorylation in this regulation, since genistein completely blocked the stimulatory effect of the hormone on the H^+-ATPase activity.

In addition, we were able to show an involvement of the p38 mitogen-activated protein kinase (MAPK) cascade in the stimulatory effect of Ang II on the H^+-ATPase activity. This effect was totally abolished after pre-incubation with the p38 MAPK inhibitor SB-203580. Similarly, other authors observed that the stimulation of NHE3 by luminal glucose (by a process involving sodium-dependent glucose co-transporters (SGLT1)) was also dependent on an activation of signaling pathways that include early activation of p38 MAPK, followed by activation of MAPK-activated protein kinase (MAPKAPK2), PI3-K, Akt2, and ezrin, which lead to stimulation of NHE3 activity due to increased exocytosis [68]. We speculated that similar pathways could be involved in our model. Unlike the possible transcriptional-independent role of p38 MAPK, Larsen *et al.* have shown that glucose regulates osteoclast H^+-ATPase expression by a mechanism involving p38 MAPK [69]. In our system, the increased activity of H^+-ATPase in chronically Ang II-treated cells was not accompanied by changes in the expression of, at least, the B2 subunit of vacuolar H^+-ATPase, suggesting that maybe P38 MAPK plays a role in the control of cytoskeleton proteins.

10.3 V-ATPase is Associated with (Pro)-Renin Receptor

Mutations in the genes encoding for Ang II, renin, Angiotensin converting enzyme (ACE), or AT_1R in humans are associated with renal tubular dysgenesis (RTD). Kidneys of patients with RTD

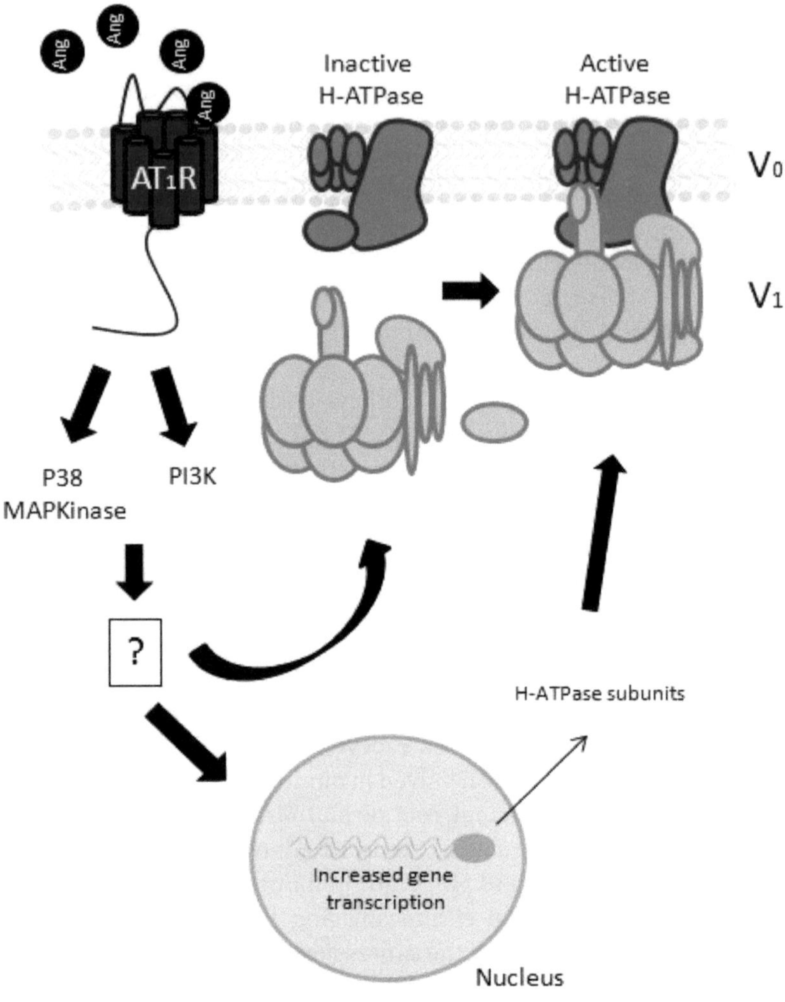

Figure 10.2 Schematic illustration of the mechanisms involved in the long-term modulation of H^+-ATPase by Angiotensin II. Ligation of Ang II to its receptor AT_1 leads to activation of P38 MAPKinase and PI3K pathways which cause the upstream assembly of V_0 and V_1 domains, ultimately forming the active complex. The same pathways are probably involved also with the stimulation of gene transcription of one or more H^+-ATPase subunits, which also contributes to the appearing of mounted H^+-ATPase at the plasma membrane.

demonstrate reduced number of PTs, collapsed collecting ducts, enlarged glomeruli, and thickened arteries [70].

Thus, Ang II is not the only member of the RAS system that plays an important role in modulating H^+-ATPase. The (pro)renin receptor [(P)-RR] is a 350 amino acid, single-transmembrane protein which binds both renin and prorenin with high affinity [71]. In the kidneys, *in situ* hybridization revealed low levels of (P)-RR expression in the glomerulus, PTs and arteries, and relatively high levels in the distal tubule. The most prominent expression was detected in the collecting duct. A detailed study showed that (P)-RR is expressed specifically at intercalated A cell types, which is a V-ATPase-rich cell and important for urinary acidification. Co-localization studies have demonstrated that (P)-RR co-localizes to B1/2 V-ATPase subunit in intercalated A cells [72].

The idea that the prorenin and renin receptor could have any relation to V-ATPase emerged from the comparison of (P)-RR protein structure to other proteins using the available databases, that (P)-RR is identical to the ATPase, H^+ transporting, lysosomal accessory protein 2 (ATP6ap2), a protein that associates with the vacuolar H^+-ATPase [73, 74]. In fact, ablation of (P)-RR in cardiomyocytes selectively suppressed expression of the V_0 subunits of V-ATPase, resulting in de-acidification of the intracellular vesicles [75]. It was determined that (P)-RR binding to V_0 depends on transmembrane and extracellular domains of the receptor, but it is not clear if this interaction is direct or not [76].

Functional studies have shown that V-ATPase inhibition attenuates renin and P-RR-induced (P)-RR-dependent extracellular signal-regulated kinase (ERK) phosphorylation in MDCK cells [72]. Besides, (P)-RR was related to Wnt receptor as an accessory adaptor between Wnt receptor and the V-ATPase complex, and both, (P)-RR and V-ATPase are required for Wnt signaling transduction [76]. A novel line of evidence suggests that both (pro)renin and angiotensin activate p38 MAPKinase, which, in turn, activates actin filaments [77]. Concerning angiotensin, it is well established that activation of this pathway promotes increasing vacuolar H^+-ATPase activity [19], which gives some evidence that (pro)renin binding to its receptor may act in a similar way.

References

1. McDonough, A.A. (2010) *Am. J. Physiol. Regul. Integr. Comp. Physiol.* 298, R851–R861.
2. Skelton, L.A., Boron, W.F., and Zhou, Y. (2010) *J. Nephrol.* 23(Suppl 16), S4–S18.
3. McDonough, A.A. (2010) *Am. J. Physiol. Regul. Integr. Comp. Physiol.* 298, R851–R861.
4. Allen, A.M., Zhuo, J., and Mendelsohn, F.A. (2000) *Am. J. Hypertens.* 13, S31S–S38.
5. Harris, P.J. and Navar, L.G. (1985) *Am. J. Physiol.* 248, F621–F630.
6. Hall, J.E., Brands, M.W., and Henegar, J.R. (1999) *J. Am. Soc. Nephrol.* 10(Suppl 12), S258–S265.
7. Tojo, A., Tisher, C.C., and Madsen, K.M. (1994) *Am. J. Physiol.* 267, F1045–F1051.
8. Horita, S., Zheng, Y., Hara, C., Yamada, H., Kunimi, M., Taniguchi, S., Uwatoko, S., Sugaya, T., Goto, A., Fujita, T., and Seki, G. (2002) *Hypertension.* 40, 707–712.
9. Turban, S., Beutler, K.T., Morris, R.G., Masilamani, S., Fenton, R.A., Knepper, M.A., and Packer, R.K. (2006) *Kidney Int.* 70, 660–668.
10. Geibel, J., Giebisch, G., and Boron, W.F. (1990) *Proc. Natl. Acad. Sci. USA* 87, 7917–7920.
11. Dixit, M.P., Xu, L., Xu, H., Bai, L., Collins, J.F., and Ghishan, F.K. (2004) *Biochim. Biophys. Acta.* 1664, 38–44.
12. Du, Z., Ferguson, W., and Wang, T. (2003) *Am. J. Physiol. Renal Physiol.* 284, F688–F692.
13. Leong, P.K., Yang, L.E., Holstein-Rathlou, N.H., and McDonough, A.A. (2002) *Am. J. Physiol. Renal Physiol.* 283, F1142–F1150.
14. Xu, L., Dixit, M.P., Nullmeyer, K.D., Xu, H., Kiela, P.R., Lynch, R.M., and Ghishan, F.K. (2006) *Biochim. Biophys. Acta.* 1758, 519–526.
15. du, C.D., Chalumeau, C., Defontaine, N., Klein, C., Kellermann, O., Paillard, M., and Poggioli, J. (2003) *Kidney Int.* 64, 939–949.
16. Houillier, P., Chambrey, R., Achard, J.M., Froissart, M., Poggioli, J., and Paillard, M. (1996) *Kidney Int.* 50, 1496–1505.
17. Reilly, A.M., Harris, P.J., and Williams, D.A. (1995) *Am. J. Physiol.* 269, F374–F380.
18. Wagner, C.A., Giebisch, G., Lang, F., and Geibel, J.P. (1998) *Proc. Natl. Acad. Sci. USA* 95, 9665–9668.

19. Carraro-Lacroix, L.R., Girardi, A.C., and Malnic, G. (2009) *Pflugers Arch.* 458, 969–979.
20. Valles, P.G., Carrizo, L., Seltzer, A., and Manucha, W. (2005) *Nephron Physiol.* 99, 90–100.
21. Valles, P., Wysocki, J., Salabat, M.R., Cokic, I., Ye, M., LaPointe, M.S., and Batlle, D. (2005) *Hypertension.* 45, 818–823.
22. Oliveira-Souza, M., Malnic, G., and Mello-Aires, M. (2002) *Kidney Int.* 62, 1693–1699.
23. Pech, V., Kim, Y.H., Weinstein, A.M., Everett, L.A., Pham, T.D., and Wall, S.M. (2007) *Am. J. Physiol Renal. Physiol.* 292, F914–F920.
24. Hakam, A.C. and Hussain, T. (2006) *Am. J. Physiol. Renal Physiol.* 290, F1430–F1436.
25. Shah, S. and Hussain, T. (2006) *Clin. Exp. Hypertens.* 28, 29–40.
26. Yingst, D.R., Massey, K.J., Rossi, N.F., Mohanty, M.J., and Mattingly, R.R. (2004) *Am. J. Physiol. Renal Physiol.* 287, F713–F721.
27. Assuncao-Miranda, I., Guilherme, A.L., Reis-Silva, C., Costa-Sarmento, G., Oliveira, M.M., and Vieyra, A. (2005) *Regul. Pept.* 127, 151–157.
28. Xu, L., Dixit, M.P., Chen, R., Dixit, N.M., Collins, J.F., and Ghishan, F.K. (2004) *Biochim. Biophys. Acta.* 1667, 114–121.
29. Beutler, K.T., Masilamani, S., Turban, S., Nielsen, J., Brooks, H.L., Ageloff, S., Fenton, R.A., Packer, R.K., and Knepper, M.A. (2003) *Hypertension.* 41, 1143–1150.
30. Peti-Peterdi, J., Warnock, D.G., and Bell, P.D. (2002) *J. Am. Soc. Nephrol.* 13, 1131–1135.
31. Sun, P., Yue, P., and Wang, W.H. (2012) *Am. J. Physiol. Renal Physiol.* 302, F679–F687.
32. Bautista, R., Manning, R., Martinez, F., Avila-Casado, M.C., Soto, V., Medina, A., and Escalante, B. (2004) *Am. J. Physiol. Renal Physiol.* 286, F127–F133.
33. Wei, Y. and Wang, W. (2003) *Am. J. Physiol. Renal Physiol.* 284, F175–F181.
34. Lu, M., Zhu, Y., Balazy, M., Reddy, K.M., Falck, J.R., and Wang, W. (1996) *J. Gen. Physiol.* 108, 537–547.
35. Ye, T., Liu, Z.Q., Sun, C.F., Zheng, Y., Ma, A.Q., and Fang, Y. (2005) *Chin. Med. J. (Engl.)* 118, 1945–1951.
36. Kovacs, G., Peti-Peterdi, J., Rosivall, L., and Bell, P.D. (2002) *Am. J. Physiol. Renal Physiol.* 282, F301–F306.
37. Alpern, R.J. (1990) *Physiol. Rev.* 70, 79–114.

38. Wagner, C.A., Finberg, K.E., Breton, S., Marshansky, V., Brown, D., and Geibel, J.P. (2004) *Physiol. Rev.* 84, 1263–1314.
39. Carraro-Lacroix, L.R. and Malnic, G. (2006) *Pflugers Arch.* 452, 728–736.
40. de, G.M., Catt, K.J., Inagami, T., Wright, J.W., and Unger, T. (2000) *Pharmacol. Rev.* 52, 415–472.
41. Miyata, N., Park, F., Li, X.F., and Cowley, A.W., Jr. (1999) *Am. J. Physiol.* 277, F437–F446.
42. Navar, L.G., Harrison-Bernard, L.M., Imig, J.D., Cervenka, L., and Mitchell, K.D. (2000) *Am. J. Hypertens.* 13, 45S–54S
43. Levine, D.Z., Iacovitti, M., Buckman, S., Hincke, M.T., Luck, B., and Fryer, J.N. (1997) *Am. J. Physiol.* 272, F799–F808.
44. Barreto-Chaves, M.L. and Mello-Aires, M. (1996) *Am. J. Physiol.* 271, F977–F984.
45. Douglas, J.G., Romero, M., and Hopfer, U. (1990) *Kidney Int. Suppl.* 30, S43–S47.
46. Wang, T. and Chan, Y.L. (1990) *J. Pharmacol. Exp. Ther.* 252, 689–695.
47. Schelling, J.R., Singh, H., Marzec, R., and Linas, S.L. (1994) *Am. J. Physiol.* 267, C1239–C1245.
48. Doan, T.N., Ali, M.S., and Bernstein, K.E. (2001) *J. Biol. Chem.* 276, 20954–20958.
49. Holliday, L.S., Lu, M., Lee, B.S., Nelson, R.D., Solivan, S., Zhang, L., and Gluck, S.L. (2000) *J. Biol. Chem.* 275, 32331–32337.
50. Vitavska, O., Wieczorek, H., and Merzendorfer, H. (2003) *J. Biol. Chem.* 278, 18499–18505.
51. Breton, S., Wiederhold, T., Marshansky, V., Nsumu, N.N., Ramesh, V., and Brown, D. (2000) *J. Biol. Chem.* 275, 18219–18224.
52. Nicoletta, J.A., Ross, J.J., Li, G., Cheng, Q., Schwartz, J., Alexander, E.A., and Schwartz, J.H. (2004) *Am. J. Physiol. Cell Physiol.* 287, C1366–C1374.
53. Banerjee, A., Shih, T., Alexander, E.A., and Schwartz, J.H. (1999) *J. Biol. Chem.* 274, 26518–26522.
54. Banerjee, A., Li, G., Alexander, E.A., and Schwartz, J.H. (2001) *Am. J. Physiol. Cell Physiol.* 280, C775–C781.
55. Levine, D.Z., Iacovitti, M., Luck, B., Hincke, M.T., Burns, K.D., and Fryer, J.N. (2000) *Am. J. Physiol. Renal Physiol.* 278, F476–F483.
56. Rothenberger, F., Velic, A., Stehberger, P.A., Kovacikova, J., and Wagner, C.A. (2007) *J. Am. Soc. Nephrol.* 18, 2085–2093.
57. Paunescu, T.G., Russo, L.M., Da, S.N., Kovacikova, J., Mohebbi, N., Van Hoek, A.N., McKee, M., Wagner, C.A., Breton, S., and Brown, D. (2007) *Am. J. Physiol. Renal Physiol.* 293, F1915–F1926.

58. Paunescu, T.G., Da, S.N., Marshansky, V., McKee, M., Breton, S., and Brown, D. (2004) *Am. J. Physiol. Cell Physiol.* 287, C149–C162.

59. Mohebbi, N., Kovacikova, J., Nowik, M., and Wagner, C.A. (2007) *Am. J. Physiol. Renal Physiol.* 293, F416–F427.

60. Breton, S. and Brown, D. (2007) *Am. J. Physiol. Renal Physiol.* 292, F1–F10.

61. Forgac, M. (2007) *Nat. Rev. Mol. Cell Biol.* 8, 917–929.

62. Marshansky, V. and Futai, M. (2008) *Curr. Opin. Cell Biol.* 20, 415–426.

63. Nanda, A., Brumell, J.H., Nordstrom, T., Kjeldsen, L., Sengelov, H., Borregaard, N., Rotstein, O.D., and Grinstein, S. (1996) *J. Biol. Chem.* 271, 15963–15970.

64. Brown, D., Sabolic, I., and Gluck, S. (1992) *J. Exp. Biol.* 172, 231–243.

65. Schwartz, J.H., Li, G., Yang, Q., Suri, V., Ross, J.J., and Alexander, E.A. (2007) *Kidney Int.* 72, 1310–1315.

66. Chen, S.H., Bubb, M.R., Yarmola, E.G., Zuo, J., Jiang, J., Lee, B.S., Lu, M., Gluck, S.L., Hurst, I.R., and Holliday, L.S. (2004) *J. Biol. Chem.* 279, 7988–7998.

67. Sautin, Y.Y., Lu, M., Gaugler, A., Zhang, L., and Gluck, S.L. (2005) *Mol. Cell Biol.* 25, 575–589.

68. Donowitz, M. and Li, X. (2007) *Physiol. Rev.* 87, 825–872.

69. Larsen, K.I., Falany, M.L., Ponomareva, L.V., Wang, W., and Williams, J.P. (2002) *J. Cell Biochem.* 87, 75–84.

70. Song, R. and Yosypiv, I.V. (2011) *Int. J. Nephrol.* 2011, 247048.

71. Nguyen, G., Delarue, F., Burckle, C., Bouzhir, L., Giller, T., and Sraer, J.D. (2002) *J. Clin. Invest.* 109, 1417–1427.

72. Advani, A., Kelly, D.J., Cox, A.J., White, K.E., Advani, S.L., Thai, K., Connelly, K.A., Yuen, D., Trogadis, J., Herzenberg, A.M., Kuliszewski, M.A., Leong-Poi, H., and Gilbert, R.E. (2009) *Hypertension* 54, 261–269.

73. Campbell, D.J. (2008) *Hypertension* 51, 1259–1264.

74. Ludwig, J., Kerscher, S., Brandt, U., Pfeiffer, K., Getlawi, F., Apps, D.K., and Schagger, H. (1998) *J. Biol. Chem.* 273, 10939–10947.

75. Kinouchi, K., Ichihara, A., Sano, M., Sun-Wada, G.H., Wada, Y., Kurauchi-Mito, A., Bokuda, K., Narita, T., Oshima, Y., Sakoda, M., Tamai, Y., Sato, H., Fukuda, K., and Itoh, H. (2010) *Circ. Res.* 107, 30–34.

76. Cruciat, C.M., Ohkawara, B., Acebron, S.P., Karaulanov, E., Reinhard, C., Ingelfinger, D., Boutros, M., and Niehrs, C. (2010) *Science* 327, 459–463.

77. Danser, A.H. (2009) *Hypertension* 54, 219–221.

Chapter 11

Vacuolar H^+-ATPase in Distal Renal Tubular Acidosis and Diabetes

Irfan Shukrullah,[a] Patricia Valles,[b] and Daniel Batlle[a]

[a] *Division of Nephrology/Hypertension, Northwestern University Feinberg School of Medicine, Chicago, Illinois 60601, USA*

[b] *Cátedra de Fisiopatología, Facultad de Ciencias Médicas, Universidad Nacional de Cuyo and Hospital H. Notti, Mendoza, Argentina*

d-batlle@northwestern.edu

11.1 Introduction

The vacuolar (H^+)-ATPases are multi-subunit enzymatic proton pump complex. It has two domains: cytoplasmic V1 domain and transmembrane V0 [1]. The cytoplasmic V1 hydrolyzes ATP while transmembrane V0 translocates protons. It is an ATP-dependent proton pump that acidifies intracellular compartment or transport protons (H^+) across the plasma membrane. It is found on the plasma membrane or in association with various intracellular organelles. V1 domain is composed of eight types of subunits (A–H), whereas V0 domain consist of six subunits (a, c, c″, d, e, and Ac45) [1]. V-ATPase subunits gene have spatiotemporal gene expression pattern and is tissue and cell type-specific [2]. V-ATPases have been identified in the plasma membranes of various cells including renal intercalated

Handbook of H^+-ATPases
Edited by Suguru Nakamura

ISBN 978-981-4411-91-2 (Hardcover), 978-981-4411-92-9 (eBook)
www.panstanford.com

cells, osteoclasts, and macrophages [3]. The important subunits reported to be involved in human diseases are A and B of the V0 and V1 domain, respectively [4]. Subunit A has four isoforms (a1–a4) and subunit B has two isoforms (B1 and B2) [3]. Isoform that are present in limited number of tissues such as the kidneys are B1 (ATP6V1B1), a4 (ATP6V0A4), G3, C2, and d2 [3]. Besides kidneys, B1 subunits are also present in epididymis, ciliary body of the eye, and in the inner ear. A4 subunits are present only in epididymis, inner ear, and kidney. In the kidney, V-ATPases are localized in the apical membrane of type A-intercalated cells, basolateral membrane of type B-intercalated cells, and both apical and basolateral in non-A and non-B type of cells [3].

V-ATPases have remarkably diverse functions in different cells [1, 3]. Inside the cells, V-ATPases help in normal cell processes of receptor-mediated endocytosis, intracellular membrane trafficking, pro-hormone synthesis, protein degradation, and neurotransmitters uptake [5]. They also facilitate cellular entry of various viruses and toxins. In the cellular plasma membrane, they help in renal H^+ secretion, bone resorption, and sperm maturation. The prototypic disease caused by alterations in the vacuolar H^+-ATPase (V-ATPase) is distal renal tubular acidosis (dRTA). As postulated decades ago, the main mechanism causing this disease involves an alteration in the active mechanisms for H^+ secretion in the distal nephron [6, 7]. With the advent of genetic studies on the mechanism of hereditary dRTA in the 1990s, it was elegantly demonstrated that mutations in various isoforms of H^+-ATPase result in loss of function leading to impaired hydrogen ion secretion and thereby, decreased net acid excretion. The resulting chronic metabolic acidosis is responsible for the associated morbidity that characterizes the dRTA syndromes. Mutations in specific subunits have been described to cause renal tubular acidosis, osteopetrosis, and tumor metastasis [1, 8]. Mutations in ATP6V0A4 and ATP6V0B1 gene encoding A4 and B1 subunit have been linked to human inherited dRTA [4, 9]. In addition to this prototype disease of H^+-ATPase function, it is now increasingly recognized that alterations occur in common disease such as diabetes, cardiovascular diseases, and cancer [10–14]. In this chapter, we provide an overview on this topic.

11.2 Renal Tubular Acidosis (RTA)

Vaculaor H^+-ATPase mutations involving the ATP6V1B1 and the ATP6V0A4 genes account for an important number of cases of hereditary dRTA [4]. In children, dRTA is mostly observed as a primary entity. Prominent clinical features include impairment of growth, polyuria, hypercalciuria, nephrocalcinosis, lithiasis, and K^+ depletion [15]. Progression of nephrocalcinosis may lead to development of chronic renal failure. If detected early in life, therapeutic correction of the acidosis by continuous alkali administration may induce resumption of normal growth, arrest of nephrocalcinosis, and preservation of renal function. Primary dRTA may be observed sporadically or with autosomal dominant or recessive transmission [16]. Patients with recessive dRTA are typically severely affected, presenting either with acute illness or with growth failure in the early years of life. Two types of recessive dRTA have been differentiated by the presence or absence of sensorineural hearing loss (SNHL), but appear otherwise clinically similar. Karet *et al.* have recently demonstrated that most patients with dRTA and nerve deafness present mutations in the *ATP6V1B1* gene encoding the B1subunit of H^+-ATPase [17].

A genome wide linkage screen of a set of 31 kindreds with recessive dRTA (rdRTA) resulted in the identification of *ATP6V1B1* as the first gene associated with this disorder [18]. This gene is encoded in 14 exons, yielding the B1 subunit of 513 amino acids. It is expressed by interdental cells and endolymphatic sac epithelia, accounting for the associated hearing impairment. *ATP6V1B* mutations were identified that produce highly disruptive changes likely to result in loss of function of the encoded ATP6V1B1 protein; these mutations introduce premature termination codons, frameshift mutations, and splice site mutations. In other kindreds, missense mutations were detected; these missense and other mutations are dispersed along *ATP6V1B1*, rather than being clustered in a particular region of the protein, indicating that alterations in several regions of this protein led to a loss of B1 subunit, and by inference, V-ATPase function in intercalated cells [17].

Sporadic or autosomal rdRTA without sensorineural deafness is the most frequently encountered primary form. Karet *et al.* also demonstrated that this form may be caused by mutations in the gene *ATP6V0A4* encoding the 116 kDa subunit of V-ATPase [19]. Characterization of the *ATP6V0A4* genomic locus revealed that it comprises 23 exons; 20 of these exons encode the 840 amino acid a4 subunit of the V-ATPase [18].

Screening of the entire *ATP6V0A4* coding region as well as all intron–exon boundary sequences in nine rdRTA kindreds supporting linkage to chromosome 7q33–34 led to the identification of different *ATP6V0A4* mutations [20]. Several *ATP6V0A4* mutations were identified that produce highly disruptive changes predicted to result in loss of function of the encoded ATP6V0A4 protein; these mutations include a premature termination codon, frameshift mutations, and splice site mutations.

However, several patients with *ATP6V0A4* mutations have developed hearing loss, usually in young adulthood [21]. It is now known that *ATP6V0A4* is also expressed within the human inner ear. These findings provide further evidence for genetic heterogeneity in dRTA; extend the spectrum of disease causing mutations in *ATP6V1B1* and *ATP6V0A4*.

In contrast, autosomal dominant dRTA is typically a milder disease. Autosomal dominant dRTA was found to be associated in several kindred with mutations in the gene encoding the Cl^-–HCO_3^- exchanger AE1 or band three proteins [22], affecting the ability of α-intercalated cell to transport protons into the lumen of the collecting duct, and thus reducing net acid secretion. Alterations in H$^+$-ATPase expression have also been described in acquired forms of dRTA [15]. This has been best documented in patients with Sjogren's syndrome (Fig. 11.1) [23].

11.2.1 *ATP6VIB1 Mutations in dRTA*

As noted before, Karet *et al.* were the first to identify ATP6V1B1 gene mutations in a subset of patients with rdRTA [4]. They also described in detail the clinical characteristics of the 31 unrelated kindred with rdRTA [17]. Of the 31, 27 had family history of consanguineous marriage. All reported cases were diagnosed at an early age with

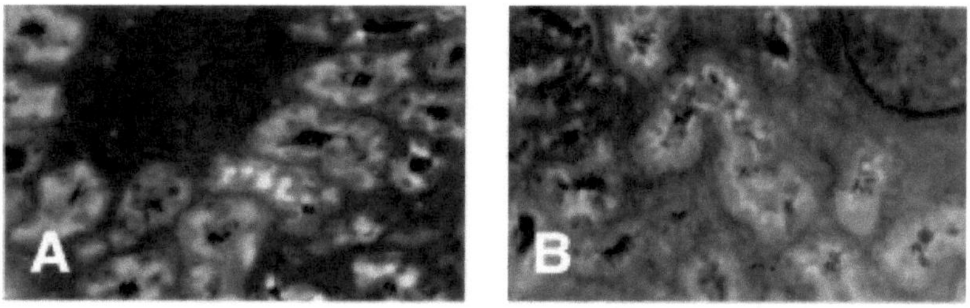

Figure 11.1 (A) Immunofluorescence staining with a vacuolar H^+-ATPase antibody in a normal kidney. (B) A renal biopsy from a patient with dRTA. Intense staining of collecting tubules is observed in normal kidney but not in the patient's biopsy specimen [15]. Reprinted from Ref. [23] with permission from the American Society of Nephrology, copyright 2013.

majority diagnosed by age of one year of which 19 cases were identified with ATP6B1 mutation. About 87% of the cases had hearing impairment with the mean blood pH of 7.2; bicarbonate, 10.9 mmol/L; and K^+ was 3.1 mmol/L. Mean urinary pH was 7.3 suggesting acidosis with inability to acidify urine.

Joshua *et al.* studied hearing loss in five children with dRTA from three unrelated families [24]. All families had different ATP6V1B1 mutations. Hearing loss was diagnosed between three months and two years. Audiometric and Imaging techniques were used to characterize the hearing loss. The hearing loss was bilateral, asymmetrical, progressive, and sometimes had conductive component. High resolution computed tomography performed in four patients showed the only radiological abnormality of bilaterally enlarged vestibular aqueduct (EVA) [24].

11.2.2 *ATP6V0A4 Mutations in dRTA*

Karet *et al.* besides identifying ATP6V0A4 gene mutation in a subset of recessive dRTA patients with normal hearing as mentioned before described their relevant phenotypic features [4, 19]. All except one had family history of consanguinity. All of them had presented in early childhood and had severe metabolic acidosis and hypokalemia with normal renal function. Urinary pH was reported to be greater than 6.5. All had nephrocalcinosis, and with the exception of two had elevated urine calcium. Ricket was present in seven of them. Audiometric evaluation revealed normal hearing.

Stover *et al.* later described several novel mutations in ATP6V0A4 gene causing autosomal rdRTA [18]. The clinical and biochemical parameters were similar to those described earlier. A follow-up of these patients found hearing loss in some at a later age.

Saito *et al.* described two-month old Japanese male infant who had presented with vomiting and failure to thrive [25]. The patient had hyperchloremic metabolic acidosis, hypokalemia, a normal serum anion gap, a positive urine anion gap, nephrocalcinosis, and high urine pH. Auditory brain stem response test revealed normal hearing bilaterally.

Both B1 and A4 subunits have been localized at the apical membrane of type B-intercalated cells [3]. Several mutations causing dRTA have been located to the gene encoding these subunits. Premature termination codons, frameshift mutations, and splice site mutations have been described in some, while in others missense mutations were identified.

Urinary acidification is primarily controlled in the distal nephron. H^+ ion secretion is coupled to bicarbonate absorption in the polarized type A-intercalated cells [9]. The V-ATPase present on the apical membrane of these cells is the main H^+ ion secretory protein [9]. In response to a fall in the blood pH, ATP-powered V-ATPase can increase H^+-ion secretion by several thousand folds. This secretion is coupled to bicarbonate absorption by the anion exchanger 1 (AE1) at the basolateral membrane of these cells. A defect in V-ATPase, therefore, disrupts the ability of the kidney to correct acidosis through acidifying urine [9, 19].

Su *et al.* found that G820R mutation caused complete loss of phosphofructokinase-1 (PFK-1) binding to a4 subunit without affecting PFK-1 activity whereas, R807Q mutation reduced a4 subunit production rendering V-ATPase inactive [26]. They also found that G820R resulted in loss of proton transport by 78% but caused only 36% loss in ATPase activity which suggested the importance of a4/PFK-1 binding to couple the ATPase activity to proton transport [26].

A study on rat inner medullary collecting duct (IMCD) cells has revealed another possible mechanism whereby mutant subunit did not combine with partially active V-ATPase complexes, but trafficked to the apical membrane in response to cellular acidification and inhibited proton pump activity [27]. It has been suggested that this mechanism might be involved in B1 missense mutation R394Q wherein, unlike the usual autosomal recessive pattern of V-ATPase mutations, dominant inheritance pattern was described [28, 29].

Finberg *et al.* generated an ATP6V1B1$^{-/-}$ deficient mice and reported that these mice produced alkaline urine [30]. Phenotypically, the mutant mice grew normally and did not develop systemic acidosis on a normal rodent diet. Apical expression of B2 subunit was increased suggesting compensation, although it failed to correct acidosis when challenged with oral acids [30]. Unlike humans, hypercalciuria, nephrocalcinosis, and any skeletal abnormality were absent. Mice also express the a4 subunit but knock-out mice (ATP6V0A4$^{-/-}$), to our knowledge, has not been reported yet [29].

11.3 Hearing Impairment in Hereditary dRTA

The kidney-specific a4 and B1 subunits of H^+-ATPase have been identified in inner ear as well. It is known that the pH of endolymph plays an important role in hearing [31]. In the inner ear, endolymphatic pH is maintained by H^+-ATPase and Cl–HCO_3 exchanger including pendrin. Disruption of their function such as through mutations leads to hearing loss [31].

Hearing loss is seen in autosomal recessive type of dRTA. Based on the type of hearing loss, the type of mutations can be suspected. Conductive deafness is seen with CAII mutations whereas; SNHL is seen with ATP6V1B1 and ATP6V0A4 mutations. Studies have reported that early SNHL (before the age of 10) occurs in most patients with ATP6V1B1 mutations, whereas late-onset SNHL is seen with ATP6V0A4 mutations, although there have been exceptions [18, 21, 32]. Vargas-Poussou *et al.* studied 39 families with autosomal rdRTA and identified V-ATPase mutations in 31 families [33] of which 21 of them had ATP6V0A4 mutations while only 10 had ATP6V1B1 mutations. In eight cases, these two mutations were not identified. Early-onset SNHL was observed in

70% of cases with ATP6V1B1 gene mutations and in 39% of cases with ATP6V0A4 gene mutations. Overall, 53.8% of the cases had ATP6V0A4 mutations and 25.6% of them had ATP6V1B1 mutations. Only 41% had hearing loss at the time of diagnosis. Based on these findings, they suggested that a genetic screening strategy based on early or late-onset hearing loss would be misleading [33].

Vargas-Poussou *et al.* proposed a new algorithm for the genetic screening of autosomal rdRTA [33]. They suggested that gene be investigated based on the presence or absence of consanguinity rather than hearing loss. In consanguineous families, homozygous mapping using microsatellite markers at the ATP6V1B1 and ATP6V0A4 loci could identify the mutations, and if inconclusive then ATP6V0A4 gene should be analyzed for mutation as it accounts for most cases of autosomal rdRTA [33].

11.4 Diabetes

V-ATPase has two functional domains V0 and V1, which are involved in proton translocation and ATP hydrolysis, respectively. The V1 domain contains eight subunits (A–H, with stoichiometry $A_3B_3C_1D_1E_3F_1G_3H_1$), which form a peripheral complex responsible for the hydrolysis of ATP. One of the important and interesting regulating subunit of V1 is V1H [34]. The V1H subunit, encoded by ATP6V1H, has been shown to have an important regulatory effect on V-ATPase activity. V1H has a helical structure consisting of two domains: the *N*-domain with seventeen consecutive a-helices (amino acids 2–352) and the C-domain with eight a-helices (amino acids 353–478). Multiple studies demonstrate that the *N*-terminal domain is important for the ATP hydrolysis function of the V1 domain of V-ATPase, and the C-terminal domain is needed for proper communication between the V1 and V0 domains of V-ATPase [35, 36]. In addition, it has also been shown that the C-terminal domain of V1H is necessary in inhibiting ATPase activity of free V1 domain complexes by interacting with subunit F *via* a cysteine residue at position 1 [37]. These above-mentioned studies demonstrated the critical role of V1H in maintaining proper V-ATPase function.

Downregulation of V1H gene expression has been reported in patients with type 2 diabetes (T2D) [38]. Olson *et al.* has shown that pancreatic islets from patients with T2D showed decreased expression of a set of oxidative phosphorylation (OXPHOS) genes (NDUFA5, NDUFA10, and COX11) in the ATP6V1H gene, which may lead to impaired insulin secretion. All four OXPHOS genes were significantly downregulated in islets from patients with T2D compared with non-diabetic donors. In addition, HbAlc levels associated negatively with gene expression of NDUFA5, COX11, and ATP6V1H but correlated positively with glucose-stimulated insulin secretion [13]. Thus, impaired V1H expression in diabetes could be important in the development of T2D and its complications. Recently, one cohort study compared the expression of ATP6V1H in peripheral white blood cells pre- and post-development diabetes in nine subjects [39]. Interestingly, the decrease in expression of ATP6V1H was associated with the development of T2D. This novel observation points out new mechanisms involved in development of diabetes and clues to the processes associated with onset and progress of the disease state and its complications [39] (Fig. 11.2). This was a prospective longitudinal study comparing gene expression in each individual over time associated with actual progression from pre-diabetes to diabetes. Interestingly, it also examined the changes in peripheral blood white cell expression associated with development of T2D. Not all these subjects were under any treatment for diabetes at the time of their conversion. Thus, a decrease in expression of ATP6V1H in white blood cells can be a potential marker for the development of T2D.

The main function of V-ATPase is to acidify intracellular compartments, which serve a vital role in receptor-mediated endocytosis, intracellular trafficking processes and protein degradation [34, 39] (Fig. 11.3). The V-ATPase pumps protons into the endosome lumen, creating an acidic environment that not only enables ligand-receptor complexes (brought in from the plasma membrane) to dissociate [39], but also allows for the vesicular trafficking of released ligands from early endosomes to late endosomes. These processes are important in the dissociation of insulin and its receptor. Benzi *et al.* has demonstrated the abnormal processing of insulin in T2D by examining the effects of the ionophore monensin, which inhibits

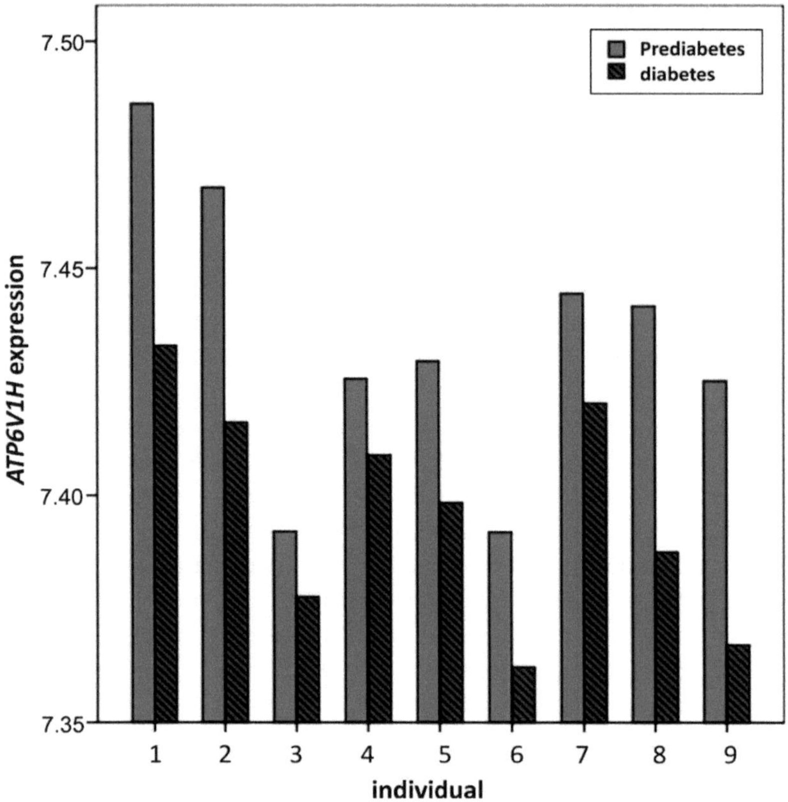

Figure 11.2 The expression of ATP6V1H in each participant. Dramatic variations were observed among the different individuals, which was concordant with PCA findings of the whole transcriptome profiles. However, compared with pre-diabetes status, the expression of ATP6V1H in diabetes decreased in each individual without exception, with pairwise Z test $P = 7.18 \times 10^{-11}$. Reprinted from Ref. 39, copyright 2011, with permission from Elsevier.

endosomal acidification, on the cellular processing of insulin and insulin receptor in monocytes from patients with and without diabetes [40, 41]. The acidification of endocytic vesicles may be the mechanism by which internalized insulin is dissociated from its receptor and properly sorted and processed. Therefore, V-ATPase plays an important role in the development of insulin resistance and diabetes.

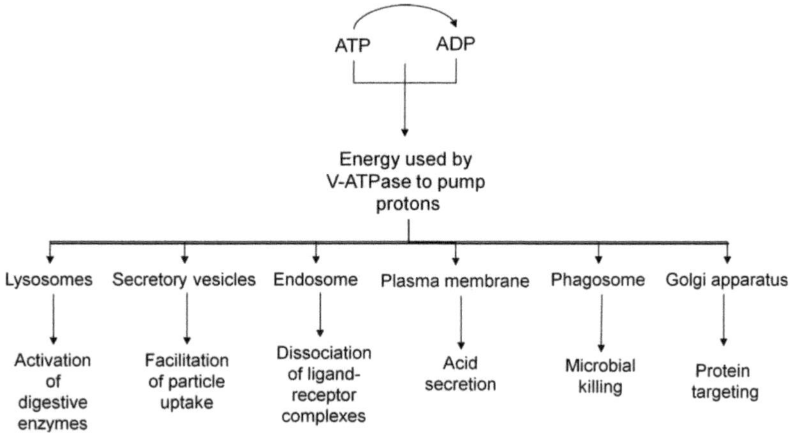

Figure 11.3 Cellular localizations and functions of V-ATPase: V-ATPase is present in a variety of intracellular organelles. In the lysosomes and central vacuoles of eukaryotic organisms, the action of the V-ATPase creates an acidic environment which facilitates the degradation of macromolecules by activating digestive enzymes. In synaptic vesicles and chromaffin granules, the proton gradient generated by V-ATPase acts as a driving force, which promotes the uptake of neurotransmitters and other small molecules destined for secretion. In some cases, a low pH is needed in secretory vesicles to activate enzymes which can convert hormones such as insulin from their precursor to mature form. At the plasma membrane in a variety of highly specialized cells, the V-ATPase is involved in acid secretion; an important process in urine formation bone resorption, and sperm maturation and storage. The enzyme also plays an important role in the lysosomal targeting of enzymes in the Golgi apparatus, giving rise to a low pH within late endsomes that allows the enzymes to dissociate from the mannose-6-phosphate receptor. Finally, in the Golgi apparatus, the acidification provided by V-ATPase ensures proper sorting of proteins to their destined location within the cell. Reprinted from Ref. 39, copyright 2011, with permission from Elsevier.

One study found that one of the *a*isoforms, *a*3, is localized to the membranes of insulin-containing secretory granules in pancreatic beta-cells, whereas the other subunit isoforms, *a*1 and *a*2, are localized to the compartments negative for insulin [35]. They found out that the secretion of insulin from isolated islets in response to glucose or depolarizing stimulation was impaired. Therefore, this study suggests that the *a*3 isoform of V-ATPase also has a regulatory function in the exocytosis of insulin secretion [35]. In undifferentiated cells, the *a*3 isoform localizes to late endosomes and lysosomes [42] and participates in the acidification of these lytic compartments. Davies *et al.* showed the human V-ATPase *a*3 isoform, encoded by the *ATP6i* gene, was mapped to chromosome 11q13, a region that had been associated to insulin-dependent diabetes mellitus (IDDM) [43]. It is possible that a genetic alteration of the gene encoding the human *a*3, could contribute to type 1 or insulin-dependant diabetes [44].

V-ATPase may also be involved in the increased risk of tuberculosis in patients with diabetes since it participates in endocytosis through the acidification of the phagosome [45] which is a critical process in intracellular microbial killing [46]. One study showed that the inhibition of V-ATPase in alveolar macrophages led to a rise in phagosomal pH and a decrease in reactive oxygen species (ROS) production that, in turn, decreases the ability to kill microbial at intracellular level [46]. Abnormal function of V-ATPase in diabetes may worsen the susceptibility to mycobacterium tuberculosis and possibly other microorganisms.

11.5 Connection with the (Pro)renin Receptor and Possible Role in Hypertension, Cardiovascular, and Renal Diseases

The main biologic peptide within the renin-angiotensin system (RAS), Ang II, has been shown to upregulate H-ATPase within the kidney. In addition to its main effect on the proximal tubules, Ang II when infused to adrenalectomized (ADX) rats, increases H^+-ATPase B1 subunit protein expression in medullary collecting ducts (Fig. 11.4) [47]. Selective aldosterone deficiency created by adrenalectomy with glucocorticoid replacement also resulted in downregulation in the expression of H^+-ATPase B1 subunit in medullary collecting duct [47]. Such an action may be responsible, in part, for an increase in H^+ secretion and thereby contribute to metabolic alkalosis in various hypertensive syndromes associated with Ang II overactivity. A more direct and intriguing connection between the RAS system and H^+-ATPase comes from the finding that

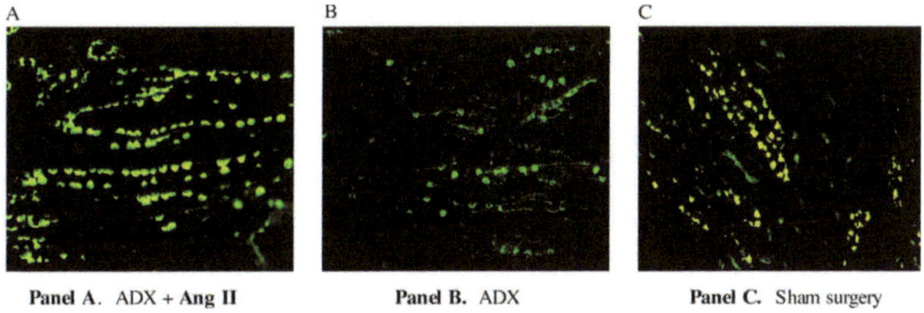

Figure 11.4 Immunofluorescence staining using antibodies against the 56 kDa H^+-ATPase B1 subunit in sections of rat kidney medulla. Magnification 400×. (A) In collecting ducts from adrenalectomized (ADX) rat administered Ang II, cells with intense apical immunoreactivity corresponding to α-intercalated cells were frequently seen. (B) In collecting ducts from ADX rats, cells had a marked attenuation of staining. (C) Cells from medullary collecting ducts from sham surgery rats showed intense staining on the apical membrane similar to that seen in tubules from Ang II infused rats, although staining does not appear to be as widespread within the tubules. Reprinted from Ref. 47 with permission from Wolters Kluwer Health, copyright 2005.

the (P)RR, a multi-functional molecule with functions beyond renin and prorenin [48], is associated with V-ATPase. Deletion of (P)RR, also called ATP6AP2, results in V-ATPase dysfunction, suggesting that the (P)RR is essential for its role as a proton pump [46]. ATP6AP2 has recently been found to be identical to the (P)RR and has a dual role: (i) acting within the RAS also regulates V-ATPase activity; (ii) acting as an adapter by binding to both the V-ATPase and the Wnt receptor complex, recruits the receptor complex into an acidic microenvironment [49].

Much of the supporting evidence to support a role of (P)RR in hypertension, cardiovascular, and renal disease came from animal over expression of (P)RR or from the use of presumed (P)RR blockers. Many groups have tried to produce (P)RR knockout. Recently, a study has demonstrated that cardiomyocyte specific (P)RR knockout mice develop fulminant heart failure and early death [12]. Connelly *et al.* showed that (P)RR is mainly found at the Z-disc and dyad of cardiac myocytes, coinciding closely with the distributions of the vacuolar H^+-ATPase and ryanodine receptor, known to be located within T-tubules and the sarcoplasmic reticulum's terminal cisternae, respectively. They also found that (P)RR mRNA/protein abundance were increased approximately three-fold in the hearts of diabetic rats in association with diastolic dysfunction, myocyte hypertrophy, and interstitial fibrosis. Direct renin inhibition reduced cardiac (P)RR expression in association with improved cardiac structure and function [35]. Several studies used a (P)RR antagonist, handle region peptide (HRP), mainly in diabetes setting as in this condition prorenin is high and renin levels are low. After treatment with HRP, beneficial effects were seen on the development of diabetic nephropathy and retinopathy and on cardiac fibrosis without effect on blood pressure [36]. It has been described that blocking the prorenin and (P)RR has a beneficial effect on tissue damage in the setting of high prorenin levels [50].

It has shown that (P)RR is expressed at high levels in neurons of nuclei which are important for central blood pressure and that it binds renin and is involved with ERK1/2 phosphorylation. (P)RR is also present in synaptic vesicles where V-ATPase is essential for concentration and maturation of neurotransmitter suggesting a link between (P)RR and V-ATPase [51]. Shan *et al.* demonstrated that central (P)RR might have a role in the control of blood pressure. They also showed that the level of expression of (P)RR in the supraoptic nucleus (SON) was higher in spontaneously hypertensive rats (SHRs) than in Wistar Kyoto control animals and that cultured neuronal cells from SHRs have a higher susceptibility to prorenin stimulation and displayed 50% higher ERK1/2 phosphorylation in response to prorenin [36]. Furthermore, downregulation of (P)RR expression in the SON of SHRs by local injection of an adenovirus coding for PRR small-hairpin (sh) RNA reduced the increase of blood pressure in the SHRs [36–48].

(P)RR is also expressed in podocytes, where it is involved in both tissue angiotensin II production and (P)RR-mediated intracellular signaling. Nevertheless, the physiological role of the PRR in podocyte has not been elucidated. (P)RR is important to maintain the normal structure and function of normal podocyte. In addition, it has been suggested that (P)RR-null podocyte develops proteinuria because of impaired function of V-ATPase and autophagia [48]. Overexpression of human (P)RR in rats caused slowly progressive proteinuria and glomerular sclerosis, which suggests that (P)RR-mediated signaling is involved in the development of glomerular diseases [52].

11.6 Cancer

Cancer cells show a metabolic shift that makes them overproduce protons; this has the potential to disturb cellular acid–base homeostasis [53]. However, these cells show cytoplasmic alkalinization, increased acid extrusion, and endosome-dependent drug resistance. V-ATPases, together with other transporters, may be responsible to a great extent for these features. These multi-subunit proton pumps are involved in the control of cytosolic pH and the generation of proton gradients (positive inside) across endocellular membrane systems like Golgi, endosomes, or lysosomes. In addition, in tumors, they have been shown to play an important role in the acidification of the intercellular medium [4]. Scientific evidence suggests that the acidic tumor microenvironment is a key to managing cancer progression and metastasis. V-ATPases play a major role in metastasis tumor development because many tumor cells secrete lysosomal enzymes that participate in the extracellular matrix degradation necessary form etastatic invasion. These enzymes are most active at low optimal pH; moreover, V-ATPases are responsible for microenvironment acidification [15]. Among the many mechanisms that regulate the tumor microenvironment, V-ATPases are especially significant because they can be inhibited by proton pumpinhibitors [23].

V-ATPase mediates multiple pH-dependent intracellular processes including vesicular trafficking, coupled ion gradient-molecular transport, and protease activation [47, 54]. The V-ATPase was originally characterized in association with intracellular organelles, but is now also recognized as a component of plasma membranes that mediates proton extrusion in some normal and malignant cells [14, 55]. In breast cancer cells, the abundance of V-ATPases on the plasma membrane correlates with an invasive phenotype [14]. Further, V-ATPase inhibitors reduce cell migration in cancer cells with high levels of plasma membrane V-ATPase [14]. It has been postulated that cell surface V-ATPase activity can provide a localized proton flux creating an acidic extracellular microenvironment [56]. One potential effect of this activity is to optimize pH for extracellular protease activation, thereby facilitating matrix degradation and cellular invasion. In breast cancer cells, the V_0a3 and V_0a4 isoforms were also recently shown to be essential for cellular invasion [8]. These findings demonstrate that specific V_0a isoforms provide a mechanism to target V-ATPase activities to the cell surface and affect cell function. These findings also suggest that differences in the V-ATPase localization assessed by immuno-labeling could be used to predict the behavior of cancer types, and that its use in clinical specimens could serve as a histological marker of aggressiveness. A recent study demonstrated that the cellular distribution of V-ATPase in human pancreatic cancer tissues may influence cancer cell activity since polarity is lost and expression is increased with advancing malignant features. In fact, striking differences in V-ATPase polarity and staining intensity distinguished early from advanced pancreatic intraepithelialneoplasia (PanIN) lesions. Invasive pancreatic cancers and metastatic lesions meanwhile demonstrated uniformly diffuse and intense V-ATPase staining. These findings indicate that elevated expression and loss of V-ATPase polarity may be key steps in modulating the tumor microenvironment, thereby providing a clinical correlate to previous *in vitro* work in breast cancer cells that identified V-ATPase expression as a marker of cancer cell aggressiveness. V-ATPase staining of human pancreatic cancer, ranging from PanIN lesions to pancreatic ductal adenocarcinoma (PDAC), demonstrated a marked loss of polarity and increased intensity with increasing tumor invasiveness. Inhibiting V-ATPase function decreased metalloprotein as, matrix metallopeptidase-9 (MMP-9) activities, but increased MMP-2 activation *in vitro*. These results indicate that the V-ATPase plays a complex role in regulating MMPs and influences the development of pancreatic cancer [55].

ATP6V1C1, one of the main genes regulating V-ATPase activity, has been also implicated in metastasis and multiple drug resistance [57]. One study suggests that inhibition of the development of cancer-associated acidic environments by suppressing a3 V-ATPase could be a novel therapeutic approach for the treatment of cancer metastasis [58].

References

1. Toei, M., Saum, R., and Forgac, M. (2010). Regulation and isoform function of the V-ATPases. *Biochemistry* 49, pp. 4715–4723.
2. Miranda, K.C., Karet, F.E., and Brown, D. (2010). An extended nomenclature for mammalian V-ATPase subunit genes and splice variants. *PLoS One* 5, p. e9531.
3. Nishi, T., and Forgac, M. (2002). The vacuolar (H+)-ATPases–nature's most versatile proton pumps. *Nat. Rev. Mol. Cell. Biol.* 3, pp. 94–103.
4. Batlle, D., and Haque, S.K. (2012). Genetic causes and mechanisms of distal renal tubular acidosis. *Nephrol. Dial. Transplant* 27, pp. 3691–3704.
5. Gu, F., and Gruenberg, J. (2000). ARF1 regulates pH-dependent COP functions in the early endocytic pathway. *J. Biol. Chem.* 275, pp. 8154–8160.
6. Batlle, D., and Kurtzman, N.A. (1982). Distal renal tubular acidosis: Pathogenesis and classification. *Am. J. Kidney Dis.* 1, pp. 328–344.
7. Batlle, D.C. (1986). Segmental characterization of defects in collecting tubule acidification. *Kidney Int.* 30, pp. 546–554.
8. Hinton, A., Sennoune, S.R., Bond, S., Fang, M., Reuveni, M., Sahagian, G.G., Jay, D., Martinez-Zaguilan, R., and Forgac, M. (2009). Function of a subunit isoforms of the V-ATPase in pH homeostasis and *in vitro* invasion of MDA-MB231 human breast cancer cells. *J. Biol. Chem.* 284, pp. 16400–16408.
9. Wagner, C.A., Finberg, K.E., Breton, S., Marshansky, V., Brown, D., and Geibel, J.P. (2004). Renal vacuolar H^+-ATPase. *Physiol. Rev.* 84, pp. 1263–1314.
10. Finberg, K.E., Wagner, C.A., Bailey, M.A., Wang, T., Mentone, S.A., Kashgarian, M., Giebisch, G., Geibel, J.P., and Lifton, R.P. (2002). Loss of plasma membrane H^+-ATPase activity from cortical collecting duct intercalated cells of H^+-ATPase B1-subunit deficient mice: A mouse model of distal renal tubular acidosis (Abstract). *J. Am. Soc. Nephrol.* 13, p. 4.
11. Finberg, K.E., Wang, T., Wagner, C.A., Geibel, J.P., Dou, H., and Lifton, R.P. (2001). Generation and characterization of H^+-ATPase B1 subunit deficient mice (Abstract). *J. Am. Soc. Nephrol.* 12, p. 15.
12. Kinouchi, K., Ichihara, A., Sano, M., Sun-Wada, G.H., Wada, Y., Kurauchi-Mito, A., Bokuda, K., Narita, T., Oshima, Y., Sakoda, M., *et al.* (2010).

The (pro)renin receptor/ATP6AP2 is essential for vacuolar H^+-ATPase assembly in murine cardiomyocytes. *Circ. Res.* 107, pp. 30–34.

13. Olsson, A.H., Yang, B.T., Hall, E., Taneera, J., Salehi, A., Nitert, M.D., and Ling, C. (2011). Decreased expression of genes involved in oxidative phosphorylation in human pancreatic islets from patients with type 2 diabetes. *Eur. J. Endocrinol.* 165, pp. 589–595.
14. Sennoune, S.R., Bakunts, K., Martinez, G.M., Chua-Tuan, J.L., Kebir, Y., Attaya, M.N., and Martinez-Zaguilan, R. (2004). Vacuolar H^+-ATPase in human breast cancer cells with distinct metastatic potential: Distribution and functional activity. *Am. J. Physiol. Cell Physiol.* 286, pp. C1443–C1452.
15. Batlle, D., and Flores, G. (1996). Underlying defects in distal renal tubular acidosis: New understandings. *Am. J. Kidney Dis.* 27, pp. 896–915.
16. DuBose, T.D., and Alpern, R.J., Jr. (2001). Renal tubular acidosis, in: *The Metabolic and Molecular Bases of Inherited Disease*, 8thEd.(Scriver, C.R., Beaudet, A.L., Sly, W.S., Valle, D., eds), McGraw-Hill, New York, pp. 4983–5021.
17. Karet, F.E., Finberg, K.E., Nelson, R.D., Nayir, A., Mocan, H., Sanjad, S.A., Rodriguez-Soriano, J., Santos, F., Cremers, C.W., Di Pietro, A., *et al.* (1999). Mutations in the gene encoding B1 subunit of H^+-ATPase cause renal tubular acidosis with sensorineural deafness. *Nat. Genet.* 21, pp. 84–90.
18. Stover, E.H., Borthwick, K.J., Bavalia, C., Eady, N., Fritz, D.M., Rungroj, N., Giersch, A.B., Morton, C.C., Axon, P.R., Akil, I., *et al.* (2002). Novel ATP6V1B1 and ATP6V0A4 mutations in autosomal recessive distal renal tubular acidosis with new evidence for hearing loss. *J. Med. Genet.* 39, pp. 796–803.
19. Karet, F.E., Finberg, K.E., Nayir, A., Bakkaloglu, A., Ozen, S., Hulton, S.A., Sanjad, S.A., Al-Sabban, E.A., Medina, J.F., and Lifton, R.P. (1999). Localization of a gene for autosomal recessive distal renal tubular acidosis with normal hearing (rdRTA2) to 7q33-34. *Am. J. Hum. Genet.* 65, pp. 1656–1665.
20. Chaabani, H., Hadj-Khlil, A., Ben-Dhia, N., and Braham, H. (1994). The primary hereditary form of distal renal tubular acidosis: Clinical and genetic studies in 60-member kindred. *Clin. Genet.* 45, pp. 194–199.
21. Smith, A.N., Skaug, J., Choate, K.A., Nayir, A., Bakkaloglu, A., Ozen, S., Hulton, S.A., Sanjad, S.A., Al-Sabban, E.A., Lifton, R.P., *et al.* (2000).

Mutations in ATP6N1B, encoding a new kidney vacuolar proton pump 116-kD subunit, cause recessive distal renal tubular acidosis with preserved hearing. *Nat. Genet.* 26, pp. 71–75.

22. Karet, F.E., Gainza, F.J., Gyory, A.Z., Unwin, R.J., Wrong, O., Tanner, M.J., Nayir, A., Alpay, H., Santos, F., Hulton, S.A., *et al.* (1998). Mutations in the chloride-bicarbonate exchanger gene AE1 cause autosomal dominant but not autosomal recessive distal renal tubular acidosis. *Proc. Natl. Acad. Sci. USA* 95, pp. 6337–6342.
23. Cohen, E.P., Bastani, B., Cohen, M.R., Kolner, S., Hemken, P., and Gluck, S.L. (1992). Absence of H(+)-ATPase in cortical collecting tubules of a patient with Sjogren's syndrome and distal renal tubular acidosis. *J. Am. Soc. Nephrol.* 3, pp. 264–271.
24. Joshua, B., Kaplan, D.M., Raveh, E., Lotan, D., and Anikster, Y. (2008). Audiometric and imaging characteristics of distal renal tubular acidosis and deafness. *J. Laryngol. Otol.* 122, pp. 193–198.
25. Saito, T., Hayashi, D., Shibata, S., Jogamoto, M., and Kamoda, T. (2010). Novel compound heterozygous ATP6V0A4 mutations in an infant with distal renal tubular acidosis. *Eur. J. Pediatr.* 169, pp. 1271–1273.
26. Su, Y., Blake-Palmer, K.G., Sorrell, S., Javid, B., Bowers, K., Zhou, A., Chang, S.H., Qamar, S., and Karet, F.E. (2008). Human H^+-ATPase a4 subunit mutations causing renal tubular acidosis reveal a role for interaction with phosphofructokinase-1. *Am. J. Physiol. Renal Physiol.* 295, pp. F950–F958.
27. Yang, Q., Li, G., Singh, S.K., Alexander, E.A., and Schwartz, J.H. (2006). Vacuolar H^+-ATPase B1 subunit mutations that cause inherited distal renal tubular acidosis affect proton pump assembly and trafficking in inner medullary collecting duct cells. *J. Am. Soc. Nephrol.* 17, pp. 1858–1866.
28. Carboni, I., Andreucci, E., Caruso, M.R., Ciccone, R., Zuffardi, O., Genuardi, M., Pela, I., and Giglio, S. (2009). Medullary sponge kidney associated with primary distal renal tubular acidosis and mutations of the H^+-ATPase genes. *Nephrol. Dial. Transplant* 24, pp. 2734–2738.
29. Alper, S.L. (2010). Familial renal tubular acidosis. *J. Nephrol.* 23(Suppl 16), pp. S57–S76.
30. Finberg, K.E., Wagner, C.A., Bailey, M.A., Paunescu, T.G., Breton, S., Brown, D., Giebisch, G., Geibel, J.P., and Lifton, R.P. (2005). The B1-subunit of the H(+) ATPase is required for maximal urinary acidification. *Proc. Natl. Acad. Sci. USA* 102, pp. 13616–13621.

31. Lang, F., Vallon, V., Knipper, M., and Wangemann, P. (2007). Functional significance of channels and transporters expressed in the inner ear and kidney. *Am. J. Physiol. Cell Physiol.* 293, pp. C1187–C1208.

32. Ruf, R., Rensing, C., Topaloglu, R., Guay-Woodford, L., Klein, C., Vollmer, M., Otto, E., Beekmann, F., Haller, M., Wiedensohler, A., *et al.* (2003). Confirmation of the ATP6B1 gene as responsible for distal renal tubular acidosis. *Pediatr. Nephrol.* 18, pp. 105–109.

33. Vargas-Poussou, R., Houillier, P., Le Pottier, N., Strompf, L., Loirat, C., Baudouin, V., Macher, M.A., Dechaux, M., Ulinski, T., Nobili, F., *et al.* (2006). Genetic investigation of autosomal recessive distal renal tubular acidosis: Evidence for early sensorineural hearing loss associated with mutations in the ATP6V0A4 gene. *J. Am. Soc. Nephrol.* 17, pp. 1437–1443.

34. Geyer, M., Fackler, O.T., and Peterlin, B.M. (2002). Subunit H of the V-ATPase involved in endocytosis shows homology to beta-adaptins. *Mol. Biol. Cell* 13, pp. 2045–2056.

35. Lu, M., Vergara, S., Zhang, L., Holliday, L.S., Aris, J., and Gluck, S.L. (2002). The amino-terminal domain of the E subunit of vacuolar H(+)-ATPase (V-ATPase) interacts with the H subunit and is required for V-ATPase function. *J. Biol. Chem.* 277, pp. 38409–38415.

36. Liu, M., Tarsio, M., Charsky, C.M., and Kane, P.M. (2005). Structural and functional separation of the N- and C-terminal domains of the yeast V-ATPase subunit H. *J. Biol. Chem.* 280, pp. 36978–36985.

37. Jefferies, K.C., Cipriano, D.J., and Forgac, M. (2008). Function, structure and regulation of the vacuolar (H+)-ATPases. *Arch. Biochem. Biophys.* 476, pp. 33–42.

38. Lu, H., Yang, Y., Allister, E.M., Wijesekara, N., and Wheeler, M.B. (2008). The identification of potential factors associated with the development of type 2 diabetes: A quantitative proteomics approach. *Mol. Cell. Proteomics* 7, pp. 1434–1451.

39. Molina, M.F., Qu, H.Q., Rentfro, A.R., Nair, S., Lu, Y., Hanis, C.L., McCormick, J.B., and Fisher-Hoch, S.P. (2011). Decreased expression of ATP6V1H in type 2 diabetes: A pilot report on the diabetes risk study in Mexican Americans. *Biochem. Biophys. Res. Commun.* 412, pp. 728–731.

40. Benzi, L., Cecchetti, P., Ciccarone, A.M., Nardone, A., Merola, E., Maggiorelli, R., Campi, F., Di Cianni, G., and Navalesi, R. (1997). Inhibition of endosomal acidification in normal cells mimics the derangements of cellular insulin and insulin-receptor metabolism observed in non-insulin-dependent diabetes mellitus. *Metabolism* 46, pp. 1259–1265.

41. Hamm, L.L., and Alpern, R.J. (2000). Cellular mechanisms of renal tubular acidification, in: *The kidney: Physiology and Pathophysiology* (3rd ed), pp. 1935–1979, New York, Raven Press.

42. Toyomura, T., Murata, Y., Yamamoto, A., Oka, T., Sun-Wada, G.H., Wada, Y., and Futai, M. (2003). From lysosomes to the plasma membrane: Localization of vacuolar-type H^+-ATPase with the a3 isoform during osteoclast differentiation. *J. Biol. Chem.* 278, pp. 22023–22030.

43. Davies, J.L., Kawaguchi, Y., Bennett, S.T., Copeman, J.B., Cordell, H.J., Pritchard, L.E., Reed, P.W., Gough, S.C., Jenkins, S.C., Palmer, S.M., *et al.* (1994). A genome-wide search for human type 1 diabetes susceptibility genes. *Nature* 371, pp. 130–136.

44. Sun-Wada, G.H., Toyomura, T., Murata, Y., Yamamoto, A., Futai, M., and Wada, Y. (2006). The a3 isoform of V-ATPase regulates insulin secretion from pancreatic beta-cells. *J. Cell Sci.* 119, pp. 4531–4540.

45. Lee, B.Y., Jethwaney, D., Schilling, B., Clemens, D.L., Gibson, B.W., and Horwitz, M.A. (2010). The Mycobacterium bovis bacille Calmette-Guerin phagosome proteome. *Mol. Cell. Proteomics* 9, pp. 32–53.

46. Ip, W.K., Sokolovska, A., Charriere, G.M., Boyer, L., Dejardin, S., Cappillino, M.P., Yantosca, L.M., Takahashi, K., Moore, K.J., Lacy-Hulbert, A., *et al.* (2010). Phagocytosis and phagosome acidification are required for pathogen processing and MyD88-dependent responses to Staphylococcus aureus. *J. Immunol.* 184, pp. 7071–7081.

47. Valles, P., Wysocki, J., Salabat, M.R., Cokic, I., Ye, M., LaPointe, M.S., and Batlle, D. (2005). Angiotensin II increases H^+-ATPase B1 subunit expression in medullary collecting ducts. *Hypertension* 45, pp. 818–823.

48. Ichihara, A., and Kinouchi, K. (2011). Current knowledge of (pro)renin receptor as an accessory protein of vacuolar H^+-ATPase. *J. Renin Angiotensin Aldosterone Syst.* 12, pp. 638–640.

49. Jansen, E.J., and Martens, G.J. (2013). Novel insights into V-ATPase functioning: Distinct roles for its accessory subunits ATP6AP1/Ac45 and ATP6AP2/(pro)renin receptor. *Curr. Protein Pept. Sci.* 13(2): 124–33.

50. Susic, D., Zhou, X., Frohlich, E.D., Lippton, H., and Knight, M. (2008). Cardiovascular effects of prorenin blockade in genetically spontaneously hypertensive rats on normal and high-salt diet. *Am. J. Physiol. Heart. Circ. Physiol.* 295, pp. H1117–H1121.

51. Contrepas, A., Walker, J., Koulakoff, A., Franek, K.J., Qadri, F., Giaume, C., Corvol, P., Schwartz, C.E., and Nguyen, G. (2009). A role of the (pro)renin

receptor in neuronal cell differentiation. *Am. J. Physiol. Regul. Integr. Comp. Physiol.* 297, pp. R250–R257.

52. Ichihara, A. (2012). (Pro)renin receptor and autophagy in podocytes. *Autophagy* 8, pp. 271–272.
53. Spugnini, E.P., Citro, G., and Fais, S. Proton pump inhibitors as anti vacuolar-ATPases drugs: A novel anticancer strategy. *J. Exp. Clin. Cancer Res.* 29, p. 44.
54. Beyenbach, K.W., and Wieczorek, H. (2006). The V-type H^+-ATPase: Molecular structure and function, physiological roles and regulation. *J. Exp. Biol.* 209, pp. 577–589.
55. Chung, C., Mader, C.C., Schmitz, J.C., Atladottir, J., Fitchev, P., Cornwell, M.L., Koleske, A.J., Crawford, S.E., and Gorelick, F. (2011). The vacuolar-ATPase modulates matrix metalloproteinase isoforms in human pancreatic cancer. *Lab. Invest.* 91, pp. 732–743.
56. Sennoune, S.R., and Martinez-Zaguilan, R. (2007). Plasmalemmal vacuolar H^+-ATPases in angiogenesis, diabetes and cancer. *J. Bioenerg. Biomembr.* 39, pp. 427–433.
57. Perez-Sayans, M., Reboiras-Lopez, M.D., Somoza-Martin, J.M., Barros-Angueira, F., Diz, P.G., Rey, J.M., and Garcia-Garcia, A. (2010). Measurement of ATP6V1C1 expression in brush cytology samples as a diagnostic and prognostic marker in oral squamous cell carcinoma. *Cancer Biol. Ther.* 9, pp. 1057–1064.
58. Nishisho, T., Hata, K., Nakanishi, M., Morita, Y., Sun-Wada, G.H., Wada, Y., Yasui, N., and Yoneda, T. (2011). The a3 isoform vacuolar type H-ATPase promotes distant metastasis in the mouse B16 melanoma cells. *Mol. Cancer Res.* 9, pp. 845–855.

Chapter 12

Vacuolar H^+-ATPase in Cancer and Diabetes

Souad R. Sennoune and Raul Martínez-Zaguilán
Department of Cell Physiology and Molecular Biophysics, Texas Tech University Health Sciences Center, Lubbock, Texas 79430, USA
Souad.Sennoune@ttuhsc.edu

12.1 Introduction

Glucose is the preferred carbon source for most cell types and a major determinant of cell growth. Cancer and diabetes mellitus are metabolism disorders, where glucose homeostasis plays an important role on the initiation and the progression of these diseases.

In first half of the 20th century, Otto Warburg proposed that cancer was caused by altered cellular metabolism. He made the fundamental observation, that to obtain energy, cancer cells relied on glycolysis [1–4]. In his view, the up-regulation of glycolysis caused normal cells to become cancerous. However, glucose transport inside of the cells of many tissues is a process that requires the hormone, insulin. Insulin promotes the uptake and the utilization of glucose. Therefore, in order to increase glucose uptake and tumor

Handbook of H^+-ATPases
Edited by Suguru Nakamura

ISBN 978-981-4411-91-2 (Hardcover), 978-981-4411-92-9 (eBook)
www.panstanford.com

growth, cancer cells trigger a change in the hormonal environment, that is, increasing insulin levels. Another hormone that has been shown to be involved in cancer progression and development is Insulin Growth Factor (IGF) [5–10].

Diabetes mellitus is classified into two subtypes that have fundamentally different metabolic and hormonal characteristics: type 1 and type 2. Type 1 diabetes is characterized by the cessation of insulin biosynthesis due to the autoimmune destruction of insulin-producing beta cells in the pancreas. To overcome the absence of insulin, type 1 diabetes patients need exogenously administered insulin. Type 2 diabetes is characterized by impaired insulin sensitivity, and cells do not respond normally to insulin, known as insulin resistance. It is well known that insulin resistance is a central defect in metabolic syndromes such as type 2 diabetes mellitus. However, there are patients that exhibit the characteristics of both type 1 and type 2 and are defined as type 1.5 or latent autoimmune diabetes in adults (LADA). Although this classification is an oversimplification, it has allowed researchers to better understand the basis of the disease. Regardless of the classification, all type of diabetes involve altered glucose metabolism. These disorders have also been associated with an activation of cellular stress responses caused by high levels of triglycerides or an elevation of glucose or insulin [11–13]. Not surprisingly, exogenous insulin also may be used in type 2 diabetes to overcome the resistance of the cells to insulin.

In both diseases, cancer and diabetes, insulin acts as a critical player for cell survival. Cancer cells predominantly generate energy *via* glycolysis even in the presence of oxygen, and produce lactic acid in the cytosol. This process is known as Warburg effect [1–3, 14]. This should be contrasted with normal cells that typically generate their energy by glycolysis followed by oxidation of pyruvate in the mitochondria under aerobic conditions *via* the Tricarboxylic acid (TCA) cycle or Krebbs cycle. This latter process is far more efficient than glycolysis that only generates four molecules of ATP for each glucose molecule Oxidation of pyruvate in the mitochondria *via* the TCA cycle generates 36 molecules of ATP. Not surprisingly, cancer cells need a huge amount of glucose to maintain cell proliferation. Interestingly, in type 1 diabetes, there are elevated levels of glucose in the blood circulation; however, no glucose is transported into the cell unless exogenous insulin is administered. Without the addition of insulin, the cells undergo programmed cell death or apoptosis. These seminal observation have lead to the prediction that patients with type 1 diabetes, where insulin levels are deficient, would be associated with a lower cancer risk. In contrast, one would also predict that in patients with type 2 diabetes, the risk of cancer should be greater due to excess of insulin in blood [15–16]. Indeed, among type 2 diabetes patients, increased circulating levels of insulin precursors have been implicated as mediators of the elevated risks of cancer of the liver, pancreas, kidney, and endometrium [17–22].

12.1.1 *Elevated Glycolysis in Cancer Induces Significant Acid Production*

Due to preferential utilization of glucose *via* glycolysis, there is an excessive amount of lactic acid being produced. The first step in glycolysis, that is, the production of glucose 1,6 bisphosphate *via* hexokinase/glucokinase generates acid because of the pK_a of glucose 1 phosphate is acidic (pKa 1.1 and 6.1), similar to that of glucose 6 phosphate (pKa 0.9 and 6.1) [23]. Excessive acid is deleterious to cell function and it is known to cause apoptosis. To survive this cytosolic acidification caused by excessive glycolysis, cells must use pH regulatory mechanisms to extrude acid from the cytosol. There are several pH regulatory mechanisms that all eukaryotic cells have evolved, including a variety of Na^+–H^+ exchangers and HCO_3 based H^+-transporting mechanisms. Although they are sufficient to extrude acid in most normal cells, these pH^{cyt} regulatory mechanisms, typically do not work at pH^{cyt} larger than 7.1; and most tumor cells maintain a pH^{cyt} >7.1 that is favorable for growth. Thus, tumor cells require additional pH^{cyt} regulatory mechanisms. We hypothesize that a primary H^+-transporting mechanism is the vacuolar H^+-ATPase (V-ATPase) that typically resides in acidic vesicles inside of the cell. However, in tumors and metastatic cells, the V-ATPase is also located at the cell surface, where it plays an important role in extruding acid generated by excessive glycolysis (Fig. 12.1). The V-ATPase is composed of two

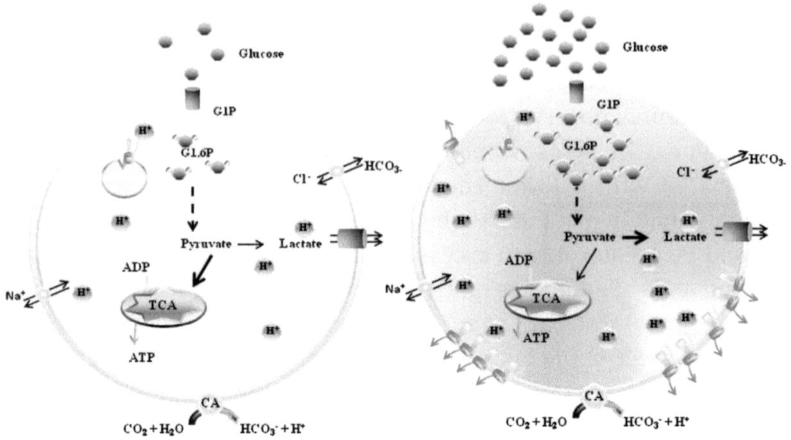

Figure 12.1 Cancer cells exhibit increase glycolysis than normal cells. To generate energy, tumor cells preferentially use glycolysis rather than oxidative phosphorylation, even in the presence of adequate supply of oxygen. This process is known as the Warburg effect. Consequently, tumor cells consume excessive amounts of glucose. Metabolism of glucose *via* glycolysis generates only 4 moles of ATP per mole of glucose, whereas, glucose metabolism *via* oxidative phosphorylation generates 36 moles of ATP. Further, glycolysis generates lactic acid and excessive accumulation of acid is deleterious to cells. To extrude this excessive acid, tumor cells use V-ATPase located at the cell surface. The V-ATPase allows them to maintain an alkaline intracellular pH optimal for growth and survival, while maintaining an acidic extracellular pH, optimal for the activity of proteases needed for invasion and metastasis.

structural domains containing multiple subunits associated in the V0 and V1 domain. The V0 domain is transmembrane and is part of the H^+ conductance channel; the V1 domain is located in the cytosolic side and contains the catalytic sector for ATP hydrolysis To detect and respond to changes in glucose, cells employ multiple signaling pathways. We know that the V-ATPase activity is sensitive to glucose availability. The regulation of V-ATPase by glucose is a well known, but poorly understood mechanism. Glucose withdrawal initiates the reversible dissociation of the V-ATPase into V0 and V1 domains, shutting down its activity. Addition of glucose triggers the assembly of V0 and V1 domains, thus activating the enzyme [24–26]. The addition of glucose, after a brief glucose deprivation in yeast, resulted in V-ATPase activation probably due to its reassembly [27–28]. We have recently shown that at 10 h of glucose deprivation, there is dissociation of V-ATPase in human prostate cancer cell lines (our unpublished observations). The pathway of glucose activation of V-ATPase reassembly and its significance for cancer progression remains unclear. Understanding or resolving this pathway will help to understand the contribution of V-ATPases in cellular signaling pathways.

12.2 Structure and Functions of Vacuolar H^+-ATPases

The V-ATPase belongs to a class of pumps that also includes the F_0-F_1-ATP synthase (F-ATPase). The V- and F-ATPases share structural and functional similarities [29–30]. They are distinguished from other membrane pumps by being composed of two multi-subunit domains that are operated by a rotary mechanism energized by ATP hydrolysis, and not having a phospho-enzyme intermediate characteristic of the P-type ATPases. The V-ATPase is therefore an ATP-driven enzyme that transforms the energy of ATP hydrolysis to generate electrochemical potential differences of protons across diverse biological membranes *via* the primary active transport of H^+. In turn, the transmembrane electrochemical potential of H^+ is used to drive a variety of secondary active transport systems *via* H^+-dependent symporters and antiporters, and channel-mediated transport systems [31–32].

V-ATPase is a giant multi-subunits nanomotor present in all eukaryotic cells. It is composed of two reversibly-associated multisubunit domains called V1 and V0. The extra membranous V1 domain contains the catalytic site and the membranous V0 domain translocates protons.

There are eight different subunits (A to H) in the detachable V1 catalytic domain. Three copies each of the A and B subunits are organized in an alternating fashion to form a hexamer. ATP hydrolysis occurs at the interface between the A and B subunits. The other V_1 subunits constitute peripheral and central stalks that have distinct functions in the rotary mechanism by which the V-ATPases couple ATP hydrolysis to proton pumping. One copy each of the C and H subunits and three copies each of the G and E subunits form peripheral stalks that act as stators. The D and F subunits form a central stalk that serves as a rotor to couple the energy generated by ATP hydrolysis to the actual rotation of the proteolipid ring in the V0 domain, to transport protons.

In mammals, the transmembrane V0 domain contains six hydrophobic subunits (five c, and one c″), and single copies of a, d, and e subunits. During rotation, protons are captured and translocated at the interface of the a- and c-subunits by consecutive binding to acidic residues in a- and c-subunits [30, 33–34].

12.2.1 *Accessory Proteins in Vacuolar H^+-ATPase*

In addition to the core subunits, two accessory proteins were found to be associated with V-ATPase: ATP6ap1 and ATP6ap2. The highly glycosylated type-I transmembrane protein Ac45 (ATP6ap1), a 45 kDa polypeptide, has been reported to co-purify with the V0 domain of V-ATPases from bovine adrenal chromaffin granules [35]. Ac45 is a neuroendocrine-enriched accessory subunit of the V-ATPase and found as an integral component of the V-ATPase V0 domain [36–40]. Ac45 was found to be specifically associated with the V0 domain subunits a3, c, and c″. The Ac45 appears to closely associate with the c″ subunit through its C-terminal domain. The Ac45 *N*-terminal region may serve to anchor Ac45 with a3 isoform [35]. The cytoplasmic terminus of ATP6ap1 is required for its proper interaction with V0 domain subunits and plays an important role in osteoclastic bone resorption [41]. Ac45 appears to be involved in regulating the expression of the plasma membrane V-ATPase and increasing the efficiency of calcium-dependent secretion in neuro-endocrine cell [38]. These data suggest that the transmembrane protein Ac45 may contribute to the membrane fusion event during exocytosis.

The human ATP6ap2 gene codes for a protein with 350 amino acids and a predicted mass of approximately 37 kDa that can be divided into four different domains: an *N*-terminal signal peptide, an extracellular domain binding (pro)renin, a single transmembrane domain, and a short cytoplasmic domain [42–43]. ATP6ap2 (also called (P)RR/ATP6ap2, ac8-9, or M8-9) was never found as part of the V-ATPase V0 domain as Ac45. These accessory proteins are processed by furin which is a prerequisite for Ac45 to function [44]. A furin cleavage site on the extracellular part of (P)RR/ATP6ap2 allows the release of a 28 kDa fragment during passage through the Golgi apparatus [45]. The cleavage leaves behind a short protein consisting of the transmembrane and the intracellular domains, which corresponds to the M8-9 fragment of (P)RR/ATP6ap2 found to be associated with V-ATPase [42, 45]. The significance of these accessory proteins is emphasized by the fact that Ac45 knockout in mice is lethal to embryos [46]; and ATP6ap2 knockout mice could not be generated [43, 47–48]. A further discussion on the significance of ATP6ap2 and the regulation of V-ATPase is addressed later in this chapter.

12.2.2 *Inhibitors of Vacuolar H^+-ATPase*

Analysis of the role and the presence of the VATPase were facilitated by the discovery of specific inhibitors [49]. The first inhibitors identified in the 1980s were the plecomacrolides, bafilomycins and concanamycin [50–52]. Since, new inhibitors were identified including: the benzolactone enamides salicylihalamide, lobatamide, apicularen, oximidine, cruentaren [53–57], and the indolyls [58–59]. One of the novel inhibitors is the recently discovered macrolactone archazolid [60–61]. The plecomacrolides were shown to binds to the c-subunit of the V-ATPase at nanomolar concentrations [62–63]. Studies have also shown a minor contribution of the V0 subunit-a to the binding site [64–65]. The new archazolid appeared to be an exclusive and highly potent inhibitor of V-ATPases with IC50 values in the nanomolar range [66]. It binds with the V0 subunit-c as it competes with concanamycin for its binding site within this subunit. All of these inhibitors bind to the V0 subunit-c, albeit the salicylhamides appear to bind to a distinct site than the plecomacrolides [67]. The toxicity of the plecomacrolides bafilomycins and concanamycins when administered to animals, apparently are higher than the salicylhamides [68]. This has limited the use of the plecomacrolides and opened new opportunities for the development of novel V-ATPase for therapeutic purposes [69].

12.2.3 *Physiological Significance of Vacuolar H^+-ATPase*

The primary function of the V-ATPases is proton pumping. In 1905, Eli Metchnikoff has shown the first evidence for the physiological significance [70]. He demonstrated vacuolar acidification in digestive vacuoles of unicellular organisms [70]. Later, about 40 years ago, the V-ATPase was discovered, in various endomembranes of eukaryotic cells [71–78]. It is now recognized that the V-ATPase is a key enzyme acidifying intracellular organelles and some extracellular milieus, localized in membranes of organelles of exocytic and endocytic pathways and cell surface. Maintenance of a luminal acidic environment in the exocytic and endocytic pathways is required at various stages in the processing and sorting of vesicle contents for a variety of signaling molecules, and ligand-receptor complexes [79–81]. Many enzymes and growth factors, for example, processing of insulin from pro-insulin, are produced as latent inactive forms and undergo several modifications before being converted to mature active forms within the secretory vesicles. It has been suggested that the proton gradients might be involved in the fusion of secretory vesicles to the target membrane [82–83] and that acidic pH might regulate priming of the vesicles for secretion [84]. These data indicate that the V-ATPase is also a key regulator of membrane fusion. In addition, recent studies indicated a role of V-ATPases in signal transduction in multiple systems [85–89].

The physiological significance of V-ATPases for cell growth and survival was first shown in yeast. Disruption of single-copy genes encoding the V1 subunit-b or the V0 subunit-c resulted in the inability of yeast cells to survive at pH values higher than 6.5 and they only grew well at pH values of 5.5, indicating that V-ATPase was essential for survival at pH >6.5 [90]. The V-ATPase appears to be the major electrogenic pump of vacuolar membranes. Although first found in endo-membranes, the V-ATPase has also been found in the plasma membranes of various organisms, where it functions both in physiological and pathological processes [27, 33, 79, 91–104].

This distinct subcellular localization must be involved in regulating distinct cellular functions.

12.3 Vacuolar H^+-ATPase in Endomembranous Compartments and Vesicular Trafficking

V-ATPase is ubiquitous on intracellular membranes (vacuoles, lysosomes, endosomes, secretory, Golgi and synaptic vesicles) where it plays a major role in luminal acidification to provide optimal pH for multiple functions (degradation, protein sorting, receptor retrieval) [72, 105]. It not only acidifies intracellular organelles, but also confers a voltage gradient. The proton potential across the membrane generated by V-ATPases provides a driving force to transport small molecules and ions into the lumen. The acidic luminal pH of intracellular organelles is a prerequisite for proper vesicular trafficking. It is essential for the formation of endosome carrier vesicles. In this process, V-ATPase itself may serve as a binding scaffold for proteins involved in budding, such as ADP ribosylation factor, a small GTPase (ARF6) and ADP ribosylation factor nucleotide site opener (ARNO), a guanine nucleotide exchange factor for ARF6 [106–108]. V-ATPase recruits ARF6 and ARNO to endosomal membranes *via* interaction of c-subunit with ARF6 and the a2 isoform with ARNO [106]. Functionally, both ARF6 and ARNO have been implicated in regulation of endocytotic pathway, organelle biogenesis and actin cytoskeleton remodeling [108–109]. Therefore, perturbation of these interactions with V-ATPase subunits leads to the inhibition of endocytosis.

Independent of its role in acidification, the V-ATPase transmembrane V0 domain has been proposed to be involved in membrane fusion, independent of the V1 domain [110–113]. There is evidence that V0 domain participates in vesicular membrane fusion that is critical for both exocytotic and endocytotic pathways; and important for various cellular activities, including neurotransmitter release and hormone secretion. Altogether, these data suggest that V-ATPase subunits play a regulatory role in vesicular trafficking and in other physiological phenomena involving homotypic and heterotypic membrane–membrane fusion.

12.3.1 Vacuolar H^+-ATPase in Membrane Trafficking

Independently from its role in proton pumping activity, several reports have converged to support the idea that the V0 domain may be involved in membrane fusion and neurotransmitter release. About 30 years ago, experiments designed to identify proteins involved in acetylcholine release from synaptosomes isolated from Torpedo, uncovered the role of a hydrophobic oligomer, called, mediatophore. Using artificial membrane vesicles, Israel *et al.* found that the mediatophore conferred calcium-dependent release of acetylcholine [114], suggesting that it has pore-forming properties. Later, the mediatophore protein components were found to be identical to the c-subunit of the V0 domain [115]. It was suggested that two V0 sectors in trans at the membrane interface between the two vacuoles destined to fuse, constitute a gap junction-like channel. Studies from invertebrates showed that V0 c-subunit participates in gap junction organization [116–117]. Peters *et al.* uncovered a crucial role of V0 c-subunit in yeast vacuole fusion while searching for molecular partners of calmodulin, which acts as a Ca^{2+}-sensor [118]. Analysis of V0 function showed that it acts downstream of SNARE complex assembly. In order to further understand if the interaction of V-ATPase subunits with SNAREs could influence neurotransmitter release, a study addressed the possibility of direct binding between V0 c-subunit and VAMP2 using a recombinant c-subunit and VAMP2 expressed in *Escherichia coli*. The authors demonstrated a direct interaction between these two proteins and identified loop 3.4 of V0 csubunit as a binding domain for the membrane proximal region of VAMP2 [119] which is an important determinant for Ca^{2+}/CaM binding [120–121].

Zhang *et al.* identified and characterized a CaM-binding site in subunit a1 isoform of Drosophila [122]. CaM interacts with V0 a-subunit in a Ca^{2+}-dependent, high affinity manner. Ablation of this binding site results in a loss of CaM recruitment to synapses. Expression of V0 a-subunit with a mutated CaM-binding site at low levels leads to incomplete rescue of the neuronal function, and high levels of expression lead to cell death. These data uncover a critical regulatory function of CaM that controls the V0 a-subunit and possibly V-ATPase functions [120, 122].

The V0 a1 isoform has also been identified in Drosophila as an essential component in evoked synaptic transmission, acting downstream of SNAREs [123]. Direct interactions between a1-isoform and t-SNAREs were demonstrated [123]. These data are consistent with a role for the a1-isoform at a late step in exocytosis. It has been suggested that when neurotransmitter filling is completed, the vesicular transporter can no longer dissipate the proton gradient and signals the dissociation of V1 [124]. A further role for V-ATPase subunits in exocytosis is supported by the observation that V0 interacts with SNAREs, and it has been specifically shown that the a- and c-subunits of V-ATPase bind syntaxin [123] and VAMP2 [119].

The V0 a2-isoform has been shown to interact with the guanine nucleotide-exchange factor ARNO in a pH-dependent manner, suggesting its function as a pH sensor for controlling the recruitment of the GTPase activated by ARNO, ARF6, which in turn directly interacts with the V0 sector [106, 108, 125]. Moreover, the V0 a-subunit has been shown to play a crucial role in exosome-mediated apical secretion of Hedgehog-related proteins in *Caenorhabditis elegans* [112] and the V0 a3-isoform to regulate insulin secretion in pancreatic β-cells [126]. Altogether, these data supports an important role for V-ATPases in the formation of the SNARE complex and in exocytosis in several cell models.

12.4 Physiological and Pathological Significance of Vacuolar H^+-ATPase at the Plasma Membrane

In addition to the localization of V-ATPase in endomembranous compartments, V-ATPase is also present at the plasma membrane of some specialized cells, where they perform coupled transport of molecules to carry out cell specific functions, such as, bone resorption [127], maturation and sperm storage in epididymal lumen [128], bicarbonate re-absorption in renal proximal tubes [129], renal acidification in distal nephron [130], pH regulation in the inner ear [131], and as well as cytoplasmic pH regulation in microvascular endothelial cells and cancer metastasis [95–99, 132–135]. The functional significance of V-ATPase expression on pre-synaptic plasma membranes is still a matter of debate [124], although it has recently been shown to be involved in stimulation-dependent alkalinization of the cytoplasm which may be involved in regulating endocytosis [136].

We have characterized a number of primary human tumor cells in terms of V-ATPase activity, and determined that a subset of them translocated V-ATPases to the cell surface [96–99, 137]. This was determined using bafilomycin, and following cytosolic pH recovery to acid loads in the absence of Na^+ and HCO_3, to inhibit two major cytosolic pH regulatory systems, that is, Na^+/H^+ exchanger (NHE) and HCO_3-based H^+-transporting mechanisms. In the absence of Na^+ and HCO_3, only cells exhibiting an alternate cytosolic pH regulatory mechanism should recover from an acid load. Treatment of these cells with SCH28080, to block H^+/K^+-ATPase does not affect cytosolic pH regulation. Highly metastatic human melanoma cells exhibit higher pmV-ATPase than lowly metastatic human melanoma cells [97–99]. We have made similar observation in human breast cancer cells with high and low metastatic potential [135] as well as in pancreatic carcinoma and prostate cancer cells (unpublished data).

12.4.1 *Vacuolar H^+-ATPases are Essential for Enveloped Virus Replication in Host Cells*

In neutrophils, it appears that the V-ATPase is important in regulating the cytosolic pH in response to activation by the "switching on" of host defense systems upon microbial infection [138]. The acidic environment of endosomes is crucial for the entry of various enveloped viruses [118, 138–140], such as influenza A virus [139, 141] into host cells by triggering the formation of membrane pores. The enveloped viruses rely on vesicular acidification created by the V-ATPase to escape the endosome and release their RNA into the cytoplasm, where viral replication occurs. Examples of these enveloped viruses include influenza, hepatitis C, West Nile viruses, Dengue fever virus, Ebola fever virus, HIV, Japanese encephalitis virus, among others [141–151]. Not surprisingly, inhibition of V-ATPase is a target for antiviral therapies. Most recently, 23 cellular factors, including members of the V-ATPases, have been identified as essential for virus entry [152].

V-ATPase activity is elevated during infection of cell monolayers with influenza A virus, as measured by intracellular pH change, *via* a mechanism mediated by extracellular signal-regulated kinase (ERK) and phosphatidylinositol 3-kinase (PI3K). Inhibition of influenza A virus entry into host cells decreased the activation of ERK and PI3K and V-ATPase activity, therefore the intracellular compartment is less acidic [147]. Influenza A virus-activated ERK and PI3K appear to interact directly, and they co-localize with the VATPase E subunit [147]. This is just one example of the potential mechanisms by which V-ATPase is involved in virus replication in host cells that involves both vesicular acidification and the involvement of V-ATPase in signaling pathways.

12.5 Significance of Vacuolar H^+-ATPase in Diabetes and Cancer

In a model of diabetes type 1, we found that V-ATPase activity was significantly decreased in coronary microvascular endothelial cells [133]. In this study, we isolated coronary microvascular endothelial cells from diabetic BB and non-diabetes-prone rat heart. This rat model develops spontaneously diabetes as in human type 1 diabetes. The rats are treated daily with insulin, otherwise they will die. These data corroborated that glucose is crucial for V-ATPase activity [113]. In this study, we reported a decreased ATP hydrolytic activity. Whether this was due to dislocation of V-ATPase from plasma membrane to intracellular endo-membranes related to glucose deficiency is unclear. As shown in other experimental models, glucose levels change the trafficking/localization of V-ATPase. We have observed in streptozotocin (STZ)-induced diabetic nude mice, that the tumor growth of the inoculated human melanoma cells was significantly decreased (unpublished data). Other studies in STZ-induced diabetic nude mice and STZ-induced diabetic hamster [153–154] showed that the induction and the establishment of human pancreatic cancer implants were decreased. However, in db/db mice (type 2 diabetes), the chemical induction of colon cancer is exacerbated, when compared to heterozygous db/+ or controls [155–156]. These studies support our hypothesis that glucose homeostasis is an important mechanism for controlling V-ATPase both *in vitro* and *in vivo*; and emphasize the significance of V-ATPase in diabetes and cancer.

The a-subunits are present in four isoforms (a1, a2, a3, and a4) that contain information to target V-ATPase to different membranous locations, and are expressed in a tissue-specific manner [31, 149, 157–159]. The a1-, a2-, and a3-isoforms have been shown by different investigators to be ubiquitously expressed in mammalian cells, albeit their levels of expression are tissue-specific [103–104, 159]. The a4-isoform is expressed in renal intercalated cells and epididymal clear cells, where it is localized in the apical membrane [91, 160]; while the a3-isoform is expressed in osteoclasts, where it is translocated to the plasma membrane from the lysosome upon activation of osteoclasts [103, 159]. In the brain, the a1-isoform is present in both synaptic vesicles and the presynaptic plasma membrane of presynaptic nerve terminals; whereas the a2-isoform localizes only to the apical endosomes of the renal proximal tubule cells. It is also clear that a given cell type can express more than one a-isoform in the plasma membrane, as it has been demonstrated for rat vas deferens and epididymal cells [160]. In our studies, we have shown that although all four isoforms are detectable in both highly (MB231) and lowly (MCF7) metastatic breast cells, the levels of a3 and a4 are much higher in MB231 than in MCF7 cells [95]. Cytosolic pH was decreased only on knockdown of a3. The knockdown of either a3- or a4-isoforms significantly inhibited invasion of MB231 cells. We have found that the isoforms a1 is downregulated and a4 is upregulated in highly metastatic cells compared to the lowly metastatic cells. Altogether, these data emphasize the significance of a3 and a4 for the acquisition of a more metastatic phenotype.

12.6 Regulation of Vacuolar H^+-ATPase by Reversible Dissociation of V0 and V1 Domains

It might be expected that V-ATPase activity would be tightly regulated at different levels since it is located in several endomembranes and at the cell surface. However, the exact mechanisms that regulate V-ATPase activity remain to be elucidated.

Little is known about the exact mechanism that modulates the differential coupling efficiency between ATP hydrolysis and proton translocation resulting in differential acidification along the exocytosis and endocytosis pathways. In yeast, it has been shown that the coupling efficiency is controlled by the C-terminal hydrophobic domain of a-subunit [161]. However, the best characterized mechanism of regulating V-ATPase activity described to date is the reversible dissociation of the V1 and V0 domains (Fig. 12.2).

The reversible dissociation of V0 and V1 domains for the regulation of V-ATPase was first observed in *Saccharomice scerevisiae* and *Manduca sexta* following glucose deprivation and starvation, respectively [24–26, 162]. The V0 free of V1 is incompetent for proton translocation [24–26, 163]. Similarly, cytoplasmic pools of V1 appear to be unable to hydrolyze ATP. The dissociation of V0 and V1 domain is, therefore, an efficient mechanism of controlling the V-ATPase activity. In these organisms, however, dissociation

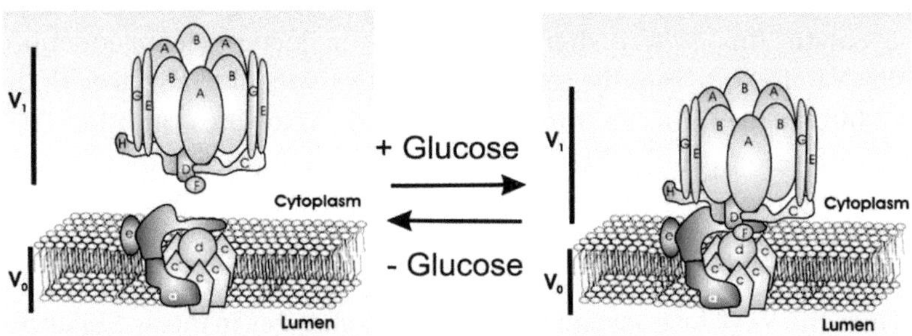

Figure 12.2 Regulation of V-ATPase by glucose. The V-ATPase is composed of multi-subunit domains called the V0 and V1 domains. The V1 domain contains the catalytic sector responsible for ATP hydrolysis. The V0 domain is embedded in the plasma membrane and is responsible for the translocation of H^+ *via* a rotary mechanism that is coupled to ATP hydrolysis. The presence of glucose triggers the association of the V0/V1 domains, thus activating V-ATPase. The absence of glucose triggers the reversible dissociation of the V0 and V1 shutting down V-ATPase activity. This mechanism effectively regulates the V-ATPase.

appears as a survival mechanism to conserve the ATP stock upon conditions of energy limitation by glucose deprivation or starvation [33, 92]. When glucose becomes available again pre-existing V0 and V1 domain re-associate. As expected, this process does not require *de novo* synthesis of proteins. Disassembly–reassembly process appears to be a way of regulating V-ATPase activity, and therefore regulating acidification.

Change in V1–V0 association was also reported during maturation of murine dendritic cells [164]. This association was aimed at controlling endosomal pH, where increased acidification was shown to increase the activity of hydrolytic enzymes and thus the efficiency of antigen processing [164]. These data suggest that, as in maturing dendritic cells, increased assembly of the V-ATPase sectors along the endocytic pathway of fibroblast-like cells contributes to increasing the acidity of the organelles. Changes in V1–V0 association raise the questions of the mechanisms that regulate this process.

Several signaling pathways that regulate reversible dissociation of the V-ATPase have been proposed. By monitoring the subcellular localization of GFP-tagged subunits using microfluidic chips, Dechant *et al.* assessed V-ATPase assembly in living cells exposed to rapid shifts in glucose concentrations. During conditions of high glycolytic flux, where there is increase in lactic acid production, the V1 domain rapidly relocalizes to the vacuolar surface. Upon a shift to low glucose, the V1 domain disassociates, limiting ATP consumption [165]. Their results suggest that glucose metabolism and intracellular pH regulates the assembly of the V-ATPase. Moreover, V-ATPase functions as an activator of protein Kinase A (PKA) in response to acidic pH caused by increased in glycolytic flux. Thus, the V-ATPase assembly responds to changes in intracellular pH and acts as a sensor. In addition, it has been shown that dissociation requires an intact microtubule network [166]; and reassembly of V1 onto V0 appears to be mediated by the regulator of H^+-ATPase of vacuolar and endosomal membranes (RAVE complex) and aldolase [167–172]. RAVE appears to stabilize the dissociated V1 complex in an assembly competent form; and mediates assembly in both the normal biosynthetic pathway and the glucose-regulated process. In contrast, the glycolytic enzyme aldolase appears to be important in glucose mediated assembly. Mutations in aldolase prevent assembly of the V-ATPase *in vivo* [169].

12.7 Regulation of Vacuolar H^+-ATPase by Glucose in Cancer

Glucose is the preferred carbon source for most cell types and a major determinant of cell growth. Cells employ multiple signaling pathways to detect and respond to changes in glucose level [173]. The reversible dissociation of V0 and V1 domains has also been described in renal cells [105, 174] where reversible dissociation is, as in yeast, coupled to glucose levels. Glucose metabolism is required for maintaining V-ATPase assembly [163]. Further, one of the hallmarks in cancer cells is a shift in glucose metabolism from oxidative phosphorylation to aerobic glycolysis (Warburg effect). The interactions between glucose homeostasis and V-ATPase activity in tumors are unclear. The reversible dissociation mechanism by glucose availability may be a mechanism of regulating V-ATPase activity in tumor cells. Therefore, we hypothesized that glucose withdrawal in cancer cells will initiate the reversible dissociation of the V-ATPase, shutting down its activity. To address this hypothesis, we incubated lowly and highly prostate cancer cells in medium without glucose, and after 6 h, we observed a significant decrease of the V-ATPase activity, acidifying the cytosol and leading to cell death (our unpublished observations).

V-ATPases are activated by glucose as well. Martinez-Munoz and Kane have shown in yeast that the change in pH is sensitive to glycolytic flux [27]. Paradoxically, they showed that cells in the presence of high glucose concentration have a more alkaline intracellular pH, whereas cells starved for glucose experience a more acidic intracellular pH. These data suggest that in response to elevated glucose that would cause increased in lactic acid production upregulate a pH regulatory mechanism to cope with excess acid that could kill the cells. Interestingly, the same phenomenon is observed in tumor cells. The high level of glycolysis leads the tumor cells to maintain a more alkaline intracellular pH (7.2). In glucose starvation, the prostate cancer cells experience also a more acidic intracellular pH (6.8) (our unpublished observations). Furthermore, the high level of glycolysis leads also to an acidification of the tumor micro-environment that is caused by extrusion of H^+ *via* the V-ATPase at the cell surface in tumors.

It has been well documented that the tumor micro-environment pH is more acidic (pH 6.6–6.8) than the normal tissues, due to the high rate of glycolysis and large amount of lactic acid produced in tumor areas [175–178]. Importantly, the tumor cells are known to maintain a more alkaline intracellular pH (7.1–7.8) than normal tissue (6.9–7.1) and to be efficient at maintaining pH gradients [97]. Since mammalian cells can only survive at neutral intracellular pH, these data suggest that tumor cells have evolved mechanisms that allowed them to proliferate in a hostile acidic environment strictly regulating their intracellular pH to allow for optimal functioning of glycolytic pathways. We hypothesize that such a mechanism is the V-ATPase. It has been shown that in kidney, the V-ATPase E subunit interacts directly with the glycolytic enzyme aldolase [167–169] and the a-subunit interacts with phosphofructokinase-1 [179], thereby providing a functional and spatial coupling of V-ATPase with the glycolytic pathway. It is unclear; however whether glucose or any of its metabolites is responsible for the regulation of V-ATPase. It has been shown that pyruvate also activates V-ATPase activity in the absence of glucose. These results extend those on the interaction of the V-ATPase with aldolase or glucose metabolites that provide further evidence for coupling between V-ATPase activity and glycolysis [105]. However, due to the complexity of mammalian systems, it is expected that the regulation of V-ATPase by glucose is not limited to only its interaction with the glycolytic enzymes.

12.8 Regulation of Vacuolar H^+-ATPase by Phosphorylation

Although the complete mechanism regulating assembly–disassembly of V-ATPase is unclear, in yeast, glucose causes assembly of the V-ATPase by activating the Ras/cAMP/PKA pathway [110]. An elevated glucose level inhibits the activity of the Ras GTPase activating proteins (GAPs) Ira1p and Ira2p, which leads to an increase in the amount of GTP-bound Ras; and hence increased Ras activity. Activated Ras, in turn, activates adenylate cyclase, leading to increased cellular levels of cAMP. Elevated cAMP levels causes dissociation of the regulatory subunits of PKA that in turn is free to phosphorylate several protein targets, including the V-ATPase.

Transient phosphorylation of the c-subunit (V1 domain) during reassembly has been observed in *Manduca sexta*, and this may well be a critical target in the signaling pathways [180]. These studies suggest that c-subunit binds to and serves as a substrate for PKA. These data also suggest that phosphorylation may be a regulatory switch for the formation of the active V0/V1 holoenzyme [180].

Other studies have demonstrated that activation of protein kinase C (PKC) augments V-ATPase-mediated proton translocation across the plasma membrane of macrophages [181], and inhibition of PKC decreased V-ATPase acid secretion in osteoclasts [182]. In ischemia preconditioning, Carini *et al.* proposed that the development of ischemic preconditioning in liver parenchymal cells might involve the PKC-mediated activation of V-ATPase. After preconditioning, the stimulation of PKC might activate proton extrusion through V-ATPase, thus, limiting intracellular acidosis and Na^+ overload promoted by Na^+-dependent acid buffering systems [183].

Lui *et al.* identified the phosphorylation site of V-ATPase A-subunit in plants. Their data indicated that A-subunit can generate 61 peptide fragments after tryptic digestion and only one fragment, F56 can be phosphorylated at Ser525 in maize roots [184]. Furthermore, they confirmed the presence of a calcium dependent protein kinase [185] that might phosphorylate the A-subunit in maize root at the serine site and consequently regulate V-ATPase activity.

Hong-Hermesdorf *et al.* reported that a member of the Arabidopsis WNK (with no lysine (K)) protein kinase family, AtWNK8, interacts with subunit C of V-ATPase [186]. WNK protein kinases are found in many eukaryotes and share a unique active site. AtWNK8 is shown to autophosphorylate intermolecularly and to phosphorylate subunit C at multiple sites in *Arabidopsis thaliana*. Furthermore, the authors show that the subunit C and other V-ATPase subunits are phosphorylated when V1-complexes are used as substrates for

AtWNK8. These data provide evidence that V-ATPases are potential targets of WNK kinases and their associated signaling pathways.

12.9 Insulin Signaling and Vacuolar H^+-ATPase

Insulin and IGFs belong to the insulin-like peptide super-family, and have the same evolutionary origin. Secretion of insulin is essential for metabolic homeostasis where glucose plays a prominent role. Insulin and IGFs have distinct physiological functions despite their high homology. Insulin regulates metabolite uptake and energy homeostasis, whereas IGFs promotes cell growth, proliferation, and differentiation. Insulin and IGF exert their physiological functions by interacting with their specific receptors. The insulin receptor (IR) and the IGF receptor (IGFR) both represent fundamental regulators of glucose metabolism.

Upon insulin stimulation, two signaling pathways are activated: the Ras/Raf/MEK/extracellular signal-regulated kinase 1 and 2 (ERK1/2) pathway and the phosphatidylinositol 3-kinase (PI3K)/phosphoinositide dependent kinase 1 (PDK-1)/Akt signaling pathway [187].

The insulin receptor is able to recruit and phosphorylate insulin receptor substrate proteins after ligand binding, leading to the activation of the PI3K/Akt pathway (Fig. 12.3). Tyrosine-phosphorylated insulin receptor substrate proteins, in turn, bind src homology 2 domain-containing signaling proteins, such as the p85 regulatory subunit of the type IA phosphatidylinositol 3-kinase (PI3K) [188] which activates the p110 catalytic subunit of PI3K.

PI3K is an important component in the signal transduction cascade initiated by various receptor tyrosine kinase specific growth factors. This kinase is a heterodimeric protein with a 110 kDa catalytic subunit (p110) and an 85 kDa regulatory adaptor subunit (p85) with two Src homology (SH2) domains that interact with the cytosolic portion of the tyrosine kinase domain of the receptor [189]. PI3K catalyzes the phosphorylation of the 3-OH position of inositol head groups of phosphoinositide lipids, namely phosphatidylinositol (PI), phosphatidylinositol-4-phosphate (PI-4-P), and phosphatidylinositol-4,5-bisphosphate (PI-4,5-P2). This

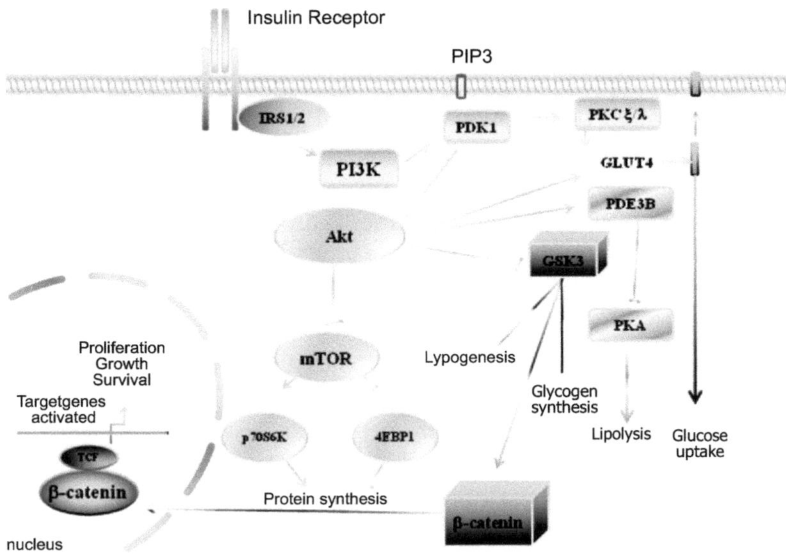

Figure 12.3 Insulin signaling. Insulin activates the insulin receptor tyrosine kinase which phosphorylates and recruits different substrates. The activated tyrosine-phosphorylated insulin receptor substrate (IRS1/2) allows binding of PI3 kinase that activates PDK1 and subsequently Akt pathway. Akt then induces glucose uptake *via* the translocation of GLUT4 vesicles to the plasma membrane, lipolysis, glycogen synthesis, lipogenesis, and gluconeogenesis, by targeting GLUT4, PDE3B, and GSK3. Akt also induces protein synthesis *via* mammalian target of rapamycin (mTOR). The activation of the PI3K/Akt pathway is responsible for glucose metabolism.

results in the formation of PI-3-P, PI-3,4-P2, and PI-3,4,5-P3. These are collectively termed 3′-phosphoinositide lipids [189]. Several studies have highlighted the importance of PI3K-generated phosphoinositides in the regulation of cell proliferation, cellular differentiation, cytoskeletal reorganization, membrane trafficking, glucose transport, survival and metabolism, and the prevention of apoptosis [189]. Akt and p70 S6K are two major protein kinases involved in signal transduction downstream to PI3K. They are directly activated by phosphatidylinositol 3-phosphate–dependent protein kinase (PDK1) in response to activation of PI3K and synthesis of 3-PI [190]. The activation of the PI3K/Akt pathway is responsible for glucose metabolism, including glucose uptake and glycogenesis (Fig. 12.3) [191–193].

Glucose metabolism is required for maintaining V-ATPase activity. Although the regulation of V-ATPase by glucose is a well-known phenomenon; however the underlying extra- and intracellular signaling mechanisms are unclear.

Sautin *et al.* have demonstrated an involvement of PI3K-dependent signaling in the control of V-ATPase trafficking, assembly, and function by glucose in renal epithelial cells [174]. In this study, glucose removal diminished V-ATPase dependent acidification, decreased ATP hydrolytic activity of V-ATPase, induced its disassembly and led to translocation of both V1 and V0 domains from plasma membrane and sub-membrane vesicles to intracellular vesicles (V1 and V0) and the cytoplasm (V1). Glucose deprivation or replacement dramatically changed the pattern of V-ATPase localization. The effects of glucose on V-ATPase trafficking and assembly can be abolished by pretreatment with the PI3K inhibitor LY294002; and can be reproduced in glucose-deprived cells by adenoviral expression of the constitutively active catalytic subunit p110 of PI3K. These data provide evidence that, in renal epithelial cells, glucose plays an important role in the control of V-ATPase-dependent acidification of intracellular compartments and V-ATPase assembly and trafficking. These data also indicate that the effects of glucose are mediated by PI3K-dependent signaling [174]. The glucose effects on V-ATPase trafficking can be mediated by changes in actin dynamics in the formation of microfilaments promoted by PI3K.

Further, involvement for PI3K in the regulation of V-ATPase has been obtained in other cell types. In the ruffled border of osteoclasts, the V-ATPase co-localizes with the p85 regulatory subunit of PI3K [194]. Wortmannin, a specific inhibitor of PI3K, inhibited the formation of ruffled border and osteoclastic bone resorption [194–195], supporting a role for PI3K-dependent signaling in the control of V-ATPase. A recent study showed activation of V-ATPase regardless of glucose levels (deprivation or stimulation) in cells infected with influenza A virus [147]. Here, V-ATPase is mainly responsible of the acidification of the organelle and its activation seems to be regulated, at least in part, through interaction of E subunit with virus-activated ERK and PI3K signaling molecules.

To address the role of glucose in the regulation of V-ATPase in highly and lowly prostate cancer cells, we starved these cells from glucose. After about 6 h, we observed that the activity of V-ATPase *via* proton fluxes was significantly decreased (our unpublished observations). We do not know yet if the effects of glucose changed the pattern of V-ATPase localization or if its effects are mediated by PI3K-dependent signaling. This is currently being investigated.

12.10 Vacuolar H^+-ATPase and Insulin Secretion

There are studies suggesting that V-ATPase is important for the appropriate processing and maturation of insulin [196–199]. Sun-Wada *et al.* found that the a3-isoform was specifically targeted to the membranes of insulin-containing secretory granules; whereas, a1, and a2-isoforms, were localized in the compartments negative for insulin [126]. These results indicate that the a3-isoform of V-ATPase has a regulatory role in the secretion of insulin containing granules. The a-subunit, encoded by multiple genes, is present in several isoforms (a1, a2, a3, and a4) that contain information to target V-ATPases to different membranous locations, which are expressed in a tissue-specific manner [32, 103–104, 126, 161]. It has been shown that mutant mice without the a3-isoform exhibited a significantly lower level of plasma insulin than the wild-type mice, despite the fact they produced and contained substantial amounts of insulin in the pancreas. Insulin secretion in response to glucose or depolarizing stimulation was significantly impaired in oc/oc mutant islets. These observations suggest that the mutants are defective in the step of exocytosis of secretory granules containing insulin.

Several studies have identified regions on human chromosomes that encode a-subunit isoforms of V-ATPase. The human V-ATPase a3-isoform was mapped to chromosome 11q13, a region that has been linked to IDDM4 [200–201]. Several positional candidates including Fas-associated death domain (FADD) protein, galanin (GALN) [202] and lipoprotein receptor-related protein 5 (LRP5)

[203] have been investigated, although the functional candidate gene has not been identified yet. The a3-isoform gene ATP6i is located approximately 200 kb apart from the LRP5 locus, which shows strong linkage to the disease [203]. Sun-Wada *et al.* data raise the possibility that genetic alteration(s) of the gene encoding human a3 could also contribute to diabetes [126]. There are other proteins that appear to play a role in insulin secretion. The furin processing appears to be a prerequisite for Ac45 function, as reduced proteolytic processing of Ac45 impairs acidic vesicle formation and is responsible for defective insulin secretion in pancreatic β-cells [44]. In type 2 diabetes, increased insulin might be due to a3-isoform overexpression as well as in cancer cells. These studies emphasize the interrelationship between insulin secretion, glucose metabolism and V-ATPase.

12.11 ATP6ap2 and its Relationship with Vacuolar H$^+$-ATPase

The (pro)renin receptor ((P)RR) was initially characterized as a component of the renin-angiotensin system (RAS) known to play a key role in the control of blood pressure and salt balance. However, it has also been involved in other physiological and pathological processes. (P)RR/ATP6ap2 [42–43, 204] was first discovered not as a (pro)renin binding protein but as a protein associated with the V-ATPase in adrenal chromaffin cells. Ludwig *et al.* in 1998, described a truncated form of (P)RR/ATP6ap2 composed of the C-terminal part and the transmembrane domain, which co-purified with the V-ATPase in chromaffin cells and was later named ATP6ap2, for V-ATPase associated protein 2 [204]. In 2002, Nguyen *et al.* discovered the full-length (P)RR/ATP6ap2 [205]. The identity of the two proteins was not immediately obvious leading to two different names for the same protein.

(P)RR/ATP6ap2 is a small protein (35–39 kDa) organized into a large extracellular domain, a single transmembrane domain and a short intracellular domain. The truncated form of (P)RR initially described to be associated with the V-ATPase in bovine chromaffin granules [204] is composed of the transmembrane intracellular domains and a short portion of the extracellular domain. A furin cleavage site on the extracellular part of (P)RR/ATP6ap2 allows the release of a 28 kDa fragment during passage through the Golgi apparatus. A short protein consisting of the transmembrane domain and the intracellular domain released from the furin cleavage corresponds to the M8-9 fragment of (P)RR/ATP6ap2 that has been found associated with V-ATPase. The furin cleavage site may allow the separation of the M8-9 fragment from the rest of the molecule and confers functional independence. The subcellular localization of (P)RR/ATP6ap2 is unusual for a receptor. Sequence analysis predicted motifs in the cytosolic domain of (P)RR/ATP6ap2 targeting the protein to distinct intracellular vesicle compartments [48]. Indeed, the majority of (P)RR/ATP6ap2 is located on intracellular vesicles, although a significant amount of the protein is also found on the plasma membrane, possibly due to vesicle recycling. Interestingly, this distribution concurs with the localization of the V-ATPase in metastatic tumors that exhibit this enzyme both in acidic vesicles and at the cell surface [96–99, 135]. The cellular distribution of (P)RR/ATP6ap2 in tumors is unclear.

Until recently, the only function assigned to renin and its precursor prorenin was the cleavage of angiotensinogen as a first step in the RAS, finally resulting in several angiotensin peptides [42–43, 48, 205–208]. The binding of prorenin to the receptor has two important consequences. First, prorenin undergoes non-proteolytic activation, displays full enzymatic activity, and initiates Angiotensin II-dependent effects.

The (P)RR/ATP6AP2 function in regulating the RAS pathway is well established; however its role in regulating V-ATPase has not been established. The V-ATPase is critical to supply the driving force for vesicle trafficking, neurotransmitter uptake, and exocytosis and the lack of (P)RR/ATP6ap2 may impair neurotransmission. Contrepas *et al.* showed that the D4-(P)RR mutant have altered trafficking of this receptor to the neurite tips *in vitro* [208]. V-ATPases and endosomal acidification are also involved in the entry process of viruses into cells; and (P)RR/ATP6ap2, among other V-ATPase subunits, has recently been discovered to be pivotal for influenza virus infection [139, 152, 209–210]. The signaling pathways involved in the regulation of (P)RR/ATP6ap2 and V-ATPase interactions are unclear.

The biogenesis of V-ATPase complex requires the coordinated association of V1 with V0 domains. Studies in yeast have shown that several genes Vma12p, Vma21p, and Vma22p are required for V-ATPase assembly. This suggests that mammalian cells may have similar assembly mechanism. It is unclear; however, what is the assembly chaperone of mammalian V-ATPase. A functional link between the accessory protein 2 (ATP6ap2) and assembly of V0/V1 to form V-ATPase has been suggested.

Kinouchi *et al.* suggested that ATP6ap2 is essential for V-ATPase assembly in murine cardiomyocytes. Interestingly, there is no yeast homolog to mammalian ATP6ap2 [211]. However, ATP6ap2 showed sequence and structural similarity to the yeast chaperone, Vma21p, needed for the V-ATPase assembly [201]. Assembly of V0 and V1 domains in yeast relies on glucose and needs chaperones. Because cancer cells also rely on glucose, they may need a chaperone to assemble V-ATPase. Whether (P)RR/ATP6ap2 operates as a chaperone or glucose sensor, and accordingly assembles/disassembles and targets the V-ATPase to specific compartments, is unclear and warrants further investigation.

12.11.1 *Wnt Signaling Pathway Crosstalk with Vacuolar H^+-ATPase*

Prominent in the signaling pathways leading to cell growth and differentiation is the Wnt pathway [212–213]. Currently, three different pathways appear to be activated upon Wnt receptor activation: the canonical Wnt/β-catenin cascade, the non-canonical planar cell polarity (PCP) pathway, and the Wnt/Ca^{2+} pathway [214–219]. The best understood Wnt signal transduction cascade is the Wnt/β-catenin pathway. Wnt genes occur throughout the animal kingdom and encode a large family of secreted glycoproteins [214–215, 217, 220].

The Wnt signaling pathway involves proteins that directly participate in both gene transcription and cell adhesion. Wnt glycoproteins act as ligands for receptor mediated signaling pathways. The key to the transmission of canonical Wnt signals is the intracellular protein β-catenin, which is a transcriptional co-activator that also binds to cadherin proteins to form part of adherens junctions. The membrane-associated, cadherin-bound pool of β-catenin is highly stable. The cytosolic β-catenin is usually found in a protein complex with glycogen synthase kinase 3 (GSK3), Axin and adenomatous polyposis coli (APC). This leads to the phosphorylation of β-catenin by GSK3, which targets it for rapid ubiquitin-mediated degradation, thereby maintaining low levels of cytosolic β-catenin [221–222].

Activation of the canonical Wnt pathway involves the stabilization of β-catenin through the binding of Wnt ligands to cell surface receptors: Frizzled (Fz) family receptors and LRP5 and 6 [221–222]. The Fz proteins are members of the seven transmembrane domain cell surface receptors that belong phylogenetically to the large family of G protein-coupled receptors. LRP5 and LRP6 are co-receptors that are members of the family of low-density LRP single transmembrane receptors. The binding of Wnts to cell-surface Fz receptors and LRP5 and 6 co-receptors results in a functional change in this complex such that GSK3 no longer phosphorylates LRP-catenin. The resultant accumulation of β-catenin in the cytoplasm leads to nuclear accumulation and binding to T cell/lymphoid enhancer-binding transcription factors to induce the expression of specific target genes [214–217, 220]. Many of these genes are involved in regulating cell growth and differentiation.

The Wnt-signaling pathway traditionally is associated with regulation of development and differentiation, with defects in this pathway strongly associated with tumorigenesis [214–217]. It is known that tumors have greatly increased levels of glucose uptake, insulin–insulin growth factor and their respective receptors. It is becoming clear that the Wnt/β-catenin pathway is involved in the mechanisms regulating energy metabolism [223] and glucose metabolism [224].

Interest in the crosstalk of Wnt signaling pathway with diabetes has risen since the identification of variants of the transcription factor TCF7L2, a β-catenin effector gene, were shown to be involved in β-cell dysfunction and the etiology of type 2 diabetes [225–226]. Earlier to these studies, suggestion for a role of Wnt β-catenin signaling pathway in the pathogenesis of type 2 diabetes came from a study by Kanazawa *et al.* which reported a single nucleotide polymorphism locus in the Wnt5b gene in a Japanese population [227–228]. In addition, a link between cellular glucose sensing and the Wnt/β-catenin pathway was recently reported in macrophages [229], indicating that this pathway could be activated in diabetes. A more recent study demonstrated that Wnt and insulin signaling pathways crosstalk at multiple levels [230]. Wnt induces phosphorylation of Akt, ERK1/2, and GSK3β, and this is dependent on insulin/IGF-1 receptors. Insulin signaling also involves the Wnt co-receptor LRP5, which has a positive effect on insulin signaling. Thus, altered Wnt and LRP5 activity can serve as modifiers of insulin action and insulin resistance in the pathophysiology of diabetes and metabolic syndrome [230]. These studies suggest that Wnt signaling is involved in the regulation of glucose homeostasis.

In the Wnt pathway, the endosomal acidic environment, provided by V-ATPase, is essential for the phosphorylation of low-density LRP6 involved in signal transduction leading to activation of β-catenin [85]. Since the V-ATPases are also located at the plasma membrane in certain cell types, including highly metastatic cells, several interesting questions emerge regarding the relationship between V-ATPase and Wnt signaling pathways; and the factors upstream/downstream that regulate proton transporters or *vice versa*. As we know, in cancer cells, there is a shift in energy production. Evidence suggests that the activation of glycolysis favors tumor growth as well as tumor invasion and metastasis *via* acidification of the tumor microenvironment. As a consequence of the switch to glycolysis, cancer cells rapidly take up glucose and convert most of it to lactate and proton leading to a more acidic cytosol. Changes in cytosolic pH are critical for cell function/survival.

12.11.2 *(P)RR/ATP6ap2 Crosstalk with Vacuolar H^+-ATPase*

There are recent studies suggesting a role for the V-ATPase *via* (P)RR/ATP6ap2 in canonical and PCP signaling pathways [85–87]. PRR/ATP6ap2 seems to be required for the assembly of the V-ATPase [211]. (P)RR/ATP6ap2 was shown to associate with the Wnt receptor complex, consisting of Fz and LRP6, in HEK293T cells and in *Xenopus laevis* tadpoles [85]. (P)RR/ATP6ap2 and other subunits of the V-ATPase complex also interacted physically with Fz and LRP6, this suggests a model that characterizes (P)RR as an adaptor that links Fz and LRP6 to the V-ATPase complex (Fig. 12.4) [86–87]. This

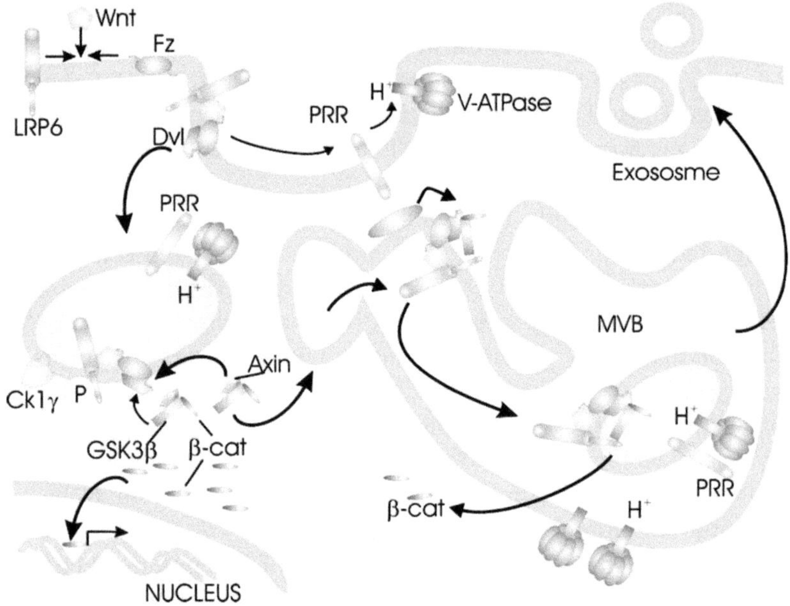

Figure 12.4 V-ATPase signaling *via* (P)RR/ATP6ap2 in Wnt β-catenin pathway. At the cell surface, the V-ATPase forms a complex with the (P)RR/ATP6ap2 that associates with the Wnt receptor complex (LRP6, Zf, and Dvl). This suggests that PRR is an adaptor that links Fz and LRP6 to the V-ATPase. The Wnt signaling triggers GSK3 sequestration in multivescular bodies (MVB) containing V-ATPase. The MVB represents a late endocytic compartment that contains intralumenal vesicles highly enriched with V-ATPase. The interaction between V-ATPase and PRR constitutes a poorly understood signaling pathway for V-ATPase regulation in signal transduction. The Wnt signaling triggers GSK3 sequestration in multivesicular bodies. This process requires endocytosis of Wnt receptor complexes that could be regulated by the acidification provided by the V-ATPase (a more complete description of these events is in the text).

adaptor function is predominantly carried out by the extracellular domain of (P)RR, which is also regarded as the prorenin and renin binding domain, and therefore seems to serve multiple purposes. Intriguingly, renin showed no effect on Wnt signaling [86–87].

Cruciat *et al.* characterized V-ATPase and (P)RR/ATP6ap2 as essential components of Wnt signaling in Xenopus embryos, which is crucial for several processes in embryonic development [85]. The authors show that phosphorylation of the Wnt co-receptor LRP6, and thereby activation of intracellular Wnt signaling, is dependent on its sequestration in vesicles, which are acidified by the V-ATPases. Thus, it appears that the crucial link between LRP6 and V-ATPase is formed by (P)RR/ATP6ap2 complexing with V-ATPase (Fig. 12.4)

A functional link between (P)RR and V-ATPase has been suggested earlier by Advani *et al.* in renal intercalated cells [231]. More recently, Kinouchi *et al.* suggested that (P)RR/ATP6ap2 is essential for V-ATPase assembly in murine cardiomyocytes [211]. In their study, they generated a mouse with a cardiac-specific deficiency in ATP6ap2. The ATP6ap2-disrupted cardiomyocytes showed extensive vacuolization, a phenotype that could be reproduced by pharmacological inhibition of intracellular acidification. These studies demonstrated for the first time that ATP6ap2 might be an essential assembly chaperone of mammalian V-ATPase; and that genetic ablation of ATP6ap2 created a loss-of-function model for V-ATPase representing a function that is unique to mammalian cells. We have shown that ATP6ap2 mRNA levels are up-regulated in highly metastatic cancer cells compared to the lowly metastatic (unpublished data). Therefore, we hypothesize that (P)RR/ATP6ap2 is the pH sensor/glucose sensor, and accordingly assembles V-ATPase and targets it to specific compartments to regulate the pH gradient and/or initiates signaling pathways. Whether this is a direct effect of the (P)RR/ATP6ap2, or an indirect effect due to interactions with a-subunit, a putative pH sensor in V-ATPase [125], requires further investigation.

More recently, Taelman *et al.* showed that Wnt signaling triggers GSK3 sequestration in multivesicular bodies, a late endocytic compartment that harbors intralumenal vesicles highly enriched with V-ATPase [232]. The sequestration of GSK3 in multivesicular bodies is thus one reason canonical Wnt signaling requires endocytosis of Wnt receptor complexes that could be regulated by the acidification provided by the V-ATPase [222]. Another recent work shows that vesicle acidification, which is mediated by V-ATPase and occurs during endosomal traffic, is required for canonical and non-canonical Wnt signaling [85]. Thus, endocytosis and V-ATPase may play other roles in Wnt signaling beyond multivesicular bodies' formation. Furthermore, experiments with a fusion protein of LRP6 and a pH reporter suggested that the V-ATPase mediates acidification of vesicles specialized for signalosome formation [233–234]. This makes V-ATPase a unique component of the signalosome complex that could regulate activity and function of the complex *via* localized regulation of the pH microenvironment.

In addition to its function in the Wnt signaling pathway, β-catenin also binds tightly to the cytoplasmic domain of type I cadherins [235]. Thus, β-catenin plays an essential role in the structural organization and function of cadherins by linking cadherins through β-catenin to the actin cytoskeleton. Members of the catenin family were first isolated complexed with the cytoplasmic domains of cadherins at cell–cell junctions [235]. The structural and functional integrity of the cadherin catenin complex is regulated by phosphorylation [235]. Serine–threonine phosphorylation of β-catenin resulted in increased stabilization of the cadherin–catenin complex. However, tyrosine phosphorylation of β-catenin disrupts binding of β-catenin to α-catenin whereas phosphorylation by Src or the epidermal growth factor receptor disrupts binding of β-catenin to cadherin. In general, activation of tyrosine kinases results in a loss of cadherin-mediated cell–cell adhesion; and increase in the level of cytoplasmic β-catenin, either by direct release of β-catenin into the cytoplasm or by activating cadherin endocytosis. It is important to consider that phosphorylation-dependent release of β-catenin from the cadherin complex not only regulates the integrity and function of the adhesion complex, but also may be an alternative mechanism for activating β-catenin signaling. The significance of V-ATPase and the regulation of this signaling pathway are unclear.

12.12 Hypoxia-Induced Factor (HIF) Pathway Crosstalk with Vacuolar H^+-ATPase

In eukaryotic cells, the major routes of energy production in cells are glycolysis and oxidative phosphorylation. However, as indicated earlier, the biochemical hallmark of cancer cells is a shift in glucose metabolism to aerobic glycolysis that, in bioenergetic grounds, is less effective than oxidative phosphorylation. This conundrum raises the question of how cancer cells grow and survive. To further complicate this issue, as the tumor growth, the neovascularization is often inadequate, leading to an insufficient supply of oxygen to the tumor, that is, hypoxia. Hypoxia causes the induction of various genes that contribute to tumor cell invasion and metastasis [236–241]. Several studies suggest that activation of the hypoxia inducible factor (HIF) is a common consequence of cancer.

HIF has been shown as the sensor of oxygen and respond to various cellular stimuli by upregulating genes involved in glycolytic energy metabolism, angiogenesis, cell proliferation and survival [242–244]. HIF-1 is a transcription factor composed of the subunits HIF-1α and HIF-1β. Both genes are ubiquitously expressed. They work as heterodimers to form the active HIF-1 that activates gene transcription by binding to hypoxia-responsive element (HRE) in promoter target genes [238, 243, 245]. The activity of HIF-1 is predominantly regulated at the posttranslational level by regulating HIF-1α protein stability. At normal oxygen tension, HIF-1α is hydroxylated by prolyl hydroxylases (PHD) in the oxygen dependent degradation domain. Hydroxylated HIF-1α is recognized by the Von Hippel–Lindau (VHL) protein, ubiquitinated, and destined for degradation by proteasomes [182]. At low oxygen pressure, stabilized HIF-1α subunits heterodimerize with β subunits to transactivate target genes after nuclear translocation.

In addition to directly upregulating glycolytic enzyme expression, HIF-1 directly contributes to downregulating the TCA cycle and oxidative phosphorylation (OXPHOS). OXPHOS is governed by the availability of its two major substrates: oxygen and pyruvate. Pyruvate is the end product of glycolysis, after which it enters the mitochondria and is converted to acetyl-CoA by the pyruvate dehydrogenase (PDH) complex, allowing it to enter the TCA cycle. PDH activity is inhibited through phosphorylation by pyruvate dehydrogenase kinase (PDK). Pyruvate conversion to acetyl-CoA in the mitochondria is thought to be an irreversible step, making this a critical control point in cellular metabolism. Studies found that PDK is a HIF-1 target gene [245–247]. By inhibiting PDH *via* PDK, cells with activated HIF-1 will accumulate pyruvate, which is then converted into lactate by another HIF-1 target: lactate dehydrogenase (LDH). Lactate is then released into the extracellular space by the mono carboxylic transporter (MCT), ultimately regenerating NAD^+ for another cycle of glycolysis. The MCT4 isoform was recently demonstrated to be a HIF-1 target [245, 248], suggesting that activation of HIF-1 drives expression of every enzyme in the pathway from glucose import to lactate production and release. HIF also induces over-expression of carbonic anhydrase 9 that catalyzes the extracellular conversion of CO_2 and H_20 to HCO_3^-, and H^+. The HCO_3^- exchangers import extracellular HCO_3^- into the cytosol to decrease intracellular acidification. To prevent intracellular acidification during glycolysis, HIFs also drive the expression of the NHE [236]. However, as indicate earlier, in metastatic tumors we hypothesize that the main H^+-transporting mechanism is the V-ATPase at the cell surface. Given the significance of V-ATPase for pH regulation, we hypothesize that HIF interacts with V-ATPase and possibly regulates transcription of V-ATPase subunits, since regulated energy metabolism and pH are central for cell survival (Fig. 12.5).

HIF-1 stabilization may also occur under oxygen independent conditions, including infection with oncogenic viruses, loss-of-function mutations in tumor suppressor genes such as VHL, or signaling by receptor tyrosine kinases, prostaglandin E2 receptor, or nitric oxide [245, 248]. Sonveaux *et al.* showed that human cancer cells cultured under low oxygen conditions convert glucose to lactate and extrude it *via* MCT4, whereas aerobic cancer cells take up lactate *via* MCT1 and utilize it for oxidative phosphorylation. When MCT1 is inhibited, aerobic cancer cells take up glucose rather than lactate, and anaerobic cancer cells die due to glucose deprivation [249].

Since V-ATPase assembly relies on glucose, these data suggested that under conditions of low glucose, V-ATPase may be disassembled, leading to cell death. Indeed, V-ATPase over-expression is an anti-apoptotic signal and inhibition of V-ATPase triggers apoptosis (Fig. 12.5).

Many tumors demonstrate high HIF activity even under normoxic conditions. Importantly, HIF stabilization has been observed under conditions of cytosolic acidification. Mekhail *et al.* show that normoxic acidosis neutralizes the function of VHL by triggering its nucleolar sequestration. Nucleolar sequestration of VHL enables HIF to evade destruction in the presence of oxygen and activate its target

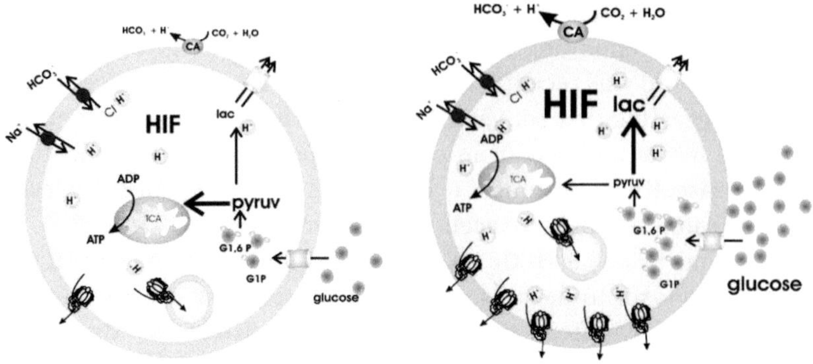

Figure 12.5 Crosstalk between HIF and V-ATPase. In addition to extracellular acidosis caused by the excessive amount of glucose that tumor cells need to survive, the tumor environment is also hypoxic. To survive these conditions, cells overexpress HIF that in turn regulate the expression of many genes involved in glycolysis and pH regulation. Due to enhanced glycolysis, the production of glucose 1 phosphate (G1P) and glucose 1,6 bisphoshate (G1, 6P; with pKa of 1.1 and 6.1) contribute to a large glucose-induced acid load. We hypothesize that one of these key pH regulatory mechanism is overexpression of V-ATPases at the plasma membrane. This pump helps to prevent acid loads that could be deleterious for cell function and survival. The c-subunit of V-ATPase, in turn help to maintain HIF stabilized in the nucleus by shuttling HIF into the nucleous (a more complete description of the interaction between V-ATPase and HIF is provided in the text).

genes [250]. Their findings suggest that an increase in hydrogen ions elicits a transient and reversible loss of VHL function by promoting its nucleolar sequestration. Therefore, it is likely that increased acid production will further induce HIF activity and glycolysis. In this case, V-ATPase is suitable positioned at the plasma membrane of tumor cells to maintain extracellular acidosis, thus creating a positive feedback loop to enhance the tumor phenotype.

Interestingly, a study of the growth inhibition of tumor cells by bafilomycin revealed an induction of HIF-1α expression, subsequent p21 induction, and cell-cycle arrest [251]. In this study, the effect of V-ATPase inhibition by bafilomycin on tumor survival does not appear to be at the level of intracellular pH regulation as previously proposed. This interaction was confirmed in a follow-up study where it was demonstrated that the effect of bafilomycin on cell survival was not a result of intracellular pH acidification [252]. Instead it was found that bafilomycin enhances the binding of the V-ATPase V0 c-subunit to the *N*-terminus of HIF-1, altering protein structure to prevent pVHL binding and thus results in HIF-1 stabilization. Thus, it appears that in addition to its role in regulating pH, V-ATPase may have alternative important signaling mechanisms. The significance of V-ATPase in signal transduction pathways is clearly an emerging field of study in cell physiology and pathophysiology.

In contrast to what occurs in cancer, insulin-dependent diabetes mellitus induces a rapid reduction of glucose uptake and utilization, and an inadequate neovascularization in certain capillary beads (e.g., lower extremities and microvascular endothelium in the heart). Thus, glucose metabolism is downregulated. Studies have shown that in diabetes HIF1-α stability and function are decreased [253–255].

Reduced glucose utilization in diabetic heart is associated with decreased activity of glucose transporter such as glucose transporter 1 (GLUT-1) and glycolytic enzymes, such as hexokinase II (HK-II) [256]. The inadequate neovascularization in diabetic heart is associated with decreased of vascular endothelial growth factor (VEGF) and its receptors [257–260]. Chou *et al.* have shown that the expression of mRNA and protein of VEGF and its receptors Flt-1 and Flk-1 in the myocardium was decreased significantly in both diabetic and insulin-resistant non-diabetic rats, as well as in diabetic patients [258].

The downregulation of myocardial VEGF expression preceded all other features of diabetic cardiomyopathy and was followed by an increased apoptosis of endothelial cells, decreased numbers of circulating endothelial progenitor cells, decreased capillary density, and impaired myocardial perfusion. VEGF, VEGF receptor, GLUT-1, and HK-II are the transcriptional targets of HIF-1 in various cell types [258–259]. Thangarajah *et al.* demonstrated that decreased hypoxia-induced VEGF expression in diabetes is the result of impaired transactivation of HIF-1 [258].

Studies attempting to increase HIF-1 activity have shown beneficial effects in protecting diabetic heart from ischemic insult [261]. Xue *et al.* demonstrated that over-expression of HIF-1 made the heart highly resistant to diabetes-induced injury by restoring GLUT-1, HK-II and VEGF expression and activity, suggesting an important role of HIF-1 in glucose metabolism and angiogenesis in diabetic heart [262]. Thus, HIF-1 plays a critical role in regulating cardiac protective mechanism against diabetic pathogenesis. The significance of the interrelation between V-ATPase and HIF-1 deserves further investigation.

12.13 Crosstalk between Diabetes and Cancer

Due to overlapping functional similarities between the insulin-IR and IGF-IGFR systems, the relationship between diabetes and cancer has been intensively investigated. However, due to the complexity of both diseases, studies characterizing the cross-talk between the two conditions are difficult and the underlying mechanisms of these diseases are still ambiguous. Both IR and IGFR are over-expressed in many human cancers [5, 8–10]. Recent evidence reveals the role of IGFR in metastasis of human cancers. Epidemiological studies have demonstrated a positive correlation between the levels of components from the IGF system and the relative risk of certain cancer in humans [17–18, 20, 22]. The majority of tumors have been observed in type 2 diabetic patients. In these patients, increased circulating levels of insulin precursors have been implicated as mediators of the elevated risks of cancer of the liver, pancreas, colorectal, breast, bladder and endometrium [21, 263–267]. Most studies report a reduced risk of prostate cancer. Diabetes appears to protect the patients with diabetes against prostate cancer [15, 268]. Type 2 diabetic patients in early stages of the disease have significantly elevated circulating insulin levels, potentially leading to binding to the IGFR and activation of IGF signaling. The consequent mitogenic effects of IGF signaling may lead to the development of malignant tumors. This scenario is supported by epidemiological studies demonstrating positive correlations between type 2 diabetes and cancer risk [269], particularly in insulin treated patients [5, 7, 270]. The few studies that attempted to address cancer incidence in type 1 diabetic patients suffer from poor diabetes type assessment. A study, on a Swedish population, evaluating cancer incidence in nearly 30,000 type 1 diabetic patients diagnosed in the period between 1965 and 1999 has identified 355 cases of cancer (about 1.8%) [16]. In contrast to type 2 diabetic patients, no increased risk of breast, pancreatic, and colorectal cancer was found in this cohort of patients. However, type 1 diabetic patients had an increased risk for stomach, endometrial, and cervical cancer that have been attributed to the high prevalence of *Helicobacter pylori* infection or of pernicious anemia for gastric carcinomas [271–272] and to the higher incidence of nulliparity, irregular menses, and fertility disorders in type 1 diabetic women (for uterine malignancies). In contrast with this report, a recent meta-analysis including three cohort studies and six case–control studies, has found that the risk for pancreatic cancer was doubled in type 1 diabetic patients and young-onset diabetics in comparison with non-diabetics [273].

Although the positive side effect of preventing cancer is rarely reported for anti-diabetic drugs, anti-cancer drugs targeting the IGF-IGFR survival pathway may cause abnormal glucose homeostasis, progression of diabetes or deterioration of undiagnosed pre-diabetes. The diabetogenic effect for those anti-cancer molecules may result from the inhibitory actions on the intracellular substrates shared between the insulin and IGF signaling pathways. Evans *et al.* in 2005, were first to report a potential association of metformin (a drug primarily used to treat diabetes type 2) use, with reduced cancer incidence [274]. Since then, several epidemiological studies have reported similar findings [275–278]. Evidence has emerged that metformin use may reverse the increased cancer risk associated with administration of insulin or insulin secretagogues [279] decreases glucose in the blood, and as a secondary effect, decreases insulin levels [8]. The anti-diabetic drug metformin is emerging as a potential anti-cancer agent [280–281] and may be useful in the prevention and treatment of cancer. The complex interdependencies between mechanisms of diabetes and cancer need further investigation.

12.14 Conclusion

To conclude, over the past four decades, we have seen the emergence of V-ATPase as an important physiological cellular mechanism not just to pump acid inside of vesicles and to the extracellular space when located at the cell surface, but more recently in the organization and participation of signaling mechanisms both at the cell surface and in vesicular trafficking. No doubt that given the significance of V-ATPase for diabetes and cancer, understanding of the molecular mechanisms that dictate its function in pH regulation and signal transduction is of paramount importance in the developing of strategies to halt these diseases.

References

1. Abaci, A., Oguzhan, A., Kahraman, S., Eryol, N.K., Unal, S., Arinc, H., and Ergin, A. (1999). Effect of diabetes mellitus on formation of coronary collateral vessels, *Circulation*, 99, pp. 2239–2242.
2. Adami, H.O., McLaughlin, J., Ekbom, A., Berne, C., Silverman, D., Hacker, D., and Persson, I. (1991). Cancer risk in patients with diabetes mellitus. *Cancer Causes Control*, 2, pp. 307–314.
3. Advani, A., Kelly, D.J., Cox, A.J., White, K.E., Advani, S.L., Thai, K., Connelly, K.A., Yuen, D., Trogadis, J., Herzenberg, A.M., Kuliszewski, M.A., Leong-Poi, H., and Gilbert, R. E. (2009). The (pro)renin receptor: Site-specific and functional linkage to the vacuolar H-ATPase in the kidney, *Hypertension*, 54, pp. 261–269.
4. Al-Awqati, Q. (1986). Proton-translocating ATPases, *Annu. Rev. Cell Biol.*, 2, pp. 179–199.
5. Al-Awqati, Q. (1996). Plasticity in epithelial polarity of renal intercalated cells: Targeting of the $H(^+)$-ATPase and band 3, *Am. J. Physiol.*, 270, pp. C1571–C1580.
6. Amsterdam, A., Nissen, R.M., Sun, Z., Swindell, E.C., Farrington, S., and Hopkins, N. (2004). Identification of 315 genes essential for early zebrafish development, *Proc. Natl. Acad. Sci. USA*, 101, pp. 12792–12797.
7. Anagnostou, S.H. and Shepherd, P.R. (2008). Glucose induces an autocrine activation of the Wnt/beta-catenin pathway in macrophage cell lines, *Biochem. J.*, 416, pp. 211–218.

8. Apps, D.K. and Reid, G.A. (1977). Adenosine triphosphatase and adenosine diphosphate/adenosine triphosphate isotope-exchange activities of the chromaffin granule membrane, *Biochem. J.*, 167, pp. 297–300.
9. Aragonés, J., Fraisl, P., Baes, M., and Carmeliet, P. (2009). Oxygen sensors at the crossroad of metabolism, *Cell Metab.*, 9, pp. 11–22.
10. Bader, M. (2007). The second life of the (pro)renin receptor, *J. Renin. Angiotensin. Aldosterone. Syst.*, 8, pp. 205–208.
11. Barg, S., Huang, P., Eliasson, L., Nelson, D.J., Obermuller, S., Rorsman, P., Thevenod, F., and Renstrom, E. (2001). Priming of insulin granules for exocytosis by granular Cl^- uptake and acidification, *J. Cell Sci.*, 114, pp. 2145–2154.
12. Barone, B.B., Yeh, H.C., Snyder, C.F., Peairs, K.S., Stein, K.B., Derr, R.L., Wolff, A. C., and Brancati, F.L. (2008). Long-term all-cause mortality in cancer patients with preexisting diabetes mellitus: A systematic review and meta-analysis, *JAMA*, 300, pp. 2754–2764.
13. Belfiore, A. and Malaguarnera, R. (2011). Insulin receptor and cancer, *Endocr. Relat. Cancer*, 18, pp. R125–R147.
14. Belham, C., Wu, S., and Avruch, J. (1999). Intracellular signalling: PDK1–a kinase at the hub of things, *Curr. Biol.*, 9, pp. R93–R96.
15. Bell, G.I. and Polonsky, K.S. (2001). Diabetes mellitus and genetically programmed defects in beta-cell function, *Nature*, 414, pp. 788–791.
16. Bengali, Z., Townsley, A.C., and Moss, B. (2009). Vaccinia virus strain differences in cell attachment and entry, *Virology*, 389, pp. 132–140.
17. Bertout, J.A., Patel, S.A., and Simon, M.C. (2008). The impact of O_2 availability on human cancer, *Nat. Rev. Cancer*, 8, pp. 967–975.
18. Beyenbach, K.W. and Wieczorek, H. (2006). The V-type H^+ ATPase: Molecular structure and function, physiological roles and regulation, *J. Exp. Biol.*, 209, pp. 577–589.
19. Bhattacharyya, S., Hope, T.J., and Young, J.A.T. (2011). Differential requirements for clathrin endocytic pathway components in cellular entry by Ebola and Marburg glycoprotein pseudovirions, *Virology*, 419, pp. 1–9.
20. Biddinger, S.B. and Kahn, C.R. (2006). From mice to men: Insights into the insulin resistance syndromes, *Annu. Rev. Physiol.*, 68, pp. 123–158.
21. Bilic, J., Huang, Y.L., Davidson, G., Zimmermann, T., Cruciat, C.M., Bienz, M., and Niehrs, C. (2007). Wnt induces LRP6 signalosomes and promotes dishevelled-dependent LRP6 phosphorylation, *Science*, 316, pp. 1619–1622.

22. Bond, S. and Forgac, M. (2008). The Ras/cAMP/protein kinase A pathway regulates glucose-dependent assembly of the vacuolar (H^+)-ATPase in yeast, *J. Biol. Chem.*, 283, pp. 36513–36521.
23. Bonifacino, J.S. and Glick, B.S. (2004). The mechanisms of vesicle budding and fusion, *Cell*, 116, pp. 153–166.
24. Bowman, E.J., Siebers, A., and Altendorf, K. (1988). Bafilomycins: A class of inhibitors of membrane ATPases from microorganisms, animal cells, and plant cells, *Proc. Natl. Acad. Sci. USA*, 85, pp. 7972–7976.
25. Bowman, B.J. and Bowman, E.J. (2002). Mutations in subunit c of the vacuolar ATPase confer resistance to bafilomycin and identify a conserved antibiotic binding site, *J. Biol. Chem.*, 277, pp. 3965–3972.
26. Braithwaite, S.S., Palazuk, B., Colca, J.R., Edwards, C.W. 3rd, and Hofmann, C. (1995). Reduced expression of hexokinase II in insulin-resistant diabetes, *Diabetes*, 44, pp. 43–48.
27. Breton, S., Smith, P.J., Lui, B. and Brown, D. (1996). Acidification of the male reproductive tract by a proton pumping (H^+)-ATPase, *Nat. Med.*, 2, pp. 470–472.
28. Brochier, G. and Morel, N. (1993). The same 15 kDa proteolipid subunit is a constituent of two different proteins in Torpedo, the acetylcholine releasing protein mediatophore and the vacuolar H^+ ATPase, *Neurochem. Int.*, 23, pp. 525–539.
29. Brown, D. and Marshansky, V. (2004). Renal V-ATPase: Physiology and pathophysiology, in: *Handbook of ATPases: Biochemistry, Cell Biology, Pathophysiology* (Futai, M., Wada, Y., Kaplan, J.H., eds.), Wiley-VCH, Weinheim, pp. 414–442.
30. Buechling, T., Bartscherer, K., Ohkawara, B., Chaudhary, V., Spirohn, K., Niehrs, C., and Boutros, M. (2010). Wnt/Frizzled signaling requires dPRR, the Drosophila homolog of the prorenin receptor, *Curr. Biol.*, 20, pp. 1263–1268.
31. Burckle, C. and Bader, M. (2006). Prorenin and its ancient receptor, *Hypertension*, 48, pp. 549–551.
32. Burgering, B.M. and Coffer, P.J. (1995). Protein kinase B (c-Akt) in phosphatidylinositol-3-OH kinase signal transduction, *Nature*, 376, pp. 599–602.
33. Burgess, T.L. and Kelly, R.B. (1987). Constitutive and regulated secretion of proteins, *Annu. Rev. Cell Biol.*, 3, pp. 243–293.
34. Cadigan, K.M. and Nusse, R. (1997). Wnt signaling: A common theme in animal development, *Genes Dev.*, 11, pp. 3286–3305.

35. Carini, R., De Cesaris, M.G., Splendore, R., Bagnati, M., and Albano, E. (2000). Ischemic preconditioning reduces Na^+ accumulation and cell killing in isolated rat hepatocytes exposed to hypoxia, *Hepatology*, 31, pp. 166–172.

36. Catrina, S.B., Okamoto, K., Pereira, T., Brismar, K., and Poellinger, L. (2004). Hyperglycemia regulates hypoxia-inducible factor-1alpha protein stability and function, *Diabetes*, 53, pp. 3226–3232.

37. Cauchi, S., El Achhab, Y., Choquet, H., Dina, C., Krempler, F., Weitgasser, R., Nejjari, C., Patsch, W., Chikri, M., Meyre, D. and Froguel, P. (2007). TCF7L2 is reproducibly associated with type 2 diabetes in various ethnic groups: A global meta-analysis, *J. Mol. Med.*, 85, pp. 777–782.

38. Chen, J.X. and Stinnett, A. (2008). Ang-1 gene therapy inhibits hypoxia-inducible factor-1alpha (HIF-1alpha)-prolyl-4-hydroxylase-2, stabilizes HIF-1alpha expression, and normalizes immature vasculature in db/db mice, *Diabetes*, 57, pp. 3335–3343.

39. Chen, S., Chen, J., and Wang, X-C. (2002). Existence and characteristics of tonoplast-bound protein kinase in the tip cell of maize root, *Acta Bot. Sin.*, 44, pp. 661–666.

40. Chou, E., Suzuma, I., Way, K.J., Opland, D., Clermont, A.C., Naruse, K., Suzuma, K., Bowling, N.L., Vlahos, C.J., Aiello, L.P., and King, G.L. (2002). Decreased cardiac expression of vascular endothelial growth factor and its receptors in insulin-resistant and diabetic States: A possible explanation for impaired collateral formation in cardiac tissue, *Circulation*, 105, pp. 373–379.

41. Chow, W.H., Gridley, G., Nyren, O., Linet, M.S., Ekbom, A., Fraumeni, J.F., Jr., and Adami, H.O. (1995). Risk of pancreatic cancer following diabetes mellitus: A nationwide cohort study in Sweden, *J. Natl. Cancer Inst.*, 87, pp. 930–931.

42. Cidon, S. and Nelson, N. (1983). A novel ATPase in the chromaffin granule membrane, *J. Biol. Chem.*, 258, pp. 2892–2898.

43. Cipriano, D.J., Wang, Y., Bond, S., Hinton, A., Jefferies, K.C., Qi, J., and Forgac, M. (2008). Structure and regulation of the vacuolar ATPases, *Biochim. Biophys. Acta*, 1777, pp. 599–604.

44. Clevers, H. (2006). Wnt/beta-catenin signaling in development and disease, *Cell*, 127, pp. 469–480.

45. Contrepas, A., Walker, J., Koulakoff, A., Franek, K.J., Qadri, F., Giaume, C., Corvol, P., Schwartz, C.E., and Nguyen, G. (2009). A role of the (pro)renin receptor in neuronal cell differentiation, *Am. J. Physiol. Regul. Integr. Comp. Physiol.*, 297, pp. R250–R257.

46. Cosset, F.L., Marianneau, P., Verney, G., Gallais, F., Tordo, N., Pécheur, E.I., ter Meulen, J., Deubel, V., and Bartosch, B. (2009). Characterization of Lassa virus cell entry and neutralization with Lassa virus pseudoparticles, *J. Virol.*, 83, pp. 3228–3237.
47. Cousin, C., Bracquart, D., Contrepas, A., Corvol, P., Muller, L., and Nguyen, N. (2009). Soluble form of the (pro)renin receptor generated by intracellular cleavage by furin is secreted in plasma, *Hypertension*, 53, pp. 1077–1082.
48. Cruciat, C.M., Ohkawara, B., Acebron, S.P., Karaulanov, E., Reinhard, C., Ingelfinger, D., Boutros, M., and Niehrs, C. (2010). Requirement of prorenin receptor and vacuolar H^+-ATPase-mediated acidification for Wnt signaling, *Science*, 327, pp. 459–463.
49. Currie, C.J., Poole, C.D., Gale, E.A.M., Evans, J.M.M., Donnelly, L.A., Emslie-Smith, A.M., Alessi, D.R., and Morris, A. D. (2005). Metformin and reduced risk of cancer in diabetic patients. *Br. Med. J.*, 330, pp. 1304–1305.
50. Currie, C.J., Poole, C.D., and Gale, E.A.M. (2009). The influence of glucose-lowering therapies on cancer risk in type 2 diabetes, *Diabetologia*, 52, pp. 1766–1777.
51. Cusi, K., Maezono, K., Osman, A., Pendergrass, M., Patti, M.E., Pratipanawatr, T., DeFronzo, R.A., Kahn, C.R., and Mandarino, L.J. (2000). Insulin resistance differentially affects the PI 3-kinase- and MAP kinase-mediated signaling in human muscle, *J. Clin. Invest.*, 105, pp. 311–320.
52. Czyzyk, A. and Szczepanik, Z. (2000). Diabetes mellitus and cancer, *Eur. J. Intern. Med.*, 11, pp. 245–252.
53. Davies, J., Kawaguchi, Y., Bennett, S.T., Copeman, J.B., Cordell, H.J., Pritchard, L.E., Reed, P.W., Gough, S.C., Jenkins, S.C., Palmer, S.M., Balfour, K.M., Rowe, B.R., Farrall, M., Barnett A.H., Bain, S.C., and Todd, J.A. (1994). A genome-wide search for human type 1 diabetes susceptibility genes, *Nature*, 371, pp. 130–136.
54. Dawson, R.M.C., Elliot, D.C., Elliot, W.H., and Jones, K.M. (1986). *Data for Biochemical Research*, 3rd Ed., Oxford Science Publications, Oxford.
55. De Block, C.E., De Leeuw, I.H., and Van Gaal, L.F. (1999). High prevalence of manifestations of gastric autoimmunity in parietal cell antibody-positive type 1 (insulin-dependent) diabetic patients. The Belgian Diabetes Registry, *J. Clin. Endocrinol. Metab.*, 84, pp. 4062–4067.
56. De Milito, A., Canese, R., Marino, M.L., Borghi, M., Iero, M., Villa, A., Venturi, G., Lozupone, F., Iessi, E., Logozzi, M., Della Mina, P.,

Santinami, M., Rodolfo, M., Podo, F., Rivoltini, L., and Fais, S. (2010). pH-dependent antitumor activity of proton pump inhibitors against human melanoma is mediated by inhibition of tumor acidity, *Int. J. Cancer*, 127, pp. 207–219.

57. Decensi, A., Puntoni, M., Goodwin, P., Cazzaniga, M., Gennari, A., Bonanni, B., and Gandini, S. (2010). Metformin and cancer risk in diabetic patients: A systematic review and meta-analysis, *Cancer Prev. Res.*, 3, pp. 1451–1461.
58. Dechant, R., Binda, M., Lee, S.S., Pelet, S., Winderickx, J., and Peter, M. (2010). Cytosolic pH is a second messenger for glucose and regulates the PKA pathway through V-ATPase, *EMBO J.*, 29, pp. 2515–2526.
59. Di Giovanni, J., Boudkkazi, S., Mochida S., Bialowas, A., Samari, N., Leveque, C., Youssouf, F., Brechet, A., Iborra, C., Maulet, Y., Moutot, N., Debanne, D., Seargar, M., and El Far, O. (2010). V-ATPase membrane sector associates with synaptobrevin to modulate neurotransmitter release, *Neuron*, 67, pp. 268–279.
60. Di Giovanni, J., Iborra, C., Maulet, Y., Leveque, C., El Far, O., and Seagar, M. (2010). Calcium-dependent regulation of SNARE-mediated membrane fusion by calmodulin. *J. Biol. Chem.*, 285, pp. 23665–23675.
61. Dowling, R.J., Zakikhani, M., Fantus, I.G., Pollak, M., and Sonenberg, N. (2007). Metformin inhibits mammalian target of rapamycin-dependent translation initiation in breast cancer cells. *Cancer Res.*, 67, pp. 10804–10812.
62. Dröse, S., Bindseil, K.U., Bowman, E.J., Siebers, A., Zeeck, A., and Altendorf, K. (1993). Inhibitory effect of modified bafilomycins and concanamycins on P- and V-type adenosinetriphosphatases, *Biochemistry*, 32, pp. 3902–3906.
63. Dröse, S. and Altendorf, K. (1997). Bafilomycins and concanamycins as inhibitors of V-ATPases and P-ATPases, *J. Exp. Biol.*, 200, pp. 1–8.
64. D'Souza-Schorey, C. and Chavrier, P. (2006). ARF proteins: Roles in membrane traffic and beyond, *Nat. Rev. Mol. Cell Biol.*, 7, pp. 347–358.
65. Duan, X., Lu, X., Li, J., and Liu, Y. (2008). Novel binding between pre-membrane protein and vacuolar ATPase is required for efficient dengue virus secretion, *Biochem. Biophys. Res. Commun.*, 373, pp. 319–324.
66. Eckenrode, S., Marron, M.P., Nicholls, R., Yang, M.C., Yang, J.J., Guida Fonseca, L.C., and She, J.X. (2000). Fine-mapping of the type 1 diabetes locus (IDDM4) on chromosome 11q and evaluation of two candidate

genes (FADD and GALN) by affected sibpair and linkage disequilibrium analyses, *Hum. Genet.*, 106, pp. 14–18.

67. El Far, O. and Seagar, M. (2011). A role for V-ATPase subunits in synaptic vesicle fusion? *J. Neurochem.*, 117, pp. 603–612.
68. Elwing, J.E., Gao, F., Davidson, N.O., and Early, D.S. (2006). Type 2 diabetes mellitus: The impact on colorectal adenoma risk in women, *Am. J. Gastroenterology*, 101(8), pp. 1866–1871.
69. Erickson, K.L., Beutler, J.A., Cardellina, I.J., and Boyd, M.R. (1997). Salicylihalamides A and B, novel cytotoxic macrolides from the marine sponge Haliclona sp, *J. Org. Chem.*, 62, pp. 8188–8192.
70. Evans, J.M., Donnelly, L.A., Emslie-Smith, A.M., Alessi, D.R., and Morris, A.D. (2005). Metformin and reduced risk of cancer in diabetic patients, *Br. Med. J.*, 330, pp. 1304–1305.
71. Fais, S., De Milito, A., You, H., and Qin, W. (2007). Targeting vacuolar H^+-ATPases as a new strategy against cancer, *Cancer Res.*, 67, pp. 10627–10630.
72. Feng, H., Cheng, T., Pavlos, N.J., Yip, K.H., Carrello, A., Seeber, R., Eidne, K., Zheng, M.H., and Xu, J. (2008). Cytoplasmic terminus of vacuolar type proton pump accessory subunit Ac45 is required for proper interaction with V(0) domain subunits and efficient osteoclastic bone resorption, *J. Biol. Chem.*, 283, pp. 13194–13204.
73. Fisher, W.E., Muscarella, P., Boros, L.G., and Schirmer, W.J. (1998). Variable effect of streptozotocin-diabetes on the growth of hamster pancreatic cancer (H2T) in the Syrian hamster and nude mouse, *Surgery*, 123, pp. 315–320.
74. Forgac, M., Cantley, L., Wiedenmann, B., Altstiel, L., and Branton, D. (1983). Clathrin-coated vesicles contain an ATP-dependent proton pump, *Proc. Natl. Acad. Sci. USA*, 80, pp. 1300–1303.
75. Forgac, M. (2007). Vacuolar ATPases: Rotary proton pumps in physiology and pathophysiology, *Nat. Rev. Mol. Cell Biol.*, 8, pp. 917–929.
76. Franke, T.F., Yang, S.I., Chan, T.O., Datta, K., Kazlauskas, A., Morrison, D.K., Kaplan, D.R., and Tsichlis, P.N. (1995). The protein kinase encoded by the Akt proto-oncogene is a target of the PDGF-activated phosphatidylinositol 3-kinase, *Cell*, 81, pp. 727–736.
77. Frayn, K.N., Tan, G.D., and Karpe, F. (2007). Adipose tissue: A key target for diabetes pathophysiology and treatment, *Horm. Metab. Res.*, 39, pp. 739–742.

78. Fresno Vara, J.A., Casado, E., de Castro, J., Cejas. P., Belda-Iniesta, C., and González-Barón, M. (2004). PI3K/Akt signalling pathway and cancer, *Cancer Treat. Rev.*, 30, pp. 193–204.

79. Fujino, T., Asaba, H., Kang, M.J., Ikeda, Y., Sone, H., Takada, S., Kim, D.H., Ioka, R.X., Ono, M., Tomoyori, H., Okubo, M., Murase, T., Kamataki, A., Yamamoto, J., Magoori, K., Takahashi, S., Miyamoto, Y., Oishi, H., Nose, M., Okazaki, M., Usui, S., Imaizumi, K., Yanagisawa, M., Sakai, J., and Yamamoto, T.T. (2003). Low-density lipoprotein receptor-related protein 5 (LRP5) is essential for normal cholesterol metabolism and glucose-induced insulin secretion, *Proc. Natl. Acad. Sci. USA*, 100, pp. 229–234.

80. Futai, M., Oka, T., Sun-Wada, G., Moriyama, Y., Kanazawa, H., and Wada, Y. (2000). Luminal acidification of diverse organelles by V-ATPase in animal cells, *J. Exp. Biol.*, 203, pp. 107–116.

81. Gagliardi, S., Nadler, G., Consolandi, E., Parini, C., Morvan, M., Legave, M.N., Belfiore, P., Zocchetti, A., Clarke, G.D., James, I. Nambi, P., Gowen, M., and Farina, C. (1998). 5-(5,6-Dichloro- 2-indolyl)-2-methoxy-2,4-pentadienamides: Novel and selective inhibitors of the vacuolar H^+-ATPase of osteoclasts with bone antiresorptive activity, *J. Med. Chem.* 41, pp. 1568–1573.

82. Galinis, D.L., McKee, T.C., Pannell, L.K., Cardellina, J.H., and Boyd, M.R. (1997). Lobatamides A and B, novel cytotoxic macrolides from the tunicate Aplidium lobatum, *J. Org. Chem.*, 62, pp. 8968–8969.

83. Gao, W., Ferguson, G., Connell, P., Walsche, T., Murphy, R., Birney, Y.A., O'Brien, C., and Cahill, P. A. (2006). High glucose concentrations alter hypoxia-induced control of vascular smooth muscle cell growth via a HIF-1alpha-dependent pathway, *J. Mol. Cell Cardiol.*, 42, pp. 609–619.

84. Getlawi, F., Laslop, A., Schägger, H., Ludwig, J., Haywood, J., and Apps, D. (1996). Chromaffin granule membrane glycoprotein IV is identical with Ac45, a membrane-integral subunit of the granule's $H(^+)$-ATPase. *Neurosci. Lett.*, 219, pp. 13–16.

85. Gillies, R.J., Martínez-Zaguilán, R., Martinez, G.M., Serrano, R., and Perona, R. (1990). Tumorigenic 3T3 cells maintain an alkaline intracellular pH under physiological conditions, *Proc. Natl. Acad. Sci. USA*, 87, pp. 7414–7418.

86. Gillies, R.J., Robey, I. and Gatenby, R.A. (2008). Causes and consequences of increased glucose metabolism of cancers, *J. Nucl. Med.*, 49, pp. 24S–42S.

87. Giovannucci, E. (2001). Insulin, insulin-like growth factors and colon cancer: A review of the Evidence, *J. Nutr.*, 131, pp. 3109S–3120S.

88. Giovannucci, E., Harlan, D.M., Archer, M.C., Bergenstal, R.M., Gapstur, S.M., Habel, L.A., Pollak, M., Regensteiner, J.G., and Yee, D. (2010). Diabetes and cancer: A consensus report, *Diabetes Care*, 33, pp. 1674–1685.

89. Grant, S.F., Thorleifsson, G., Reynisdottir, I., Benediktsson, R., Manolescu, A., Sainz, J., Helgason, A., Stefansson, H., Emilsson, V., Helgadottir, A. *et al.* (2006) Variant of transcription factor 7-like 2 (TCF7L2) gene confers risk of type 2 diabetes, *Nat. Genet.*, 38, pp. 320–323.

90. Gruenberg, J. and van der Goot, F.G. (2006). Mechanisms of pathogen entry through the endosomal compartments, *Nat. Rev. Mol. Cell Biol.*, 7, pp. 495–504.

91. Guinea, R. and Carrasco, L. (1995). Requirement for vacuolar proton-ATPase activity during entry of influenza virus into cells, *J. Virol.*, 69, pp. 2306–2312.

92. Harikumar, P. and Reeves, J.P. (1983). The lysosomal proton pump is electrogenic, *J. Biol. Chem.*, 258, pp. 10403–10410.

93. Harrison. M., Durose, L., Song, C.F., Barratt, E., Trinick, J., Jones, R., and Findlay, J.B. (2003). Structure and function of the vacuolar H^+-ATPase: Moving from low-resolution models to high resolution structures, *J. Bioenerg. Biomembr.*, 35, pp. 337–345.

94. Hashimoto, L., Habita, C., Beressi, J.P., Delepine, M., Besse, C., Cambon-Thomsen, A., Deschamps, I., Rotter, J.I., Djoulah, S., James, M.R., Froguel, P., Weissenbach, J., Lathrop, G. M., and Julier, C. (1994). Genetic mapping of a susceptibility locus for insulin-dependent diabetes mellitus on chromosome 11q, *Nature*, 371, pp. 104–105.

95. He, X., Semenov, M., Tamai, K., and Zeng, X. (2004). LDL receptor-related proteins 5 and 6 in Wnt/beta-catenin signaling: Arrows point the way, *Development*, 131, pp. 1663–1677.

96. Hermle, T., Saltukoglu, D., Grunewald, J., Walz, G., and Simons, M. (2010). Regulation of Frizzled-dependent planar polarity signaling by a V-ATPase subunit. *Curr. Biol.*, 20, pp. 1269–1276.

97. Hermle, T., Petzoldt, A.G., and Simons, M. (2011). The role of proton transporters in epithelial Wnt signaling pathways, *Pediatr. Nephrol.*, 26, pp. 1523–1527.

98. Hiesinger, P.R., Fayyazuddin, A., Mehta, S.Q., Rosenmund, T., Schulze, K.L., Zhai, R.G., Verstreken, P., Cao, Y., Zhou, Y., Kunz, J., and Bellen, H.J.

(2005). The V-ATPase V(0) subunit a1 is required for a late step in synaptic vesicle exocytosis in drosophila, *Cell*, 121, pp. 607–620.

99. Hinton, A., Sennoune, S.R., Bond, S., Fang, M., Reuveni, M., Sahagian, G.G., Jay, D., Martínez-Zaguilán, R., and Forgac, M. (2009). Function of a subunit isoforms of the V-ATPase in pH homeostasis and in vitro invasion of MDA-MB231 human breast cancer cells, *J. Biol. Chem.*, 284, pp. 16400–16408.
100. Holthuis, J.C., Jansen, E.J., Schoonderwoert, V.T., Burbach, J.P., and Martens, G.J. (1999). Biosynthesis of the vacuolar H^+-ATPase accessory subunit Ac45 in Xenopus pituitary, *Eur. J. Biochem.*, 262, pp. 484–491.
101. Holzenburg, A., Jones, P.C., Franklin, T., Pali, T., Heimburg, T., Marsh, D., Findlay, J.B., and Finbow, M.E. (1993). Evidence for a common structure for a class of membrane channels, *Eur. J. Biochem.*, 213, pp. 21–30.
102. Hong-Hermesdorf, A., Brüx, A., Grüber, A., Grüber, G., and Schumacher, K. (2006). A WNK kinase binds and phosphorylates V-ATPase subunit C, *FEBS Lett.*, 580, pp. 932–939.
103. Hurtado-Lorenzo, A., Skinner, M., El Annan, J., Futai, M., Sun-Wada, G.H., Bourgoin, S., Casanova, J., Wildeman, A., Bechoua, S., Ausiello, D.A., Brown, D., and Marshansky, V. (2006). V-ATPase interacts with ARNO and Arf6 in early endosomes and regulates the protein degradative pathway, *Nat. Cell Biol.*, 8, pp. 124–136.
104. Huss, M., Ingenhorst, G., König, S., Gassel, M., Dröse, S., Zeeck, A., Altendorf, K. and Wieczorek, H. (2002). Concanamycin A, the specific inhibitor of V-ATPases, binds to the V0 subunit c, *J. Biol. Chem.*, 277, pp. 40544–40548.
105. Huss, M., Sasse, F., Kunze, B., Jansen, R., Steinmetz, H., Ingenhorst, G., Zeeck, A., and Wieczorek, H. (2005). Archazolid and apicularen: Novel specific V-ATPase inhibitors, *BMC Biochem.*, 6, pp. 13.
106. Huss, M. and Wieczorek, H. (2009). Inhibitors of V-ATPases: Old and new players, *J. Exp. Biol.*, 212, pp. 341–346.
107. Hutton, J.C. (1994). Insulin secretory granule biogenesis and the proinsulin-processing endopeptidases, *Diabetologia*, 37, pp. S48–S56.
108. Israel, M., Morel, N., Lesbats, B., Birman, S., and Manaranche, R. (1986). Purification of a presynaptic membrane protein that mediates a calcium-dependent translocation of acetylcholine, *Proc. Natl Acad. Sci. USA*, 83, pp. 9226–9230.

109. Jansen, E.J., Scheenen, W.J., Hafmans, T.G., and Martens, G.J. (2008). Accessory subunit Ac45 controls the V-ATPase in the regulated secretory pathway, *Biochim. Biophys. Acta*, 1783, pp. 2301–2310.

110. Jansen, E.J., van Bakel, N.H., Coenen, A.J., van Dooren, S.H., van Lith, H.A., and Martens, G.J. (2010). An isoform of the vacuolar ($H(^+)$)-ATPase accessory subunit Ac45, *Cell Mol. Life Sci.*, 67, pp. 629–640.

111. Jefferies, K.C., Cipriano, D.J. and Forgac, M. (2008). Function, structure and regulation of the vacuolar (H^+) ATPases, *Arch. Biochem. Biophys.*, 476, pp. 33–42.

112. Kaelin, W.G., Jr. and Ratcliffe, P.J. (2008). Oxygen sensing by metazoans: The central role of the HIF hydroxylase pathway, *Mol. Cell*, 30, pp. 393–402.

113. Kaelin, W.G., Jr. and Thompson, C.B. (2010). Q&A: Cancer: Clues from cell metabolism, *Nature*, 465, pp. 562–564.

114. Kammoun, H.L., Chabanon, H., Hainault, I., Luquet, S., Magnan, C., Koike, T., Ferré, P., and Foufelle, F. (2009). GRP78 expression inhibits insulin and ER stress-induced SREBP-1c activation and reduces hepatic steatosis in mice, *J. Clin. Invest.*, 119, pp. 1201–1215.

115. Kanazawa, A., Tsukada, S., Sekine, A., Tsunoda, T., Takahashi, A., Kashiwagi, A., Tanaka, Y., Babazono, T., Matsuda, M., Kaku, K., Iwamoto, Y., Kawamori, R., Kikkawa, R., Nakamura, Y., and Maeda S. (2004). Association of the gene encoding wingless-type mammary tumor virus integration-site family member 5B (WNT5B) with type 2 diabetes, *Am. J. Hum. Genet.*, 75, pp. 832–843.

116. Kane, P.M. (1995). Disassembly and reassembly of the yeast vacuolar H^+-ATPase in vivo, *J. Biol. Chem.*, 270, 17025–17032.

117. Kane, P.M. (2000). Regulation of V-ATPases by reversible disassembly, *FEBS Lett.*, 469, pp. 137–141.

118. Kane, P.M. and Parra, K.J. (2000). Assembly and regulation of the yeast vacuolar $H(^+)$-ATPase, *J. Exp. Biol.*, 203, pp. 81–87.

119. Kane, P.M. (2006). The where, when, and how of organelle acidification by the yeast vacuolar H^+-ATPase, *Microbiol. Mol. Biol. Rev.*, 70, pp. 177–191.

120. Kapeller, R. and Cantley, L.C. (1994). Phosphatidylinositol 3-kinase, *Bioessays*, 16, pp. 565–576.

121. Karlas, A., Machuy, N., Shin, Y., Pleissner, K.P., Artarini, A., Heuer, D., Becker, D., Khalil, H., Ogilvie, L.A., Hess, S., Mäurer, A.P., Müller, E., Wolff, T., Rudel, T., and Meyer, T.F. (2010). Genome-wide RNAi screen identifies human host factors crucial for influenza virus replication, *Nature*, 463, pp. 818–822.

122. Kasper, J.S. and Giovannucci, E. (2006). A meta-analysis of diabetes mellitus and the risk of prostate cancer. *Cancer Epidemiol. Biomarkers Prev.*, 15, pp. 2056–2062.

123. Katoh, M. (2005). WNT/PCP signaling pathway and human cancer, *Oncol. Rep.*, 14, pp. 1583–1588.

124. Katoh, M. (2008). WNT signaling in stem cell biology and regenerative medicine, *Curr. Drug Targets*, 9, pp. 565–570.

125. Kawasaki-Nishi, S., Bowers, K., Nishi, T., Forgac, M., and Stevens, T.H. (2001). The amino-terminal domain of the V-ATPase a subunit controls targeting and in vivo dissociation and the carboxyl-terminal domain affects coupling of proton transport and ATP hydrolysis, *J. Biol. Chem.*, 276, pp. 47411–47420.

126. Kawasaki-Nishi, S., Nishi, T., and Forgac, M. (2003). Proton translocation driven by ATP-hydrolysis in V-ATPases, *FEBS Lett.*, 545, pp. 76–85.

127. Kikuchi, A. and Yamamoto, H. (2007). Regulation of Wnt signalling by receptor-mediated endocytosis, *J. Biochem.*, 141, pp. 443–451.

128. Kim, J.W., Shin-Ya, K., Furihata, K., Hayakawa, Y., and Seto, H. (1999). Oximidines I and II: Novel antitumor macrolides from Pseudomonas sp, *J. Org. Chem.*, 64, pp. 153–155.

129. Kim, J.W., Tchernyshyov, I., Semenza, G.L., and Dang, C.V. (2006). HIF-1-mediated expression of pyruvate dehydrogenase kinase: A metabolic switch required for cellular adaptation to hypoxia, *Cell Metab.*, 3, pp. 177–185.

130. Kim, J.W., Gao, P., Liu, Y.C., Semenza, G.L., and Dang, C.V. (2007). Hypoxia-inducible factor 1 and dysregulated c-Myc cooperatively induce vascular endothelial growth factor and metabolic switches hexokinase 2 and pyruvate dehydrogenase kinase 1, *Mol. Cell Biol.*, 27, pp. 7381–7393.

131. Kinashi, H., Someno, K., and Sakaguchi, K. (1984) Isolation and characterization of concanamycins A, B and C, *J. Antibiot. (Tokyo)*, 37, pp. 1333–1343.

132. Kinouchi, K., Ichihara, A., Sano, M., Sun-Wada, G.H., Wada, Y., Kurauchi-Mito, A., Bokuda, K., Narita, T., Oshima, Y., Sakoda, M., Tamai, Y., Sato, H., Fukuda, K. and Itoh, H. (2010). The (Pro)renin receptor/ATP6AP2 is essential for vacuolar H^+-ATPase assembly in murine cardiomyocytes, *Circ. Res.*, 107, pp. 30–34.

133. Kirshner, N. (1962). Uptake of catecholamines by a particulate fraction of the adrenal medulla, *J. Biol. Chem.*, 237, pp. 2311–2317.

134. Kohn, A.D. and Moon, R.T. (2005). Wnt and calcium signaling: Beta-catenin-independent pathways, *Cell Calcium*, 38, pp. 439–446.

135. König, R., Stertz, S., Zhou, Y., Inoue, A., Hoffmann, H.H., Bhattacharyya, S., Alamares, J.G., Tscherne, D.M., Ortigoza, M.B., Liang, Y., Gao, Q., Andrews, S.E., Bandyopadhyay, S., De Jesus, P., Tu, B. P., Pache, L., Shih, C., Orth, A., Bonamy, G., Miraglia, L., Ideker, T., García-Sastre, A., Young, J.A., Palese, P., Shaw, M.L., and Chanda, S.K. (2010). Human host factors required for influenza virus replication, *Nature*, 463, pp. 813–817.

136. Kourelis, T.V. and Siegel, R.D. (2012). Metformin and cancer: New applications for an old drug. *Med. Oncol.*, 29, pp. 1314–1327.

137. Kunze, B., Jansen, R., Sasse, F., Höfle, G., and Reichenbach, H. (1998). Apicularens A and B, new cytostatic macrolides from Chondromyces species (myxobacteria): Production, physico-chemical and biological properties, *J. Antibiot.*, 51, pp. 1075–1080.

138. Kunze, B., Steinmetz, H., Hofle, G., Huss, M., Wieczorek, H., and Reichenbach, H. (2006). Cruentaren, a new antifungal salicylate-type macrolide from Byssovorax cruenta (myxobacteria) with inhibitory effect on mitochondrial ATPase activity. Fermentation and biological properties, *J. Antibiot.*, 59, pp. 664–668.

139. Larsson, S.C., Orsini, N., and Wolk, A. (2005). Diabetes mellitus and risk of colorectal cancer: A meta-analysis, *J. Nat. Cancer Inst.*, 97, pp. 1679–1687.

140. Larsson, S. C., Orsini, N., Brismar, K., and Wolk, A. (2006). Diabetes mellitus and risk of bladder cancer: A meta-analysis, *Diabetologia*, 49, pp. 2819–2823.

141. Larsson, S. C., Mantzoros, C. S., and Wolk, A. (2007). Diabetes mellitus and risk of breast cancer: A meta-analysis, *Int. J. Cancer*, 121, pp. 856–862.

142. Lecot, S., Belouzard, S., Dubuisson, J., and Rouille, Y. (2005). Bovine viral diarrhea virus entry is dependent on clathrin-mediated endocytosis, *J. Virol.*, 79, pp. 10826–10829.

143. Lerman, O.Z., Galiano, R.D., Armour, M., Levine, J.P., and Gurtner, G.C. (2003). Cellular dysfunction in the diabetic fibroblast: Impairment in migration, vascular endothelial growth factor production, and response to hypoxia, *Am. J. Pathol.*, 162, pp. 303–312.

144. Levin, M. (2002). Isolation and community: A review of the role of gap junctional communication in embryonic patterning, *J. Membr. Biol.*, 185, pp. 177–192.

145. Li, D., Yeung, S.C., Hassan, M.M., Konopleva, M., and Abbruzzese, J.L. (2009). Antidiabetic therapiesaffect risk of pancreatic cancer, *Gastroenterology*, 137, pp. 482–488.

146. Li, G., Yang, Q., Krishnan, S., Alexander, E.A., Borkan, S.C., and Schwartz, J.H. (2006). A novel cellular survival factor—the B2 subunit of vacuolar H^+-ATPase inhibits apoptosis. *Cell Death Differ.*, 13, pp. 2109–2117.

147. Li, S.L., Goko, H., Xu, Z.D., Kimura, G., Sun, Y., Kawachi, M.H., Wilson, T.G., Wilczynski, S. and Fujita-Yamaguchi, Y. (1998). Expression of insulin-like growth factor (IGF)-II in human prostate, breast, bladder, and paraganglioma tumors, *Cell Tissue Res.*, 291, pp. 469–479.

148. Libby, G., Donnelly, L.A., Donnan, P.T., Alessi, D.R., Morris, A.D., and Evans, J.M.M. (2009). New users of metformin are at low risk of incident cancer: A cohort study among people with type 2 diabetes, *Diabetes Care*, 32, pp. 1620–1625.

149. Liegeois, S., Benedetto, A., Garnier, J.M., Schwab, Y., and Labouesse, M. (2006). The V_0-ATPase mediates apical secretion of exosomes containing Hedgehog-related proteins in Caenorhabditis elegans, *J. Cell Biol.*, 173, pp. 949–961.

150. Lim, J.H., Park, J.W., Kim, M.S., Park, S.K., Johnson, R.S., and Chun, Y.S. (2006). Bafilomycin induces the p21-mediated growth inhibition of cancer cells under hypoxic conditions by expressing hypoxia-inducible factor-1 alpha, *Mol. Pharmacol.*, 70, pp. 1856–1865.

151. Lim, J.H., Park, J.W., Kim, S.J., Kim, M.S., Park, S.K., Johnson, R.S., and Chun, Y.S. (2007). ATP6V0C competes with von Hippel-Lindau protein in hypoxia-inducible factor 1 alpha (HIF-1 alpha) binding and mediates HIF-1alpha expression by bafilomycin A1, *Mol. Pharmacol.*, 71, pp. 942–948.

152. Limburg, P.J., Vierkant, R.A., Fredericksen, Z.S., Leibson, C.L., Rizza, R.A., Gupta, A.K., Ahlquist, D.A., Melton, L.J., III, Sellers, T.A., and Cerhan, J.R. (2006). Clinically confirmed type 2 diabetes mellitus and colorectal cancer risk: A populationbased, retrospective cohort study, *Am. J. Gastroenterol.*, 101, pp. 1872–1879.

153. Lipscombe, L.L., Goodwin, P.J., Zinman, B., McLaughlin, J.R., and Hux, J.E. (2008). The impact ofdiabetes on survival following breast cancer, *Breast Cancer Res. Treat.*, 109, pp. 389–395.

154. Louagie, E., Taylor, N.A., Flamez, D., Roebroek, A.J., Bright, N.A., Meulemans, S., Quintens, R., Herrera, P.L., Schuit, F., Van de Ven, W.J., and Creemers, J.W. (2008). Role of furin in granular acidification in the endocrine pancreas: Identification of the V-ATPase subunit Ac45 as a candidate substrate, *Proc. Natl. Acad. Sci. USA*, 105, pp. 12319–12324.

155. Lu, M., Holliday, L.S., Zhang, L., Dunn, W.A., Jr., Gluck, S.L. (2001). Interaction between aldolase and vacuolar H^+-ATPase: Evidence for

direct coupling of glycolysis to the ATP-hydrolyzing proton pump, *J. Biol. Chem.*, 276, pp. 30407–30413.

156. Lu, M., Sautin, Y.Y., Holliday, L.S., Gluck, S.L. (2004). The glycolytic enzyme aldolase mediates assembly, expression, and activity of vacuolar H^+-ATPase, *J. Biol. Chem.*, 279, pp. 8732–8739.

157. Lu, M., Ammar, D., Ives, H., Albrecht, F., and Gluck, S.L. (2007). Physical interaction between aldolase and vacuolar H^+-ATPase is essential for the assembly and activity of the proton pump, *J. Biol. Chem.*, 282, pp. 24495–24503.

158. Ludwig, J., Kerscher, S., Brandt, U., Pfeiffer, K., Getlawi, F., Apps, D.K., and Schägger, H. (1998). Identification and characterization of a novel 9.2-kDa membrane sector-associated protein of vacuolar proton-ATPase from chromaffin granules, *J. Biol. Chem.*, 273, pp. 10939–10947.

159. Lui, G-S., Chen, S., Chen, J., and Wang, X-C. (2004). Identification of the phosphoryaltion site of the V-ATPase subunit A in maize roots, *Acta Bot. Sin.*, 46, pp. 428–435.

160. Lustig, B. and Behrens, J. (2003). The Wnt signaling pathway and its role in tumor development, *J . Cancer Res. Clin. Oncol.*, 129, pp. 199–221.

161. Marjuki, H., Gornitzky, A., Marathe, B.M., Ilyushina, N.A., Aldridge, J.R., Desai, G., Webby, R.J., and Webster, R.G. (2011). Influenza A virus-induced early activation of ERK and PI3K mediates V-ATPase-dependent intracellular pH change required for fusion, *Cell Microbiol.*, 13, pp. 587–601.

162. Marshansky, V. (2007). The V-ATPase a2-subunit as a putative endosomal pH-sensor, *Biochem. Soc. Trans.*, 35, pp. 1092–1099.

163. Marshansky, V. and Futai, M. (2008). The V-type $H(^+)$-ATPase in vesicular trafficking: Targeting, regulation and function, *Curr. Opin. Cell Biol.*, 20, pp. 415–426.

164. Martínez-Munoz, G.A. and Kane, P. (2008). Vacuolar and plasma membrane proton pumps collaborate to achieve cytosolic pH homeostasis in yeast, *J. Biol. Chem.*, 283, pp. 20309–20319.

165. Martínez-Zaguilán, R. and Gillies, R.J. (1992). A plasma membrane V-type $H(^+)$-ATPase may contribute to elevated intracellular pH (pHin) in some human tumor cells, *Ann. N. Y. Acad. Sci.*, 671, pp. 478–480.

166. Martínez-Zaguilán, R., Lynch, R.M., Martinez, G.M., and Gillies, R.J. (1993). Vacuolar-type $H(^+)$-ATPases are functionally expressed in plasma membranes of human tumor cells, *Am. J. Physiol.*, 265, pp. C1015–C1029.

167. Martínez-Zaguilán, R., Martinez, G.M., Gomez, A., Hendrix, M.J., and Gillies, R.J. (1998). Distinct regulation of pHin and [Ca^{2+}] in human melanoma cells with different metastatic potential, *J. Cell Physiol.*, 176, pp. 196–205.

168. Martínez-Zaguilán, R., Raghunand, N., Lynch, R.M., Bellamy, W., Martinez, G.M., Rojas, B., Smith, D., Dalton, W.S., and Gillies, R.J. (1999). pH and drug resistance I. Functional expression of plasmalemmal V-type H^+-ATPase in drug-resistant human breast carcinoma cell lines, *Biochem. Pharmacol.*, 57, pp. 1037–1046.

169. Maxwell, P.H. (2005). The HIF pathway in cancer, *Semin. Cell Dev. Biol.*, 16, pp. 523–530.

170. Mekhail, K., Gunaratnam, L., Bonicalzi, M.E., and Lee, S. (2004). HIF activation by pH-dependent nucleolar sequestration of VHL, *Nat. Cell Biol.*, 6, pp. 642–647.

171. Menche, D., Hassfeld, J., Steinmetz, H., Huss, M., Wieczorek, H., and Sasse, F. (2007). The first hydroxylated archazolid from the myxobacterium Cystobacter violaceus: Isolation, structural elucidation and V-ATPase inhibition, *J. Antibiot.*, 60, pp. 328–331.

172. Merkulova, M., Hurtado-Lorenzo, A., Hosokawa, H., Zhuang, Z., Brown, D., Ausiello, D.A., and Marshansky, V. (2011). Aldolase directly interacts with ARNO and modulates cell morphology and acidic vesicle distribution, *Am. J. Physiol. Cell Physiol.*, 300, pp. C1442–C1455.

173. Metchnikoff, E. (1905). *Immunity in Infective Diseases*, Cambridge University Press, Cambridge.

174. Moon, R.T., Bowerman, B., Boutros, M., and Perrimon, N. (2002). The promise and perils of Wnt signaling through beta-catenin, *Science*, 296, pp. 1644–1646.

175. Morel, N., Dedieu, J.C., and Philippe, J.M. (2003). Specific sorting of the a1 isoform of the V-H^+-ATPase a subunit to nerve terminals where it associates with both synaptic vesicles and the presynaptic plasma membrane, *J. Cell Sci.*, 116, pp. 4751–4762.

176. Müller, K.H., Kainov, D.E., El Bakkouri, K., Saelens, X., De Brabander, J.K., Kittel, C., Samm, E., and Muller, C.P. (2011). The proton translocation domain of cellular vacuolar ATPase provides a target for the treatment of influenza A virus infections, *Br. J. Pharmacol.*, 164, pp. 344–357.

177. Nadler, G., Morvan, M., Delimoge, I., Belfiore, P., Zocchetti, A., James, I., Zembryki, D., Lee-Rycakzewski, E., Parini, C., Consolandi, E., Gagliardi, S., and Farina, C. (1998). (2Z,4E)-5-(5,6-dichloro-2-indolyl)-2-methoxy-*N*-(1,2,2,6,6-pentamethylpiperidin-4-yl)-2,4-pentadienamide, a

novel, potent and selective inhibitor of the osteoclast V-ATPase, *Bioorg. Med. Chem. Lett.*, 8, pp. 3621–3626.

178. Nakamura, I., Takahashi, N., Sasaki, T., Tanaka, S., Udagawa, N., Murakami, H., Kimura, K., Karubawa, T., and Suda, T. (1995). Wortmannin, a specific inhibitor of Phosphatidylinositol-3-kinase, blocks osteoclast resorption, *FEBS Lett.*, 361, pp. 79–84.
179. Nakamura, I., Sasaki, T., Tanaka, S., Takahashi, N., Jimi, E., Kurokawa, T., Kita, Y., Ihara, S., Suda, T., and Fukui, Y. (1997). Phosphatidylinositol-3-kinase is involved in ruffled border formation in osteoclasts, *J. Cell. Physiol.*, 172, pp. 230–239.
180. Nakamura, S. (2004). Glucose activates H^+-ATPase in kidney epithelial cells, *Am. J. Physiol. Cell Physiol.*, 287, pp. C97–C105.
181. Nanda, A., Brumell, J.H., Nordstrom, T., Kjeldsen, L., Sengelov, H., Borregaard, N., Rotstein, O.D. and Grinstein, S. (1996). Activation of proton pumping in human neutrophils occurs by exocytosis of vesicles bearing vacuolar-type H^+-ATPases, *J. Biol. Chem.*, 271, pp. 15963–15970.
182. Nawa, M. (1998). Effects of bafilomycin A1 on Japanese encephalitis virus in C6/36 mosquito cells, *Arch. Virol.*, 143, pp. 1555–1568.
183. Nawa, M., Machida, S., Takasaki, T., and Kurane, I. (2007). Plaque formation by Japanese encephalitis virus bound to mosquito C6/36 cells after low pH exposure on the cell surface, *Jpn. J. Infect. Dis.*, 60, pp. 118–120.
184. Nelson, H. and Nelson, N. (1990). Disruption of genes encoding subunits of yeast vacuolar $H(^+)$-ATPase causes conditional lethality, *Proc. Natl. Acad. Sci. USA*, 87, pp. 3503–3507.
185. Nelson, W.J. and Nusse, R. (2004). Convergence of Wnt, beta-catenin, and cadherin pathways, *Science*, 303, pp. 1483–1487.
186. Nguyen, G., Delarue, F., Burckle, C., Bouzhir, L., Giller, L. and Sraer, J.D. (2002). Pivotal role of the renin/prorenin receptor in angiotensin II production and cellular responses to renin, *J. Clin. Invest.*, 109, pp. 1417–1427.
187. Nguyen, G. and Muller, D.N. (2010). The biology of the (pro)renin receptor, *J. Am. Soc. Nephrol.*, 21, pp. 18–23.
188. Nguyen, G. (2011). Renin, (pro)renin and receptor: An update, *Clin. Sci. (Lond.),* 120, pp. 169–178.
189. Nicolucci, A. (2010). Epidemiological aspects of neoplasms in diabetes, *Acta Diabetol.*, 47, pp. 87–95.

190. Niehrs, C. and Acebron, S.P. (2010). Wnt signaling: Multivesicular bodies hold GSK3 captive, *Cell*, 143, pp. 1044–1046.

191. Nishi, T. and Forgac, M. (2000). Molecular cloning and expression of three isoforms of the 100-kDa a subunit of the mouse vacuolar proton-translocating ATPase, *J. Biol. Chem.*, 275, pp. 6824–6830.

192. Nishi, T. and Forgac, M. (2002). The vacuolar (H^+)-ATPases–nature's most versatile proton pumps, *Nat. Rev. Mol. Cell Biol.*, 3, pp. 94–103.

193. Nishikawa, T., Edelstein, D., and Brownlee, M. (2000). The missing link: A single unifying mechanism for diabetic complications, *Kidney Int. Suppl.*, 77, pp. S26–S30.

194. Nordström, T., Grinstein, S., Brisseau, G.F., Manolson, M.F., and Rotstein, O.D. (1994). Protein kinase C activation accelerates proton extrusion by vacuolar type H1-ATPase in murine peritoneal macrophages, *FEBS Lett.*, 350, pp. 82–86.

195. Nusse, R. and Varmus, H.E. (1982). Many tumors induced by the mouse mammary tumor virus contain a provirus integrated in the same region of the host genome, *Cell*, 31, pp. 99–109.

196. Ochiai, H., Sakai, S., Hirabayashi, T., Shimizu, Y., and Terasawa, K. (1995). Inhibitory effect of bafilomycin A1, a specific inhibitor of vacuolar-type proton pump, on the growth of influenza A and B viruses in MDCK cells, *Antiviral Res.*, 27, pp. 425–430.

197. Ohkuma, S., Moriyama, Y., and Takano, K. (1982). Identification and characterization of a proton pump on lysosomes by fluoresceiniso-thiocyanate-dextran fluorescence, *Proc. Natl. Acad. Sci. USA*, 79, pp. 2758–2762.

198. Oldenburg, B., Diepersloot, R.J., and Hoekstra, J.B. (1996). High seroprevalence of Helicobacter pylori in diabetes mellitus patients, *Dig. Dis. Sci.*, 41, pp. 458–461.

199. Orci, L., Ravazzola, M., Storch, M.J., Anderson, R.G., Vassalli, J.D., and Perrelet, A. (1987). Proteolytic maturation of insulin is a post-Golgi event which occurs in acidifying clathrin-coated secretory vesicles, *Cell*, 49, pp. 865–868.

200. Orci, L., Halban, P., Perrelet, A., Amherdt, M., Ravazzola, M., and Anderson, R.G. (1994). pH-independent and -dependent cleavage of proinsulin in the same secretory vesicle, *J. Cell Biol.*, 126, pp. 1149–1156.

201. Palsgaard, J., Emanuelli, B., Winnay, J.N., Sumara, G., Karsenty, G., and Kahn, C.R. (2012). Cross-talk between insulin and Wnt signaling in preadipocytes: Role of Wnt co-receptor low density lipoprotein

receptor-related protein-5 (LRP5), *J. Biol. Chem.*, 287, pp. 12016–12026.

202. Papa, V., Pezzino, V., Costantino, A., Belfiore, A., Giuffrida, D., Frittitta, L., Vannelli, G. B., Brand, R., Goldfine, I.D., and Vigneri, R. (1990). Elevated insulin receptor content in human breast cancer, *J. Clin. Invest.*, 86, pp. 1503–1510.
203. Parra, K.J. and Kane, P.M. (1998). Reversible association between the V1 and V0 domains of yeast vacuolar H-ATPase is an unconventional glucose-induced effect, *Mol. Cell. Biol.*, 18, pp. 7064–7074.
204. Perez, L. and Carrasco, L. (1994). Involvement of the vacuolar H^+-ATPase in animal virus entry, *J. Gen. Virol.*, 75, pp. 2595–2606.
205. Pérez-Sayáns, M., Somoza-Martín, J.M., Barros-Angueira, F., Diz, P.G., Rey, J.M., and García-García, A. (2010). Multidrug resistance in oral squamous cell carcinoma: The role of vacuolar ATPases, *Cancer Lett.*, 295, pp. 135–143.
206. Perona, R. and Serrano, R. (1988). Increased pH and tumorigenicity of fibroblasts expressing a yeast proton pump, *Nature*, 334, pp. 438–340.
207. Peters, C., Bayer, M.J., Buhler, S., Andersen, J.S., Mann, M., and Mayer, A. (2001). Trans-complex formation by proteolipid channels in the terminal phase of membrane fusion, *Nature*, 409, pp. 581–588.
208. Pietrement, C., Sun-Wada, G.H., Silva, N.D., McKee, M., Marshansky, V., Brown, D., Futai, M., and Breton, S. (2006). Distinct expression patterns of different subunit isoforms of the V-ATPase in the rat epididymis, *Biol. Reprod.*, 74, pp. 185–194.
209. Pollak, M. (2008). Insulin and insulin-like growth factor signalling in neoplasia, *Nat. Rev. Cancer*, 8, pp. 915–928.
210. Prestwich, T.C. and Macdougald, O.A. (2007). Wnt/beta-catenin signaling in adipogenesis and metabolism, *Curr. Opin. Cell Biol.*, 19, pp. 612–617.
211. Qi, J. and Forgac, M. (2007). Cellular environment is important in controlling V-ATPase dissociation and its dependence on activity, *J. Biol. Chem.*, 282, pp. 24743–24751.
212. Quetglas, S., Leveque, C., Miquelis, R., Sato, K., and Seagar, M. (2000). Ca^{2+}-dependent regulation of synaptic SNARE complex assembly via a calmodulin- and phospholipid-binding domain of synaptobrevin, *Proc. Natl Acad. Sci. USA*, 97, pp. 9695–9700.
213. Quintero, M., Brennan, P.A., Thomas, G.J., Moncada, S. (2006). Nitric oxide is a factor in the stabilization of hypoxia-inducible factor-1 in cancer: Role of free radical formation, *Cancer Res.*, 66, pp. 770–774.

214. Racker, E. (1972). Bioenergetics and the problem of tumor growth, *Am. Sci.*, 60, pp. 56–63.

215. Raghunand, N., Martínez-Zaguilán, R., Wright, S.H. and Gillies, R.J. (1999). pH and drug resistance. II. Turnover of acidic vesicles and resistance to weakly basic chemotherapeutic drugs, *Biochem. Pharmacol.*, 57, pp. 1047–1058.

216. Robey, I.F., Lien, A.D., Welsh, S.J., Baggett, B.K., and Gillies, R.J. (2005). Hypoxia-inducible factor-1alpha and the glycolytic phenotype in tumors, *Neoplasia*, 7, pp. 324–330.

217. Robey, I.F., Stephen, R.M., Brown, K.S., Baggett, B.K., Gatenby, R.A., and Gillies, R.J. (2008). Regulation of the Warburg effect in early-passage breast cancer cells, *Neoplasia*, 10, pp. 745–756.

218. Rojas, J.D., Sennoune, S.R., Maiti, D., Martinez, G.M., Bakunts, K., Wesson, D.E. and Martínez-Zaguilán, R. (2004). Plasmalemmal V-H($^+$)-ATPases regulate intracellular pH in human lung microvascular endothelial cells, *Biochem. Biophys. Res. Commun.*, 320, pp. 1123–1132.

219. Rojas, J.D., Sennoune, S.R., Martinez, G.M., Bakunts, K., Meininger, C.J., Wu, G., Wesson, D.E., Seftor, E.A., Hendrix, M.J., and Martínez-Zaguilán, R. (2004). Plasmalemmal vacuolar H^+ATPase is decreased in microvascular endothelial cells from a diabetic model, *J. Cell Physiol.*, 201, pp. 190–200.

220. Rojas, J.D., Sennoune, S.R., Maiti, D., Bakunts, K., Reuveni, M., Sanka, S.C., Martinez, G.M., Seftor, E.A., Meininger, C.J., Wu, G., Wesson, D.E., Hendrix, M.J., and Martínez-Zaguilán, R. (2006). Vacuolar-type H^+-ATPases at the plasma membrane regulate pH and cell migration in microvascular endothelial cells, *Am. J. Physiol. Heart Circ. Physiol.*, 291, pp. H1147–H1157.

221. Rubenstein, E.M. and Schmidt, M.C. (2010). The glucose signal and metabolic p[H^+]lux, *EMBO J.*, 29, pp. 2473–2474.

222. Samani, A.A., Yakar, S., LeRoith, D., and Brodt, P. (2007). The role of the IGF system in cancer growth and metastasis: Overview and recent insights, *Endocr. Rev.*, 28, pp. 20–47.

223. Sasse, F., Steinmetz, H., Höfle, G., and Reichenbach, H. (2003). Archazolids, new cytotoxic macrolactones from Archangium gephyra (Myxobacteria). Production, isolation, physico-chemical and biological properties, *J. Antibiot.*, 56, pp. 520–525.

224. Sautin, Y.Y., Lu, M., Gaugler, A., Zhang, L., and Gluck, S.L. (2005). Phosphatidylinositol 3-kinase-mediated effects of glucose on vacuolar H^+-ATPase assembly, translocation, and acidification of intracel-

lular compartments in renal epithelial cells, *Mol. Cell Biol.*, 25, pp. 575–589.

225. Schlesinger, P.H., Blair, H.C., Teitelbaum, S.L., and Edwards, J.C. (1997). Characterization of the osteoclast ruffled border chloride channel and its role in bone resorption, *J. Biol. Chem.*, 272, pp. 18636–18643.
226. Schoonderwoert, V.T. and Martens G.J. (2002). Targeted disruption of the mouse gene encoding the V-ATPase accessory subunit Ac45, *Mol. Membr. Biol.*, 19, pp. 67–71.
227. Schoonderwoert, V.T., Jansen, E.J., and Martens, G.J. (2002). The fate of newly synthesized V-ATPase accessory subunit Ac45 in the secretory pathway, *Eur. J. Biochem.*, 269, pp. 1844–1853.
228. Scott, B.B. and Chapman, C.G. (1998). The putative 116 kDa osteoclast specific vacuolar proton pump subunit has ubiquitous tissue distribution, *Eur. J. Pharmacol.*, 346, pp. R3–R4.
229. Semenza, G.L. and Wang, G.L. (1992). A nuclear factor induced by hypoxia via de novo protein synthesis binds to the human erythropoietin gene enhancer at a site required for transcriptional activation, *Mol. Cell Biol.*, 12, pp. 5447–5454.
230. Semenza, G.L., Jiang, B.H., Leung, S.W., Passantino, R., Concordet, J.P., Maire, P., and Giallongo, A. (1996). Hypoxia response elements in the aldolase A, enolase 1, and lactate dehydrogenase A gene promoters contain essential binding sites for hypoxia-inducible factor 1, *J. Biol. Chem.*, 271, pp. 32529–32537.
231. Semenza, G.L. (2010). HIF-1: Upstream and downstream of cancer metabolism, *Curr. Opin. Genet. Dev.*, 20, pp. 51–56.
232. Sennoune, S.R., Luo, D., and Martínez-Zaguilán, R. (2004) Plasmalemmal vacuolar-type H^+-ATPase in cancer biology, *Cell Biochem. Biophys.*, 40, pp. 185–206.
233. Sennoune, S.R., Bakunts, K., Martínez, G.M., Chua-Tuan, J.L., Kebir, Y., Attaya, M.N., and Martínez-Zaguilán, R. (2004). Vacuolar H^+-ATPase in human breast cancer cells with distinct metastatic potential: Distribution and functional activity, *Am. J. Physiol. Cell Physiol.*, 286, pp. C1443–C1452.
234. Sennoune, S.R. and Martínez-Zaguilán, R. (2007). Plasmalemmal vacuolar H^+-ATPases in angiogenesis, diabetes and cancer, *J. Bioenerg. Biomembr.*, 39, pp. 427–433.
235. Seol, J.H., Shevchenko, A., and Deshaies, R.J. (2001). Skp1 forms multiple protein complexes, including RAVE, a regulator of V-ATPase assembly, *Nat. Cell Biol.*, 3, pp. 384–391.

236. Shepherd, P.R., Withers, D.J., and Siddle, K. (1998). Phosphoinositide 3-kinase: The key switch mechanism in insulin signaling, *Biochem. J.*, 333, pp. 471–490.

237. Sihn, G., Rousselle, A., Vilianovitch, L., Burckle, C., and Bader, M. (2010). Physiology of the (pro)renin receptor: Wnt of change? *Kidney Int.*, 78, pp. 246–256.

238. Smardon, A.M., Tarsio, M., Kane, P.M. (2002). The RAVE complex is essential for stable assembly of the yeast V-ATPase, *J. Biol. Chem.*, 277, pp. 13831–13839.

239. Smardon, A.M. and Kane, P.M. (2007). RAVE is essential for the efficient assembly of the C subunit with the vacuolar H-ATPase, *J. Biol. Chem.*, 282, pp. 26185–26194.

240. Smeekens, S.P., Montag, A.G., Thomas, G., Albiges-Rizo, C., Carroll, R., Benig, M., Phillips, L.A., Martin, S., Ohagi, S., Gardner, P., Swift, H.H., and Steiner, D.F. (1992). Proinsulin processing by the subtilisin-related proprotein convertases furin, PC2, and PC3, *Proc. Natl. Acad. Sci. USA*, 89, pp. 8822–8826.

241. Sonveaux, P., Végran, F., Schroeder, T., Wergin, M.C., Verrax, J., Rabbani, Z.N., De Saedeleer, C.J., Kennedy, K.M., Diepart, C., Jordan, B.F., Kelley, M.J., Gallez, B., Wahl, M.L., Feron, O., and Dewhirst, M.W. (2008). Targeting lactate-fueled respiration selectively kills hypoxic tumor cells in mice, *J. Clin. Invest.*, 118, pp. 3930–3942.

242. Sørensen, M.G., Henriksen, K., Neutzsky-Wulff, A.V., Dziegiel, M.H., and Karsdal, M.A. (2007). Diphyllin, a novel and naturally potent V-ATPase inhibitor, abrogates acidification of the osteoclastic resorption lacunae and bone resorption, *J. Bone Miner. Res.*, 22, pp. 1640–1648.

243. Sørensen, M.G., Karsdal, M.A., Dziegiel, M.H., Boutin, J.A., Nosjean, O., and Henriksen, K. (2010). Screening of protein kinase inhibitors identifies PKC inhibitors as inhibitors of osteoclastic acid secretion and bone resorption, *BMC Musculoskelet. Disord.*, 11, pp. 250.

244. Stankovic, K.M., Brown, D., Alper, S.L., and Adams, J.C. (1997). Localization of pH regulating proteins H^+-ATPase and Cl^-/HCO_3^- exchanger in the guinea pig inner ear, *Hear. Res.*, 114, pp. 21–34.

245. Stevens, R.J., Roddam, A.W., and Beral, V. (2007). Pancreatic cancer in type 1 and young-onset diabetes: Systematic review and meta-analysis, *Br. J. Cancer*, 96, pp. 507–509.

246. Su, Y., Zhou, A., Al-Lamki, R.S., and Karet, F.E. (2003). The a-subunit of the V-type H^+-ATPase interacts with phosphofructokinase-1 in humans, *J. Biol. Chem.*, 278, pp. 20013–20018.

247. Sun-Wada, G.H., Toyomura, T., Murata, Y., Yamamoto, A., Futai, M., and Wada, Y. (2006), The a3 isoform of V-ATPase regulates insulin secretion from pancreatic beta-cells, *J. Cell Sci.*, 119, pp. 4531–4540.

248. Sun-Wada, G.H. and Wada, Y. (2010). Vacuolar-type proton pump ATPases: Roles of subunit isoforms in physiology and pathology, *Histol. Histopathol.*, 25, pp. 1611–1620.

249. Supek, F., Supekova, L., Mandiyan, S., Pan, Y.C., Nelson, H., and Nelson, N. (1994). A novel accessory subunit for vacuolar $H(^+)$-ATPase from chromaffin granules, *J. Biol. Chem.*, 269, pp. 24102–24106.

250. Taelman, V.F., Dobrowolski, R., Plouhinec, J.L., Fuentealba, L.C., Vorwald, P.P., Gumper, I., Sabatini, D.D., and De Robertis, E.M. (2010). Wnt signaling requires sequestration of glycogen synthase kinase 3 inside multivesicular endosomes, *Cell*, 143, pp. 1136–1148.

251. Taubes, G. (2012). Cancer research. Ravenous for glucose, *Science*, 335, pp. 29–32.

252. Thangarajah, H., Yao, D., Chang, E.I., Shi, Y., Jazayeri, L., Vial, I.N., Galiano, R.D., Du, X.L., Grogan, R., Galvez, M.G., Januszyk, M., Brownlee, M., and Gurtner, G.C. (2009). The molecular basis for impaired hypoxia-induced VEGF expression in diabetic tissues, *Proc. Natl. Acad. Sci. USA*, 106, pp. 13505–13510.

253. Toei, M., Saum, R., and Forgac, M. (2010). Regulation and isoform function of the V-ATPases, *Biochemistry*, 49, pp. 4715–4723.

254. Toyomura, T., Oka, T., Yamaguchi, C., Wada, Y., and Futai, M. (2000). Three subunit a isoforms of mouse vacuolar H^+-ATPase. Preferential expression of the a3 isoform during osteoclast differentiation, *J. Biol. Chem.*, 275, pp. 8760–8765.

255. Toyomura, T., Murata, Y., Yamamoto, A., Oka, T., Sun-Wada, G.H., Wada, Y., and Futai, M. (2003). From lysosomes to the plasma membrane: Localization of vacuolar-type H^+-ATPase with the a3 isoform during osteoclast differentiation, *J. Biol. Chem.*, 278, pp. 22023–22030.

256. Trombetta, E.S., Ebersold, M., Garrett, W., Pypaert, M., and Mellman, I. (2003). Activation of lysosomal function during dendritic cell maturation, *Science*, 299, pp. 1400–1403.

257. Twells, R.C., Mein, C.A., Payne, F., Veijola, R., Gilbey, M., Bright, M., Timms, A., Nakagawa, Y., Snook, H., Nutland, S., Rance, H.E., Carr, P., Dudbridge, F., Cordell, H.J., Cooper, J., Tuomilehto-Wolf, E., Tuomilehto, J., Phillips, M., Metzker, M., Hess, J.F., and Todd, J.A. (2003). Linkage and association mapping of the LRP5 locus on chromosome 11q13 in type 1 diabetes, *Hum. Genet.*, 113, pp. 99–105.

258. Vander Heiden, M.G., Cantley, L.C. and Thompson, C.B. (2009). Understanding the Warburg effect: The metabolic requirements of cell proliferation, *Science*, 324, pp. 1029–1033.

259. Vigneri, P., Frasca, F., Sciacca, L., Pandini, G., and Vigneri, R. (2009). Diabetes and cancer, *Endocr. Relat. Cancer*, 16, pp. 1103–1123.

260. Voss, M., Vitavska, O., Walz, B., Wieczorek, H., and Baumann, O. (2007). Stimulus-induced phosphorylation of vacuolar H($^+$)-ATPase by protein kinase A, *J. Biol. Chem.*, 282, pp. 33735–33742.

261. Wagner, C.A., Finberg, K.E., Breton, S., Marshansky, V., Brown, D., and Geibel, J.P. (2004). Renal vacuolar H^+-ATPase, *Physiol. Rev.*, 84, pp. 1263–1314.

262. Wang, H. and Jiang, C. (2009). Influenza A virus H5N1 entry into host cells is through clathrin-dependent endocytosis, *Sci. China C. Life Sci.*, 52, pp. 464–469.

263. Wang, Y., Inoue, T., and Forgac, M. (2005). Subunit a of the yeast V-ATPase participates in binding of bafilomycin, *J. Biol. Chem.*, 280, pp. 40481–40488.

264. Warburg, O., Wind, F., and Negelein, E. (1927). The Metabolism of tumors in the body, *J. Gen. Physiol.*, 8, pp. 519–530.

265. Warburg, O., Wind, F., and Neglers, E. (1930). On the metabolism of tumors in the body, in: *Metabolism of Tumors* (Warburg, O., ed.), Constable & Co, London, pp. 254–270.

266. Warburg, O. (1956). On the origin of cancer cells, *Science*, 123, pp. 309–314.

267. Weisz, O.A. (2003). Acidification and protein traffic, *Int. Rev. Cytol.*, 226, pp. 259–319.

268. Welters, H.J. and Kulkarni, R.N. (2008). Wnt signaling: Relevance to beta-cell biology and diabetes, *Trends Endocrinol. Metab.*, 19, pp. 349–355.

269. Werner, H. and Bruchim, I. (2009). The insulin-like growth factor-I receptor as an oncogene, *Arch. Physiol. Biochem.*, 115, pp. 58–71.

270. Wideroff, L., Gridley, G., Mellemkjaer, L., Chow, W.H., Linet, M., Keehn, S., Borch-Johnsen, K., and Olsen, J.H. (1997). Cancer incidence in a population-based cohort of patients hospitalized with diabetes mellitus in Denmark, *J. Natl. Cancer Inst.*, 89, p. 136.

271. Wilkens, S. and Forgac, M. (2001). Three-dimensional structure of the vacuolar ATPase proton channel by electron microscopy, *J. Biol. Chem.*, 276, pp. 44064–44068.

272. Wilson, P.W. (1998). Diabetes mellitus and coronary heart disease, *Am. J. Kidney Dis.*, 32, pp. S89–S100.

273. Xie, X.S., Stone, D.K., and Racker, E. (1983). Determinants of clathrin-coated vesicle acidification, *J. Biol. Chem.*, 258, pp. 14834–14838.

274. Xie, X.S., Padron, D., Liao, X., Wang, J., Roth, M.G., and De Brabander, J.K. (2004). Salicylihalamide A inhibits the VO sector of the V-ATPase through a mechanism distinct from bafilomycin A1, *J. Biol. Chem.*, 279, pp. 19755–19763.

275. Xu, T. and Forgac, M. (2001). Microtubules are involved in glucose dependent dissociation of the yeast vacuolar H$^+$-ATPase in vivo, *J. Biol. Chem.*, 276, pp. 24855–24861.

276. Xue, W., Liu, Y., Zhao, J., Cai, L., Li, X., and Feng, W. (2012). Activation of HIF-1 by metallothionein contributes to cardiac protection in diabetic heart, *Am. J. Physiol. Heart Circ. Physiol.,* 302(12), pp. H2528–H2535.

277. Yan, Y., Denef, N., and Schupbach, T. (2009). The vacuolar proton pump, V-ATPase, is required for notch signaling and endosomal trafficking in Drosophila, *Dev. Cell*, 17, pp. 387–402.

278. Zendehdel, K., Nyrén, O., Ostenson, C.G., Adami, H.O., Ekbom, A., and Ye, W. (2003). Cancer incidence in patients with type 1 diabetes mellitus: A population-based cohort study in Sweden, *J. Natl. Cancer. Inst.*, 95, pp. 1797–1800.

279. Zhang, J., Feng, Y. and Forgac, M. (1994). Proton conduction and bafilomycin binding by the V0 domain of the coated vesicle V-ATPase, *J. Biol. Chem.,* 269, pp. 23518–23523.

280. Zhang, W., Wang, D., Volk, E., Bellen, H.J., Hiesinger, P.R., and Quiocho, F.A. (2008). V-ATPase V0 sector subunit a1 in neurons is a target of calmodulin, *J. Biol. Chem.*, 283, pp. 294–300.

281. Zhang, Z., Nguyen, K.T., Barrett, E.F., and David, G. (2010). Vesicular ATPase inserted into the plasma membrane of motor terminals by exocytosis alkalinizes cytosolic pH and facilitates endocytosis, *Neuron*, 68, pp. 1097–1108.

Chapter 13

The a2 Isoform of Vacuolar ATPase and Cancer-Related Inflammation

Kenneth Beaman, Christina Kwong, Mukesh Jaiswal, and Alice Gilman-Sachs
Department of Microbiology and Immunology, Clinical Immunology Laboratory, Rosalind Franklin University of Medicine and Science, North Chicago, IL, USA
kenneth.beaman@rosalindfranklin.edu

13.1 Introduction

Vacuolar ATPase (V-ATPase), specifically the a2 isoform, which is discussed in this chapter, has been shown to be important in tumor progression and metastases. V-ATPase is found as a multi-protein complex found on the surface of tumors (Martinez-Zaguilan *et al.* 1993, Sennoune *et al.* 2004b). Its role in tumorigenesis involves acidification of the extracellular environment via its proton pumping activity. This activity enables maturation of pH-sensitive degradative enzymes, such as matrix metalloproteinases, which lead to tumor metastasis (Kubota & Seyama 2000, Rofstad *et al.* 2006). Increased cell surface expression of V-ATPases is linked to cell invasion (Sennoune *et al.* 2004a). V-ATPase has at least three independent activities that support cancer growth by controlling

Handbook of H^+-ATPases
Edited by Suguru Nakamura

ISBN 978-981-4411-91-2 (Hardcover), 978-981-4411-92-9 (eBook)
www.panstanford.com

related inflammatory processes. The first is its function in the control of hydrogen ion production. There is mounting evidence that it also functions as a surface ATPase released from damaged and necrotic cells (Derks & Beaman 2004a). It may be that the a2 isoform has the capacity to redirect its activity and to be either an ATPase or control acid hydrolysis on the cell surface. The second activity is that of as a modulator of chemokines and cytokine expression through a released peptide a2NTD (Jaiswal *et al.* 2012, Kwong *et al.* 2011). Finally, and perhaps most importantly, is its expression in tumor macrophage where it can regulate IL-1 directly leading to inflammation and subsequent angiogenesis.

13.2 V-ATPase in Tumors

The normal function of V-ATPase is to maintain the correct cytoplasmic pH and to acidify intracellular organelles such as endosomes, Golgi, and lysosomes. In tumor cells the activity of V-ATPase increases to accommodate tumor cell survival in an acidified environment (Kubota & Seyama 2000, Martinez-Zaguilan *et al.* 1993, Rofstad *et al.* 2006, Sennoune *et al.* 2004a, Sennoune *et al.* 2004b). V-ATPase is an intracellular multi-protein complex that translocates to the cell surface in a functional form under appropriate conditions such as lymphocyte activation and trophoblast invasion (Ntrivalas *et al.* 2007a, Ntrivalas *et al.* 2007b).

Recently, we have shown the role of the a2 isoform of V-ATPase (a2V) in tumor progression via a novel mechanism. a2V located in the plasma membrane of human tumor cell lines stimulated immune cells by cleaving the *N*-terminus region of the a2V protein and secreting the resulting peptide out of the cell (Kwong *et al.* 2011). When incubated with peripheral blood mononuclear cells (PBMCs), this peptide (a2NTD) caused monocytes to produce IL-1β (Ntrivalas *et al.* 2007a, Ntrivalas *et al.* 2007b). A more detailed investigation showed that a2NTD induced anunique phenotype: IL-12^{low}, IL-23^{low}, and IL-10^{high}, in primary monocytes post stimulation. This phenotype is referred to as M2-like and will be discussed later. Thus, a2NTD was the first non-cytokine described to induce the alternative activation pathway in macrophages (Kwong *et al.* 2011).

The a2NTD peptide plays a novel role in the tumor microenvironment. While functioning to promote a localized inflammatory reaction, it also prevents immune surveillance. Its specific interaction with monocytes is ideal because monocytes and macrophages are well characterized in their ability to promote tumorigenesis (Kwong *et al.* 2011). The stimulation of both $CD16^{-/+}$ monocytes indicates there is more than one responder cell type to a2NTD. In the future, identifying a2NTD's receptors as well as understanding the differences between responder cells will aid in finding new antitumor therapies.

We have detected the expression of the $\alpha 2$ subunit of a V-ATPase in the malignant cell population of a panel of human ovarian carcinoma tumors by immunohistochemistry. It appears that in tumors from later stage disease, there is a significant increase in the expression of a2NTD compared to earlier stage disease. Using confocal microscopy examination we have also demonstrated that in high density culture of ovarian carcinoma cell lines, greater levels of membrane bound form of a2V are expressed. *In vivo* mice injected with ovarian carcinoma cells and then given anti-a2V antibody did not develop tumors or developed tumors greatly decreased in volume compared to control mice not given antibody. Given the increased expression of a2V in the malignant cell population of the tumors, and based on our *in vitro* and *in vivo* studies we suggest that a2Vmay be a possible target for ovarian carcinoma cell killing. We hypothesize that a2V promotes ovarian carcinoma cell survival because of the predominant expression of a2V on the malignant cell population. Based on our *in vitro* model we hypothesized that the mechanism to promote tumor cell survival involves dephosphorylation of extracellular ATP. Unlike other chemical reagents, such as concanamycin A and bafilomycin A1, that block the function of intracellular V-ATPase, we have created an antibody that recognizes the functional cell membrane form of V-ATPase present on malignant cells. Therefore, we conclude that the cell surface form of the V-ATPase $\alpha 2$ subunit is a potential tumor associated antigen candidate to target for the diagnosis and treatment of ovarian cancer. We suggest that anti-a2V antibody may be superior to chemical reagents for targeting V-ATPase because it will target the cell surface form of V-ATPase (a potential tumor cell marker) on malignant cells and spare normal cells.

13.3 V-ATPase and Cytokine Induction and Macrophage Development

A novel peptide from the a2 isoform V-ATPase called a2NTD has been shown to have an immunoregulatory role in the tumor microenvironment by affecting monocyte maturation into a tumor-associated macrophage phenotype. Tumors secrete cytokines and other proteins that dictate the type of immune response that will occur including the activation and polarization of specific macrophage subsets (Mantovani *et al.* 2002). Cytokines, such as IFN-γ, skew macrophages toward a classically activated phenotype called M1, which participates in tumor destruction. In comparison, cytokines, such as IL-4, IL-13, and IL-10, secreted by tumors induce the alternatively activated M2 macrophage (Goerdt & Orfanos 1999, Stein *et al.* 1992). An M2 macrophage promotes tumor survival by aiding in proliferation and progression and stimulating angiogenesis and lymphangiogenesis (Albini *et al.* 2005). They promote extracellular matrix remodeling (Wynn 2004) while at the same time regulating a complex chemokine network (Mantovani *et al.* 2004).

We verified the expression of a2NTD in multiple human ovarian carcinoma cell lines of epithelial origin. In order to determine the consistency of staining within a cell line, we used immunofluorescent staining. For higher resolution, images were collected by confocal microscopy. Cells from a non-synchronous population grown on glass chamber slides were stained with 2C1 conjugated to FITC. To identify cells that were negative for the expression of a2NTD, cells were counter stained with the nuclear dye propidium iodine. We found that all cell lines examined A2780, PTX10, PTX22 and NIH OVCAR 3, had a high expression of a2NTD. However, in the NIH OVCAR 3 cell line we were able to identify cells that were negative for the expression of a2NTD. Previously, by western blot we have found that cells from low-density culture have a low level of expression of a2NTD (data not shown). The heterogeneous staining pattern in the NIH OVCAR 3 cell line suggests that a2NTD expression may be correlated with the cell cycle.

Intracellular V-ATPase has been and is presently under investigation in the treatment of cancer because it has been implicated in anti-cancer drug resistance (Martinez-Zaguilan *et al.* 1993, Sennoune *et al.* 2004b). Normal and malignant rapidly regenerating tissues must survive under adverse conditions such as nutrient deprivation and hypoxia. V-ATPase is thought to facilitate tumor cell survival under such conditions both by its expression in cancer cells and perhaps more importantly by tumor macrophage.

13.4 Vacuolar ATPase in Normal Cellular Processes and Human Disease

There are three models of plasma membrane associated V-ATPases that have been well characterized: intercalated cells of the kidney, osteoclasts in bone, and clear cells in the epididymus and vas deferens (Hinton *et al.* 2009). *Kidney*: Intercalated cells of the late distal tubule and collecting duct contain a high density of V-ATPases at their apical surfaces and participate in pumping protons out of the cytoplasm into the renal tubules (Wagner *et al.* 2004). *Osteoclasts:* Bone density is maintained in a balance between osteoblasts that form bone and osteoclasts that degrade it. Osteoclasts are specialized macrophages that resorb bone by creating a closed-off extracellular space to which acid and digestive enzymes are secreted to dissolve bone matrix (Nordstrom *et al.* 1997, Li *et al.* 1999). *Vas deferens/epididymus*: Clear cells secrete acid into the vas deferens and epididymus to maintain low pH. Macrophage and neutrophils are often recruited to environments that have a low pH, such as the site of infection or a tumor. In addition to the Na^+/H^+ pump, these cells have V-ATPases at the plasma membrane surface (Nanda *et al.* 1996, Nanda & Grinstein 1995). Unlike macrophages, which have a basal expression level of V-ATPases, neutrophils up-regulate cell surface expression due to chemoattractants (Nanda & Grinstein 1995). The expression of V-ATPase and calcineurin, a protein phosphatase, constitutively act upon macrophage as negative regulators of pro-inflammatory cytokine production by inhibiting transcription factor NF-κB (Conboy *et al.* 1999).

A handful of genetic diseases have been linked to mutations in V-ATPase genes. A genetic defect directly linked to mutations in V-ATPase subunits genes is osteopetrosis. The defect traces back to mutations in the a3 isoform causing an inability of osteoclast to secrete acid and degrade bone (Frattini *et al.* 2000). Some non-genetic diseases linked to plasma membrane V-ATPases include angiogenesis, tumor invasion, and HIV infection. Microvascular endothelial cell invasion and migration are essential processes during angiogenesis. Normal microvascular cells have been shown to express plasma membrane V-ATPase at the leading edge of migrating cells and this process can be effectively reduced by the addition of V-ATPase inhibitor bafilomycin (Rojas *et al.* 2004, Rojas *et al.* 2006). This correlates well with diabetic animal models that have problems with angiogenesis and reduced plasma membrane V-ATPases (Rojas *et al.* 2004). The first direct evidence for a role of V-ATPases in tumor cell invasiveness came from a study on breast tumor cells. These experiments demonstrated that highly invasive breast tumor cells exhibited significant plasma membrane V-ATPase activity compared to less invasive breast tumor cells (Martinez-Zaguilan *et al.* 1993, Sennoune *et al.* 2004a). Increased V-ATPase expression is also seen in increasing grades of pancreatic cancer where it co-localizes with cortactin, a protein involved in actin rearrangement and used in matrix metalloproteinase secretion (Chung *et al.* 2011). V-ATPases are also involved in the control of HIV infection in cells (Geyer *et al.* 2002). Subunit H binds HIV-encoded proteins involved with cell-surface receptor internalization, especially the HIV receptor, CD4. It appears that subunit H is required for this internalization process to take place; and by reducing the number of CD4 available on the cell surface, multiple viral particles are prevented from infecting the cell.

Interestingly, antibodies specific to a2 administered during implantation in mice can ablate pregnancy (Beaman & Hoversland 1988). Subsequent experiments demonstrate the presence of a2 on PBMC, specifically B-cells in pregnant women but not in non-pregnant women (Nichols *et al.* 1994). Exploring the expression of a2 on PBMC in women shows that CD56+ cells also exhibit high levels of a2 and this correlates with failure to deliver live births. CD56 is a cell surface marker for natural killer (NK) cells and their presence in the decidua is up-regulated during pregnancy (Mincheva-Nilsson *et al.* 1994). Correlating a2 expression on NK cells and B-cells in peripheral blood with pregnancy outcome proves to have high prognostic value. Specifically, there is more than 90% expression of a2 on B-cells and less than 50% expression of a2 on NK cells in association with viable pregnancies. Conversely, a2 expression at less than 90% in B-cells and greater than 50% in NK cells results in non-viable pregnancies (Coulam & Beaman 1995). The immunological profile of a2V-ATPase has also been investigated in other cellular and disease processes, including HIV infection (Boomer *et al.* 2000, Derks & Beaman 2004a, DuChateau *et al.* 1999, Givens *et al.* 1999, Sung *et al.* 2000), chronic lymphocytic leukemia (Aslakson *et al.* 1999), hepatitis C (Aslakson *et al.* 1999, Chedid *et al.* 2001, Derks & Beaman 2004a, DuChateau *et al.* 1999, Givens *et al.* 1999, Sung *et al.* 2000), T-cell activation and apoptosis (Boomer *et al.* 2000, Boomer *et al.* 2001, Ntrivalas *et al.* 2007b, Roth *et al.* 2006), glioblastoma (Roth *et al.* 2006), and modulation of IL-1β (Derks & Beaman 2004b, Ntrivalas *et al.* 2007a, Ntrivalas *et al.* 2007b) and IL-10 (Lee *et al.* 2001) production.

13.5 The *N*-Terminus Domain of the a2 Isoform of Vacuolar ATPase (a2NTD) and its Role as an Immune Modulator

a2NTD induces an M2-like phenotype in monocytes. It was not until studies that demonstrated the IL-1β-inducing capability of the *N*-terminal domain did researchers start looking at the domain as a post-translationally cleaved region of subunit a. Research shows that the human choriocarcinoma cell line, JEG-3, and human PBMC co-cultures induce pro-inflammatory cytokine IL-1β secretion (Ntrivalas *et al.* 2007b). An analysis of the PBMC after culture with ra2NTD indicates monocytes as the source of the IL-1.

Several reports have shown the importance of V-ATPases in inflammatory conditions (Brisseau *et al.* 1996, Nordstrom *et al.* 1997). In the cited studies, a2NTD clearly has an effect on monocytes; and though monocytes are normally thought of as agents of the inflammatory reaction during an innate immune response, in this instance a2NTD is initiating inflammatory cytokine production without microbial antigenic involvement. In this case, a2NTD may provide the link between V-ATPases and inflammation.

13.6 Monocyte-Macrophage Polarization and Cancer Related Inflammation

Heterogeneity and plasticity are hallmarks of cells of the monocyte-macrophage lineage (Mantovani *et al.* 2002, Mantovani *et al.* 2005). Mononuclear phagocytes express specialized and polarized functional properties in response to cytokines and microbial products (Gordon 2003, Mantovani *et al.* 2004). In an effort to mimic the T-cell classifications, many researchers refer to polarized macrophages as M1 and M2 cells. Extremes on a linear scale, M1 cells are thought to be the classically activated macrophages and M2 cells the alternatively activated macrophages (Gordon 2003).

A multitude of factors produced by neoplastic cells recruit the blood monocytes to tumors. Inflammatory chemokines macrophage colony stimulating factor (M-CSF/CSF-1) and vascular endothelial growth factor A (VEGFA) also participate in recruitment (Solinas *et al.* 2009). The unique monocytes recruited to tumor sites are called tumor-associated macrophages/monocytes (TAMs) (Mantovani *et al.* 2008, Pollard 2004). In the tumor microenvironment, TAMs are skewed toward an M2 phenotype (Montovani *et al.* 2002, Pollard J. 2004). More specifically, due to the production of IL-1 (Giavazzi *et al.* 1990, Porta *et al.* 2007), TAMs most resemble the M2b phenotype. M2 cells are present in established tumors, promote tumor proliferation and progression (Wyckoff *et al.* 2004), and stimulate angiogenesis and lymphangiogenesis (Albini *et al.* 2005, De Palma *et al.* 2007).

In a human papillomavirus-driven carcinogenesis model, genetic elimination of T and B lymphocytes blocks recruitment of innate immune cells, tissue remodeling and angiogenesis. It is postulated that B cells, which do not infiltrate lesions, act as remote controls of innate immune cells *in situ* by depositing immunoglobulins in the extracellular matrix. Innate inflammation is driven mainly by cells of the innate immune system, predominantly TAMs (Mantovani *et al.* 2008, Mantovani *et al.* 2009). Other investigators have described the fundamental role of V-ATPase during cytokine trafficking and secretion (Hinoki *et al.* 2006, Huang *et al.* 2007).

13.7 Innate Immune System, Inflammation and ATPase

An inflammatory response is a general term for the local accumulation of fluid, plasma proteins, and leukocytes that is initiated by physical injury, infection, or local immune response. Inflammation ranges from milder processes, such as those seen in some chronic infections (Crowe 2005, Eksteen *et al.* 2005, Wouters 2005), to more vigorous acute inflammatory conditions, such as pathogen clearance or allograft rejection (Bowen & Walker 2005). Acute inflammation is used to describe early and often transient episodes, whereas chronic inflammation occurs during persistent infection, autoimmune disease, or tumorigenesis (Cassel *et al.* 2008). Sites of chronic inflammation usually cycle through rounds of tissue damage and attempt to repair it. The type of inflammatory response can be disease specific.

The immune system has in its arsenal a protein complex, located in monocytes and macrophages, that responds to both pathogen-associated molecular patterns (PAMPs) and danger-associated molecular patterns (DAMPs) called the inflammasome. It is composed of an NLR molecule (nucleotide-binding domain leucine-rich repeat containing receptors), an adaptor molecule ASC (apoptosis associated speck-like protein containing CARD), and the enzyme caspase-1. An important proinflammatory mediator is IL-1α, which is localized to areas of injury or inflammation. While, proinflammatory stimuli will upregulate IL-1α gene and subsequent pro-IL-1α protein expression, the cleavage of the pro-cytokine and its release are regulated by the inflammasome (Dinarello 2010).

The NLRP3 inflammasome is the most fully characterized inflammasome and consists of the NLRP3 scaffold, the ASC adaptor, and caspase-1 (Martinon *et al.* 2009). Environmental irritants include silica and asbestos (Yamasaki *et al.* 2009). The mechanism(s) leading to NLRP3 inflammasome activation are not understood but several models exist. Due to the sheer number of stimuli that can activate the NLRP3 inflammasome, a general consensus is that these stimuli share a pathway that leads to the NLRP3 inflammasome but a single direct ligand is responsible for activating it (Stutz *et al.* 2009). The first model describes extracellular ATP as stimulating the purinogenic P2X7 ATP-gated ion channel. Pore formation in the cell membrane allows NLRP3 agonists access to the cytosol to activate NLRP3. A second model describes the method crystalline NLRP3 activators use. The third model claims all NLRP3 agonists trigger the generation of reactive oxygen species (ROS) and it is this common volatile metabolite that activates the NLRP3 inflammasome (Cassel *et al.* 2008, Cruz *et al.* 2007).

Septic inflammation is the most characterized response whereby PAMPs of extracellular microbial components are recognized by PRR on phagocytic cells that then stimulate inflammation. Because Janeway's model did not explain how immune responses are generated to tumors, transplants, and autologous tissue (in autoimmunity), Matzinger has described a new model called the "danger hypothesis" utilized in sterile inflammation. She states that the immune system evolved to discriminate DAMPs from non-dangerous ones (Matzinger 2002a, Matzinger 2002b, Matzinger 2007, Seong & Matzinger 2004). This theory describes the danger signal arising from injured cells of the host and then activating antigen presenting cells (APCs), such as monocytes and macrophages, to become effective stimulators of T-cells. Several proteins are now known to be endogenous danger signals including heat-shock proteins (Quintana & Cohen 2005, Tsan & Gao 2004) and high-mobility group box-1 proteins (HMGB-1) (Harris & Raucci 2006, Li *et al.* 2004, Rao *et al.* 2007), which are released during tissue injury and necrosis, and oxidized low-density lipoprotein (Freeman 1997, Tian *et al.* 2006), which is important in the pathogenesis of atherosclerosis. Breakdown products of the extracellular matrix are also danger signals. Inflammation causes an enzymatic reaction that produces small fragments of hyaluronic acid (Jiang *et al.* 2007, Jiang *et al.* 2010) and activated monocytes secrete MMPs that cleave fibronectin into small pieces (Barilla & Carsons 2000, Trial *et al.* 2004). Macrophages and other cells that have been stimulated with lipopolysaccharide (LPS) are accompanied by secretion of ATP (Kahlenberg *et al.* 2005). It is possible that it is this ATP that actually activates V-ATPase to initiate the inflammatory cascade that drives tumor metastasis and that, if checked, can be used to stop tumor cell growth and spread.

References

Albini A, Tosetti F, Benelli R & Noonan DM 2005 Tumor inflammatory angiogenesis and its chemoprevention. *Cancer Research* 65, 10637–10641.

Aslakson CJ, Lee G, Boomer JS, Gilman-Sachs A, Kucuk O & Beaman KD 1999 Expression of regeneration and tolerance factor on B cell chronic lymphocytic leukemias: a possible mechanism for escaping immune surveillance. *American Journal of Hematology* 61, 46–52.

Barilla ML & Carsons SE 2000 Fibronectin fragments and their role in inflammatory arthritis. *Seminars in Arthritis and Rheumatism* 29, 252–265.

Beaman KD & Hoversland RC 1988 Induction of abortion in mice with a monoclonal antibody specific for suppressor T-lymphocyte molecules. *Journal of Reproduction and Fertility* 82, 691–696.

Boomer JS, Derks RA, Lee GW, DuChateau BK, Gilman-Sachs A & Beaman KD 2001 Regeneration and tolerance factor is expressed during T-lymphocyte activation and plays a role in apoptosis. *Human Immunology* 62, 577–588.

Boomer JS, Lee GW, Givens TS, Gilman-Sachs A & Beaman KD 2000 Regeneration and tolerance Factor's potential role in T-cell activation and apoptosis. *Human Immunology* 61, 959–971.

Bowen DG & Walker CM 2005 Adaptive immune responses in acute and chronic hepatitis C virus infection. *Nature* 436, 946–952.

Brisseau GF, Grinstein S, Hackam DJ, Nordstrom T, Manolson MF, Khine AA & Rotstein OD 1996 Interleukin-1 increases vacuolar-type H^+-ATPase activity in murine peritoneal macrophages. *The Journal of Biological Chemistry* 271, 2005–2011.

Cassel SL, Eisenbarth SC, Iyer SS, Sadler JJ, Colegio OR, Tephly LA, Carter AB, Rothman PB, Flavell RA & Sutterwala FS 2008 The Nalp3 inflammasome is essential for the development of silicosis. *Proceedings of the National Academy of Sciences of the United States of America* 105, 9035–9040.

Chedid A, Sung CC, Lepe MR, Ahmed SA, Iftikhar SA, Feller A & Beaman KD 2001 Expression of a novel protein by regenerating hepatocytes and peripheral blood lymphocytes. *Clinical and Diagnostic Laboratory Immunology* 8, 1292–1294.

Chung C, Mader CC, Schmitz JC, Atladottir J, Fitchev P, Cornwell ML, Koleske AJ, Crawford SE & Gorelick F 2011 The vacuolar-ATPase modulates matrix metalloproteinase isoforms in human pancreatic cancer. *Investigation; A Journal of Technical Methods and Pathology* 91, 732–743.

Conboy IM, Manoli D, Mhaiskar V & Jones PP 1999 Calcineurin and vacuolar-type H^+-ATPase modulate macrophage effector functions. *Proceedings of the National Academy of Sciences of the United States of America* 96, 6324–6329.

Coulam CB & Beaman KD 1995 Reciprocal alteration in circulating TJ6+ CD19+ and TJ6+ CD56+ leukocytes in early pregnancy predicts success or miscarriage. *American Journal of Reproductive Immunology* 34, 219–224.

Crowe SE 2005 Helicobacter infection, chronic inflammation, and the development of malignancy. *Current Opinion in Gastroenterology* 21, 32–38.

Cruz CM, Rinna A, Forman HJ, Ventura AL, Persechini PM & Ojcius DM 2007 ATP activates a reactive oxygen species-dependent oxidative stress response and secretion of proinflammatory cytokines in macrophages. *The journal of Biological Chemistry* 282, 2871–2879.

De Palma M, Murdoch C, Venneri MA, Naldini L & Lewis CE 2007 Tie2-expressing monocytes: regulation of tumor angiogenesis and therapeutic implications. *Trends in Immunology* 28, 519–524.

Derks R & Beaman K 2004 Regeneration and tolerance factor modulates the effect of adenosine triphosphate-induced interleukin 1 beta secretion in human macrophages. *Human Immunology* 65, 676–682.

Derks RA & Beaman KD 2004 Regeneration and tolerance factor prevents bystander T-cell death associated with human immunodeficiency virus infection. *Clinical and Diagnostic Laboratory Immunology* 11, 835–840.

Dinarello CA 2010 IL-1: discoveries, controversies and future directions. *European Journal of Immunology* 40, 599–606.

DuChateau BK, Lee GW, Westerman MP & Beaman KD 1999 Increased expression of regeneration and tolerance factor in individuals with human immunodeficiency virus infection. *Clinical and Diagnostic Laboratory Immunology* 6, 193–198.

Eksteen B, Walker LS & Adams DH 2005 Immune regulation and colitis: suppression of acute inflammation allows the development of chronic inflammatory bowel disease. *Gut* 54, 4–6.

Frattini A, Orchard PJ, Sobacchi C, Giliani S, Abinun M, Mattsson JP, Keeling DJ, Andersson AK, Wallbrandt P, Zecca L *et al.* 2000 Defects in TCIRG1 subunit of the vacuolar proton pump are responsible for a subset of human autosomal recessive osteopetrosis. *Nature Genetics* 25, 343–346.

Freeman MW 1997 Scavenger receptors in atherosclerosis. *Current Opinion in Hematology* 4, 41–47.

Geyer M, Yu H, Mandic R, Linnemann T, Zheng YH, Fackler OT & Peterlin BM 2002 Subunit H of the V-ATPase binds to the medium chain of adaptor protein complex 2 and connects Nef to the endocytic machinery. *The Journal of Biological Chemistry* 277, 28521–28529.

Giavazzi R, Garofalo A, Bani MR, Abbate M, Ghezzi P, Boraschi D, Mantovani A & Dejana E 1990 Interleukin 1-induced augmentation of experimental metastases from a human melanoma in nude mice. *Cancer Research* 50, 4771–4775.

Givens TS, DuChateau BK, Boomer JS, Westerman MP, Gilman-Sachs A & Beaman KD 1999 Regeneration and tolerance factor: a correlate of human immunodeficiency virus-associated T-cell activation. *Clinical and Diagnostic Laboratory Immunology* 6, 872–877.

Goerdt S & Orfanos CE 1999 Other functions, other genes: alternative activation of antigen-presenting cells. *Immunity* 10, 137–142.

Gordon S 2003 Alternative activation of macrophages. *Nature Reviews Immunology* 3, 23–35.

Harris HE & Raucci A 2006 Alarmin(g) news about danger: workshop on innate danger signals and HMGB1. *EMBO Reports* 7, 774–778.

Hinoki A, Yoshimura K, Fujita K, Akita M, Ikeda R, Nagashima M, Nomura M & Satomi A 2006 Suppression of proinflammatory cytokine production in macrophages by lansoprazole. *Pediatric Surgery International* 22, 915–923.

Hinton A, Bond S & Forgac M 2009 V-ATPase functions in normal and disease processes. *Pflugers Archiv: European Journal of Physiology* 457, 589–598.

Huang X, Masselli A, Frisch SM, Hunton IC, Jiang Y & Wang JY 2007 Blockade of tumor necrosis factor-induced Bid cleavage by caspase-resistant Rb. *The Journal of Biological Chemistry* 282, 29401–29413.

Jaiswal MK, Mallers TM, Larsen B, Kwak-Kim J, Chaouat G, Gilman-Sachs A & Beaman KD 2012 V-ATPase upregulation during early pregnancy: A possible link to establishment of an inflammatory response during preimplantation period of pregnancy. *Reproduction* 143(5), 713–725.

Jiang D, Liang J & Noble PW 2007 Hyaluronan in tissue injury and repair. *Annual Review of Cell and Developmental Biology* 23, 435–461.

Jiang D, Liang J & Noble PW 2010 Regulation of non-infectious lung injury, inflammation, and repair by the extracellular matrix glycosaminoglycan hyaluronan. *Anatomical Record* 293, 982–985.

Kahlenberg JM, Lundberg KC, Kertesy SB, Qu Y & Dubyak GR 2005 Potentiation of caspase-1 activation by the P2X7 receptor is dependent on TLR signals and requires NF-kappaB-driven protein synthesis. *Journal of Immunology* 175, 7611–7622.

Kubota S & Seyama Y 2000 Overexpression of vacuolar ATPase 16-kDa subunit in 10T1/2 fibroblasts enhances invasion with concomitant induction of matrix metalloproteinase-2. *Biochemical and Biophysical Research Communications* 278, 390–394.

Kwong C, Gilman-Sachs A & Beaman K 2011 Tumor-associated a2 vacuolar ATPase acts as a key mediator of cancer-related inflammation by inducing pro-tumorigenic properties in monocytes. *Journal of Immunology* 186, 1781–1789.

Lee GW, Boomer JS, Gilman-Sachs A, Chedid A, Gudelj L, Rukavina D & Beaman KD 2001 Regeneration and tolerance factor of the human placenta induces IL-10 production. *European Journal of Immunology* 31, 687–691.

Li J, Wang H, Mason JM, Levine J, Yu M, Ulloa L, Czura CJ, Tracey KJ & Yang H 2004 Recombinant HMGB1 with cytokine-stimulating activity. *Journal of Immunological Methods* 289, 211–223.

Li YP, Chen W, Liang Y, Li E & Stashenko P 1999 Atp6i-deficient mice exhibit severe osteopetrosis due to loss of osteoclast-mediated extracellular acidification. *Nature Genetics* 23, 447–451.

Mantovani A, Romero P, Palucka AK & Marincola FM 2008 Tumour immunity: effector response to tumour and role of the microenvironment. *Lancet* 371, 771–783.

Mantovani A, Sica A, Allavena P, Garlanda C & Locati M 2009 Tumor-associated macrophages and the related myeloid-derived suppressor cells as a paradigm of the diversity of macrophage activation. *Human Immunology* 70, 325–330.

Mantovani A, Sica A & Locati M 2005 Macrophage polarization comes of age. *Immunity* 23, 344–346.

Mantovani A, Sica A, Sozzani S, Allavena P, Vecchi A & Locati M 2004 The chemokine system in diverse forms of macrophage activation and polarization. *Trends in Immunology* 25, 677–686.

Mantovani A, Sozzani S, Locati M, Allavena P & Sica A 2002 Macrophage polarization: tumor-associated macrophages as a paradigm for polarized M2 mononuclear phagocytes. *Trends in Immunology* 23, 549–555.

Martinez-Zaguilan R, Lynch RM, Martinez GM & Gillies RJ 1993 Vacuolar-type H(+)-ATPases are functionally expressed in plasma membranes of human tumor cells. *The American Journal of Physiology* 265, C1015–C1029.

Martinon F, Mayor A & Tschopp J 2009 The inflammasomes: guardians of the body. *Annual Review of Immunology* 27, 229–265.

Matzinger P 2002a The danger model: a renewed sense of self. *Science* 296, 301–305.

Matzinger P 2002b An innate sense of danger. *Annals of the New York Academy of Sciences* 961, 341–342.

Matzinger P 2007 Friendly and dangerous signals: is the tissue in control? *Nature Immunology* 8, 11–13.

Mincheva-Nilsson L, Baranov V, Yeung MM, Hammarstrom S & Hammarstrom ML 1994 Immunomorphologic studies of human decidua-associated lymphoid cells in normal early pregnancy. *Journal of Immunology* 152, 2020–2032.

Nanda A, Brumell JH, Nordstrom T, Kjeldsen L, Sengelov H, Borregaard N, Rotstein OD & Grinstein S 1996 Activation of proton pumping in human neutrophils occurs by exocytosis of vesicles bearing vacuolar-type H^+-ATPases. *The Journal of Biological Chemistry* 271, 15963–15970.

Nanda A & Grinstein S 1995 Chemoattractant-induced activation of vacuolar H+ pumps and of an H(+)-selective conductance in neutrophils. *Journal of Cellular Physiology* 165, 588–599.

Nichols TC, Kang JA, Angkachatchai V, Beer AE & Beaman KD 1994 Expression of a membrane form of the pregnancy-associated protein TJ6 on lymphocytes. *Cellular Immunology* 155, 219–229.

Nordstrom T, Shrode LD, Rotstein OD, Romanek R, Goto T, Heersche JN, Manolson MF, Brisseau GF & Grinstein S 1997 Chronic extracellular acidosis induces plasmalemmal vacuolar type H+ ATPase activity in osteoclasts. *The Journal of Biological Chemistry* 272, 6354–6360.

Ntrivalas E, Derks R, Gilman-Sachs A, Kwak-Kim J, Levine R & Beaman K 2007a Novel role for the *N*-terminus domain of the a2 isoform of vacuolar ATPase in interleukin-1beta production. *Human Immunology* 68, 469–477.

Ntrivalas E, Gilman-Sachs A, Kwak-Kim J & Beaman K 2007b The *N*-terminus domain of the a2 isoform of vacuolar ATPase can regulate interleukin-1beta production from mononuclear cells in co-culture with JEG-3 choriocarcinoma cells. *American Journal of Reproductive Immunology* 57, 201–209.

Pollard JW 2004 Tumour-educated macrophages promote tumour progression and metastasis. *Nature Reviews Cancer* 4, 71–78.

Porta C, Subhra Kumar B, Larghi P, Rubino L, Mancino A & Sica A 2007 Tumor promotion by tumor-associated macrophages. *Advances in Experimental Medicine and Biology* 604, 67–86.

Quintana FJ & Cohen IR 2005 Heat shock proteins as endogenous adjuvants in sterile and septic inflammation. *Journal of Immunology* 175, 2777–2782.

Rao DA, Tracey KJ & Pober JS 2007 IL-1alpha and IL-1beta are endogenous mediators linking cell injury to the adaptive alloimmune response. *Journal of Immunology* 179, 6536–6546.

Rofstad EK, Mathiesen B, Kindem K & Galappathi K 2006 Acidic extracellular pH promotes experimental metastasis of human melanoma cells in athymic nude mice. *Cancer Research* 66, 6699–6707.

Rojas JD, Sennoune SR, Maiti D, Bakunts K, Reuveni M, Sanka SC, Martinez GM, Seftor EA, Meininger CJ, Wu G *et al.* 2006 Vacuolar-type H^+-ATPases at the plasma membrane regulate pH and cell migration in microvascular endothelial cells. *American Journal of Physiology Heart and Circulatory Physiology* 291, H1147–H1157.

Rojas JD, Sennoune SR, Martinez GM, Bakunts K, Meininger CJ, Wu G, Wesson DE, Seftor EA, Hendrix MJ & Martinez-Zaguilan R 2004 Plasmalemmal vacuolar H^+-ATPase is decreased in microvascular endothelial cells from a diabetic model. *Journal of Cellular Physiology* 201, 190–200.

Roth P, Aulwurm S, Gekel I, Beier D, Sperry RG, Mittelbronn M, Meyermann R, Beaman KD, Weller M & Wischhusen J 2006 Regeneration and tolerance factor: a novel mediator of glioblastoma-associated immunosuppression. *Cancer Research* 66, 3852–3858.

Sennoune SR, Bakunts K, Martinez GM, Chua-Tuan JL, Kebir Y, Attaya MN & Martinez-Zaguilan R 2004a Vacuolar H^+-ATPase in human breast cancer cells with distinct metastatic potential: distribution and functional activity. *American Journal of Physiology Cell Physiology* 286, C1443–1452.

Sennoune SR, Luo D & Martinez-Zaguilan R 2004b Plasmalemmal vacuolar-type H^+-ATPase in cancer biology. *Cell Biochemistry and Biophysics* 40, 185–206.

Seong SY & Matzinger P 2004 Hydrophobicity: an ancient damage-associated molecular pattern that initiates innate immune responses. *Nature Reviews Immunology* 4, 469–478.

Solinas G, Germano G, Mantovani A & Allavena P 2009 Tumor-associated macrophages (TAM) as major players of the cancer-related inflammation. *Journal of Leukocyte Biology* 86, 1065–1073.

Stein M, Keshav S, Harris N & Gordon S 1992 Interleukin 4 potently enhances murine macrophage mannose receptor activity: a marker of alternative immunologic macrophage activation. *The Journal of Experimental Medicine* 176, 287–292.

Stutz A, Golenbock DT & Latz E 2009 Inflammasomes: too big to miss. *The Journal of Clinical Investigation* 119, 3502–3511.

Sung CC, Boomer JS, Givens TS, DuChateau BK, Lepe MR, Feller A, Westerman MP, Gilman-Sachs A, Chedid A & Beaman KD 2000 Expression of regeneration and tolerance factor correlates directly with human immunodeficiency virus infection and inversely with hepatitis C virus infection. *Clinical and Diagnostic Laboratory Immunology* 7, 200–205.

Tian Y, Kelemen SE & Autieri MV 2006 Inhibition of AIF-1 expression by constitutive siRNA expression reduces macrophage migration, proliferation, and signal transduction initiated by atherogenic stimuli. *American Journal of Physiology Cell Physiology* 290, C1083–C1091.

Trial J, Rubio JA, Birdsall HH, Rodriguez-Barradas M & Rossen RD 2004 Monocyte activation by circulating fibronectin fragments in HIV-1-infected patients. *Journal of Immunology* 173, 2190–2198.

Tsan MF & Gao B 2004 Heat shock protein and innate immunity. *Cellular & Molecular Immunology* 1, 274–279.

Wagner CA, Finberg KE, Breton S, Marshansky V, Brown D & Geibel JP 2004 Renal vacuolar H^+-ATPase. *Physiological Reviews* 84, 1263–1314.

Wouters EF 2005 Local and systemic inflammation in chronic obstructive pulmonary disease. *Proceedings of the American Thoracic Society* 2, 26–33.

Wyckoff J, Wang W, Lin EY, Wang Y, Pixley F, Stanley ER, Graf T, Pollard JW, Segall J & Condeelis J 2004 A paracrine loop between tumor cells and macrophages is required for tumor cell migration in mammary tumors. *Cancer Research* 64, 7022–7029.

Wynn TA 2004 Fibrotic disease and the T(H)1/T(H)2 paradigm. *Nature Reviews Immunology* 4, 583–594.

Yamasaki K, Muto J, Taylor KR, Cogen AL, Audish D, Bertin J, Grant EP, Coyle AJ, Misaghi A, Hoffman HM *et al.* 2009 NLRP3/cryopyrin is necessary for interleukin-1beta (IL-1beta) release in response to hyaluronan, an endogenous trigger of inflammation in response to injury. *The Journal of Biological Chemistry* 284, 12762–12771.

Chapter 14

V-ATPases in Oral Squamous Cell Carcinoma

Mario Pérez-Sayáns García and Abel García García
University of Santiago de Compostela, Department of Dentistry
Entrerríos s/n Santiago de Compostela 15782, Spain
perezsayans@gmail.com

14.1 Introduction

Oral squamous cell carcinoma (OSCC) is the most common malignancy of the oral cavity. Accounting for between 90% and 95% of all malignant lesions of the mouth, OSCC has become practically synonymous with oral cancer [1, 2]. The main characteristics of OSCC, a solid tumor, are acidity and hypoxia, both of which have been implicated in tumor metastasis [3], sensitivity to chemotherapy agents [4, 5], and proliferation [6].

Cytosolic pH appears to be strictly regulated by four mechanisms: sodium/proton exchangers (NHEs), bicarbonate transporters (BCTs), monocarboxylate transporters (MCTs), and proton pumps (ATPases) [7, 8]. Lactate production has commonly been considered to be the main mechanism underlying the acidification of the tumor microenvironment [9]. Nonetheless, solid tumors such

Handbook of H^+-ATPases
Edited by Suguru Nakamura

ISBN 978-981-4411-91-2 (Hardcover), 978-981-4411-92-9 (eBook)
www.panstanford.com

as OSCC are capable of creating an acidic environment, even in conditions of limited lactate production [10, 11]. To survive in this microenvironment, tumor cells need a cytosolic pH regulation system to help protect themselves from harmful H^+ ions. This may explain why V-ATPases, which are normally located in acidic organelles, may also reside on the cell surface, where they regulate pH and increase the migratory capacity of metastatic cells [12–14].

14.2 Role of V-ATPases in Oral Cancer

V-ATPases have been related in solid tumors in terms of pH control, metastasis, tumor cell growth and survival, multidrug resistance (MDR), and the possible therapeutic applications associated with the use of specific V-ATPase inhibitors.

14.2.1 *Role of V-ATPases in the Control of pH*

Extracellular pH in patients with OSCC is slightly acidic (around 6.8–7.0), as has been previously described for solid tumors in animal models. Becelli *et al.* [15] found that reversed pH gradient was directly related to resistance to chemotherapy agents. To survive in this microenvironment, tumor cells need to be able to regulate cytosolic pH. This may explain why V-ATPases, which are normally located in acidic organelles, may also reside on the cell surface, where they regulate pH and increase the migratory capacity of metastatic cells [14, 16]. This is accompanied by a parallel increase in intracellular pH in addition to an increase in DNA synthesis [17–19], cell cycle progression [20–22], serum- and substrate-independent growth [17], and *in vivo* tumor growth [17, 23], leading to a pathological increase and a disruption in cell density and number. This, in turn, complicates access to the circulatory system [24], causing inefficient elimination of metabolic products such as carbonic acid. Together with an increase in aerobic glycolysis [17, 25], this provides the tumor with a metabolic microenvironment characterized by low levels of serum, hypoxia, and acidic extracellular pH. Such an environment increases the invasive capacity of the tumor and the expression of growth and angiogenic factors/receptors, and promotes the transition to a more aggressive, metastatic phenotype [26].

14.2.2 *Role of V-ATPases in Metastasis*

The development and maintenance of the proton gradient present in tumors is directly due to the ability of tumor cells to secrete protons (H^+) [27, 28], acidify the extracellular medium [26, 29], and keep the cytosolic pH alkaline [30]. This ability also increases with tumor aggressiveness [31, 32]. In addition, low pH may cause extracellular matrix (ECM) degradation and remodeling through activation of proteolytic enzymes which contribute to invasion and cancer metastasis [3, 33]. Proteases need low extracellular pH to optimize their activation, including metalloproteinases (MMP), morphogenetic bone metalloproteinases (protein type 1), tissue serine proteases, and adamalysin-related proteinases. Among them, MMPs are the proteases basically involved in degradation and remodeling of all ECM structural components [34–38].

Sennoune *et al.* assessed the effect of bafilomycin A1 in breast tumor cells and found that cytoplasmic pH recovery was inhibited in response to acid load, in both highly and lowly metastatic cells, although to a greater extent in highly metastatic cells [39] (Fig. 14.1). This suggests that V-ATPases in the plasma membrane are involved in the acquisition of a more metastatic phenotype and that the use of V-ATPase inhibitors allows distant metastasis to be minimized.

14.2.3 *Role of V-ATPases in Tumor Cell Growth and Survival*

V-ATPases may also play a significant role in tumor cell survival by regulating pH and preventing apoptosis. As previously reported, plasma membrane V-ATPases help regulate cytosolic pH in macrophages and neutrophils [40]. This mechanism may also be used by tumor cells, which produce more H^+ due to high glycolytic activity [41]. Treatment with V-ATPase inhibitors lowers H^+ extrusion, both *in vitro* and *in vivo* [42, 43]).

Bafilomycin A1 was assessed as a potential anticancer agent because it inhibits cell proliferation and tumor growth. Although this effect has been attributed to the inhibition of intracellular acidosis

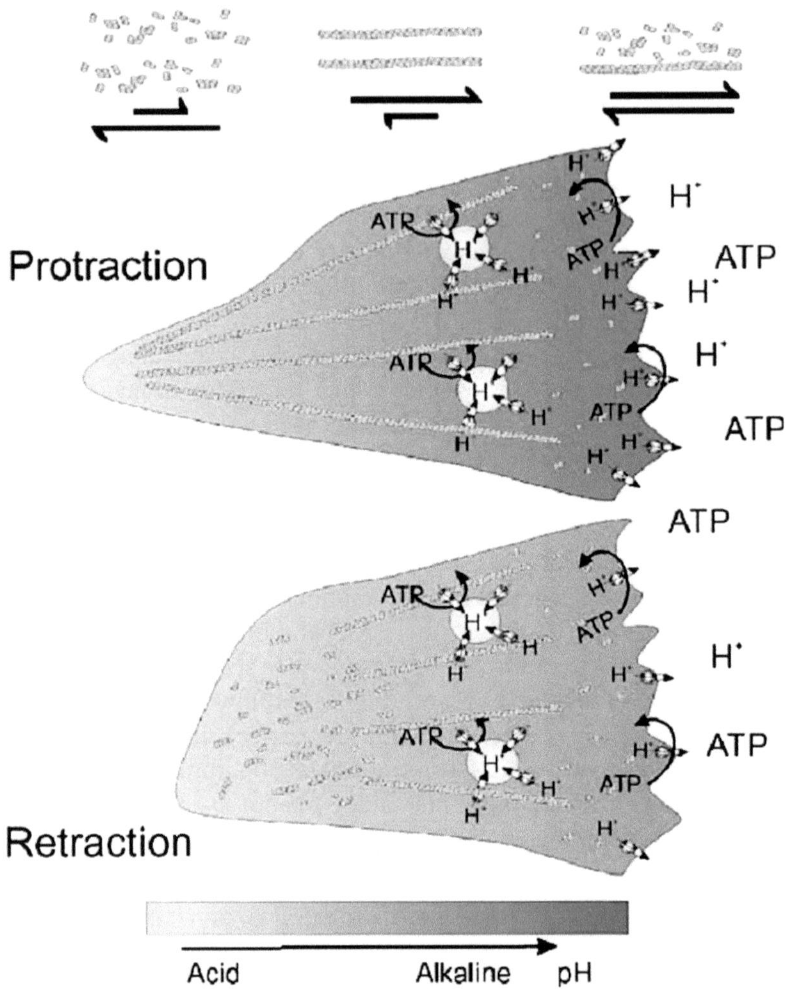

Figure 14.1 Proposed mechanism by which overexpression of pmV-ATPase at the leading edge of the cell modulates cell migration/invasion. The proposed model should be viewed as a framework to explain how pmV-ATPases determine the acquisition of an invasive phenotype needed for angiogenesis and metastasis. Changes in pHcyt are critical for establishing cell polarity needed for cell movement. A critical step in directed motility and migration is the asymmetric actin polymerization at the leading edge. Reprinted from Ref. 30, with kind permission from Springer Science and Business Media.

by blocking V-ATPases, the precise mechanism remains unknown [44]. A study conducted by Lim *et al.*, hypothesized that bafilomycin A1 and its analog, concanamycin A, stimulate a tumor growth factor, hypoxia-inducible 1α (HIF-1α) [45]. The interaction of bafilomycin with HIF-1α increases with hypoxia, causing strong induction of the p21 gene which, in turn, leads to cell cycle detection in cancer cells [46].

V-ATPase inhibition has also been shown to trigger apoptosis through caspase-dependent and caspase-independent mechanisms [47, 48], and bafilomycin and concanamycin induce apoptosis in other types of cells, including neutrophils [49] and osteoclasts [50].

Morimura *et al.* described the growth-inhibiting effect of apoptosis stimulation in human hepatoblastomas using bafilomycin A1. In particular, electron microscopy, morphological observations, and flow cytometry showed higher apoptotic cell ratios and diminished cell reproduction. Cell growth inhibition in normal liver cells was insignificant [51].

A study conducted by Wu *et al.* has shown that bafilomycin A1 suppresses macroautophagy by preventing lysosome acidification [52]. Macroautophagy is a protein degradation pathway that allows increased cell survival under stress and in cancer [53, 54]. Macroautophagic inhibition in HT-29, HCT-116, and SW1116 colon cancer cells is accompanied by downregulation of cyclin D and E and upregulation of $p21^{Cip1}$ and various caspases, causing an antiproliferative effect [52].

Cancer cells are more likely to express V-ATPase than normal cells, causing abnormalities in the acidic microenvironment and affecting cancer cell growth and infiltration significantly [14, 26, 29].

14.2.4 *Role of V-ATPases in MDR*

Extracellular pH in solid tumors is significantly more acidic than in normal tissue. This increased acidity interferes with the absorption of basic chemotherapy drugs, reducing their effect on tumors [55, 56]. Becelli *et al.* [15] found that reversed pH gradient was directly related to drug resistance. Recent studies have suggested that V-ATPases, which secrete protons through the plasma membrane, may play a key role in the acidification of the tumor environment.

Several human tumor cells are characterized by increased V-ATPase expression and activity, and pretreatment with proton pump inhibitors (PPIs) has been found to sensitize tumor cell lines to the effect of different chemotherapy drugs [5, 57–59].

Murakami *et al.* found that all V-ATPase subunits, and *ATP6C* in particular, are overexpressed in cisplatin-resistant tumors [60]. They also found significantly higher levels of cellular pH in cisplatin-resistant tumor cells than in cells sensitive to vincristine and etoposide. Torigoe *et al.* showed that treatment with anticancer agents increased expression of the *ATP6L* gene (c subunit) [61]. They argued that V-ATPase overexpression was a response to cellular acidosis and that c-subunit promoter activity was activated by anticancer agents, and above all by DNA-topoisomerase II inhibitor (TAS-103) [62]. These data suggest that the induction of V-ATPase expression in MDR is an anti-apoptotic defense [61]. Treatment with PPIs has also been found to sensitize tumor cells to cisplatin, 5-fluoracil, and vinblastine through changes in cellular pH gradients, with retention of the drugs in the cytoplasm, and in the nucleus in the case of doxorubicin [5, 57, 63]. It is also known that low pH levels are suitable for the complete activation of PPIs [48], suggesting that tumor alkalinization may be an extremely interesting target for future anticancer treatments [5, 57, 58].

There is scientific evidence that suggests that the acidic tumor microenvironment holds the key to cancer management in terms of disease progression and metastasis. Of all the mechanisms that regulate this microenvironment, V-ATPases are key targets as they can be inhibited by RNA interference and PPIs [64]. While there are many pH regulator inhibitors, V-ATPase inhibitors have proven to be the most efficient as V-ATPases are the main regulators of pH [12, 14, 27]. Specific V-ATPase inhibitors such as concanamycin and bafilomycin are other candidates for investigation, not only to treat cancer but also to reduce MDR in tumors [12].

14.2.5 *Role of V-ATPases as Therapeutic Target*

Scientific evidence suggests that the acidic tumor microenvironment is the key to managing cancer progression and metastasis. In particular, V-ATPases play a major role in metastasis tumor development because many tumor cells secrete lysosomal enzymes that participate in the ECM degradation necessary for metastatic invasion. These enzymes are most active at low optimal pH; moreover, V-ATPases are responsible for microenvironment acidification [27, 65]. Among the many mechanisms that regulate the tumor microenvironment, V-ATPases are especially significant because they can be inhibited by PPIs [64].

Initial attempts to block V-ATPases were made after bafilomycin and concanamycin were discovered in 1988 [66]. New molecules capable of inhibiting V-ATPase to a greater or lesser extent via different mechanisms of action were later discovered. Such molecules include benzolactone enamides salicylihalamide [67], lobatamide A and B [68], apicularen [69], indolyls [70, 71], oximidine [72], macrolactone archazolid [73], lobatamide C [74], and cruentaren [75]. The latest generation of inhibitors include NiK12192 [14, 76], FR202126 [77], and PPI SB 242784 [78] (Table 14.1).

14.3 Role of ATP6V1C1 and Subunit C1 in V-ATPases and OSCC

V-ATPase is composed of a cytosolic V1 domain and a transmembrane V_0 domain. The former consists of 3 A subunits, 2 B subunits, 2 G subunits, and 1 C, D, E, F, and H subunits. Two alternative transcript variants encoding different isoforms have been found in the gene that controls the expression of the C subunit: *ATP6V1C1* and *ATP6V1C2a,b.* While *ATP6V1C1* is expressed continuously in all tissues, *ATP6V1C2a,b* is found only in the lungs, kidneys, and epididymis, where it has an actin-binding function [79, 80] (Fig. 14.2).

The C subunit is a 40-kDa protein located in the V_1 domain of V-ATPase [81–85]. This subunit, which is essential for the proton secretion function of V-ATPases [86], is intimately involved in the reversible dissociation of the V_1 and V_0 domains [87–89] and is considered to be solely responsible for regulating the dissociative mechanism of the enzyme [90, 91]. According to Otero *et al.* [80] the V-ATPase C1 subunit appears to be responsible for the assembly of the V_0 membrane component and the cytosolic catalytic component.

Table 14.1 Differences and similarities of V-ATPase inhibitors

	Classes of V-ATPase Inhibitors			
	Chemistry	Provenience	Binding site	Action
Plecomacrolide				
Concanamycin & Bafilomycin	Macrolide antibiotics with 18-membered lactone rings	*Streptomyces*	Unknown	V-ATPases inhibition ionophoric properties
Benzolactone enamides				
Salicylihalamide A	Macrocyclic salicylate	Sponge *Haliclona* sp.	VO complex	Animal V-ATPases inhibition cytotoxin
Apicularens	Lactone ring	*Chondromyces*	VO complex	Highly toxic for human and animal cell
Lobatamides Oximidines	Substitution of enamide NH, salicylate, and phenyl salicylate	*Tunicate Aplidium lobatum*	VO complex	Animal and mammalian V-ATPases inhibition
Cruentaren	Lactone ring	*Pseudomonas* sp.	VO complex	Animal and mammalian V-ATPases inhibition
	Lactone ring	*Byssovorax cruenta*	VO complex	Cytotoxicity on mammalian and fungal cells at mitochondrial F-ATPases
Archazolid	Macrocyclic lactone ring with a thiazole side	*Archangium gephyra Cystobacter violaceus*	VO subunit c	Cytotoxicity on mammalian cell line
Indolyls	Bafilomycin-based	Synthesis	VO subunit c	V-ATPase inhibitor
Late-generation V-ATPase inhibitors	NiK12192, SB 242784, FR202126, 3-bromopyruvate (3-Br PA), Tributyltin chloride (TBTCl), FR177995, FR167356			

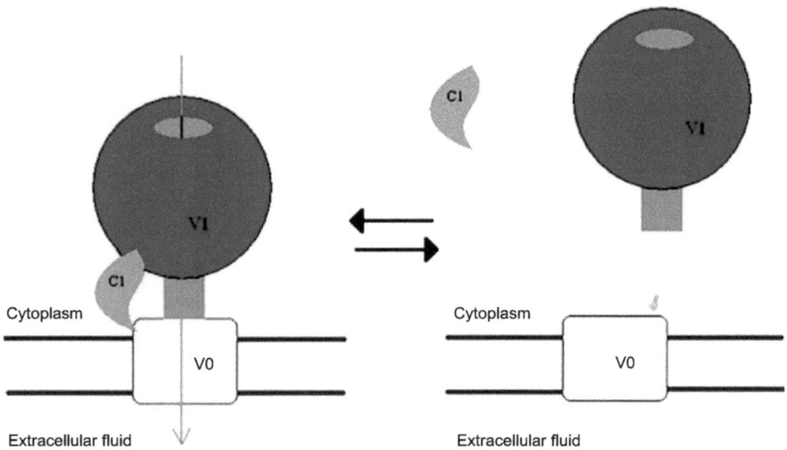

Figure 14.2 Structural model of the vacuolar H^+-ATPase, comprising a membrane component (V_0) and a catalytic cytosolic component (V_1). Dissociation of the VATPase into the V_1 and V_0 components is regulated by the C1 subunit [80].

The results regarding the role of ATP6V1C1 in OSCC as diagnostic and prognostic marker are a follow-up of previous work by our group analyzing differential gene expression in OSCC and normal oral mucosa cells using DNA microarray analysis of biopsy samples [92]. Of 322 genes found to be overexpressed in tumor tissue compared to normal tissue, we observed, using real-time quantitative polymerase chain reaction (qRT-PCR), that *ATP6V1C1* was overexpressed in 100% of OSCC samples [80]. In view of the important biological functions of this gene, we wished to measure *ATP6V1C1* levels in brush cytology OSCC samples and assess their value as tumor diagnostic and prognostic markers.

We designed a study where we analyzed patients with OSCC and a control group of healthy individuals. The clinical and demographic variables analyzed included age, sex, smoking, tumor location, and tumor stage. Brush cytology samples were obtained using a cytology brush and analyzed by real-time quantitative polymerase chain reaction for *ATP6V1C1* expression. The differences in *ATP6V1C1* expression between patients and controls were statistically significant (Mann-Whitney U test=26, $p < 0.001$). Receiver operating

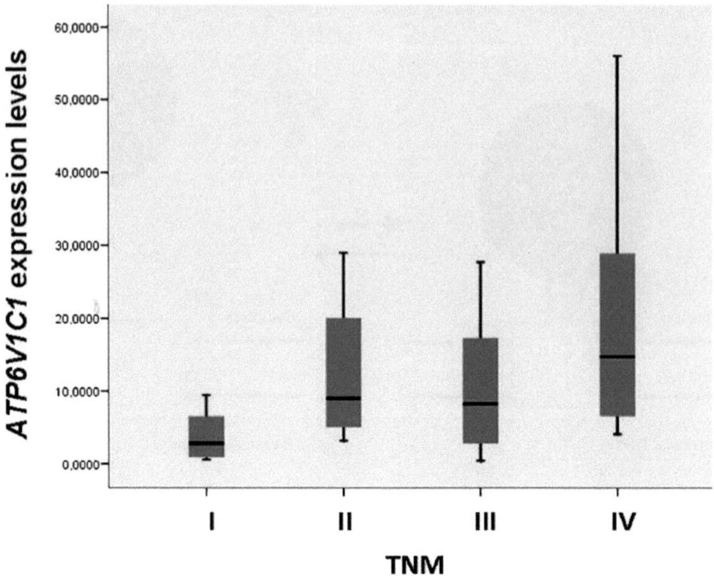

Figure 14.3 *ATP6V1C1* expression levels according to tumor stage.

characteristic (ROC) curve analysis showed an area under the curve of 0.9476, with the following diagnostic indices: sensitivity, 81.25%, specificity, 93.75%; accuracy, 87.50%; positive predictive value, 92.86%; negative predictive value, 83.33%; positive likelihood ratio, 30; and negative likelihood ratio, 0.06. It was confirmed that *ATP6V1C1* levels were significantly higher in patients with OSCC than in healthy controls, with expression increasing with higher tumor stage (Fig. 14.3). ROC analysis showed that the measurement of *ATP6V1C1* expression levels is a highly sensitive and specific diagnostic method [93].

Regarding the expression of subunit C1, we developed a study where we analyzed the expression of the C1 subunit in eight OSCC samples and two normal oral mucosa samples using polyclonal V-ATPase C1 antibody (clone H-300). In the normal oral mucosa samples, C1 subunit staining was observed in the basal and intermediate layers of the epithelium. No staining was visible in the keratinized superficial layers. More intense staining was observed in the OSCC samples, with the predominant expression at the

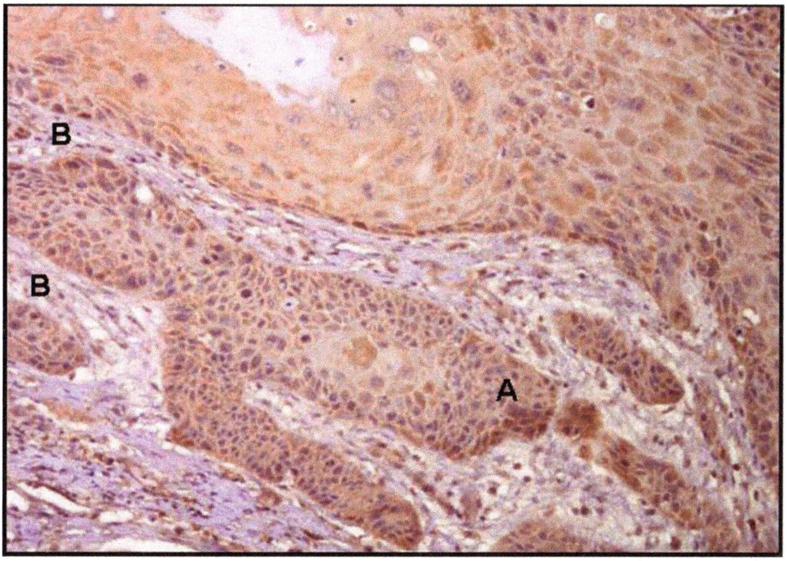

Figure 14.4 APTase expression in well-differentiated oral carcinoma. (A) Predominant expression at the periphery of tumor nests. (B) Absence of expression in dyskeratotic areas (20×).

periphery of tumor nests and absence of expression in dyskeratotic areas. C1 subunit expression in tumor cells was predominantly cytoplasmic, although there was perinuclear and nuclear expression in some samples (Fig. 14.4). These findings demonstrate that V-ATPase is necessary for proper epithelial functioning and show its importance in the development of OSCC as evidenced by the over-expression of *ATP6V1C1* in OSCC [94].

14.4 Conclusions

We confirmed that the *ATP6V1C1* gene was significantly over-expressed in patients with OSCC compared to healthy controls, characterized by an increasing gradient, between *ATP6V1C1* levels and tumor stage. The measurement of *ATP6V1C1* expression is a highly sensitive and specific means of discriminating between OSCC samples and normal oral mucosa samples. There is sufficient evidence to suggest that V-ATPase, which is believed to be largely responsible for regulating the acidity of the tumor microenvironment, may also be inherently responsible for MDR in solid tumors such as OSCC via the chemical control of the absorption of chemotherapy drugs. The findings suggest that the induced expression of V-ATPases in MDR is an anti-apoptotic defense and that the combined use of PPIs or V-ATPase inhibitors and low chemotherapy doses might be a possible treatment target. We believe that the future of these molecules in cancer treatment involves measuring the overexpression of specific V-ATPase subunits in tumors to be treated and then using inhibitors specific for the subunits being expressed. This will allow clinicians to provide more specific treatment, while also minimizing adverse effects.

References

1. Kosicki DM, Riva C, Pajarola GF, Burkhardt A, Gratz KW. OralCDx brush biopsy—a tool for early diagnosis of oral squamous cell carcinoma. *Schweiz Monatsschr Zahnmed* 2007; 117(3): 222–227.
2. Robinson PN, Mickelson AR. Early diagnosis of oral cavity cancers. *Otolaryngol Clin North Am* 2006 Apr; 39(2): 295–306.
3. Martinez-Zaguilan R, Seftor EA, Seftor RE, Chu YW, Gillies RJ, Hendrix MJ. Acidic pH enhances the invasive behavior of human melanoma cells. *Clin Exp Metastasis* 1996 Mar; 14(2): 176–186.
4. Raghunand N, Mahoney B, van Sluis R, Baggett B, Gillies RJ. Acute metabolic alkalosis enhances response of C3H mouse mammary tumors to the weak base mitoxantrone. *Neoplasia* 2001 May–June; 3(3): 227–235.
5. De Milito A, Fais S. Proton pump inhibitors may reduce tumour resistance. *Expert Opin Pharmacother* 2005 June; 6(7): 1049–1054.
6. Morita T, Nagaki T, Fukuda I, Okumura K. Clastogenicity of low pH to various cultured mammalian cells. *Mutat Res* 1992 Aug; 268(2): 297–305.
7. Izumi H, Torigoe T, Ishiguchi H, Uramoto H, Yoshida Y, Tanabe M, *et al.* Cellular pH regulators: potentially promising molecular targets for cancer chemotherapy. *Cancer Treat Rev* 2003 December; 29(6): 541–549.

8. Torigoe T, Izumi H, Ise T, Murakami T, Uramoto H, Ishiguchi H, *et al.* Vacuolar H(+)-ATPase: functional mechanisms and potential as a target for cancer chemotherapy. *Anticancer Drugs* 2002 March; 13(3): 237–243.
9. Gillies RJ, Raghunand N, Karczmar GS, Bhujwalla ZM. MRI of the tumor microenvironment. *J Magn Reson Imaging* 2002 October; 16(4): 430–450.
10. Newell K, Franchi A, Pouyssegur J, Tannock I. Studies with glycolysis-deficient cells suggest that production of lactic acid is not the only cause of tumor acidity. *Proc Natl Acad Sci USA* 1993 February 1; 90(3): 1127–1131.
11. Yamagata M, Hasuda K, Stamato T, Tannock IF. The contribution of lactic acid to acidification of tumours: studies of variant cells lacking lactate dehydrogenase. *Br J Cancer* 1998 June; 77(11): 1726–1731.
12. Perez-Sayans M, Somoza-Martin JM, Barros-Angueira F, Rey JM, Garcia-Garcia A. V-ATPase inhibitors and implication in cancer treatment. *Cancer Treat Rev* 2009 September 14; 35(8): 707–713.
13. Nelson N. A journey from mammals to yeast with vacuolar H^+-ATPase (V-ATPase). *J Bioenerg Biomembr* 2003 August; 35(4): 281–289.
14. Saroussi S, Nelson N. Vacuolar H(+)-ATPase-an enzyme for all seasons. *Pflugers Arch* 2008 March 5.
15. Becelli R, Renzi G, Morello R, Altieri F. Intracellular and extracellular tumor pH measurement in a series of patients with oral cancer. *J Craniofac Surg* 2007 September; 18(5):1051–1054.
16. Nelson N, Harvey WR. Vacuolar and plasma membrane proton-adenosine triphosphatases. *Physiol Rev* 1999 April 1; 79(2): 361–385.
17. Reshkin SJ, Bellizzi A, Caldeira S, Albarani V, Malanchi I, Poignee M, *et al.* Na^+/H^+ exchanger-dependent intracellular alkalinization is an early event in malignant transformation and plays an essential role in the development of subsequent transformation-associated phenotypes. *FASEB J* 2000 November; 14(14): 2185–2197.
18. Hagag N, Lacal JC, Graber M, Aaronson S, Viola MV. Microinjection of ras p21 induces a rapid rise in intracellular pH. *Mol Cell Biol* 1987 May; 7(5): 1984–1988.
19. Ober SS, Pardee AB. Intracellular pH is increased after transformation of Chinese hamster embryo fibroblasts. *Proc Natl Acad Sci USA* 1987 May; 84(9): 2766–2770.
20. Reshkin SJ, Bellizzi A, Albarani V, Guerra L, Tommasino M, Paradiso A, *et al.* Phosphoinositide 3-kinase is involved in the tumor-specific

activation of human breast cancer cell Na(+)/H(+) exchange, motility, and invasion induced by serum deprivation. *J Biol Chem* 2000 February 25; 275(8): 5361–5369.

21. Siczkowski M, Davies JE, Ng LL. Activity and density of the Na^+/H^+ antiporter in normal and transformed human lymphocytes and fibroblasts. *Am J Physiol* 1994 September; 267(3, Pt. 1): C745–C752.
22. Doppler W, Jaggi R, Groner B. Induction of v-mos and activated Ha-ras oncogene expression in quiescent NIH 3T3 cells causes intracellular alkalinisation and cell-cycle progression. *Gene* 1987; 54(1): 147–153.
23. Pouyssegur J, Franchi A, Pages G. pHi, aerobic glycolysis and vascular endothelial growth factor in tumour growth. *Novartis Found Symp* 2001; 240: 186–196; discussion 196–198.
24. Hanahan D, Folkman J. Patterns and emerging mechanisms of the angiogenic switch during tumorigenesis. *Cell* 1996 August 9; 86(3): 353–364.
25. Raghunand N, Gillies RJ. pH and drug resistance in tumors. *Drug Resist Updat* 2000 February; 3(1): 39–47.
26. Cardone RA, Casavola V, Reshkin SJ. The role of disturbed pH dynamics and the Na^+/H^+ exchanger in metastasis. *Nat Rev Cancer* 2005 October; 5(10): 786–795.
27. Martinez-Zaguilan R, Lynch RM, Martinez GM, Gillies RJ. Vacuolar-type H(+)-ATPases are functionally expressed in plasma membranes of human tumor cells. *Am J Physiol Cell Physiol* 1993 October 1; 265(4): C1015–C1029.
28. McLean LA, Roscoe J, Jorgensen NK, Gorin FA, Cala PM. Malignant gliomas display altered pH regulation by NHE1 compared with nontransformed astrocytes. *Am J Physiol Cell Physiol* 2000 April; 278(4): C676–C688.
29. Perona R, Serrano R. Increased pH and tumorigenicity of fibroblasts expressing a yeast proton pump. *Nature* 1988 August 4; 334(6181): 438–440.
30. Sennoune SR, Martinez-Zaguilan R. Plasmalemmal vacuolar H^+-ATPases in angiogenesis, diabetes and cancer. *J Bioenerg Biomembr* 2007 December; 39(5/6): 427–433.
31. Montcourrier P, Silver I, Farnoud R, Bird I, Rochefort H. Breast cancer cells have a high capacity to acidify extracellular milieu by a dual mechanism. *Clin Exp Metastasis* 1997 July; 15(4): 382–392.

32. Parkins CS, Stratford MR, Dennis MF, Stubbs M, Chaplin DJ. The relationship between extracellular lactate and tumour pH in a murine tumour model of ischaemia-reperfusion. *Br J Cancer* 1997; 75(3): 319–323.
33. Rofstad EK, Mathiesen B, Kindem K, Galappathi K. Acidic extracellular pH promotes experimental metastasis of human melanoma cells in athymic nude mice. *Cancer Res* 2006 July 1; 66(13): 6699–6707.
34. Montcourrier P, Mangeat PH, Valembois C, Salazar G, Sahuquet A, Duperray C, *et al.* Characterization of very acidic phagosomes in breast cancer cells and their association with invasion. *J Cell Sci* 1994 September; 107 (Pt. 9): 2381–2391.
35. Rozhin J, Sameni M, Ziegler G, Sloane BF. Pericellular pH affects distribution and secretion of cathepsin B in malignant cells. *Cancer Res* 1994 December 15; 54(24): 6517–6525.
36. Johnson LL, Pavlovsky AG, Johnson AR, Janowicz JA, Man CF, Ortwine DF, *et al.* A rationalization of the acidic pH dependence for stromelysin-1 (Matrix metalloproteinase-3) catalysis and inhibition. *J Biol Chem* 2000 April 14; 275(15): 11026–11033.
37. Kato Y, Lambert CA, Colige AC, Mineur P, Noel A, Frankenne F, *et al.* Acidic extracellular pH induces matrix metalloproteinase-9 expression in mouse metastatic melanoma cells through the phospholipase D-mitogen-activated protein kinase signaling. *J Biol Chem* 2005 March 25; 280(12): 10938–10944.
38. Gocheva V, Joyce JA. Cysteine cathepsins and the cutting edge of cancer invasion. *Cell Cycle* 2007 January 1; 6(1): 60–64.
39. Sennoune SR, Bakunts K, Martinez GM, Chua-Tuan JL, Kebir Y, Attaya MN, *et al.* Vacuolar H^+-ATPase in human breast cancer cells with distinct metastatic potential: distribution and functional activity. *Am J Physiol Cell Physiol* 2004 June 1; 286(6): C1443–C1452.
40. Nanda A, Brumell JH, Nordstrom T, Kjeldsen L, Sengelov H, Borregaard N, *et al.* Activation of proton pumping in human neutrophils occurs by exocytosis of vesicles bearing vacuolar-type H^+-ATPases. *J Biol Chem* 1996 July 5; 271(27): 15963–15970.
41. Gatenby RA, Gillies RJ. Why do cancers have high aerobic glycolysis? *Nat Rev Cancer* 2004 November; 4(11): 891–899.
42. Volk C, Albert T, Kempski OS. A proton-translocating H^+-ATPase is involved in C6 glial pH regulation. *Biochim Biophys Acta* 1998 June 24; 1372(1): 28–36.

43. McSheehy PM, Troy H, Kelland LR, Judson IR, Leach MO, Griffiths JR. Increased tumour extracellular pH induced by Bafilomycin A1 inhibits tumour growth and mitosis in vivo and alters 5-fluorouracil pharmacokinetics. *Eur J Cancer* 2003 March; 39(4): 532–540.
44. Bowman EJ, Graham LA, Stevens TH, Bowman BJ. The bafilomycin/concanamycin binding site in subunit c of the V-ATPases from Neurospora crassa and Saccharomyces cerevisiae. *J Biol Chem* 2004 August 6; 279(32): 33131–33138.
45. Lim JH, Park JW, Kim MS, Park SK, Johnson RS, Chun YS. Bafilomycin induces the p21-mediated growth inhibition of cancer cells under hypoxic conditions by expressing hypoxia-inducible factor-1alpha. *Mol Pharmacol* 2006 December; 70(6): 1856–1865.
46. Zhong H, De Marzo AM, Laughner E, Lim M, Hilton DA, Zagzag D, *et al.* Overexpression of hypoxia-inducible factor 1alpha in common human cancers and their metastases. *Cancer Res* 1999 November 15; 59(22): 5830–5835.
47. Aiko K, Tsujisawa T, Koseki T, Hashimoto S, Morimoto Y, Amagasa T, *et al.* Involvement of cytochrome c and caspases in apoptotic cell death of human submandibular gland ductal cells induced by concanamycin A. *Cell Signal* 2002 August; 14(8): 717–722.
48. De Milito A, Iessi E, Logozzi M, Lozupone F, Spada M, Marino ML, *et al.* Proton pump inhibitors induce apoptosis of human B-cell tumors through a caspase-independent mechanism involving reactive oxygen species. *Cancer Res* 2007 June 1; 67(11): 5408–5417.
49. Gottlieb RA, Giesing HA, Zhu JY, Engler RL, Babior BM. Cell acidification in apoptosis: granulocyte colony-stimulating factor delays programmed cell death in neutrophils by up-regulating the vacuolar H(+)-ATPase. *Proc Natl Acad Sci USA* 1995 June 20; 92(13): 5965–5968.
50. Xu J, Feng HT, Wang C, Yip KH, Pavlos N, Papadimitriou JM, *et al.* Effects of Bafilomycin A1: an inhibitor of vacuolar H (+)-ATPases on endocytosis and apoptosis in RAW cells and RAW cell-derived osteoclasts. *J Cell Biochem* 2003 April 15; 88(6): 1256–1264.
51. Morimura T, Fujita K, Akita M, Nagashima M, Satomi A. The proton pump inhibitor inhibits cell growth and induces apoptosis in human hepatoblastoma. *Pediatr Surg Int* 2008 October; 24(10): 1087–1094.
52. Wu YC, Wu WK, Li Y, Yu L, Li ZJ, Wong CC, *et al.* Inhibition of macroautophagy by bafilomycin A1 lowers proliferation and induces apoptosis in colon cancer cells. *Biochem Biophys Res Commun* 2009 May 1; 382(2): 451-456.

53. Meijer AJ, Codogno P. Regulation and role of autophagy in mammalian cells. *Int J Biochem Cell Biol* 2004 December; 36(12): 2445–2462.
54. Mortimore GE, Hutson NJ, Surmacz CA. Quantitative correlation between proteolysis and macro- and microautophagy in mouse hepatocytes during starvation and refeeding. *Proc Natl Acad Sci USA* 1983 April; 80(8): 2179–2183.
55. Griffiths JR. Are cancer cells acidic? *Br J Cancer* 1991 September; 64(3): 425–427.
56. Negendank W. Studies of human tumors by MRS: a review. *NMR Biomed* 1992 September–October; 5(5): 303–324.
57. Luciani F, Spada M, De Milito A, Molinari A, Rivoltini L, Montinaro A, *et al.* Effect of proton pump inhibitor pretreatment on resistance of solid tumors to cytotoxic drugs. *J Natl Cancer Inst* 2004 November 17; 96(22): 1702–1713.
58. De Milito A, Fais S. Tumor acidity, chemoresistance and proton pump inhibitors. *Future Oncol* 2005 December; 1(6): 779–786.
59. Sennoune SR, Luo D, Martinez-Zaguilan R. Plasmalemmal vacuolar-type H^+-ATPase in cancer biology. *Cell Biochem Biophys* 2004; 40(2): 185–206.
60. Murakami T, Shibuya I, Ise T, Chen ZS, Akiyama S, Nakagawa M, *et al.* Elevated expression of vacuolar proton pump genes and cellular PH in cisplatin resistance. *Int J Cancer* 2001 September; 93(6): 869–874.
61. Torigoe T, Izumi H, Ishiguchi H, Uramoto H, Murakami T, Ise T, *et al.* Enhanced expression of the human vacuolar H^+-ATPase c subunit gene (ATP6L) in response to anticancer agents. *J Biol Chem* 2002 September 27; 277(39): 36534–36543.
62. Kluza J, Lansiaux A, Wattez N, Mahieu C, Osheroff N, Bailly C. Apoptotic response of HL-60 human leukemia cells to the antitumor drug TAS-103. *Cancer Res* 2000 August 1; 60(15): 4077–4084.
63. Cianfriglia M, Cenciarelli C, Tombesi M, Barca S, Mariani M, Morrone S, *et al.* Murine monoclonal antibody recognizing a 90-kDa cell-surface determinant selectively lost by multi-drug-resistant variants of CEM cells. *Int J Cancer* 1990 January 15; 45(1): 95–103.
64. Fais S, De Milito A, You H, Qin W. Targeting vacuolar H^+-ATPases as a new strategy against cancer. *Cancer Res* 2007 November 15; 67(22): 10627–10630.
65. Nishi T, Forgac M. The vacuolar (H+)-ATPases—nature's most versatile proton pumps. *Nat Rev Mol Cell Biol* 2002 February; 3(2): 94–103.

66. Bowman EJ, Siebers A, Altendorf K. Bafilomycins: a class of inhibitors of membrane ATPases from microorganisms, animal cells, and plant cells. *Proc Natl Acad Sci USA* 1988 November; 85(21): 7972–7976.
67. Erickson KL, Beutler JA, Cardellina II JH, Boyd MR. Salicylihalamides A and B, novel cytotoxic macrolides from the marine sponge Haliclona sp. *J Org Chem* 1997 November 14; 62(23): 8188–8192.
68. Galinis DL, McKee TC, Pannell LK, Cardellina JH, Boyd MR. Lobatamides A and B, novel cytotoxic macrolides from the tunicate Aplidium lobatum. *J Org Chem* 1997; 62(26): 8968–8969.
69. Kunze B., Janse R., Sasse F., Höfle G., Reichenbach H. Apicularens A and B, New cytostatic macrolides from *Chondromyces* species (Myxobacteria): production, physico-chemical and biological properties. *J Antibiot (Tokyo)* 1998; 51(12): 1075–1080.
70. Gagliardi S, Nadler G, Consolandi E, Parini C, Morvan M, Legave MN, *et al.* 5-(5,6-Dichloro-2-indolyl)-2-methoxy-2,4-pentadienamides: novel and selective inhibitors of the vacuolar H^+-ATPase of osteoclasts with bone antiresorptive activity. *J Med Chem* 1998 May 7; 41(10): 1568–1573.
71. Nadler G, Morvan M, Delimoge I, Belfiore P, Zocchetti A, James I, *et al.* (2Z,4E)-5-(5,6-dichloro-2-indolyl)-2-methoxy-N-(1,2,2,6,6- pentamethylpiperidin-4-yl)-2,4-pentadienamide, a novel, potent and selective inhibitor of the osteoclast V-ATPase. *Bioorg Med Chem Lett* 1998 December 15; 8(24): 3621–3626.
72. Kim JW, Shin-Ya K, Furihata K, Hayakawa Y, Seto H. Oximidines I and II: novel antitumor macrolides from Pseudomonas sp. *J Org Chem* 1999 January 8; 64(1): 153–155.
73. Sasse F, Steinmetz H, Hofle G, Reichenbach H. Archazolids, new cytotoxic macrolactones from Archangium gephyra (Myxobacteria). Production, isolation, physico-chemical and biological properties. *J Antibiot (Tokyo)* 2003 June; 56(6): 520–525.
74. Shen R, Lin CT, Bowman EJ, Bowman BJ, Porco JA, Jr. Lobatamide C: total synthesis, stereochemical assignment, preparation of simplified analogues, and V-ATPase inhibition studies. *J Am Chem Soc* 2003 July 2; 125(26): 7889–7901.
75. Kunze B, Steinmetz H, Hofle G, Huss M, Wieczorek H, Reichenbach H. Cruentaren, a new antifungal salicylate-type macrolide from Byssovorax cruenta (myxobacteria) with inhibitory effect on mitochondrial ATPase activity. Fermentation and biological properties. *J Antibiot (Tokyo)* 2006 October; 59(10): 664–668.

76. Petrangolini G, Supino R, Pratesi G, Bo LD, Tortoreto M, Croce AC, *et al.* Effect of a novel vacuolar-H^+-ATPase inhibitor on cell and tumor response to camptothecins. *J Pharmacol Exp Ther* 2006 September 1; 318(3): 939–946.
77. Niikura K. Effect of a V-ATPase inhibitor, FR202126, in syngeneic mouse model of experimental bone metastasis. *Cancer Chemother Pharmacol* 2007 September; 60(4): 555–562.
78. Hesselink RW, Fedorov A, Hemminga MA, Prieto M. Membrane-bound peptides from V-ATPase subunit a do not interact with an indole-type inhibitor. *J Pept Sci* 2008 Apr; 14(4): 383–388.
79. Smith AN, Borthwick KJ, Karet FE. Molecular cloning and characterization of novel tissue-specific isoforms of the human vacuolar H(+)-ATPase C, G and d subunits, and their evaluation in autosomal recessive distal renal tubular acidosis. *Gene* 2002 September 4; 297(1/2): 169–177.
80. Otero-Rey EM, Somoza-Martin M, Barros-Angueira F, Garcia-Garcia A. Intracellular pH regulation in oral squamous cell carcinoma is mediated by increased V-ATPase activity via over-expression of the ATP6V1C1 gene. *Oral Oncol* 2008 February; 44(2): 193–199.
81. Svergun DI, Konrad S, Huss M, Koch MH, Wieczorek H, Altendorf K, *et al.* Quaternary structure of V1 and F1 ATPase: significance of structural homologies and diversities. *Biochemistry* 1998 December 22; 37(51): 17659–17663.
82. Armbruster A, Svergun DI, Coskun U, Juliano S, Bailer SM, Gruber G. Structural analysis of the stalk subunit Vma5p of the yeast V-ATPase in solution. *FEBS Lett* 2004 July 16; 570(1/3): 119–125.
83. Chaban YL, Juliano S, Boekema EJ, Gruber G. Interaction between subunit C (Vma5p) of the yeast vacuolar ATPase and the stalk of the C-depleted V(1) ATPase from Manduca sexta midgut. *Biochim Biophys Acta* 2005 June 30; 1708(2): 196–200.
84. Zhang Z, Inoue T, Forgac M, Wilkens S. Localization of subunit C (Vma5p) in the yeast vacuolar ATPase by immuno electron microscopy. *FEBS Lett* 2006 April 3; 580(8): 2006–2010.
85. Drory O, Frolow F, Nelson N. Crystal structure of yeast V-ATPase subunit C reveals its stator function. *EMBO Rep* 2004 December; 5(12): 1148–1152.
86. Peng S, Stone D, Xie X. Reconstitution of recombinant 40-kDa subunit of the clathrin-coated vesicle H(+)-ATPase. *J Biol Chem* 1993 November 5; 268(31): 23519–23523.

87. Gruber G. Structural features and nucleotide-binding capability of the C subunit are integral to the regulation of the eukaryotic V1Vo ATPases. *Biochem Soc Trans* 2005 August; 33(Pt. 4): 883–885.
88. Armbruster A, Hohn C, Hermesdorf A, Schumacher K, Borsch M, Gruber G. Evidence for major structural changes in subunit C of the vacuolar ATPase due to nucleotide binding. *FEBS Lett* 2005 March 28; 579(9): 1961–1967.
89. Voss M, Vitavska O, Walz B, Wieczorek H, Baumann O. Stimulus-induced phosphorylation of vacuolar H(+)-ATPase by protein kinase A. *J Biol Chem* 2007 November 16; 282(46): 33735–33742.
90. Beltran C, Kopecky J, Pan YC, Nelson H, Nelson N. Cloning and mutational analysis of the gene encoding subunit C of yeast vacuolar H(+)-ATPase. *J Biol Chem* 1992 January 15; 267(2): 774–779.
91. Perez-Sayans M, Garcia-Garcia A, Reboiras-Lopez MD, Gandara-Vila P. Role of V-ATPases in solid tumors: importance of the subunit C (Review). *Int J Oncol* 2009 June; 34(6): 1513–1520.
92. Somoza-Martin JM, Garcia-Garcia A, Barros-Angueira F, Otero-Rey E, Torres-Espanol M, Gandara-Vila P, *et al.* Gene expression profile in oral squamous cell carcinoma: a pilot study. *J Oral Maxillofac Surg* 2005 June; 63(6): 786–792.
93. Perez-Sayans M, Reboiras-Lopez MD, Somoza-Martin JM, Barros-Angueira F, Gayoso Diz P, Gandara Rey JM, *et al.* Measurement of ATP6V1C1 expression in brush cytology samples as a diagnostic and prognostic marker in oral squamous cell carcinoma. *Cancer Biol Ther* 2010 June 25; 9(12).
94. Garcia-Garcia A, Perez-Sayans Garcia M, Rodriguez MJ, Antunez-Lopez J, Barros-Angueira F, Somoza-Martin M, *et al.* Immunohistochemical localization of C1 subunit of V-ATPase (ATPase C1) in oral squamous cell cancer and normal oral mucosa. *Biotech Histochem* 2012 February; 87(2): 133–139.

Chapter 15

Vacuolar H^+-ATPase: Functional Mechanism and Potential as a Target for Cancer Chemotherapy

Hiroto Izumi and Kimitoshi Kohno

Department of Molecular Biology, School of Medicine, University of Occupational and Environmental Health, Kitakyushu 807-8555, Japan

h-izumi@med.uoeh-u.ac.jp

The complex biology of solid tumor development represents a major obstacle to its successful treatment. In addition to cancer cells, this is in part due to the fact that solid tumors are heterogeneous, comprises various cells, including stromal, inflammatory, and vascular cells. Although the regulation of intracellular pH has been shown to be critical for many cellular functions, pH regulation has not been fully investigated in the field of cancer. It has been shown however, that cellular pH is crucial for important biological functions including cell proliferation, invasion and metastasis, drug resistance, and apoptosis. Thus, pathways affected by pH signaling are often dysregulated in cancer.

Several cellular pH regulators have been identified in solid tumors including the vacuolar H^+-ATPase (V-ATPase) proton pump, the sodium proton exchanger family, the bicarbonate transporter

Handbook of H^+-ATPases
Edited by Suguru Nakamura

ISBN 978-981-4411-91-2 (Hardcover), 978-981-4411-92-9 (eBook)
www.panstanford.com

family, the monocarboxylate transporter family, and carbonic anhydrase. Among these pH regulators, the V-ATPase is primarily responsible for the regulation of both intra- and extracellular pH. In this paper, we describe recent progress in the field of cancer research, specifically with regard to pH regulation. We also summarize the structure and function of the V-ATPase, which is expressed in tumor tissue, and focus on the potential of specific V-ATPase inhibitors in cancer treatment. Understanding the role of pH regulation in tumor cells may provide insights into a novel mechanism contributing to cancer. Such discoveries are also likely to provide a rationale for the development and application of inhibitors to specifically target this pathway in cancer cells.

15.1 pH Regulation in Cancer

To treat cancer successfully, it is critical to understand the molecular mechanisms driving the growth of solid tumors. Hanahan and Weinberg in 2000 proposed that the complexity of cancer could be simplified to six hallmarks [1]. Since then, several additional traits have been added to this model, highlighting rapid advances in our understanding of molecular mechanisms underlying cancer [2]. However, the findings also indicate that cancer biology is more complex than the previously anticipated ones.

In normal cells, intracellular pH is approximately 7.2, and is lower than the extracellular pH. In cancer cells; however, intracellular pH is elevated to approximately 7.4, and the extracellular pH about 7.0. This inverse pH gradient induces pH-dependent signal transduction pathways, facilitating cancer development, survival and progression. As summarized in Table 15.1, cancer cell phenotypes and behaviors are regulated by proton dynamics [3].

In general, tumor cells exhibit increased glycolysis and are capable of growing in a hypoxic microenvironment. Highly proliferative cancer cells produce large quantities of metabolic acid generated by glycolysis, glucose utilization and lactic acid production, and increase proton efflux, thus preventing apoptosis by cellular acidosis [4]. Figure 15.1 summarizes the complex pathways involved in solid tumor development from pre-clinical to clinical phase, and

Table 15.1 pH-dependent pathway in cancer cells

1. Proliferation/evasion of apoptosis
2. CDK/Cyclin pathway/mitotic and meiotic enntry
3. Metabolic adaptation/glycolysis
4. Invasion and metastasis
5. Drug resistance

their associated cancer hallmarks. There are two major growth phases: the pre-clinical growth phase without angiogenesis and the clinical rapid growth phase with angiogenesis. In both cases, hypoxic glycolysis is activated, leading to the production of large amounts of acid metabolites including protons. The subsequent decrease in intracellular pH has been shown to reduce DNA repair activity, and this may play a role in the accumulation of spontaneous mutations, which are observed during the malignant progression of

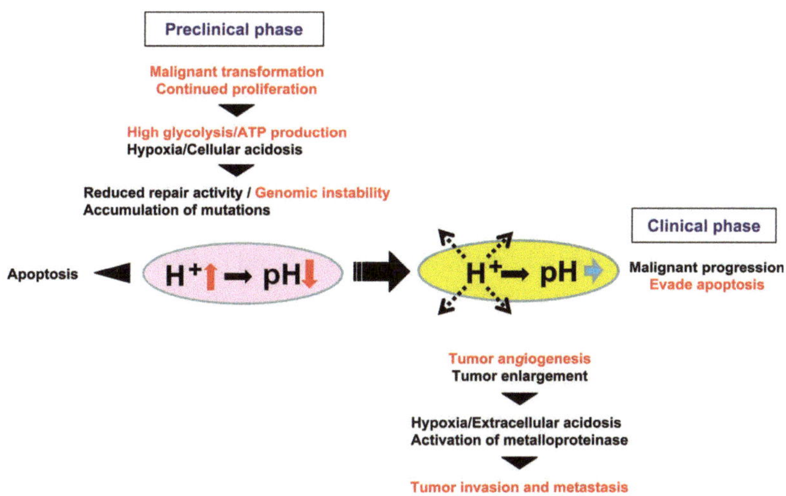

Figure 15.1 Hallmarks of cancer, cellular pH signaling, and development of solid tumor. Malignant transformation to cancer cells induces intracellular proton accumulation. Apoptosis resistance and release proton from the cells might be of great advantage to malignant progression including angiogenesis, invasion, and metastasis. Red characters indicate hallmarks of cancer.

solid tumors [5, 6]. This decrease in intracellular pH also induces genomic instability, another hallmark underlying the transformation of normal to cancer cells. Malignant progression is accompanied by novel gene expression [7, 8] and epigenetic changes, including those affecting histone modifications and DNA methylation [9, 10]. Rapid tumor growth with angiogenesis induces extracellular acidosis, which in turn leads to the activation of metalloproteinases [11]. Thus, solid tumors acquire the ability to invade tissues and to metastasize. Maintenance of pH homeostasis is crucial for solid tumor development, thus tumor cells express various pH regulators to avoid apoptosis. These data suggest that the dysregulation of pH in tumor cells is closely linked to the acquisition of properties associated with cellular transformation and cancer progression.

15.2 Structural and Functional Characteristics of V-ATPase

Proton transport is primarily mediated by the V-ATPase proton pump [12]. V-ATPase belongs to a class of pumps that includes the F-ATPase (energy-coupling factors). V-ATPases are large, multi-subunit complexes composed of two multi-subunit sectors; the proton-translocation domain, V0, and ATP hydrolysis domain, V1 (Fig. 15.2). To date, 23 genes encoding the 13 subunits of the V-ATPase in humans have been identified.

The V1 domain is a 600–650 kDa peripheral complex, composed of eight different subunits. The V0 domain is a 260 kDa integral complex, composed of five different subunits. Table 15.2 shows the characteristics of various subunits comprising the V-ATPase complex, including chromosomal location, molecular mass, gene phenotype, and the structure of the promoter including CpG content and major transcription factor binding sites.

The V-ATPase is expressed in eukaryotes from yeast to humans [13]. It is present in the membrane of intracellular organelles, in addition to the plasma membrane. The V-ATPase pumps protons from the cytoplasm to the lumen of vacuoles or into the extracellular space, using the energy produced by ATP hydrolysis. A significant feature of the V-ATPase is its involvement in a wide variety of

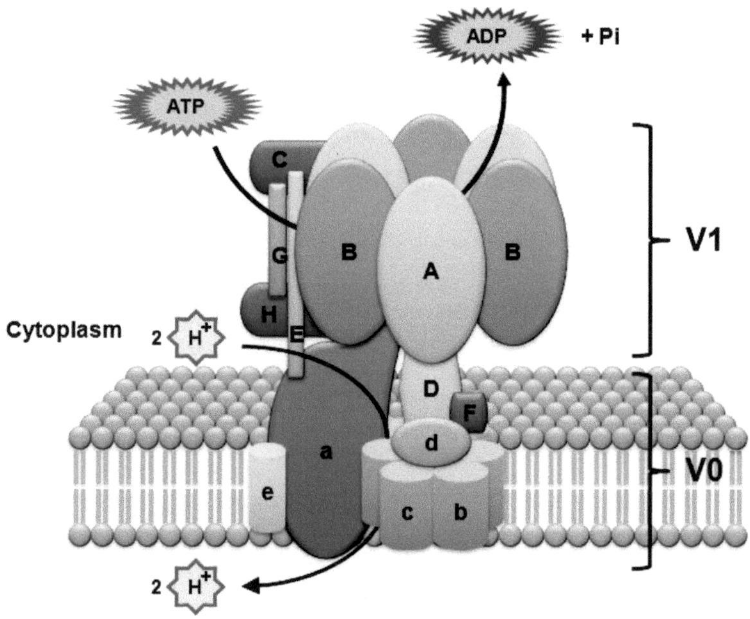

Figure 15.2 Structure of V-ATPase. V-ATPase consists of catalytic complex V1 domain and intrinsic membrane complex V0 domain. It is suggested that rotation of shaft D and six subunits containing c or d is caused by ATPase reaction to each of the subunit A in turn and drives the transport of protons through the membrane.

cellular functions (Table 15.3). V-ATPases play important roles in both endocytosis and intracellular transport. In receptor-mediated endocytosis, the dissociation of the ligand-receptor complex is initiated in acidic endosomes. After dissociation, receptors recycle to the plasma membrane. After the targeting of newly synthesized lysosomal enzymes from the Golgi to lysosomes, lysosomal enzymes are released from the mannose-6-phosphate receptor by the acidification of endosomes. Mutations in the genes encoding V-ATPases, manifest in several human diseases including renal tubular acidification, sensorineural deafness, and malignant osteopetrosis (Table 15.2). It appears that some isoforms show tissue-restricted expression. Thus, plasma membrane V-ATPase has an important role in renal acidification, pH maintenance in mechanosensory hair cells, and bone resorption.

Table 15.2 Function and subunit composition of human vacuolar proton pump (V-ATPase)

	Symbol	Subunit	Molecular weight	Previous Name	Entrez Gene ID	Location	CpG island	Element in promoter	Disease
V1 domain									
	ATP6V1A	A	68 kDa	ATP6A1	523	3q13.31	+	E-box GC box	
	ATP6V1B1	B1	57 kDa	ATP6B1	525	2p13	–		Distal renal tubular acidosis
	ATP6V1B2	B2	57 kDa	ATP6B2	526	8p21.3	+	AP-2 GC box	
	ATP6V1C1	C1	44 kDa	ATP6C, ATP6D	528	8p22.3	+	E-box(c) NF-kB GC box	
	ATP6V1C2	C2	49 kDa	ATP6C2	245973	2p25.1	+	E-box(v) GC box	
	ATP6V1D	D	28 kDa	ATP6M	51382	14q23-q24.2	+	Ets NRF2	
	ATP6V1E1	E1	26 kDa	ATP6E, ATP6V1E	529	22q11.2	+	Ets	
	ATP6V1E2	E2	26 kDa	ATP6EL2, ATP6V1EL2	90423	2p21	–	E-box(v) Ets	
	ATP6V1F	F	13 kDa	ATP6S14	9296	7q32.1	+		
	ATP6V1G1	G1	14 kDa	ATP6G1, ATP6J	9550	9q33.1	+	E-box(c) GC box	
	ATP6V1G2	G2	14 kDa	ATP6G, ATP6G2	534	6p21.3	–		
	ATP6V1G3	G3	14 kDa	ATP6G3	127124	1q32.2	–		
	ATP6V1H	H	56 kDa		51606	8q11.2	+	E-box(C) Ets	

V0 domain								
ATP6V0A1	a1	96 kDa	ATP6N1, ATP6N1A	535	17q21	+	CRE GC box	
ATP6V0A2	a2	98 kDa	ATP6a2	23545	12q24.31	+	E-box GC box	Autosomal recessive cutis laxa type 2
ATP6V0A3 (TCIRG1)	a3	93 kDa	ATP6N1C	10312	11q13.2	+	Ets GC box	Osteopetrosis autosomal recessive type 1
ATP6V0A4	a4	96 kDa	ATP6N1B, ATP6N2	50617	7q34	–		Distal renal tubular acidosis
ATP6V0B	b	21 kDa	ATP6F (subunit c")	533	1p32.3	+	E-box(c) ×2 GC box	
ATP6V0C	c	16 kDa	ATP6C, ATP6L	527	16p13.3	+	E-box(v) Oct1 GC box	
ATP6V0D1	d1	40 kDa	ATP6D, ATP6DV	9114	16q22	+		
ATP6V0D2	d2	40 kDa	ATP6D2	245972	8q21.3	–		
ATP6V0E1	e1	9 kDa	ATP6H, ATP6V0E	8992	5q35.2	+	E-box(c)	
ATP6V0E2	e2	9 kDa	ATP6V0E2L	155066	7q36.1	+	E-box(c)	

TCIRG1 (T-cell, immune regulator 1) is an approved symbol of ATP6V0A3. E-box(c) and E-box(v) indicate CACGTG and CANNTG motifs in the promoter, respectively.

Table 15.3 Functions of V-ATPase

1. Endocytosis
2. Intracellular transport
3. Cytoplasmic pH homeostasis
4. Bone resorption
5. Protein processing
6. Angiogenesis
7. Sperm maturation
8. Acid secretion
9. Epithelial-mesenchymal transition (EMT)

15.3 Interacting Molecules with V-ATPase

β1 integrin is involved in cell migration, proliferation, differentiation, cytoskeletal organization, signal transduction, and sensitivity to anti-cancer agents. It binds to the c subunit of V-ATPase on the last of four transmembrane helices. Overexpression of the c subunit or expression of a mutant subunit with the fourth helix deleted, alters the morphology of myoblasts and fibroblasts [14]. Thus, interactions of the c subunit with β1 integrin are important for the control of cell growth. Recently, the c subunit of V-ATPase has been reported to suppress β1, six branching of *N*-linked oligosaccharides to β1 integrin, and epidermal growth factor (EGF). This ability to influence glycosylation is located in the second and fourth helices of this transmembrane subunit. However, the effect of the c subunit on glycosylation is independent of its ability to bind to β1 integrin [15].

The E subunit of V-ATPase interacts with mouse Son of sevenless 1 (mSos-1), through the diffuse B-cell lymphoma (Dbl) oncoprotein homology domain. This protein has a dual role in activating Ras and Rac1 signaling pathways, suggesting that the E subunit may participate in the regulation of growth factor-mediated cell growth controlled by mSos-1-dependent Rac1 signaling [16]. The H subunit has been shown to interact with Negative regulatory factor (Nef) protein derived from human immunodeficiency virus (HIV) [17]. Nef regulates the expression of CD4, the primary HIV receptor in infected cells; it is possible that this interaction may induce endocytosis. The B subunit of V-ATPase contains an actin-binding site and interacts with cellular microfilaments suggesting that the

B subunit may be involved in the invasion and metastasis of cancer cells [18]. Future studies aimed at identifying interactions between the various V-ATPase subunits and other proteins will be important for the discovery of novel cellular functions of V-ATPase.

15.4 V-ATPase and Cancer Biology

The role of dysregulated pH in cancer biology is unclear. Although dysregulated pH is not a hallmark of cancer cells, signaling pathways involved in tumorigenesis are known to be affected by changes in intracellular and extracellular pH.

Previous studies have shown that V-ATPase is overexpressed in multidrug-resistant cells [19]. Furthermore, altered cytosolic pH has been implicated in drug resistance [20]. Multidrug-resistant cells contain more acidic organelles than their drug-sensitive counterparts, suggesting that acidic organelles may play a role in the observed resistance. In support of this hypothesis, daunomycin and several anti-cancer agents including doxorubicin and vinblastine have been shown to accumulate in acidic organelles. In cancer cells exhibiting multidrug resistance, these agents are then removed from the cell *via* exocytosis [21]. The accumulation of these compounds in acidic organelles is sensitive to the V-ATPase inhibitor bafilomycin A1, but insensitive to other transporter inhibitors and P-glycoprotein inhibitors [22–24]. In general, P-glycoprotein, an ATP-binding cassette (ABC) transporter encoded by the multidrug resistance (*MDR1)* gene, is found in cancer stem cells. It will be of interest to investigate whether cancer stem cells also contain acidic organelles. Recently, it was shown that cellular pH is higher in cisplatin-resistant cells compared with drug-sensitive cells. These studies showed that the expression of genes encoding the V-ATPase are induced by cisplatin treatment and upregulated in cisplatin-resistant cells [19]. Furthermore, cisplatin–DNA adduct formation is enhanced under acidic conditions and by a low chloride concentration. Cisplatin-resistant cells may therefore increase intracellular pH *via* induction of V-ATPase, thus avoiding the cytotoxic effects of cisplatin. We have previously shown that drug resistant cells may be established by two mechanisms, selection and induction [25], indicating that drug resistant cells are derived from cancer stem cells. The CD44 adhesion molecule is one of several cancer stem cell markers. It has been reported that cells expressing variant isoforms of CD44 exhibit increased intracellular glutathione through the interaction with a glutamate–cystine transporter [26]. High levels of glutathione are often observed in drug resistant cells. Taken together, these studies suggest that cancer stem cells possess various mechanisms for acquiring drug resistance. One interesting report revealed that V-ATPase promotes transforming growth factor β (TGF-β)-dependent epithelial–mesenchymal transition (EMT) [27]. Several transcription factors, including the E-box binding transcription factor, Snai1, are induced by TGF-β signaling. Furthermore, the Rat sarcoma (Ras) signaling pathway promotes TGF-β signaling and plays an important role in the induction of EMT [28]. Recently, recycling of V-ATPase is regulated by RhoA–RockII pathway through remodeling of the cytoskeleton [29]. Interestingly, it has been also reported that V-ATPase activates mammalian target of rapamycin complex 1 (mTORC1), a master growth regulator that induces cell growth by stimulating protein synthesis [30].

Ionizing radiation induces the development of acidic vesicular organelles in neoplastic epithelial cells [31]. Interference with acidification of vesicular organelles by bafilomycin A1, an inhibitor of V-ATPase, leads to increased radiosensitivity. Increased accumulation of p53 protein and p53-dependent transactivation of genes including p21 has also been observed in bafilomycin A1-treated cells [32]. Inhibitors of V-ATPase may produce a supra-additive effect when combined with other anticancer agents or irradiation. Thus, tumor pH plays an important role in governing the response of cancer cells to irradiation, hyperthermia, and chemotherapy. Recently, it has been shown that V-ATPase inhibitors induce EGF-dependent apoptosis in A431 cells [33]. In addition, intracellular signal transduction induced by the EGF–EGF receptor complex is blocked by V-ATPase inhibitors.

TAS-103 is a novel anticancer drug that induces apoptosis by causing cellular acidosis. It is also capable of upregulating the expression of V-ATPase subunits. The V-ATPase inhibitor, bafilomycin A1, can enhance TAS-103-induced apoptosis [34]. Intracellular acidification, an early event in apoptosis, modulates caspase activation [35]. Cellular acidosis also increases the susceptibility of cells to chemotherapeutic agents and this phenomenon has been demonstrated in HL-60 cells undergoing apoptosis in response to etoposide and camptothecin [36, 37]. This indicates that intracellular acidification activates cellular endonucleases, inducing nuclear DNA fragmentation [38]. These results indicate that V-ATPase inhibits the pH-dependent apoptosis of cancer cells and furthermore, that V-ATPase inhibitors, in combination with low-dose anticancer agents, may provide a new therapeutic approach.

Functional analysis of specific V-ATPase subunits has been reported. Targeting of V-ATPases to different cellular membranes is regulated by isoform A. Isoform-specific small interfering RNAs (siRNAs) can identify the functional differences between these isoforms. For example, cytosolic pH is specifically controlled by the A3 isoform, whereas lysosomal pH is regulated by three isoforms A1, A2, and A3 in MB231 breast cancer cells. In contrast, the A4 isoform does not affect either cytosolic or lysosomal pH [39]. Downregulation of c subunit of the V-ATPase, also known as ATP6L, has been shown to inhibit the growth and metastasis of cancer cells [40]. Furthermore, downregulation of ATP6L overcomes chemoresistance of breast cancer cells [41]. In melanoma cells, specific knockdown of the A3 isoform suppresses invasion and migration with reduced matrix metalloproteinase expression [42]. These data indicate that targeted knockdown using siRNA may provide a unique insight into subunit-specific functions of large multi-subunit enzymes, such as V-ATPase. Another feature of tumor cells is their ability to secrete lysosomal enzymes that degrade the extracellular matrix. The activity of these enzymes is markedly enhanced in an acidic extracellular environment. Thus, the plasma membrane V-ATPase may also facilitate tumor invasion and metastasis, making it an attractive potential target in cancer therapy.

15.5 V-ATPase Inhibitors

A number of V-ATPase inhibitors have been identified (Table 15.4). These inhibitors function by binding to specific sites on various subunits of the V-ATPase. Both bafilomycin A1, the first specific and

Table 15.4 V-ATPase inhibitors

1. Plecomacrolides
    - Bafilomycin A
    - Concanamycin
2. Macrolactone archazolid
3. Benzolactone enamides
    - Salicylihalamide
    - Oximidine
    - Lobatamide
    - Apicularen
    - Cruentaren
4. Indolyl
5. Macrocyclic lactams
    - Chondropsin
6. Diphyllin
7. Macrolides
    - Lejimalide
8. Others
    - NiK-12192
    - FR202126

highly potent inhibitor of V-ATPase, and concanamycin A, a microbial macrocyclic lactone, selectively inhibit V-ATPase. These agents also inhibit growth and induce apoptosis in various human cell lines at nanomolar concentrations [43, 44]. Recent analysis using bafilomycin sensitive cells, have sought to elucidate the mechanisms underlying the cellular response to low doses of this compound [45]. These studies show that mechanism of action of bafilomycin involves the regulation of cellular iron and cholesterol levels. Furthermore, it has been shown that iron uptake is critical for bafilomycin sensitivity in several cancer cell lines. V-ATPase inhibitors have been recently shown to inhibit the activity of anti-apoptotic oncoproteins, Bcl-2 and Bcl-xL [46]. Interestingly, bafilomycin A is also a potassium ionophore that impairs mitochondrial function [47]. The NCI60 cell anticancer drug discovery system allows the prediction of cellular targets for new compounds [48]. This system has been used to identify a novel antitumor benzolactone enamide that is capable of selectively inhibiting V-ATPase [49, 50]. Specific inhibitors belonging to this class include salicylihalamide, lobatamides, and oximidines, which appear promising as anticancer agents. Among these agents, salicylihalamide A can discriminate between mammalian and non-mammalian V-ATPases and can be divided into two classes: soluble-domain inhibitors and inhibitors acting at membrane sites. At low concentrations (μM), soluble-domain inhibitors, such as *N*-ethylmaleimide (NEM) and 7-chloro-4-nitrobenzo-2-oxa-1,3-diazole chloride (NBD-Cl), effectively inhibit ATP hydrolysis [44]. Inhibitors acting at membrane sites such as dicyclohexylcarbodiimide (DCCD), inhibit the c subunit of V-ATPase [51]. Another novel inhibitor is macrolactone archazolid, which is isolated from the myxobacteria [52]. Archazolid interacts with V0 subunit of V-ATPase and competes with concanamycin. The cyclohexadepsipeptide, destruxin, which is a mycotoxin, has also been shown to be a highly specific inhibitor of V-ATPase. Iejimalides are a novel family of marine macrolides that inhibit V-ATPase and have cytotoxic activity against various cancers [53]. FR202126 is a novel V-ATPase specific inhibitor, and exerts an inhibitory effect on bone absorption [54]. An indole derivative, Nik-12192 is a recently identified V-ATPase inhibitor, possessing potential anti-metastatic activity [55]. Collectively, these studies highlight the need to identify specific inhibitors for the V-ATPase and importantly, to study their effects and mechanism of action in the clinical setting.

15.6 Structure of the Promoter of V-ATPase Genes

V-ATPase genes are considered "housekeeping genes" and are positively regulated in response to growth rate. We have previously isolated and characterized several genomic clones containing the 5′ end of human V-ATPase 23 subunit genes. The promoters of all V-ATPase genes were examined with regard to their structure and transcription factor binding sites. These promoters exhibited a GC-rich region in the area of the first exon and lacked TATA and CCAAT boxes. These promoters were also characterized by the presence of putative Sp1 binding sites, which are frequently found in the regulatory regions of mammalian housekeeping genes. Interestingly, many V-ATPase subunit promoters contained E-boxes, which are capable of binding E-box binding transcription factors including Myc, USF, Twist, Snail, and Clock [56]. Taken together, the data indicate that V-ATPases may be involved in mediating drug resistance, circadian rhythm, EMT, and proliferation. We have shown that cooperative binding of Sp1 and Oct1 to the promoter region of the c subunit is required for basal expression and promoter activation by anti-cancer agents [34]. The molecular mechanisms of transcription underlying the regulation of other V-ATPase subunits remains to be elucidated. It will be of considerable interest to identify the transcription factors that regulate the V-ATPase subunit genes in cancer cells.

15.7 Prospects

In the coming decade, we envisage that significant progress toward understanding the regulation of pH, and deciphering the effect of this on metabolic and genetic reprogramming in cancer, will be made. Studies have shown that the expression profile of pH regulators differs between tumors and normal tissue and this information will be critical for the design of novel inhibitors targeting pH regulators. Epigenetic alterations, including specific chromatin modifications and the discovery of regulatory microRNAs, have led to profound changes in our understanding cancer biology [57, 58]. It is likely that these mechanisms are also involved in pH regulation in cancer cells.

The field of proton dynamics may provide the basis for selective cancer chemotherapy [57, 58]. A significant amount of work is required to analyze the functional properties of each pH regulator in maintaining physiological pH homeostasis. It has been proposed that regulation of intracellular pH may be a possible mechanism for tumor-selective therapy [59, 60]. The past decade has also seen the discovery of physiological mechanisms in which a decrease in intracellular pH is the trigger for a cascade of events leading to apoptosis. Currently, which pH regulators are expressed and may function as anti-apoptotic factors by preventing cellular acidosis in tumor tissue is unclear; however, recent reports indicate that V-ATPase is one such favorable candidate. Whether functional redundancy exists among these pH regulators also remains unclear.

Expression profiling of pH regulators in solid tumors is likely to answer some of these questions. The development of promising therapeutic agents for selective inhibition of pH regulators remains a final goal. In conclusion, further extensive investigation is required to understand the significance of tumor pH regulation in cancer development and for the design of specific therapies targeting this pathway.

Acknowledgment

This work was supported by the Ministry of Education, Culture, Sports, Science and Technology of Japan.

References

1. Hanahan, D. and Weinberg, R.A. (2000) The hallmarks of cancer, *Cell*, 100, pp. 57–70.
2. Hanahan, D. and Weinberg, R.A. (2011) Hallmarks of cancer: The next generation, *Cell*, 144, pp. 646–674.
3. Webb, B.A., Chimenti, M., Jacobson, M.P. and Barber, D.L. (2011) Dysregulated pH: A perfect storm for cancer progression, *Nat. Rev. Cancer*, 11, pp. 671–677.
4. Vaupel, P., Kallinowski, F. and Okunieff, P. (1989) Blood flow, oxygen and nutrient supply, and metabolic microenvironment of human tumors: A review, *Cancer Res.*, 49, pp. 6449–6465.
5. Yuan, J., Narayanan, L., Rockwell, S. and Glazer, P.M. (2000) Diminished DNA repair and elevated mutagenesis in mammalian cells exposed to hypoxia and low pH, *Cancer Res.*, 60, pp. 4372–4376.
6. Kondo, A., Safaei, R., Mishima, M., Niedner, H., Lin, X. and Howell, S.B. (2001) Hypoxia-induced enrichment and mutagenesis of cells that have lost DNA mismatch repair, *Cancer Res.*, 61, pp. 7603–7607.
7. Hanahan, D., Christofori, G., Naik, P. and Arbeit, J. (1996) Transgenic mouse models of tumour angiogenesis: The angiogenic switch, its molecular controls, and prospects for preclinical therapeutic models, *Eur. J. Cancer*, 32A, pp. 2386–2393.
8. Hanahan, D. and Folkman, J. (1996) Patterns and emerging mechanisms of the angiogenic switch during tumorigenesis, *Cell*, 86, pp. 353–364.

9. Brena, R.M. and Costello, J.F. (2007) Genome-epigenome interactions in cancer, *Hum. Mol. Genet.*, 16, pp. 96–105.
10. Widschwendter, M. and Jones, P.A. (2002) DNA methylation and breast carcinogenesis, *Oncogene*, 21, pp. 5462–5482.
11. Egeblad, M. and Werb, Z. (2002) New functions for the matrix metalloproteinases in cancer progression, *Nat. Rev. Cancer*, 2, pp. 161–174.
12. Nishi, T. and Forgac, M. (2002) The vacuolar (H+)-ATPases–nature's most versatile proton pumps. *Nat. Rev. Mol. Cell Biol.*, 3, pp. 94–103.
13. Forgac, M. (2007) Vacuolar ATPases: Rotary proton pumps in physiology and pathophysiology, *Nat. Rev. Mol. Cell Biol.*, 8, pp. 917–929.
14. Skinner, M.A. and Wildeman, A.G. (1999) β1 integrin binds the 16-kDa subunit of vacuolar H(+)-ATPase at a site important for human papillomavirus E5 and platelet-derived growth factor signaling, *J. Biol. Chem.*, 274, pp. 23119–23127.
15. Skinner, M.A. and Wildeman, A.G. (2001) Suppression of tumor-related glycosylation of cell surface receptors by the 16-kDa membrane subunit of vacuolar H^+-ATPase, *J. Biol. Chem.*, 276, pp. 48451–48457.
16. Miura, K., Miyazawa, S., Furuta, S., Mitsushita, J., Kamijo, K., Ishida, H., Miki, T., Suzukawa, K., Resau, J., Copeland, T.D. and Kamata, T. (2001) The Sos1-Rac1 signaling. Possible involvement of a vacuolar H(+)-ATPase E subunit, *J. Biol. Chem.*, 276, pp. 46276–46283.
17. Geyer, M., Yu, H., Mandic, R., Linnemann, T., Zheng, Y.H., Fackler, O.T. and Peterlin, B.M. (2002) Subunit H of the V-ATPase binds to the medium chain of adaptor protein complex 2 and connects Nef to the endocytic machinery, *J. Biol. Chem.*, 277, pp. 28521–28529.
18. Holliday, L.S., Bubb, M.R., Jiang, J., Hurst, I.R. and Zuo, J. (2005) Interactions between vacuolar H^+-ATPases and microfilaments in osteoclasts, *J. Bioenerg. Biomembr.*, 37, pp. 419–423.
19. Murakami, T., Shibuya, I., Ise, T., Chen, Z.S., Akiyama, S., Nakagawa, M., Izumi, H., Nakamura, T., Matsuo, K., Yamada, Y. and Kohno, K. (2001) Elevated expression of vacuolar proton pump genes and cellular pH in cisplatin resistance, *Int. J. Cancer*, 93, pp. 869–874.
20. Beck, W.T. (1987) The cell biology of multiple drug resistance, *Biochem. Pharmacol.*, 36, pp. 2879–2887.
21. Willingham, M.C., Cornwell, M.M., Cardarelli, C.O., Gottesman, M.M. and Pastan, I. (1986) Single cell analysis of daunomycin uptake and efflux in multidrug-resistant and -sensitive KB cells: Effects of verapamil and other drugs, *Cancer Res.*, 46, pp. 5941–5946.

22. Marquardt, D. and Center, M.S. (1991) Involvement of vacuolar H(+)-adenosine triphosphatase activity in multidrug resistance in HL60 cells, *J. Natl. Cancer Inst.*, 83, pp. 1098–1102.

23. Martínez-Zaguilán, R., Raghunand, N., Lynch, R.M., Bellamy, W., Martinez, G.M., Rojas, B., Smith, D., Dalton, W.S. and Gillies, R.J. (1999) pH and drug resistance. I. Functional expression of plasmalemmal V-type H^+-ATPase in drug-resistant human breast carcinoma cell lines, *Biochem. Pharmacol.*, 57, pp. 1037–1046.

24. Raghunand, N., Martínez-Zaguilán, R., Wright, S.H. and Gillies, R.J. (1999) pH and drug resistance. II. Turnover of acidic vesicles and resistance to weakly basic chemotherapeutic drugs, *Biochem. Pharmacol.*, 57, pp. 1047–1058.

25. Kohno, K., Uchiumi, T., Niina, I., Wakasugi, T., Igarashi, T., Momii, Y., Yoshida, T., Matsuo, K., Miyamoto, N. and Izumi, H. (2005) Transcription factors and drug resistance, *Eur. J. Cancer*, 41, pp. 2577–2586.

26. Ishimoto, T., Nagano, O., Yae, T., Tamada, M., Motohara, T., Oshima, H., Oshima, M., Ikeda, T., Asaba, R., Yagi, H., Masuko, T., Shimizu, T., Ishikawa, T., Kai, K., Takahashi, E., Imamura, Y., Baba, Y., Ohmura, M., Suematsu, M., Baba, H. and Saya, H. (2011) CD44 variant regulates redox status in cancer cells by stabilizing the xCT subunit of system xc(-) and thereby promotes tumor growth, *Cancer Cell*, 19, pp. 387–400.

27. Cao, X., Yang, Q., Qin, J., Zhao, S., Li, X., Fan, J., Chen, W., Zhou, Y., Mao, H. and Yu, X. (2012) V-ATPase promotes transforming growth factor-β-induced epithelial-mesenchymal transition of rat proximal tubular epithelial cells, *Am. J. Physiol. Renal Physiol.*, 302(9), F1121–F1132.

28. Miyazono, K. (2009) Transforming growth factor-beta signaling in epithelial-mesenchymal transition and progression of cancer, *Proc. Jpn. Acad., Ser. B, Phys. Biol. Sci.*, 85, pp. 314–323.

29. Shum, W.W., Da Silva, N., Belleannée, C., McKee, M., Brown, D. and Breton, S. (2011) Regulation of V-ATPase recycling via a RhoA- and ROCKII-dependent pathway in epididymal clear cells, *Am. J. Physiol. Cell Physiol.*, 301, pp. C31–C43.

30. Zoncu, R., Bar-Peled, L., Efeyan, A., Wang, S., Sancak, Y. and Sabatini, D.M. (2011) mTORC1 senses lysosomal amino acids through an inside-out mechanism that requires the vacuolar H^+-ATPase, *Science*, 334, pp. 678–683.

31. Paglin, S., Hollister, T., Delohery, T., Hackett, N., McMahill, M., Sphicas, E., Domingo, D. and Yahalom, J. (2001) A novel response of cancer cells to radiation involves autophagy and formation of acidic vesicles, *Cancer Res.*, 61, pp. 439–444.

32. Long, X., Crow, M.T., Sollott, S.J., O'Neill, L., Menees, D.S., de Lourdes Hipolito, M., Boluyt, M.O., Asai, T. and Lakatta, E.G. (1998) Enhanced expression of p53 and apoptosis induced by blockade of the vacuolar proton ATPase in cardiomyocytes, *J. Clin. Invest.*, 101, pp. 1453–1461.
33. Yoshimoto, Y. and Imoto, M. (2002) Induction of EGF-dependent apoptosis by vacuolar-type H^+-ATPase inhibitors in A431 cells overexpressing the EGF receptor, *Exp. Cell Res.*, 279, pp. 118–127.
34. Torigoe, T., Izumi, H., Ishiguchi, H., Uramoto, H., Murakami, T., Ise, T., Yoshida, Y., Tanabe, M., Nomoto, M., Itoh, H. and Kohno, K. (2002) Enhanced expression of the human vacuolar H^+-ATPase c subunit gene (ATP6L) in response to anticancer agents, *J. Biol. Chem.*, 277, pp. 36534–36543.
35. Matsuyama, S., Llopis, J., Deveraux, Q.L., Tsien, R.Y. and Reed, J.C. (2000) Changes in intramitochondrial and cytosolic pH: Early events that modulate caspase activation during apoptosis, *Nat. Cell Biol.*, 2, pp. 318–25.
36. Barry, M.A., Reynolds, J.E. and Eastman, A. (1993) Etoposide-induced apoptosis in human HL-60 cells is associated with intracellular acidification, *Cancer Res.*, 53, pp. 2349–2357.
37. Goossens, J.F., Hénichart, J.P., Dassonneville, L., Facompré, M. and Bailly, C. (2000) Relation between intracellular acidification and camptothecin-induced apoptosis in leukemia cells, *Eur. J. Pharm. Sci.*, 10, pp. 125–131.
38. Sethi, T., Rintoul, R.C., Moore, S.M., MacKinnon, A.C., Salter, D., Choo, C., Chilvers, E.R., Dransfield, I., Donnelly, S.C., Strieter, R. and Haslett, C. (1999) Extracellular matrix proteins protect small cell lung cancer cells against apoptosis: A mechanism for small cell lung cancer growth and drug resistance in vivo, *Nat. Med.*, 5, pp. 662–668.
39. Hinton, A., Sennoune, S.R., Bond, S., Fang, M., Reuveni, M., Sahagian, G.G., Jay, D., Martinez-Zaguilan, R., Forgac, M. (2009) Function of a subunit isoforms of the V-ATPase in pH homeostasis and in vitro invasion of MDA-MB231 human breast cancer cells, *J. Biol. Chem.*, 284, pp. 16400–16408.
40. Lu, X., Qin, W., Li, J., Tan, N., Pan, D., Zhang, H., Xie, L., Yao, G., Shu, H., Yao, M., Wan, D., Gu, J. and Yang, S. (2005) The growth and metastasis of human hepatocellular carcinoma xenografts are inhibited by small interfering RNA targeting to the subunit ATP6L of proton pump, *Cancer Res.*, 65, pp. 6843–6849.
41. You, H., Jin, J., Shu, H., Yu, B., De Milito, A., Lozupone, F., Deng, Y., Tang, N., Yao, G., Fais, S., Gu, J. and Qin, W. (2009) Small interfering

RNA targeting the subunit ATP6L of proton pump V-ATPase overcomes chemoresistance of breast cancer cells, *Cancer Lett.*, 280, pp. 110–119.

42. Nishisho, T., Hata, K., Nakanishi, M., Morita, Y., Sun-Wada, G.H., Wada, Y., Yasui, N. and Yoneda, T. (2011) The a3 isoform vacuolar type H^+-ATPase promotes distant metastasis in the mouse B16 melanoma cells, *Mol. Cancer Res.*, 9, pp. 845–855.
43. Bowman, E.J., Siebers, A. and Altendorf K. (1988) Bafilomycins: A class of inhibitors of membrane ATPases from microorganisms, animal cells, and plant cells, *Proc. Natl. Acad. Sci. USA*, 85, pp. 7972–796.
44. Huss. M. and Wieczorek, H. (2009) Inhibitors of V-ATPases: Old and new players, *J. Exp. Biol.*, 212, pp. 341–346.
45. Straud, S., Zubovych, I., De Brabander, J.K. and Roth, M.G. (2010) Inhibition of iron uptake is responsible for differential sensitivity to V-ATPase inhibitors in several cancer cell lines, *PLoS One*, 5, pp. e11629.
46. Sasazawa, Y., Futamura, Y., Tashiro, E. and Imoto, M. (2009) Vacuolar H^+-ATPase inhibitors overcome Bcl-xL-mediated chemoresistance through restoration of a caspase-independent apoptotic pathway, *Cancer Sci.*, 100, pp. 1460–1467.
47. Teplova, V.V., Tonshin, A.A., Grigoriev, P.A., Saris, N.E. and Salkinoja-Salonen, M.S. (2007) Bafilomycin A1 is a potassium ionophore that impairs mitochondrial functions, *J. Bioenerg. Biomembr.*, 39, pp. 321–329.
48. Shoemaker, R.H. (2006) The NCI60 human tumour cell line anticancer drug screen, *Nat. Rev. Cancer*, 6, pp. 813–823.
49. Boyd, M.R., Farina, C., Belfiore, P., Gagliardi, S., Kim, J.W., Hayakawa, Y., Beutler, J.A., McKee, T.C., Bowman, B.J. and Bowman, E.J. (2001) Discovery of a novel antitumor benzolactone enamide class that selectively inhibits mammalian vacuolar-type (H+)-ATPases, *J. Pharmacol. Exp. Ther.*, 297, pp. 114–120.
50. Xie, X.S., Padron, D., Liao, X., Wang, J., Roth, M.G. and De Brabander, J.K. (2004) Salicylihalamide A inhibits the V0 sector of the V-ATPase through a mechanism distinct from bafilomycin A1, *J. Biol. Chem.*, 279, pp. 19755–19763.
51. Pérez-Sayáns, M., García-García, A., Reboiras-López, M.D., Gándara-Vila, P. (2009) Role of V-ATPases in solid tumors: Importance of the subunit C (Review), *Int. J. Oncol.*, 34, pp. 1513–1520.
52. Menche, D., Hassfeld, J., Steinmetz, H., Huss, M., Wieczorek, H. and Sasse F. (2007) The first hydroxylated archazolid from the myxobacterium Cystobacter violaceus: Isolation, structural elucidation and V-ATPase inhibition, *J. Antibiot. (Tokyo)*, 60, pp. 328–331.

53. McHenry, P., Wang, W.L., Devitt, E., Kluesner, N., Davisson, V.J., McKee, E., Schweitzer, D., Helquist, P. and Tenniswood, M. (2010) Iejimalides A and B inhibit lysosomal vacuolar H^+-ATPase (V-ATPase) activity and induce S-phase arrest and apoptosis in MCF-7 cells, *J. Cell. Biochem.*, 109, pp. 634–642.

54. Niikura, K., Takeshita, N. and Chida, N. (2005) A novel inhibitor of vacuolar ATPase, FR202126, prevents alveolar bone destruction in experimental periodontitis in rats, *J. Toxicol. Sci.*, 30, pp. 297–304.

55. Supino, R., Petrangolini, G., Pratesi, G., Tortoreto, M., Favini, E., Bo, L.D., Casalini, P., Radaelli, E., Croce, A.C., Bottiroli, G., Misiano, P., Farina, C. and Zunino. F. (2008) Antimetastatic effect of a small-molecule vacuolar H^+-ATPase inhibitor in in vitro and in vivo preclinical studies, *J. Pharmacol. Exp. Ther.*, 324, pp. 15–22.

56. Stormo, G.D. and Zhao, Y. (2007) Putting numbers on the network connections, *Bioessays*, 29, pp. 717–721.

57. Nelson, K.M. and Weiss, G.J. (2008) MicroRNAs and cancer: Past, present, and potential future, *Mol. Cancer Ther.*, 7, pp. 3655–3660.

58. Sontheimer, E.J. and Carthew, R.W. (2005) Silence from within: Endogenous siRNAs and miRNAs, *Cell*, 122, pp. 9–12.

59. De Milito, A. and Fais, S. (2005) Tumor acidity, chemoresistance and proton pump inhibitors, *Future Oncol.*, 1, pp. 779–86.

60. Harguindey, S., Arranz, J.L., Wahl, M.L., Orive, G. and Reshkin, S.J. (2009) Proton transport inhibitors as potentially selective anticancer drugs, *Anticancer Res.*, 29, pp. 2127–2136.

Chapter 16

V-ATPase Maintains Neural Stem Cells in the Developing Mouse Cortex

Christian Lange,[a] Silvia Prenninger,[a] Philip Knuckles,[b] Verdon Taylor,[b,c] Michael Levin,[d] and Federico Calegari[a]

[a] *Vesalius Research Center, Katholieke Universiteit Leuven and the Flanders Institute for Biotechnology, Leuven, 3000, Belgium*
[b] *Department of Molecular Embryology, Max Planck Institute of Immunobiology, Stübeweg 51, 79108, Freiburg, Germany*
[c] *Centre for Stem Cell Biology, Department of Biomedical Science, University of Sheffield, Western Bank, Sheffield S10 2TN, UK*
[d] *Center for Regenerative and Developmental Biology, and Biology Department, Tufts University, Medford, MA 02155, USA*
federico.calegari@crt-dresden.de

The vacuolar H^+-ATPase (V-ATPase) is crucial for endosome acidification, endocytosis, and trafficking in essentially all eukaryotic cells. Recent studies have shown that inhibition of the V-ATPase also leads to downregulation of important signaling pathways, including Notch and Wnt, which are key regulators of cell differentiation and tissue homeostasis across the animal kingdom. However, the requirement of endosome acidification and endocytosis in the transduction of Notch signaling is still highly debated and only one study could extend the role of the V-ATPase for Notch signaling during mammalian development. Here we describe the original

Handbook of H^+-ATPases
Edited by Suguru Nakamura

ISBN 978-981-4411-91-2 (Hardcover), 978-981-4411-92-9 (eBook)
www.panstanford.com

experiments in which a dominant-negative subunit of the V-ATPase has been overexpressed in neural precursors of the developing mouse cortex, which led to depletion of neural stem cells by promoting their differentiation and the generation of neurons. In this context, inhibition of the V-ATPase reduced endogenous Notch signaling and prevented the proliferative effect of a transmembrane, γ-secretase-dependent, active Notch without blocking the effects of its cytoplasmic intracellular domain (NICD). These data are consistent with recent reports in *Drosophila* in which the V-ATPase has been suggested to be important for the transduction of Notch signaling. Extending these findings to mammalian embryos may contribute to a better understanding of the role of the V-ATPase, endosome acidification, and endocytosis in signal transduction during neural stem cell differentiation and brain development.

16.1 Introduction

The multimeric protein complex known as the vacuolar ATPase (V-ATPase) is an ancient and extremely well-conserved proton pump responsible for the acidification of intracellular compartments and the extracellular space, which are important for endocytosis and vesicle transport in essentially all eukaryotic cells, from yeast to man (Forgac, 2007; Harvey and Wieczorek, 1997; Marshansky and Futai, 2008).

More recently, three laboratories have independently shown that inhibition of the V-ATPase also affects the transduction of important signaling molecules, such as Notch (Vaccari et al., 2010; Yan et al., 2009) and Wnt (Cruciat et al., 2010), that are key determinants in controlling proliferation, tissue patterning, embryonic development, and organ homeostasis throughout the animal kingdom (Fortini, 2009; Kopan and Ilagan, 2009; Tien et al., 2009; van Amerongen and Nusse, 2009). While it is still unclear to which extent the effects of the V-ATPase on signaling are due to its direct function on endosome acidification as opposed to indirect effects on endocytosis and trafficking (Niehrs and Boutros, 2010), these pioneering studies (Cruciat et al., 2010; Vaccari et al., 2010; Yan et al., 2009) indicate that this particular proton pump is a key player in signal transduction of cell fate determinants.

Interestingly, endosome acidification and endocytosis were suggested to control Notch signaling even before manipulations of the V-ATPase were reported. This is because the γ-secretase, whose activity is essential for Notch processing, has been shown to have a higher proteolytic activity in acidic compartments (Pasternak et al., 2003) and because proteins involved in endocytosis, including dynamin and Rab5, revealed to be essential for Notch activation (Gupta-Rossi et al., 2004; Seugnet et al., 1997; Vaccari et al., 2010). On the other hand, these data may also suggest that ligand presentation by the signal-presenting cell, rather than signal-transduction in the receiving cell, may be influenced by endocytosis (Fortini and Bilder, 2009; Kopan and Ilagan, 2009). The intrinsic difficulties in distinguishing between cell autonomous *versus* extrinsic effects, and reports showing that Notch signaling may not require endocytosis (Shaye and Greenwald, 2002; Sorensen and Conner, 2010; Struhl and Adachi, 2000), have led to a long debate in the field.

Although the interplay between the V-ATPase and Notch signaling has been shown in *Drosophila*, evidence for such a link in other organisms, remarkably mammals, was still completely lacking. Considering that both the V-ATPase and Notch are well conserved across phyla, we decided to inhibit the V-ATPase in neural precursors of developing mouse embryos by tissue-specific expression of the YCHE78 peptide, a well-characterized dominant negative, truncated version of the V1e1 subunit of the V-ATPase complex (Lu et al., 2002).

YCHE78 has been successfully used to analyze the role of the proton pump in left-right asymmetry (Adams et al., 2006), appendage regeneration (Adams et al., 2007), planar cell polarity, and Wnt signaling (Cruciat et al., 2010) of non-mammalian vertebrates. In this context, YCHE78 expression is preferred to competing technologies such as RNAi because (i) it directly interferes with the existing proteins and is independent from their turnover kinetics that may delay a loss of function, (ii) it is effective despite compensation by multiple isoforms of V-ATPase subunits while multiple shRNAs would have to be combined raising serious concerns about off-targets effect, and (iii) it does not require the dicer-mediated dsRNA degradation machinery that may lead to confounding effects during development. Finally, YCHE78 expression is superior to pharmacological approaches that are difficult to perform *in vivo*, in particular while studying mammalian embryonic development.

It should also be noted that the moderate levels of YCHE78 afforded by transfection *in vivo* do not induce cell death. As shown in previous studies in *Xenopus* tadpoles (Adams et al., 2006; Adams et al., 2007; Cruciat et al., 2010), a reduction in V-ATPase activity is quite compatible with life and normal development. This is not surprising, given the many studies in which subtle patterning and regulatory functions of ion flows can be functionally manipulated and dissected separately from their housekeeping roles (Blackiston et al., 2009; Levin, 2009), which is particularly true in stem cells (Sundelacruz et al., 2008).

Thus, we decided to investigate the role of the V-ATPase, and its possible effects on Notch signaling, during mammalian development by expressing YCHE78 in neural stem cells using *in utero* electroporation, a technique widely used to acutely and tissue-specifically manipulate genes in neural stem cells of the mouse brain (LoTurco et al., 2009). This approach consists of injecting DNA plasmids into the lumen of the embryonic telencephalon followed by the delivery of electric pulses through platinum electrodes (De Pietri Tonelli et al., 2006; Lange et al., 2009). After surgery, gestation can be allowed to continue and brains collected at any desired subsequent time. Cell autonomous effects on proliferation *versus* differentiation can then be investigated in targeted cells of the developing neocortex, one of the best characterized Notch-dependent model systems in mammals (Kageyama et al., 2009; Miyata et al., 2010; Yoon and Gaiano, 2005).

Specifically, during early development of the mouse cerebral cortex, neuroepithelial and radial glial cells undergo proliferative divisions at the apical boundary of the ventricular zone (VZ), hence the name apical progenitors (AP). With the onset of differentiation, AP increasingly switch to self-renewing divisions that generate neurons or, most frequently, committed neurogenic progenitors that migrate towards the basal boundary of the VZ, hence the name basal progenitors (BP), to form the sub-ventricular zone (SVZ) (Gotz and Huttner, 2005; Miyata et al., 2010). Since Notch maintains the identity of AP while limiting the generation of BP and neurons (Kageyama et al., 2009; Miyata et al., 2010; Mizutani et al., 2007), we investigated the effects of YCHE78 expression on the generation of AP, BP, and neurons during brain development by performing immunohistochemistry for specific progenitor markers on fixed brain cryosections.

16.2 YCHE78 Expression Promotes Neurogenesis

We transfected mice at embryonic day (E) 12.5 by in utero co-electroporation with either control or YCHE78 expression vectors together with a plasmid encoding nuclear-localized RFP (RFP^{nls}) to label and reliably count targeted cells and their progeny. One day later, brains were collected and sectioned through the electroporated portion of the lateral cortex. Immunohistochemistry for the BP marker Tbr2 and/or the neuronal marker Tbr1 (Englund et al., 2005) was used to identify the VZ, SVZ, and intermediate zone (IZ) and to quantify the proportion of AP, BP, and neurons within the population of RFP^{nls}+ cells. Specifically, within the VZ AP were identified by the absence of Tbr2 immunoreactivity, which was due to difficulties in detecting Tbr2 and RFP in combination with the AP marker Pax6. This approach was legitimate despite the fact that neurons downregulate Tbr2 (Englund et al., 2005) because at this stage of development most neurons are generated by BP in the SVZ (Haubensak et al., 2004; Kowalczyk et al., 2009), thus representing a negligible proportion of all cells within the VZ (<3%) (Attardo et al., 2008; Calegari et al., 2005).

In controls we found that the majority of RFP^{nls}+ cells were Tbr2–AP (50.0 ± 5.2) and Tbr2+ newborn BP still residing in the VZ (22.6 ± 5.3%) while a minority were Tbr2+ BP in the SVZ (15.6 ± 2.2%) and neurons in the IZ (11.7 ± 2.9%) (Fig. 16.1A, C). In contrast, expression of YCHE78 induced a 50% decrease in the proportion of AP (27.4±5.1%) and newborn BP in the VZ (10.0 ± 2.8%) with a twofold increase in the proportion of Tbr2+ BP in the SVZ (30.5 ± 4.7%) and neurons in the IZ (32.2 ± 3.8%) (Fig. 16.1B, C). Importantly, upon expression of YCHE78 the vast majority of RFP^{nls}+ cells in the IZ were Tbr1+ (data not shown) indicating

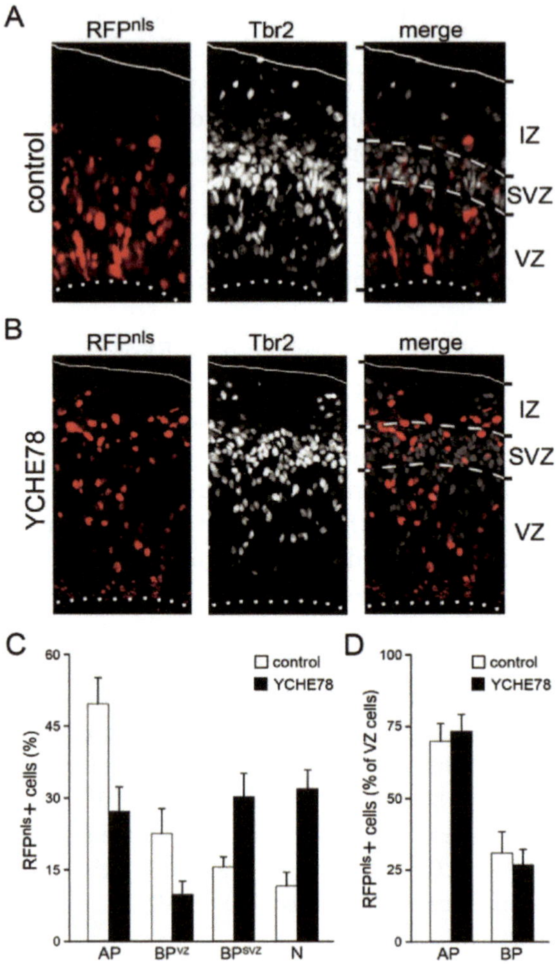

Figure 16.1 YCHE78 induces differentiation of AP. (A and B) Epifluorescence pictures of the mouse cortex 24 hours after electroporation with control (A) or YCHE78 (B) plasmids at E12.5 and immunohistochemistry for RFP (left) and Tbr2 (middle). Apical and pial boundaries of cortex and SVZ were identified by DAPI counterstaining (not shown) and Tbr2 labeling and indicated by dotted, continuous, or dashed lines, respectively. Note the increase in the proportion of RFP^{nls}+ cells in the IZ after YCHE78 expression (B) indicating an increase in neurogenesis. (C and D) Quantification of AP, BP in the VZ (BP^{VZ}), BP in the SVZ (BP^{SVZ}), and neurons (N) as a percentage of all RFP^{nls}+ cells (C) or only those within the VZ (D). Bars = SD; $p < 0.01$ for all cell populations in C; differences in D are not significant.

an increase in neurogenesis rather than just delamination of neural progenitors. No significant difference was observed with regard to the efficiency of electroporation (*ca.* 15–30% of RFP^{nls}+ cells over DAPI) in the two conditions (data not shown).

Moreover, while the proportion of RFP^{nls}+ cells in the VZ was reduced upon expression of YCHE78, the proportion of AP and newborn BP within the VZ was similar. In fact, when Tbr2+ BP in the VZ were expressed as a proportion of all RFP^{nls}+ cells in the VZ, a similar value was found in mice electroporated with control (31.1 ± 7.1%) or YCHE78 (26.7 ± 5.7%) plasmids (Fig. 16.1D). Since BP in the VZ are a transient population of migrating cells, this suggests that the phenotype induced by YCHE78 is transitory, which may be explained by the reduction in plasmid levels upon cell division and the dose-dependent effect of YCHE78 on the transduction of signaling pathways, as already reported for Wnt (Cruciat et al., 2010).

Consistent with previous results during *Xenopus* development (Adams et al., 2006; Adams et al., 2007; Cruciat et al., 2010), YCHE78 expression did not seem to have any effect neither on cell survival nor on cell migration as judged by immunostaining for cleaved caspase-3, absence of pyknotic nuclei counterstained with DAPI, and distribution of Tbr2+ or Tbr1+ nuclei in the different cortical layers, respectively (data not shown).

In essence, these data are reminiscent of the effect induced by inhibition of Notch signaling leading to an increased differentiation of AP and generation of BP and neurons (Mizutani et al., 2007; Yoon et al., 2008) and are consistent with a role of the V-ATPase in Notch signaling recently reported in flies (Vaccari et al., 2010; Yan et al., 2009).

16.3 YCHE78 Expression Inhibits Endogenous Notch Signaling

To directly investigate the effect of YCHE78 on endogenous Notch signaling we used a Notch reporter plasmid in which GFP is under the transcriptional control of the Hes5 regulatory elements (*Hes5*^{GFP}) (Basak and Taylor, 2007). *Hes5*^{GFP} was co-electroporated at E12.5, with or without YCHE78, together with a cytoplasmic RFP, which allowed us to identify somata of transfected cells. Particular attention was given to inject identical amounts of a cytoplasmic RFP and *Hes5*^{GFP} expression vectors; thus, in controls, the fluorescent reporter plasmids were supplemented with an equimolar amount of empty vectors to compensate for the lack of YCHE78. Brains were collected 16 hours later to reduce the chance of saturating the levels of GFP fluorescence at longer survival times and pictures of the individual channels taken on cryosections.

In control transfected embryos, ratio of GFP intensity relative to RFP fluorescence was variable (Fig. 16.2A, C) ranging from almost 0 (i.e., no detectable GFP) to up to 1.6 (Fig. 16.2E; white bar), likely reflecting the different levels of Notch signaling in individual VZ cells as previously described (Basak and Taylor, 2007; Kageyama et al., 2009; Mizutani et al., 2007). Moreover, GFP and RFP fluorescence intensities showed a strong and significant positive correlation (Spearman's rank correlation coefficient (r_S) = 0.71; $p < 0.001$) (Fig. 16.2C) with a median GFP/RFP ratio of 0.38 (Fig. 16.2E; white) and 33% of cells displaying values above 0.5. In contrast, when RFP and *Hes5*^{GFP} were electroporated together with YCHE78 we found a substantial overall decrease in GFP fluorescence, loss of GFP/RFP positive correlation ($r_S = 0.24$, not significantly different from $r_S = 0$ but significantly different from $r_S = 0.71$ of control, $p < 0.01$) (Fig. 16.2B, D), and a reduction by 50% in the median GFP/RFP ratio to 0.16 (Fig. 16.2E; black; $p< 0.0001$). Moreover, less than 10% of cells displayed values above 0.5.

Thus, expression of YCHE78 autonomously inhibits endogenous Notch signaling, which may, at least in part, explain the depletion of AP and increased generation of BP and neurons previously described (Fig. 16.1).

16.4 YCHE78 Expression Rescues the Phenotype Induced by Activated Notch but Not of NICD

To further characterize the effect of YCHE78 on regulation of Notch activity and neural stem cell differentiation, we expressed, with or without YCHE78, two different activators of the Notch pathway,

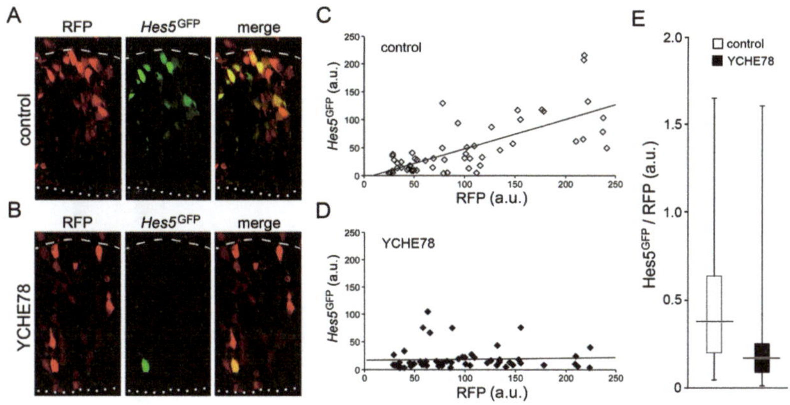

Figure 16.2 YCHE78 autonomously inhibits Notch signaling. (A and B) Confocal fluorescence pictures of the mouse cortex 16 hours after co-electroporation of *Hes5*^{GFP} with control (A) or YCHE78 (B) plasmids at E12.5 showing endogenous RFP (left) or GFP (middle) fluorescence from a confocal 0.8 μm thick optical section. Apical and basal boundaries of the VZ were identified by cell morphology and nuclear orientations after DAPI counterstaining (not shown) and indicated by dotted or dashed lines, respectively. Note the co-localization of RFP and GFP (merge; right) in control (A) but the almost complete absence of GFP in YCHE78 targeted cells (B) indicating a strong inhibition of endogenous Notch activity upon YCHE78 expression. (C and D) Plots showing the correlation between GFP and RFP fluorescence intensity (from 0 to a max of 250 a.u.) after quantification of 60, randomly chosen, RFP+ somata. Lines indicate the best-fit of the recorded values outlining a strong positive correlation of controls ($r_S = 0.71$) (C) but not YCHE78 ($r_S = 0.24$) (D) electroporated brains. (E) Box plot showing the distribution of GFP/RFP fluorescence ratio of cells shown in C and D in four quartiles. Note the decreased median (horizontal lines) in YCHE78 (black) as compared to control (white) transfected cells; U-test $p < 0.0001$.

that is, the plasma membrane, γ-secretase dependent, constitutively active Notch (ca-Notch) or its cytoplasmic, γ-secretase independent, active domain NICD. We reasoned that if the V-ATPase should be important for processing of Notch to NICD, co-expression of YCHE78 together with ca-Notch might partly rescue the phenotype induced by expression of ca-Notch alone. Conversely, co-expression of YCHE78 together with NICD may have no effect on the phenotype induced by expression of NICD alone.

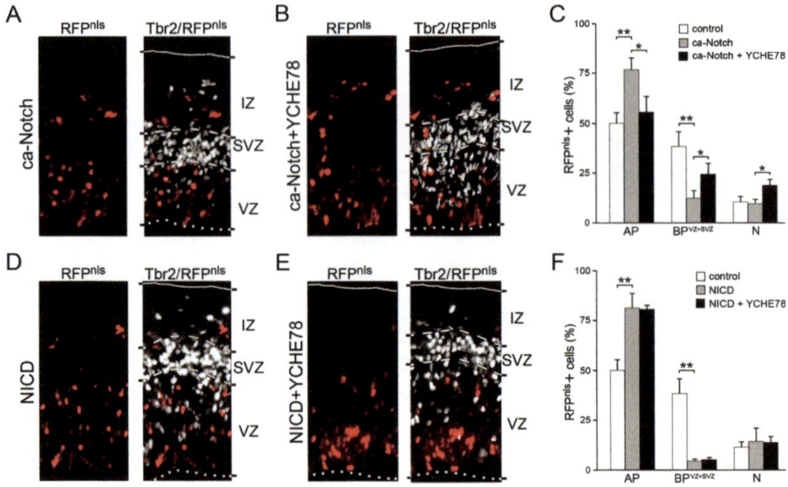

Figure 16.3 YCHE78 rescues the ca-Notch, but not NICD, induced effects. (A, B, D, and E) Epifluorescence pictures of the mouse cortex 24 hours after co-electroporation at E12.5 with ca-Notch (A and B) or NICD (D and E) with (A and D) or without (B and E) YCHE78 and immunohistochemistry for RFP (left) and Tbr2 (right; shown as merge). Apical, pial, and SVZ boundaries were identified by DAPI counterstaining (not shown) and Tbr2 immunoreactivity and identified by dotted, continuous, or dashed lines, respectively. (C and F) Quantification (%) of RFP^{nls}+ AP, BP in the VZ and SVZ (BP^{VZ+SVZ}), and neurons (N). Note the rescue of the ca-Noch (C), but not NICD (F), phenotype upon expression of YCHE78 (gray *vs.* black bars) suggesting that the V-ATPase is essential for Notch signaling downstream of ca-Notch but upstream of NICD. Values for control (white bars) are taken from Fig. 16.1C and added for better comparison. Bars = SD; * = $p < 0.05$; ** = $p < 0.005$.

One day after electroporation at E12.5, both ca-Notch and NICD expression (Fig. 16.3A, D) resulted in a similar increase in the proportion of RFP^{nls}+ AP in the VZ (76.6 ± 6.0% and 81.1 ± 7.5 for ca-Notch and NICD, respectively) as compared to control brains electroporated only with RFP^{nls} plasmids (50.0 ± 5.2%), which was concomitant to a decrease in Tbr2+ BP in the VZ and SVZ (sum in the two areas = 12.0 ± 4.2% and 4.3 ± 0.8% for ca-Notch and NICD, respectively, as compared to 38.3 ± 7.4% for control) (Fig. 16.3C, F; gray bars). Therefore, in agreement with previous reports, Notch activation inhibits the generation of the main neurogenic progenitor type, which over the long term has been shown to cause a reduction in neuronal output (Kato et al., 2010; Mizutani et al., 2007).

Compared to ca-Notch alone, co-electroporation of YCHE78 plus ca-Notch reduced the proportion of $RFP^{nls}+$ AP (55.6 $\pm$ 7.7%) and increased that of BP in the VZ and SVZ (24.2 $\pm$ 5.7%) to levels that were comparable to control (Fig. 16.3B, C). In contrast, when YCHE78 was co-electroporated together with NICD the proportion of $RFP^{nls}+$ AP (80.6 $\pm$ 2.2%) and BP in the VZ and SVZ (5.0 $\pm$ 1.2%) were undistinguishable from that observed after electroporation of NICD, or ca-Notch, alone (Fig. 16.3E, F).

Thus, the phenotype caused by activation of Notch signaling by ca-Notch was largely rescued by co-expression of YCHE78 while the NICD effect was not. This in turn suggests that the effect of the V-ATPase on neural stem cell differentiation is downstream of ligand activation of Notch but upstream of its γ-secretase-dependent cleavage and production of NICD.

16.5 Discussion

By these experiments we found that expression of YCHE78 during mouse brain development (i) triggers the differentiation of neural precursors, (ii) autonomously inhibits endogenous Notch signaling, and (iii) rescues the proliferative effect of ca-Notch, but not that of NICD, expression (Lange et al., 2011).

The fact that YCHE78 inhibited both endogenous and ectopic Notch signaling, and its effect on neurogenesis was dominant over ca-Notch, seems to suggest that, at least in our conditions, V-ATPase-independent pathways of Notch processing play a minor role in physiological Notch signaling in the embryonic mammalian cortex. Moreover, the effect of YCHE78 on other signaling molecules still need to be investigated in mammalian neural stem cells, as recently done in Xenopus for Wnt (Cruciat et al., 2010).

Beyond endosome acidification and endocytosis the V-ATPase regulates H^+ flows at the level of the plasma membrane, thereby contributing to the tissue bioelectrical gradients (Adams et al., 2007). While the bioelectrical gradients *per se* can influence cell differentiation and regeneration (Funk and Monsees, 2006; Levin,

2007), in our conditions only a minor fraction of all cells in the targeted area express YCHE78 making it unlikely that this would induce a major change in the electric potential throughout the whole neuroepithelium. Moreover, our data seem inconsistent with a γ-secretase cleavage at the inner surface of the plasma membrane because the V-ATPase pumps protons out of the cytoplasm, which would be incompatible with the optimum γ-secretase activity in an acidic milieu (Pasternak et al., 2003).

While other studies have concluded that endocytosis and endosomes are not involved in Notch signaling (Shaye and Greenwald, 2002; Sorensen and Conner, 2010; Struhl and Adachi, 2000), our data showing a reduction in both the endogenous and ectopic Notch signaling upon YCHE78 expression seem more consistent with the opposite view, which would extend recent works in flies (Seugnet et al., 1997; Vaccari et al., 2010; Yan et al., 2009) to the developing mammalian brain (Fig. 16.4). While validation of this hypothesis would still require further investigation, we hope that our observation may contribute to a better understanding of this highly debated field. Moreover, our finding that inhibition of the V-ATPase can influence the proliferation of mammalian neural stem cells and Notch signaling opens up the possibility of using V-ATPase inhibitors (Perez-Sayans et al., 2009) for therapeutic intervention in Notch-dependent brain tumors, which are thought to originate from cancer stem cells (Ischenko et al., 2008; Yang and Wechsler-Reya, 2007).

Finally, it is interesting to note that membrane potential and cellular pH can influence cell cycle progression and differentiation (Blackiston et al., 2009; Levin, 2007) and that inhibition of Notch has been shown to lengthen the cell cycle, in particular G1, of neural precursors (Borghese et al., 2010), which is probably due to its effects on cyclinD1 expression (Das et al., 2010). The fact that lengthening of G1 may alone cause the differentiation of embryonic and adult neural (Artegiani et al., 2011; Lange et al., 2009; Salomoni and Calegari, 2010), and likely also other (Lange and Calegari, 2010), stem cells reveals an elegant interplay between the cell biophysical state, signaling molecules, cell cycle progression, and differentiation during animal development, which should be better explored in future studies.

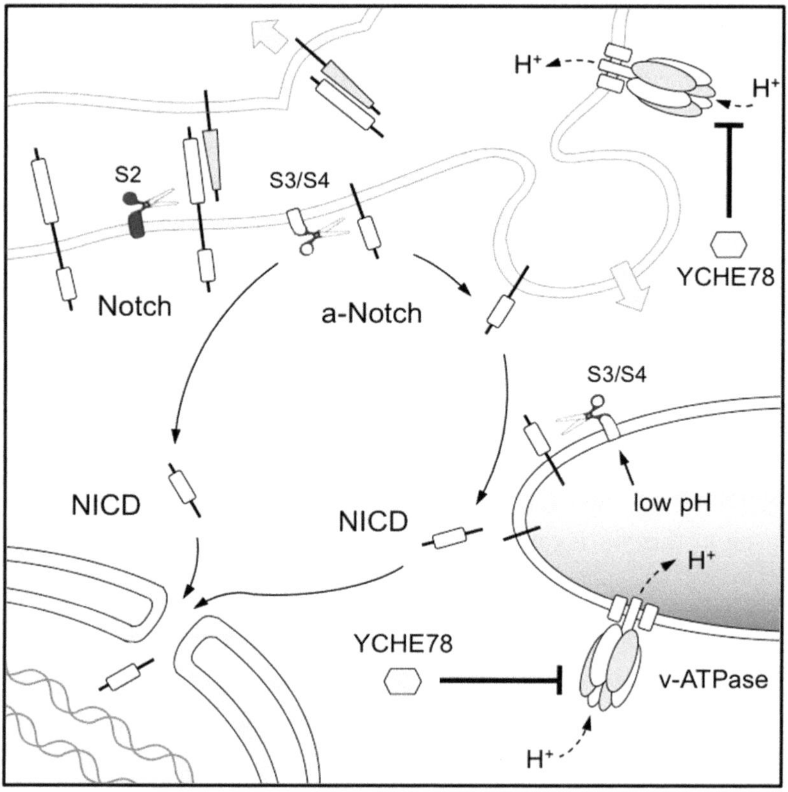

Figure 16.4 Notch pathway in a signal-receiving cell. Notch activation occurs at the plasma membrane upon ligand-mediated S2 cleavage (black scissor). Further processing of active Notch (a-Notch) occurs upon S3/S4 cleavage by γ-secretase (white scissors) resulting in the generation of NICD. While γ-secretase cleavage at the level of the plasma membrane (left route) has been originally proposed, recent works (Vaccari et al., 2010; Yan et al., 2009) suggest that an alternative route through endosomes (right route) is also possible. Our data suggest that in the mouse cortex the latter is the prominent route. Note the inhibition of the V-ATPase by YCHE78 (hexagon, bottom, right) in endosomes (bottom; right) and at the plasma membrane (top; right), which is known to affect endosome acidification and endocytosis, respectively.

Acknowledgments

We are very grateful to Dr. Masato Nakafuku and Dr. Hideyuki Okano for providing essential material to perform this study and to Änne Glass for helping us with statistical analyses. Animal experiments were approved by local authorities (249168.11-9-2007-1/2).

Authors' Disclosure Statement

This is an edited version of the original research article published in *Stem Cells and Development* by Lange et al. (2011; 20: 843–850). Reprint permission has been obtained by the corresponding author from the publisher Mary Ann Liebert, Inc., with the consent of all the other authors involved. Additional information about the methods used in this work can be found in the original publication (Lange et al., 2011).

References

Adams, D.S., Masi, A., and Levin, M. (2007). H^+ pump-dependent changes in membrane voltage are an early mechanism necessary and sufficient to induce Xenopus tail regeneration. *Development 134*, 1323–1335.

Adams, D.S., Robinson, K.R., Fukumoto, T., Yuan, S., Albertson, R.C., Yelick, P., Kuo, L., McSweeney, M., and Levin, M. (2006). Early, H^+-V-ATPase-dependent proton flux is necessary for consistent left-right patterning of non-mammalian vertebrates. *Development 133*, 1657–1671.

Artegiani, B., Lindemann, D., and Calegari, F. (2011). Overexpression of cdk4 and cyclinD1 triggers greater expansion of neural stem cells in the adult mouse brain. *J Exp Med 208*, 937–948.

Attardo, A., Calegari, F., Haubensak, W., Wilsch-Brauninger, M., and Huttner, W.B. (2008). Live imaging at the onset of cortical neurogenesis reveals differential appearance of the neuronal phenotype in apical versus basal progenitor progeny. *PLoS ONE 3*, e2388.

Basak, O., and Taylor, V. (2007). Identification of self-replicating multipotent progenitors in the embryonic nervous system by high Notch activity and Hes5 expression. *Eur J Neurosci 25*, 1006–1022.

Blackiston, D.J., McLaughlin, K.A., and Levin, M. (2009). Bioelectric controls of cell proliferation: ion channels, membrane voltage and the cell cycle. *Cell Cycle 8*, 3519–3528.

Borghese, L., Dolezalova, D., Opitz, T., Haupt, S., Leinhaas, A., Steinfarz, B., Koch, P., Edenhofer, F., Hampl, A., and Brustle, O. (2010). Inhibition of notch signaling in human embryonic stem cell-derived neural stem cells delays G1/S phase transition and accelerates neuronal differentiation in vitro and in vivo. *Stem Cells 28*, 955–964.

Calegari, F., Haubensak, W., Haffner, C., and Huttner, W.B. (2005). Selective lengthening of the cell cycle in the neurogenic subpopulation of neural progenitor cells during mouse brain development. *J Neurosci 25*, 6533–6538.

Cruciat, C.M., Ohkawara, B., Acebron, S.P., Karaulanov, E., Reinhard, C., Ingelfinger, D., Boutros, M., and Niehrs, C. (2010). Requirement of prorenin receptor and vacuolar H^+-ATPase-mediated acidification for Wnt signaling. *Science 327*, 459–463.

Das, D., Lanner, F., Main, H., Andersson, E.R., Bergmann, O., Sahlgren, C., Heldring, N., Hermanson, O., Hansson, E.M., and Lendahl, U. (2010). Notch induces cyclin-D1-dependent proliferation during a specific temporal window of neural differentiation in ES cells. *Dev Biol 348*, 153–166.

De Pietri Tonelli, D., Calegari, F., Fei, J.F., Nomura, T., Osumi, N., Heisenberg, C.P., and Huttner, W.B. (2006). Single-cell detection of microRNAs in developing vertebrate embryos after acute administration of a dual-fluorescence reporter/sensor plasmid. *Biotechniques 41*, 727–732.

Englund, C., Fink, A., Lau, C., Pham, D., Daza, R.A., Bulfone, A., Kowalczyk, T., and Hevner, R.F. (2005). Pax6, Tbr2, and Tbr1 are expressed sequentially by radial glia, intermediate progenitor cells, and postmitotic neurons in developing neocortex. *J Neurosci 25*, 247–251.

Forgac, M. (2007). Vacuolar ATPases: rotary proton pumps in physiology and pathophysiology. *Nat Rev Mol Cell Biol 8*, 917–929.

Fortini, M.E. (2009). Notch signaling: the core pathway and its posttranslational regulation. *Dev Cell 16*, 633–647.

Fortini, M.E., and Bilder, D. (2009). Endocytic regulation of Notch signaling. *Curr Opin Genet Dev 19*, 323–328.

Funk, R.H., and Monsees, T.K. (2006). Effects of electromagnetic fields on cells: physiological and therapeutical approaches and molecular mechanisms of interaction. A review. *Cells Tissues Organs 182*, 59–78.

Gotz, M., and Huttner, W.B. (2005). The cell biology of neurogenesis. *Nat Rev Mol Cell Biol 6*, 777–788.

Gupta-Rossi, N., Six, E., LeBail, O., Logeat, F., Chastagner, P., Olry, A., Israel, A., and Brou, C. (2004). Monoubiquitination and endocytosis direct gamma-secretase cleavage of activated Notch receptor. *J Cell Biol 166*, 73–83.

Harvey, W.R., and Wieczorek, H. (1997). Animal plasma membrane energization by chemiosmotic H^+ V-ATPases. *J Exp Biol 200*, 203–216.

Haubensak, W., Attardo, A., Denk, W., and Huttner, W.B. (2004). Neurons arise in the basal neuroepithelium of the early mammalian telencephalon: a major site of neurogenesis. *Proc Natl Acad Sci USA 101*, 3196–3201.

Ischenko, I., Seeliger, H., Schaffer, M., Jauch, K.W., and Bruns, C.J. (2008). Cancer stem cells: how can we target them? *Curr Med Chem 15*, 3171–3184.

Kageyama, R., Ohtsuka, T., Shimojo, H., and Imayoshi, I. (2009). Dynamic regulation of Notch signaling in neural progenitor cells. *Curr Opin Cell Biol 21*, 733–740.

Kato, T.M., Kawaguchi, A., Kosodo, Y., Niwa, H., and Matsuzaki, F. (2010). Lunatic fringe potentiates Notch signaling in the developing brain. *Mol Cell Neurosci 45*, 12–25.

Kopan, R., and Ilagan, M.X. (2009). The canonical Notch signaling pathway: unfolding the activation mechanism. *Cell 137*, 216–233.

Kowalczyk, T., Pontious, A., Englund, C., Daza, R.A., Bedogni, F., Hodge, R., Attardo, A., Bell, C., Huttner, W.B., and Hevner, R.F. (2009). Intermediate neuronal progenitors (basal progenitors) produce pyramidal-projection neurons for all layers of cerebral cortex. *Cereb Cortex 19*, 2439–2450.

Lange, C., and Calegari, F. (2010). Cdks and cyclins link G(1) length and differentiation of embryonic, neural and hematopoietic stem cells. *Cell Cycle 9*, 1893–1900.

Lange, C., Huttner, W.B., and Calegari, F. (2009). Cdk4/cyclinD1 overexpression in neural stem cells shortens G1, delays neurogenesis, and promotes the generation and expansion of basal progenitors. *Cell Stem Cell 5*, 320–331.

Lange, C., Prenninger, S., Knuckles, P., Taylor, V., Levin, M., and Calegari, F. (2011). The H(+) vacuolar ATPase maintains neural stem cells in the developing mouse cortex. *Stem Cells Dev 20*, 843–850.

Levin, M. (2007). Large-scale biophysics: ion flows and regeneration. *Trends Cell Biol 17*, 261–270.

Levin, M. (2009). Bioelectric mechanisms in regeneration: unique aspects and future perspectives. *Semin Cell Dev Biol 20*, 543–556.

LoTurco, J., Manent, J.B., and Sidiqi, F. (2009). New and improved tools for in utero electroporation studies of developing cerebral cortex. *Cereb Cortex 19*(Suppl. 1), i120–i125.

Lu, M., Vergara, S., Zhang, L., Holliday, L.S., Aris, J., and Gluck, S.L. (2002). The amino-terminal domain of the E subunit of vacuolar H(+)-ATPase (V-ATPase) interacts with the H subunit and is required for V-ATPase function. *J Biol Chem 277*, 38409–38415.

Marshansky, V., and Futai, M. (2008). The V-type H^+-ATPase in vesicular trafficking: targeting, regulation and function. *Curr Opin Cell Biol 20*, 415–426.

Miyata, T., Kawaguchi, D., Kawaguchi, A., and Gotoh, Y. (2010). Mechanisms that regulate the number of neurons during mouse neocortical development. *Curr Opin Neurobiol 20*, 22–28.

Mizutani, K., Yoon, K., Dang, L., Tokunaga, A., and Gaiano, N. (2007). Differential Notch signalling distinguishes neural stem cells from intermediate progenitors. *Nature 449*, 351–355.

Niehrs, C., and Boutros, M. (2010). Trafficking, acidification, and growth factor signaling. *Sci Signal 10*, pe26.

Pasternak, S.H., Bagshaw, R.D., Guiral, M., Zhang, S., Ackerley, C.A., Pak, B.J., Callahan, J.W., and Mahuran, D.J. (2003). Presenilin-1, nicastrin, amyloid precursor protein, and gamma-secretase activity are co-localized in the lysosomal membrane. *J Biol Chem 278*, 26687–26694.

Perez-Sayans, M., Somoza-Martin, J.M., Barros-Angueira, F., Rey, J.M., and Garcia-Garcia, A. (2009). V-ATPase inhibitors and implication in cancer treatment. *Cancer Treat Rev 35*, 707–713.

Salomoni, P., and Calegari, F. (2010). Cell cycle control of mammalian neural stem cells: putting a speed limit on G1. *Trends Cell Biol 5*, 332–342.

Seugnet, L., Simpson, P., and Haenlin, M. (1997). Requirement for dynamin during Notch signaling in Drosophila neurogenesis. *Dev Biol 192*, 585–598.

Shaye, D.D., and Greenwald, I. (2002). Endocytosis-mediated downregulation of LIN-12/Notch upon Ras activation in Caenorhabditis elegans. *Nature 420*, 686–690.

Sorensen, E.B., and Conner, S.D. (2010). Gamma-secretase-dependent cleavage initiates notch signaling from the plasma membrane. *Traffic 11*, 1234–1245.

Struhl, G., and Adachi, A. (2000). Requirements for presenilin-dependent cleavage of notch and other transmembrane proteins. *Mol Cell 6*, 625–636.

Sundelacruz, S., Levin, M., and Kaplan, D.L. (2008). Membrane potential controls adipogenic and osteogenic differentiation of mesenchymal stem cells. *PLoS One 3*, e3737.

Tien, A.C., Rajan, A., and Bellen, H.J. (2009). A Notch updated. *J Cell Biol 184*, 621–629.

Vaccari, T., Duchi, S., Cortese, K., Tacchetti, C., and Bilder, D. (2010). The vacuolar ATPase is required for physiological as well as pathological activation of the Notch receptor. *Development 137*, 1825–1832.

van Amerongen, R., and Nusse, R. (2009). Towards an integrated view of Wnt signaling in development. *Development 136*, 3205–3214.

Yan, Y., Denef, N., and Schupbach, T. (2009). The vacuolar proton pump, V-ATPase, is required for notch signaling and endosomal trafficking in Drosophila. *Dev Cell 17*, 387–402.

Yang, Z.J., and Wechsler-Reya, R.J. (2007). Hit 'em where they live: targeting the cancer stem cell niche. *Cancer Cell 11*, 3–5.

Yoon, K., and Gaiano, N. (2005). Notch signaling in the mammalian central nervous system: insights from mouse mutants. *Nat Neurosci 8*, 709–715.

Yoon, K.J., Koo, B.K., Im, S.K., Jeong, H.W., Ghim, J., Kwon, M.C., Moon, J.S., Miyata, T., and Kong, Y.Y. (2008). Mind bomb 1-expressing intermediate progenitors generate notch signaling to maintain radial glial cells. *Neuron 58*, 519–531.

Chapter 17

The Relationship between Glucose-Induced Calcium Signaling and Activation of the Plasma Membrane H^+-ATPase in *Saccharomyces cerevisiae* Cells

Rogelio Lopes Brandão
Núcleo de Pesquisas em Ciências Biológicas, Universidade Federal de Ouro Preto, LBCM-NUPEB-ICEB II, Ouro Preto, Minas Gerais 35.400-000, Brazil
rlbrand@nupeb.ufop.br

This chapter deals with several aspects related to the plasma membrane H^+-ATPases from lower eukaryotic cells, focusing on the historical findings, structural details, different mechanisms involved in regulation of activity, and the most recent data that points out for the existence of a signal transduction pathway in which the glucose-induced activation of the enzyme is dependent of the calcium availability in the cytosol.

Handbook of H^+-ATPases
Edited by Suguru Nakamura

ISBN 978-981-4411-91-2 (Hardcover), 978-981-4411-92-9 (eBook)
www.panstanford.com

17.1 Introduction

A remarkable review published by André Goffeau and Carolyn Slayman in 1981 [1] brings, just at the beginning, a very interesting Louis Pasteur's statement:

> A question naturally asked after Lavoisier is which it would be the cause of the acidity observed during the alcoholic fermentation ...Different experiments allow me to say that only succinic acid would be responsible for such acidity... Everyone will understand the consequences of these results. But, I must be more reserved than anyone in this discussion. [Louis Pasteur, *Comptes Rendus Hebdomadaires des Séances de l'Académie des Sciences* (1858), Vol. 47, p. 226]

Nowadays, one believes that the most correct answer for this phenomenon is the presence of a plasma membrane H^+-ATPase that, by pumping protons outside the cells, helps to create an electrochemical gradient across the plasma membrane, also contributing to maintain the intracellular pH in physiological ranges. In spite of the fact that this idea only began to be strengthened from data observed in the late 1960s, it is historically important to mention some observations withdrawn from experiments performed since last 150 years!

It was already known since the beginning of 1920s that alkalinization of the external medium has a strong effect on fermentation [2]. On the other hand, it was originally observed by Conway and co-workers, between 1940 and 1950, that during glucose fermentation by yeast cells there is a net efflux of protons from the cells leading to an acidification of the extracellular medium that, according to the conditions used, would be followed in parallel by an influx of K^+. Moreover, it was also observed that, depending on the external pH values, there would be a slight alkalinization of the cytosol [3–5]. In 1953, Rothstein and Denis demonstrated that an increase in the rate of the glycolysis correlates with the increase in the ph of the medium [6].

From these reports there were many attempts which try to explain the exact mechanism by which protons, and not only organic acids, are ejected outside the cells. As summarized by Goffeau and Slayman [1], three main hypothesis were suggested:

(i) Redox pump: Ejection of protons associated to the presence of organic receptors that would accept electrons like in the functioning of a cytochrome. It was admitted that, depending of the experimental conditions, there could be an influx of K^+ or an efflux of anions. However, the absence of such structures in the plasma membrane impaired the progress of such idea;
(ii) Electroneutral carrier: Exchange of extracellular K^+ and intracellular H^+ and Na^+. The main problem with this model is the fact that there is no explanation, nor on the coupling of energy to cation movements, neither with the ejection of protons observed in the absence of extracellular K^+;
(iii) ATPase pump: Net extrusion of protons outside the cells.

Two important sets of experiments were fundamental for the solidification of this last alternative. The first one performed by Cliford Slyman and co-workers by using micropipette electrodes and taking advantage of the large diameter of *Neurospora crassa* cells. This approach allowed them to observe the existence of a large membrane potential (−200 mM, negative inside) that was sensitive to metabolic inhibitors, mainly those that affect respiration (CN, azide, CO, DNP, etc.). The idea emerging from these data was that an electrogenic ion pump, coupled to the metabolism, could separate charges across the plasma membrane. Soon it was realized that protons would be the best candidates to explain the strong depolarization observed and that ATP would be the most probable source of metabolic energy. This conclusion was drawn from the strong correlation observed between a drop in the level of ATP and the parallel decline of the membrane potential [7–12].

In a parallel way, Antonio Peña and co-workers performed the second set of experiments that was important to our understanding on the presence of proton pumps strictly depending of ATP [13]. By working with yeast cells, they observed that, in the presence of glucose, the internal levels of ATP decrease when the external pH is alkalinized or when K^+ is added to external medium at low pH. By combining these results with those published before by Conway and co-workers they postulated the existence of a H^+ pump driven by ATP and inhibited by low pH, unless if, in this last condition, K^+ was added as well [13–15].

Naturally, as a consequence of the publication of Mitchell's chemiosmotic theory, many other reports on the existence of proton movements across different kinds of biological membranes (bacteria and plasma and/or organelle-membranes in eukaryotic cells) helped to strength the idea about the existence of a plasma membrane H^+-ATPase in lower eukaryotic cells.

More evidence pointing out for the existence and physiological importance of this system to the uptake of different nutrients was obtained in further experiments performed with *Saccharomyces cerevisiae* by Eddy and Nowacki [16] and by Peña [2]; in *Schizosaccharomyces pombe* by Foury and Goffeau [17]; in *Rodotorula gracilis* by Misra and Höffer [18]; and in *Neurospora crassa* by Clifford and Carolym Slayman [19]. From these studies published in the second half of the last 1970s, it was possible to reinforce the idea of the presence of a primary electrogenic H^+ pump, most probably with ATPase activity. In fact, this conclusion was facilitated by the already proposed idea that an ATPase activity existed in the plasma membrane of yeast cells [20]. By combining all the above studies with others initiated by Matile and co-workers, it became clear that in yeast cells there was an oligomycin-insensitive ATPase activity presenting a characteristic low optimum pH, since then used as a selective marker for the plasma membrane [21].

17.2 General Characteristics of Plasma Membrane H^+-ATPases

In historical point of view and by the end of 1970s, we had gone through more than 120 years since Pasteur's declaration on extracellular acidification, observed in suspensions of yeast cultures! From this moment, it soon became clear the existence of many plasma membrane H^+-ATPases in different kinds of organisms like algae, fungi, and plants. Additionally, and based in DNA sequence homology, other H^+-ATPases seem to exist in archaebacteria and protozoa [22–24].

In fact, it is known that the plasma membrane ATPase belongs to the so-called P-type cations-translocating ATPase family that transports cations against their electrochemical gradient using ATP as energy source. This family still includes K^+, Cd^{2+}, and Cu^{2+}-

ATPases from bacteria; Ca^{2+}-ATPases from animal, fungus, and plant cells; and Na^+–K^+- and H^+–K^+-ATPases from animal cells. Moreover, as all P-type ATPases, it forms an aspartylphosphate intermediate during its catalytic cycle being specifically inhibited by very low concentrations of vanadate ions. Regarding other ATPase inhibitors, it is incentive to oligomycin, azyde, and dicyclohexyl-carbodiimide. Different approaches have suggested a topological structure presenting the N- and C-terminal cytoplasmic tails and two large hydrophilic domains facing the cytoplasm, as well as 10 transmembrane spans [24].

The plasma membrane H^+-ATPase from yeast cells presents a pH optimal around 5.6 and a very high-affinity for ATP ($K_m = 0.1$ mM) being also very specific to this compound. The enzyme requires Mg^{++} for its activity and is not stimulated by K^+ or bicarbonate [25]. Two genes have been described, encoding for this enzyme: *PMA1*, highly expressed and *PMA2* that presents high homology with *PMA1* but is expressed at very low levels and still with unknown physiological function [26–28].

In plant and fungal cells, the result of its function is the creation of an electrochemical proton gradient that drives the uptake of nutrients, also contributing to control the intracellular pH. Due to its physiological relevance, the enzyme is regulated at two levels: transcriptional and post-translational. Several factors are involved in these processes. Nevertheless, the high level of the *PMA1* gene expression [29], as well as the fact that the plasma membrane H^+-ATPase is a major membrane protein, corresponding to 7–10% of the plasma membrane proteins [30] contribute to keep a constant level of this enzyme during the cellular growth. Additionally, considering that its functioning consumes up to 40% of cellular ATP [31] and that its half-life is long [32], it is relatively logical to consider that the regulatory mechanisms controlling its activity should occur mainly at a post-translational level.

Even before the Ramon Serrano's classical paper had been published [33] showing that the addition of glucose leads to an activation of the plasma membrane H^+-ATPase (accompanied by changes in the optimal pH, reduction of the K_m regarding ATP, and increase in the V_{max} for ATP hydrolysis) there was a very nice work done by McDonough and Mahler [34], who first found evidences for the phosphorylation of this enzyme by a protein kinase that would consist almost exclusively on serine residues. Together, these papers introduced the idea that the addition of glucose would lead to the activation of the plasma membrane H^+-ATPase probably through a phosphorylation mechanism controlled by an unknown signal transduction pathway. Since then many papers were published not only with different propositions on such pathway but also on different environmental factors able to provoke the post-translational activation of the plasma membrane H^+-ATPase. Before describing the glucose-induced activation of this enzyme in detail, let's briefly discuss the factors that can affect its activity. One important factor is the extracellular acidification that leads to a decrease in the intracellular pH, which, somehow, triggers the activation of the ATPase. This process was studied by at least two different groups [35, 36], but differently from the glucose induction, the acid pH-induced activation does not lead to an increase in the (V_{max}). Besides, different kinds of environmental stresses such as presence of phorbol esters, ethanol, organic acids, deprivation of nitrogen source, and higher temperatures also provoke the stimulation of the plasma membrane H^+-ATPase in yeast cells [37–42]. However, except for the acid pH, the studies of the effects of these factors on the regulation of the ATPase were not further investigated.

Interestingly, at the same period, it was observed that the plasma membrane H^+-ATPase is also activated *in vitro* by some plasma membrane phospholipids such as phosphatidylinositol phosphate, phosphatidylinositol bisphosphate, and phosphoinositol sphingolipids. On the other hand, the pre-incubation of plasma membranes with neomycin, a positively charged aminoglycoside blocks the effects observed by the addition of the plasma membrane phospholipids [43–45].

Additionally, during post-translational regulation, glucose also triggers a transcriptional regulation of the plasma membrane H^+-ATPase. Three different research groups gave an enormous contribution to the understanding of such mechanism. The phenomenon was first described by André Goffeau's group, which demonstrated the presence of two upstream activation sequences (UAS) implicated in the *PMA1* gene regulation. These regions would be recognized by a transcription factor (*TUF1*/*RAP1*/*GRF1*) that is a multifunctional DNA-binding protein indeed involved in the control of the expression of different genes [29]. In principle, these findings were in a clear contradiction with both the idea of a glucose-induced post-translational activation of the ATPase as well as with the lack of previous evidences for the constitutive behavior of the *PMA1* promoter. Nevertheless, in a paper from Carolyn Slayman's laboratory, it was demonstrated that removal of glucose from the growth medium results in a clear reduction in the expression of the *PMA1* gene. The authors used purified preparations of secretory vesicles representing a pool of *de novo* synthesized ATPase *en route* to the plasma membrane. A correlation was observed between the decay of the *PMA1* mRNA and the turn-off of ATPase synthesis [46]. Finally, the Francisco Portillo's group cloned and characterized a new gene (*APA1*) that would also be involved in the transcriptional regulation of the plasma membrane H^+-ATPase [47].

17.3 The Signal Transduction Pathway Involved in the Post-Translational Regulation of the Plasma Membrane H^+-ATPases in Yeast Cells

Classically, a signal transduction pathway must present at least three components: a signal, a receptor (generally present in plasma membrane), and a target (an enzyme, a transcriptional factor, etc.). In most sophisticated pathways, an intracellular amplificatory cascade of protein kinases, whose activity is modulated by secondary messengers, is also present and finally acts to modulate the activity of proteins, the expression of genes and/or changes in the cytoskeleton. With this general architectural concept in mind, let's review the main findings that led us to conclude that the post-translational glucose-induced activation of the plasma membrane H^+-ATPase is a typical signal transduction pathway, in this case in close relationship with calcium metabolism.

The first evidences that this enzyme could suffer modifications in response to changes in the carbon and energy provided in the medium came from the work done by Henry P. Mahler and co-workers [48, 49]. In fact, in a third paper, this group confirmed that the plasma membrane H^+-ATPase from yeast cells should phosphorylated, most probably by a protein kinase [34]. Curiously, in a classical paper that unequivocally demonstrated that addition of glucose triggers an *in vivo* activation of the yeast's plasma membrane ATPase, no changes of phosphate content on membrane proteins upon addition of glucose were observed [33]. However, it was the work of Chang and Slayman [50] that definitely demonstrated that the phosphorylation of this enzyme is associated with increase in its activity during growth on glucose, thus adjusting the ATPase activity in response to nutritional signals.

Therefore, from the ideas presented in these papers, one started to imagine a signal transduction regulatory mechanism in which the addition of glucose would trigger an internal signal that, by its turn, would be responsible for the activation of a protein kinase, leading to the activation of the plasma membrane H^+-ATPase in yeast cells. Interestingly, in early 1970s, it had already been demonstrated that addition of glucose to yeast cells increases the intracellular concentration of cyclic AMP (cAMP). This fact would be related to the activation of a protein kinase responsible for the phosphorylation of different enzymes, like trehalase and fructosebiphosphatase, provoking changes in their kinetic properties [51].

Moreover, at the same period, it was also demonstrated that the uptake of different nutrients could be stimulated by the presence of cAMP in the yeast *Schizosaccharomyces pombe* [17]. These data suggested that extracellular cAMP could activate the transport of nutrients through a possible cAMP-dependent activation of the plasma membrane H^+-ATPase. In the following decade, at least two papers reinforced the idea that cAMP-dependent protein kinase A–dependent mechanism could be involved in the glucose-induced activation of the plasma membrane H^+-ATPase. Portillo and Mazón [52] and Ulaszewski *et al.* [53], working with yeast mutants affected in cAMP metabolism, suggested its possible involvement in the regulation of the ATPase. In fact, in a paper demonstrating the presence of a similar plasma membrane H^+-ATPase in cells of *Fusarium oxysporum*, we have found evidences suggesting that the glucose-induced activation of this enzyme was indeed related to the cAMP metabolism [54]. This idea persisted until the development of the yeast genome project, which allowed us to understand, in detail, the entire pathway related to the glucose-induced cAMP–PKA activation in yeast cells. In this regard, the Johan Thevelien's group from Belgium gave a significant contribution. Thus, soon it was found that the G protein Gpa2 and the GTP–GDP exchange factor CDC25 were both required for the glucose-induced increase in the intracellular cAMP levels. Furthermore, it was also demonstrated that sugar phosphorylation was required although no further glucose metabolism would be necessary for a cAMP signal development [55–58].

Thus, by using different yeast mutants, presenting specific deletions in genes encoding for many components of this emerging cAMP–PKA transduction pathway, it was conclusively demonstrated that the post-translational glucose-induced activation of the plasma membrane H^+-ATPase was clearly independent of the cAMP metabolism [59, 60]. Nevertheless, it was also shown that glucose phosphorylation but not its further metabolization was also a pre-requisite for the glucose-induced activation of the H^+-ATPase, suggesting that sugar phosphorylation is indeed an important intracellular signal to trigger different regulatory mechanisms [59].

At that point, an obvious question was put in discussion: if the glucose-induced cAMP signaling was not involved in the activation of the H^+-ATPase, which would be the mechanism responsible for the control of its activity? In other words, if the protein kinase A is not the enzyme that phosphorylates the H^+-ATPase which protein kinase, and its corresponding signaling transduction pathway, would be responsible for the activation of the plasma membrane H^+-ATPase?

One of the first hypotheses was the involvement of a casein kinase that co-purifies with the plasma membrane H^+-ATPase *in vitro* [61, 62], but there were no further evidences suggesting that this kinase could be involved in the glucose-induced regulatory process. Interestingly, even before that hypothesis had been completely discarded, a new one arose, concerning the involvement of a phosphatidylinositol-like signaling transduction pathway in the regulation of the H^+-ATPase. It was shown that phorbol esters were able to activate this enzyme *in vivo* in a very similar way as triggered by glucose (with increase in affinity for ATP; the optimum pH is shifted to an alkaline range; higher incorporation of ^{32}Pi in the plasma membrane H^+-ATPase) [39]. The authors were inspired by the fact that the phorbol esters act in different systems as activators of protein kinase C and, by doing this, they cause the activation of Na^{+}–H^{+}-ATPases.

In the same connection, Brandão *et al.* [63] by using activators and inhibitors of the enzymes phospholipase C and protein kinase C (respectively encoded by *PLC1* and *PKC1* genes) showed that glucose-induced activation of the H^{+}-ATPase seemed to depend on a similar mechanism to the "turnover" of the mammalian phosphatidylinositol. In fact, it had already been demonstrated that the addition of glucose to yeast cells stimulates the phosphatidylinositol phosphate (PI) turnover [64]. In another study, Cocetti *et al.* using yeast mutants presenting deletion in the *PLC1* gene confirmed the requirement of phospholipase C in the glucose-induced phosphatidylinositol turnover and activation of the plasma membrane H^{+}-ATPase [65]. With these data, we gathered evidence pointing to the action of signaling transduction pathway, probably involving the metabolism of phosphatidylinositol in the activation of this enzyme.

Considering that the role of signaling molecule would be exercised by glucose, we decided and try to identify the glucose receptor that would be responsible for detecting glucose and transmitting this signal to the cells. In yeasts, three receptors seem to fulfill this role: Gpr1, Rgt2p, and Snf3p. The Gpr1 receptor is also associated with Gpa2p, which in turn binds to the enzyme phospholipase C in yeasts [66], supporting the idea of the presence of a trimeric complex [67]. By their turn, Snf3p and Rgt2p act in the generation of intracellular signals involved in transcription of genes encoding hexose transporters (HXTs) [68, 69]. Snf3p seems to function as a sensor for low glucose levels and is therefore, involved in the expression of high-affinity glucose transporter. On the other hand, Rgt2p seems to act as a sensor for high glucose levels, being involved in the control of *HXT1* (encoding the low-affinity glucose transporter) gene expression.

In a previous paper published, the only receptor that showed a real involvement with the activation of ATPase was Snf3p [70]. This study showed evidences for the need of transport and phosphorylation of glucose as a prerequisite for the activation of the enzyme, and pointed a role to the G protein Gpa2, probably by detecting the presence of phosphorylated glucose and, somehow, through this signal, activating phospholipase C. The idea of the requirement of Snf3 was contested by Kotyk *et al.* [71] where the authors suggested that indeed Rgt2 would be the sensor involved in the regulation of the plasma membrane H^+-ATPase. Very recently, and by using the EUROSCARF yeast collection, it was observed that the Gpr1 receptor seems to be also required of this regulatory process (Eduardo Perovano, unpublished data). Therefore, regarding the direct involvement of glucose sensors, some contradiction appears when all available data is put together; nevertheless, and based in other studies, we will focus our attention on the role played by Snf3 as discussed below.

From this point, it became evident that the two processes, the glucose-induced activation of the ATPase and the intracellular calcium signaling, shared many common elements. Parallel studies showed that phosphorylation of sugar and participation of the enzyme phospholipase C (Plc1) and a G protein (Gpa2) were essential for calcium signaling [72–74]. In fact, this "coincidence" led us to postulate that calcium signaling and activation of the ATPase were connected events.

Interestingly, it was also found that both the increase in the concentration of IP_3 and Ca^{2+} signaling are more pronounced in mutants with deletions in the gene encoding the polyphosphate multikinase, Arg82 [75]. Consistently, it was also showed that a mutant having a deletion in the gene *ARG82* presents a stronger activation of the ATPase enzyme upon addition of glucose. The same study also confirmed the requirement of Snf3p as an essential element to calcium signaling induced by glucose [76].

Therefore, the data accumulated in the second half of the last decade suggests that, along with the glucose transporters, also the glucose sensor Snf3p, the phospholipase C, PLC1p, the G protein, Gpa2, and an still unknown protein kinase, are components of a signaling pathway involved in the regulation of the H^+-ATPase [59, 63, 65, 70, 76]. The calcium metabolism would be the integrative event, since the activation of H^+-ATPase and the absorption of calcium and/or its internal mobilization are stimulated by the presence of glucose.

Other results demonstrate that changes in availability of calcium in the external medium affect the level of ATPase activation. However, the data also suggests that the uptake of calcium by various transport systems do not seem to be necessary for the activation of plasma membrane ATPase [76]. This apparent contradiction might be explained in different ways: first, a common feature of many experiments on the calcium uptake and/or signaling in yeasts is the fact that the external calcium concentrations are always in the milimolar range. However, in our experiments the activation of plasma membrane ATPase and detection of intracellular signals induced by calcium sugar were always performed in media containing very low concentrations of calcium (micromolar range). Second, in our experiments, yeast cells were grown in rich medium, which is a condition that causes calcineurin-mediated inhibition of the influx of calcium through the high-affinity Ca^{2+} channel, Cch1 and its subunit Mid1 [77]. Thirdly, we found that even in strains presenting mutations in the genes *MID1* and *FIG1* both the glucose-induced calcium (GIC) signaling and H^{+}-ATPase are comparable to the corresponding wild type strains [78]. Interestingly, it was also suggested the existence of an alternative mechanism that would bind the detection level of Ca^{2+}, Ca^{2+} uptake into cells, homeostasis, and signaling in yeast cells [79]. Obviously, it is still possible to speculate that the real channel has not yet been identified or a combination of different calcium carriers may be involved. Indeed, very recent data suggests the existence of another high-affinity system that was called GIC, a transporter that presents, as main characteristics, the resistance to nickel ions and verapamil. This new transporter would be the major responsible for calcium influx upon addition of glucose in rich medium [80].

Moreover, there were plenty of evidence suggesting that calcium homeostasis in yeasts would also be associated with the relative levels of glucose-6-phosphate and glucose-1-P [81]. In fact, higher level of calcium signaling is observed when 100 mM glucose is added to wild-type cells grown on glucose. However, the addition of 100 mM galactose produces no increase in cytosolic calcium level, probably because the transcription of genes encoding proteins involved in the metabolism of galactose is suppressed under these conditions. But for cells grown in galactose, the addition of glucose or galactose triggers calcium signals of lower intensity. In conclusion, the authors suggest that there is a strong correlation between the levels of these metabolites and the magnitude of sugar-induced calcium signaling.

Using strains with different levels of phosphoglucomutase enzymatic activity [81] demonstrated that there is a relationship between the intracellular level of glucose-1-P and cytosolic calcium signaling [82]. When yeast strains presenting mutations in *PGM2* gene (encoding a major isoform of phosphoglucomutase) are grown in media containing glucose as carbon source, the intracellular levels of glucose-6-phosphate and glucose-1-P are comparable to those found in the isogenic wild-type. However, when these cells are grown in medium containing galactose, there is an increase in eight times of glucose-1-P and a corresponding rate of vacuolar accumulation of calcium five times higher than in corresponding wild-type strain [81].

Moreover, while a partial GIC signaling can be observed on a *pgm2*Δ mutant strain cultivated on glucose, when those cells are grown in galactose, the re-addition of glucose or galactose leads to a quite reduced calcium signal. Therefore, the significant reduction in the phosphoglucomutase activity creates a "metabolic bottleneck", suggesting that increase in intracellular level of glucose-1-P stimulates the accumulation of calcium in intracellular compartments. Thus, this mutant is defective in calcium homeostasis and in the calcium signaling, both dependent on the carbon source in which the cells have been grown. Consistent with the results above, it has been shown that deletion of the *PMC1* gene (that encodes for a Ca^{2+}-ATPase responsible for the accumulation of calcium inside the vacuole) in a strain presenting a *pgm2*Δ mutation restores calcium homeostasis at levels comparable with the wild-type strain, suggesting that the accumulation of intracellular levels of glucose-1-P stimulates the Pmc1p activity [83].

Interestingly, the results obtained with the *pgm2*Δ mutant, regarding the lower intensity of the H^+-ATPase activation, correlate with the level of increase in the cytosolic calcium as described previously. Moreover, from experiments on the regulation of the H^+-ATPase activity using the *pgm2*Δ*pmc1*Δ double mutant there was a clear recovery of the activation pattern of the enzyme. In summary, these data seem to confirm that *Pmc1p* plays a crucial role in controlling the availability of intracellular calcium, whose activity influences the magnitude of activation of H^+-ATPase induced by sugars [76]. These results, combined with others obtained before, also indicate that sugar when phosphorylated can act as intracellular signals for H^+-ATPase activation and calcium mobilization [70, 74]. But which would be the intracellular detectors, able to perceive the phosphorylation state of sugars? Interestingly, a possible involvement of Snf3p sensor for the detection of intracellular metabolites, such as glucose-6-phosphate, has been suggested [84].

Therefore, our results on the participation of the Snf3p receptor, both in the glucose-induced activation of H^+-ATPase and the calcium signaling, lead us to propose a role for Snf3p (probably through its C-terminal tail) by which it is able to detect the intracellular levels of glucose-6-P and/or glucose-1-P, responsible for triggering an appropriate calcium signaling which, in turn, leads to activation of the ATPase [76].

However, the precise mechanism by which the Snf3p is involved in the control of the intracellular availability of calcium is unknown. Two results led us to speculate on how Snf3p participates in this signaling pathway. First, we have observed that in *snf3*Δ mutants grown in galactose there was a very high level of total cellular calcium accumulation, similar to that described in the mutant *pgm2*Δ [82]. Secondly, in the double mutant *snf3*Δ*pmc1*Δ there was also a clear recovery of the glucose-induced activation of the H^+-ATPase plasma membrane, as observed in the strain *pgm2*Δ*pmc1*Δ. These facts lead us to imagine that, somehow, after the detection of the levels of phosphorylated sugars, the Snf3p would transduce the signal by producing an inhibition of the vacuolar Ca^{2+}-ATPase, *Pmc1p* [76].

Despite the strong experimental evidence supporting the relationship between changes in intracellular calcium levels and activation of H^+-ATPase, there are some aspects that need to be elucidated before validating the existence of a complete signaling pathway. The first one would be related to the role played by IP_3: if it is indeed involved in calcium signaling (and the activation of H^+-ATPase), which would be the intracellular receptor able to detect its presence and the corresponding organelle where the receptor would be present, thus allowing the modulation of the calcium levels in the cytoplasm? On the other hand, which would be the protein kinases(s), sensitive to changes in intracellular calcium involved in the phosphorylation and activation of the H^+-ATPase?

Recently, it was demonstrated that the vacuolar calcium channel, Yvc1p (encoded by the gene *YVC1*) appears to be involved in the GIC signaling working in connection with IP_3 [78]. Evidence for this mechanism comes from experiments on GIC signaling and H^+-ATPase activation in a wild-type strain pre-incubated with aminoetoxidifenil 2-borate (2-APB), a known IP_3-receptor modulator in mammalian cells [85]. By using 2-APB, a clear reduction of both processes, the glucose-induced intracellular calcium signaling and the activation of the H^+-ATPase, were observed. Furthermore, in the strains presenting a single deletion in the gene *YVC1* or a double deletion in the genes *YVC1* and *ARG82*, there is a clear reduction in the GIC signaling and H^+-ATPase activation [78]. Nevertheless, and considering that previous results have shown that IP_3 is not directly involved with the Yvc1p calcium channel [86], it is possible that other components in the system could mediate the action of IP_3 on calcium mobilization from the vacuole.

On the other hand, and to complement the structure of this signaling pathway in yeasts, it remains to be identified the protein kinase that phosphorylates the H^+-ATPase. Since the work published by McDonough and Mahler suggesting that the activity of proton-translocating ATPase of fungal plasma membrane is subject to control by phosphorylation by a protein kinase [34], many studies have been performed to successfully identify the kinase involved in the post-translational activation of the H^+-ATPasein glucose-fermenting cells. In fact, it was also conclusively shown that this enzyme is phosphorylated on multiple Ser and Thr residues *in vivo* and the phosphorylation also occurs during the transport of newly synthesized H^+-ATPase from the endoplasmic reticulum to the cell surface [50].

A current and accepted model for the post-translational regulation of the H^+-ATPase plasma membrane suggests the occurrence of an inhibitory interaction of the C-terminus tail with the active site of the enzyme, limiting the access of ATP to its binding site.

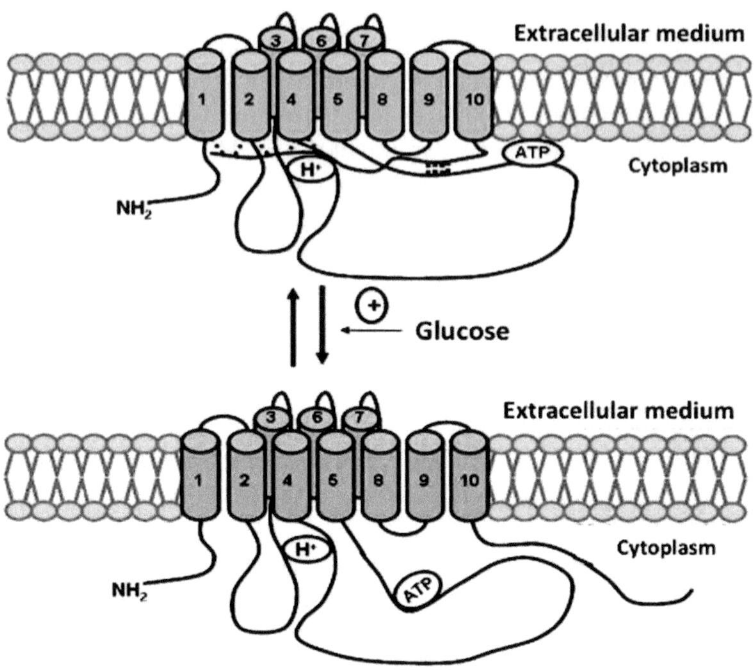

Figure 17.1 Working model for the activation of the yeast plasma membrane H^+-ATPase. Upon addition of glucose the enzyme changes its conformation after the releasing of the inhibitory interaction of the C-terminus tail with the ATP-binding domain.

After the addition of glucose, and having as a consequence the phosphorylation of specific residues, a conformational modification in the protein C-terminus would be triggered, releasing the active site and consequently activating the enzyme [24, 87–90]. By site-directed mutagenesis, it was shown that these conformational changes result from the phosphorylation of two different residues present in the C-terminus domain of this enzyme. Mutations affecting Ser-899 or Thr-912 abolish the glucose-dependent K_m or V_{max} changes, respectively [91]. Figure 17.1 illustrates this proposed model by which a conformational modification occurs in the structure of the H^+-ATPase upon addition of glucose.

However, a careful reading of the literature reveals that the identification of the phosphorylation sites involved in the activation of the plasma membrane ATPase is still a matter of controversy. In fact, the phosphorylating site Thr-912 seems to be of great importance in the regulatory process [92]; however, by using nanoflow high-pressure liquid chromatography (nHPLC) coupled with electrospray electron transfer dissociation (ETD) tandem mass spectrometry, the same group demonstrated that there is a stepwise phosphorylation of the two tandemly positioned residues Ser-911 and Thr-912. Interestingly, using this approach, the residue Ser-899 was not identified as a site of phosphorylation. This has been attributed to difficulties to detect phosphopeptides less abundant and/or with lower ionization efficiency. To add more complexity to this aspect, the data indicated that Thr-912 is a constitutive site of phosphorylation and the adjacent Ser-911 is the main target of phosphorylation, induced by the presence of glucose [93].

Indeed, considering that in the genome of *Saccharomyces cerevisiae* there are at least 127 protein kinases [94] and that the H^+-ATPase is phosphorylated at multiple sites [50], it is easy to understand the level of difficulty to correctly identify the protein kinase(s) involved in the regulatory phosphorylation of H^+-ATPase, even after more than 40 years since the first hypothesis on the existence of this enzyme [13].

Based on classical papers published by Spanish groups [33, 89, 90, 95], it has been suggested that Ptk2p would be one protein kinase involved in the phosphorylation of the residue Ser-899, triggering a reduction of the K_m value for ATP [96]. On the other hand, the *YOR137c* gene product would be involved in the glucose-induced increase of the V_{max} of the plasma membrane H^+-ATPase [97]; nevertheless, this information has not been confirmed by further and more conclusive experiments. The residue Thr-912, supposedly related to the modification in the V_{max}, has been identified as part of a classical recognition motif for both Ca^{2+}/calmodulin-dependent protein kinase II [98] and casein kinase II [93]; but again there are no evidences suggesting that these kinases could be involved in the glucose-induced regulatory process, except for the fact that the casein kinase was co-purified with the plasma membrane H^+-ATPase *in vitro* [61, 62].

In the context of the identification of a phosphatidylinositol-type signaling pathway, it was suggested that protein kinase C would be involved in the control of the H^+-ATPase activity [70] and in other physiological functions [93, 100]. However, in spite of the fact that Pkc1p possesses binding sites for diacylglycerol (DAG) and the so-called calcium-dependent C2 domain, there are evidences demonstrating that it is not activated by those signals [101].

Very recently, we have performed a detailed search in order to try to identify, among the 127 protein kinases found in yeast cells, those that could indeed be connected to the regulation of the plasma membrane H^+-ATPase. The idea was to find mutants that presented normal cellular growth but with an affected glucose-induced activation of the H^+-ATPase, by using the quantitative measurement of the proton pumping capacity, as we have used consistently in our previous works [71, 77, 79, 102]. This approach was combined with the measurement of the GIC signaling and subsequently with the determination of the kinetic parameters in each strain to determine how the absence of each kinase affects K_m and V_{max} values in glucose fermenting cells compared to glucose-starved ones.

Indeed, our results confirmed that the *ptk2*Δ mutant is the only strain where a typical change in the affinity for ATP upon glucose addition to the medium is clearly absent (Renata Pereira, personal communication), reinforcing the idea of the protein kinase Ptk2 involvement in the phosphorylation of the residue Ser-899 [90, 96, 98]. Ptk2p belongs to a family of serine/threonine protein kinases dedicated to the activation of plasma membrane transporters [98].

Interestingly, in the absence of another protein of this family, Hrk1, there is no increase in the affinity for neither ATP nor increased V_{max} for ATP hydrolysis, in an apparent contradiction with one previous work [98]. Additionally, the mutant *rim15*Δ (a serine/threonine protein kinase involved in signal transduction during cell proliferation in response to nutrients) presented a typical reduction in the K_m without increase in the V_{max} for ATP hydrolysis when cells were exposed to glucose. Furthermore, the strain deleted in the gene encoding for Pkp1 (a serine/threonine protein kinase) also did not show a normal increase of the V_{max} for ATP hydrolysis in glucose-fermenting cells. Therefore, the protein kinases Hrk1, Pkp1, and Rim15 seem be responsible and/or at least involved in the mechanism by which the increase of V_{max} by phosphorylation of the second site formed by Ser-911 and/or Thr-912 residues occurs (unplublished results). This result contradicts the observations suggesting that Thr-912, by the constitution of its recognition motif, would be a target of the Ca^{2+}/calmodulin-dependent protein kinase II [98] or casein kinase II [92]. It also has been claimed that Ser-911 could be recognized by a kinase belonging to the AGC group [92]. Nevertheless, our most recent data suggests that those kinases are not connected to the activation of the H^+-ATPase (Renata Pereira, unpublished results). Therefore, combination with the ideas expressed by [92, 93], the most obvious direction for the identification and/or confirmation of the identity of the protein kinase(s) responsible for the phosphorylation of the H^+-ATPase would be the utilization of more sophisticated techniques, like the Nhplc coupled with electrospray ETD tandem mass spectrometry, but now using mutants lacking the genes that encode for the protein kinases probably involved in the control of the activity of the H^+-ATPase (Ptk2, Hrk1, Pkp1, and Rim15). By this way, we could proceed with the identification of the residues present in the plasma membrane H^+-ATPase, whose glucose-induced phosphorylation is affected.

17.4 Conclusions and Perspectives

In conclusion, and based on all updated information available, we have proposed [78] the existence of a pathway with two branches by which the addition of glucose controls calcium availability in the cytosol having, as a consequence, the activation of the H^+-ATPase (Fig. 17.2). In the first branch, the transport of glucose and its phosphorylation by different sugar kinases generate a signal (probably the sugar phosphates themselves) that would activate the G protein Gpa2 that, by its turn, would stimulate the phospholipase C activity. Then, PIP_2 would be hydrolyzed generating DAG and inositol tri-phosphate (IP_3), the later interacting directly or indirectly with Yvc1, thus regulating the intensity of calcium signaling in the cytosol. In the second branch, the focus would be the control of the Pmc1 Ca^{2+}-ATPase activity; the detection of the levels of sugar phosphate, probably through the C-terminus tails of the glucose sensor Snf3,

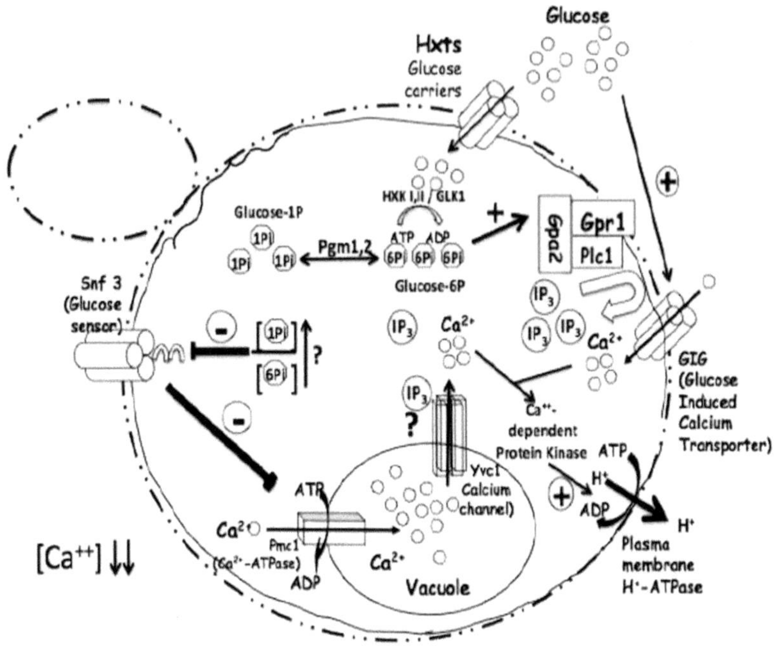

Figure 17.2 Modulation of intracellular calcium levels and the cytoplasmic membrane H^+-ATPase in yeasts.

would, in some way, signal the Pmc1, leading to an increase in its activity. In such model, the participation of a calcium transporter, sensitive to gadolinium, and nifedipine, would also connect the external calcium availability with the activity of Pmc1. In any case, the balance of these two branches would be responsible for the transient nature of calcium signaling and would also avoid the accumulation of toxic calcium concentrations in the cytosol.

Of course, this working model is incomplete and other studies must be done to clarify many different aspects. The main aspects to be deeply studied would be (i) to find the mechanism by which the G protein Gpa2 and the glucose sensor Snf3 would detect fluctuations in the levels of sugar phosphates, then triggering the respective effects on phospholipase C and the Ca^{2+}-ATPase Pmc1. Would there be other proteins involved in these processes? (ii) By which mechanism(s) would IP_3 control the opening of the vacuolar calcium channel Yvc1, thus triggering a calcium signaling? As stressed before, it was already shown that IP_3 does not act directly on Yvc1 [86]; therefore, it is quite reasonable to suppose that other proteins could be involved in the control of this calcium channel; (iii) Finally, the identity of the protein kinases directly involved in the control of the glucose-induced activation of the plasma membrane H^+-ATPase should also be known.

All these aspects are under study in many laboratories and, in the near future, a complete picture of the entire signaling transduction pathway will be available. As a eukaryotic organism easily manipulated, it is quite likely that this yeast regulatory model of a proton pump ATPase may be useful in understanding how the acid secretion transporter and the energetic metabolism would be connected, for instance, in mammalian cells. Therefore, the ideas emerging from the studies on the glucose-induced activation of the plasma membrane H^+-ATPase from yeast cells could be used and/or adapted to study pathophysiological mechanisms of several diseases.

Acknowledgments

The research developed in our laboratory has been partially financed by grants from Fundação de Amparo a Pesquisa do Estado de Minas Gerais Process CBB 824/06; from Conselho Nacional de Desenvolvimento—CNPq (Brazil) Process 300998/89-9 (research fellowship to R.L.B.), Process 475672/08-5 and Process 550195/2010-6 (research grants). The author is extremely grateful to Luis Eduardo Fernandes Rodrigues da Conceição for his critical review of this work.

References

1. Goffeau, A., and C.W. Slayman. 1981. The proton-translocating ATPase of the fungal plasma membrane. *Biochim. Biophys. Acta*, 639, pp. 197–223.

2. Peña, A., G. Cinco, A. Gómez-Puyou, and M. Tuena. 1972. Effect of the pH of the incubation medium on glycolysis and respiration in Saccharomyces cerevisiae. *Arch. Biochem. Biophys.*, 153, pp. 413–425.
3. Conway, E.J., and T. Brady. 1947. Mechanism of high acid production by yeast and its bearing on hydrochloric acid formation in the stomach. *Nature*, 159, pp. 137.
4. Conway, E.J., and T. Brady. 1950. Biological production of acid and alkali; quantitative relations of succinic and carbonic acids to the potassium and hydrogen ion exchange in fermenting yeast. *Biochem. J.*, 47, pp. 360–369.
5. Conway, E.J., T. Brady., and E. Carton. 1950. Biological production of acid and alkali; a redox theory for the process in yeast with application to the production of gastric acidity. *Biochem. J.*, 47, pp. 369–374.
6. Rothstein, A., and C. Demis. 1953. The relationship of the cell surface to metabolism; the stimulation of fermentation by extracellular potassium. *Arch. Biochem. Biophys.*, 44, pp. 18–29.
7. Mithell, P. 1961. Coupling of phosphorylation to electron and hydrogen transfer by a chemi-osmotic type of mechanism. *Nature*, 191, pp. 144–148.
8. Rothstein, A. 1974. Relationship of cation influxes and effluxes in yeast. *J. Gen. Physiol.*, 64, pp. 608–621.
9. Slayman, C.L. 1965. Electrical properties of *Neurospora crassa*. Effects of external cations on the intracellular potential. *J. Gen. Physiol.*, 49, pp. 69–92.
10. Slayman, C.L. 1965. Electrical properties of *Neurospora crassa*. Respiration and the intracellular potential. *J. Gen. Physiol.*, 49, pp. 93–116.
11. Slayman, C.L., and C.W. Slayman. 1968. Net uptake of potassium in *Neurospora*. Exchange for sodium and hydrogen ions. *J. Gen. Physiol.*, 52, pp. 424–443.
12. Slayman, C.W., and C.L. Slayman. 1970. Potassium transport in *Neurospora*. Evidence for a multisite carrier at high pH. *J. Gen. Physiol.*, 55, pp. 758–786.
13. Peña, A., G. Cinco, A.G. Puyou, and M. Tuena. 1969. Studies on the mechanism of the stimulation of glycolysis and respiration by K^+ in *Saccharomyces cerevisiae*. *Biochim. Biophys. Acta*, 180, pp. 1–8.
14. Slayman, C.L. 1970. Movement of ions and electrogenesis in microorganisms. *Am. Zool.*, 10, pp. 377–392.

15. Slayman, C.L., W.S. Long, and C.Y. Lu. 1973. The relationship between ATP and an electrogenic pump in the plasma membrane of *Neurospora crassa*. *J. Membr. Biol.*, 14, pp. 305–338.

16. Eddy, A.A., and J.A. Nowacki. 1971. Stoicheiometrical proton and potassium ion movements accompanying the absorption of amino acids by the yeast *Saccharomyces carlsbergensis*. *Biochem. J.*, 122, pp. 701–711.

17. Foury, F., and A. Goffeau. 1975. Stimulation of active uptake of nucleosides and amino acids by cyclic adenosine 3':5'-monophosphate in the yeast *Schizosaccharomyces pombe*. *J. Biol. Chem.*, 250, pp. 2354–2362.

18. Misra, P.C., and M. Höfer. 1975. An energy-linked proton-extrusion across the cell membrane *Rhodotorula gracilis*. *FEBS Lett.*, 52, pp. 95–99.

19. Slayman, C.L., and C.W. Slayman. 1974. Depolarization of the plasma membrane of *Neurospora* during active transport of glucose: Evidence for a proton-dependent cotransport system. *Proc. Natl. Acad. Sci. USA*, 71, pp. 1935–1939.

20. Schatz, G. 1965. Subcellular particles carrying mitochondrial enzymes in anaerobically-grown cells of *saccharomyces cerevisiae*. *Biochim. Biophys. Acta*, 96, pp. 342–345.

21. Matile, P., H. Moor, and K. Mühlethaler. 1967. Isolation and properties of the plasmalemma in yeast. *Arch. Mikrobiol.*, 58, pp. 201–211.

22. Bult, C.J., O. White, G.J. Olsen, L. Zhou, R.D. Fleischmann, G.G. Sutton, J.A. Blake, L.M. FitzGerald, R.A. Clayton, J.D. Gocayne, A.R. Kerlavage, B.A. Dougherty, J.F. Tomb, M.D. Adams, C.I. Reich, R. Overbeek, E.F. Kirkness, K.G. Weinstock, J.M. Merrick, A. Glodek, J.L. Scott, N.S. Geoghagen, and J.C. Venter. 1996. Complete genome sequence of the methanogenic archaeon, *Methanococcus jannaschii*. *Science*, 273, pp. 1058–1073.

23. Meade, J.C., J. Shaw, S. Lemaster, G. Gallagher, and J.R. Stringer. 1987. Structure and expression of a tandem gene pair in *Leishmania donovani* that encodes a protein structurally homologous to eucaryotic cation-transporting ATPases. *Mol. Cell Biol.*, 7, pp. 3937–3946.

24. Portillo, F. 2000. Regulation of plasma membrane H(+)-ATPase in fungi and plants. *Biochim. Biophys. Acta*, 1469, pp. 31–42.

25. Serrano, R. 1978. Characterization of the plasma membrane ATPase of Saccharomyces cerevisiae. *Mol. Cell Biochem.*, **22**(1), pp. 51–63.

26. Ghislain, M., and A. Goffeau. 1991. The pma1 and pma2 H(+)-ATPases from *Schizosaccharomyces pombe* are functionally interchangeable. *J. Biol. Chem.*, 266, pp. 18276–18279.
27. Serrano, R., and F. Portillo. 1990. Catalytic and regulatory sites of yeast plasma membrane $H(^+)$-ATPase studied by directed mutagenesis. *Biochim. Biophys. Acta*, 1018, pp. 195–199.
28. Supply, P., A. Wach, D. Thinès-Sempoux, and A. Goffeau. 1993. Proliferation of intracellular structures upon overexpression of the PMA2 ATPase in *Saccharomyces cerevisiae*. *J. Biol. Chem.*, 268, pp. 19744–19752.
29. Capieaux, E., M.L. Vignais, A. Sentenac, and A. Goffeau. 1989. The yeast H^+-ATPase gene is controlled by the promoter binding factor TUF. *J. Biol. Chem.*, 264, pp. 7437–7446.
30. Rao, R., R.K. Nakamoto, S. Verjovski-Almeida, and C.W. Slayman. 1992. Structure and function of the yeast plasmamembrane $H(^+)$-ATPase. *Ann. N. Y. Acad. Sci.*, 671, pp. 195–203.
31. Gradmann, D., U.P. Hansen, W.S. Long, C.L. Slayman, and J. Warncke. 1978. Current-voltage relationships for the plasma membrane and its principal electrogenic pump in *Neurospora crassa*: I. Steady-state conditions. *J. Membr. Biol.*, 39, pp. 333–367.
32. Benito, B., E. Moreno, and R. Lagunas. 1991. Half-life of the plasma membrane ATPase and its activating system in resting yeast cells. *Biochim. Biophys. Acta*, 1063, pp. 265–268.
33. Serrano, R. 1983. In vivo glucose activation of the yeast plasma membrane ATPase. *FEBS Lett.*, 156, pp. 11–14.
34. McDonough, J.P., and H.P. Mahler. 1982. Covalent phosphorylation of the Mg^{2+}-dependent ATPase of yeast plasma membranes. *J. Biol. Chem.*, 257, pp. 14579–14581.
35. Carmelo, V., H. Santos, and I. Sá-Correia. 1997. Effect of extracellular acidification on the activity of plasma membrane ATPase and on the cytosolic and vacuolar pH of *Saccharomyces cerevisiae*. *Biochim. Biophys. Acta*, 1325, pp. 63–70.
36. Eraso, P., and C. Gancedo. 1987. Activation of yeast plasma membrane ATPase by acid pH during growth. *FEBS Lett.*, 224, pp. 187–192.
37. Benito, B., F. Portillo, and R. Lagunas. 1992. In vivo activation of the yeast plasma membrane ATPase during nitrogen starvation. Identification of the regulatory domain that controls activation. *FEBS Lett.*, 300, pp. 271–274.

38. Piper, P.W. 1995. The heat shock and ethanol stress responses of yeast exhibit extensive similarity and functional overlap. *FEMS Microbiol. Lett.*, 134, pp. 121–127.

39. Portillo, F., and M.J. Mazón. 1985. Activation of yeast plasma membrane ATPase by phorbol ester. *FEBS Lett.*, 192, pp. 95–98.

40. Rosa, M.F., and I. Sá-Correia. 1991. In vivo activation by ethanol of plasma membrane ATPase of *Saccharomyces cerevisiae*. *Appl. Environ. Microbiol.*, 57, pp. 830–835.

41. Viegas, C.A., and I. Sá-Correia. 1991. Activation of plasma membrane ATPase of *Saccharomyces cerevisiae* by octanoic acid. *J. Gen. Microbiol.*, 137, pp. 645–651.

42. Viegas, C.A., P.B. Sebastião, A.G. Nunes, and I. Sá-Correia. 1995. Activation of plasma membrane H(+)-ATPase and expression of PMA1 and PMA2 genes in Saccharomyces cerevisiae cells grown at supraoptimal temperatures. *Appl. Environ. Microbiol.*, 61, pp. 1904–1909.

43. Chen, Q., and W.F. Boss. 1990. Short-term treatment with cell wall degrading enzymes increases the activity of the inositol phospholipid kinases and the vanadate-sensitive ATPase of carrot cells. *Plant Physiol.*, 94, pp. 1820–1829.

44. Memon, A.R., Q.Y. Chen, and W.F. Boss. 1989. Inositol phospholipids activate plasma membrane ATPase in plants. *Biochem. Biophys. Res. Commun.*, 162, pp. 1295–1301.

45. Patton, J.L., and R.L. Lester. 1992. Phosphatidylinositol phosphate, phosphatidylinositol bisphosphate, and the phosphoinositol sphingolipids are found in the plasma membrane and stimulate the plasma membrane H(+)-ATPase of Saccharomyces cerevisiae. *Arch. Biochem. Biophys.*, 292, pp. 70–76.

46. Rao, R., D. Drummond-Barbosa, and C.W. Slayman. 1993. Transcriptional regulation by glucose of the yeast PMA1 gene encoding the plasma membrane H(+)-ATPase. *Yeast*, 9, pp. 1075–1084.

47. García-Arranz, M., A.M. Maldonado, M.J. Mazón, and F. Portillo. 1994. Transcriptional control of yeast plasma membrane H(+)-ATPase by glucose. Cloning and characterization of a new gene involved in this regulation. *J. Biol. Chem.*, 269, pp. 18076–18082.

48. Mahler, H.R., P.K. Jaynes, J.P. McDonough, and D.K. Hanson. 1981. Catabolite repression in yeast: Mediation by cAMP. *Curr. Top. Cell Regul.*, 18, pp. 455–474.

49. McDonough, J.P., P.K. Jaynes, and H.R. Mahler. 1980. Partial characterization of the plasma membrane ATPase from a rho0 petite strain of Saccharomyces cerevisiae. *J. Bioenerg. Biomembr.*, 12, pp. 249–264.
50. Chang, A., and C.W. Slayman. 1991. Maturation of the yeast plasma membrane [H^+]ATPase involves phosphorylation during intracellular transport. *J. Cell Biol.*, 115, pp. 289–295.
51. van der Plaat, J.B. 1974. Cyclic 3',5'-adenosine monophosphate stimulates trehalose degradation in baker's yeast. *Biochem. Biophys. Res. Commun.*, 56, pp. 580–587.
52. Portillo, F., and M.J. Mazón. 1986. The *Saccharomyces cerevisiae* start mutant carrying the cdc25 mutation is defective in activation of plasma membrane ATPase by glucose. *J. Bacteriol.*, 168, pp. 1254–1257.
53. Ulaszewski, S., F. Hilger, and A. Goffeau. 1989. Cyclic AMP controls the plasma membrane H^+-ATPase activity from *Saccharomyces cerevisiae. FEBS Lett.*, 245, pp. 131–136.
54. Brandão, R.L., I.M. Castro, J.B. Passos, J.R. Nicoli, and J.M. Thevelein. 1992. Glucose-induced activation of the plasma membrane H(+)-ATPase in *Fusarium oxysporum. J. Gen. Microbiol.*, 138 , pp. 1579–1586.
55. Beullens, M., K. Mbonyi, L. Geerts, D. Gladines, K. Detremerie, A.W. Jans, and J.M. Thevelein. 1988. Studies on the mechanism of the glucose-induced cAMP signal in glycolysis and glucose repression mutants of the yeast *Saccharomyces cerevisiae. Eur. J. Biochem.*, 172, pp. 227–231.
56. Mbonyi, K., M. Beullens, K. Detremerie, L. Geerts, and J.M. Thevelein. 1988. Requirement of one functional RAS gene and inability of an oncogenic ras variant to mediate the glucose-induced cyclic AMP signal in the yeast *Saccharomyces cerevisiae. Mol. Cell. Biol.*, 8, pp. 3051–3057.
57. Thevelein, J.M. 1992. The RAS-adenylate cyclase pathway and cell cycle control in Saccharomyces cerevisiae. *Ant. Van Leeuw.*, 62, pp. 109–130.
58. Van Aelst, L., E. Boy-Marcotte, J.H. Camonis, J.M. Thevelein, and M. Jacquet. 1990. The C-terminal part of the CDC25 gene product plays a key role in signal transduction in the glucose-induced modulation of cAMP level in *Saccharomyces cerevisiae. Eur. J. Biochem.*, 193, pp. 675–680.
59. dos Passos, J.B., M. Vanhalewyn, R.L. Brandão, I.M. Castro, J.R. Nicoli, and J.M. Thevelein. 1992. Glucose-induced activation of plasma membrane H(+)-ATPase in mutants of the yeast Saccharomyces cerevisiae affected in cAMP metabolism, cAMP-dependent protein phosphorylation and the initiation of glycolysis. *Biochim. Biophys. Acta*, 1136, pp. 57–67.

60. Mazón, M.J., M.M. Behrens, F. Portillo, and R. Piñón. 1989. cAMP- and RAS-independent nutritional regulation of plasma-membrane H^+-ATPase activity in *Saccharomyces cerevisiae*. *J. Gen. Microbiol.*, 135, pp. 1453–1460.

61. Kolarov, J., J. Kulpa, M. Baijot, and A. Goffeau. 1988. Characterization of a protein serine kinase from yeast plasma membrane. *J. Biol. Chem.*, 263, pp. 10613–10619.

62. Yanagita, Y., M. Abdel-Ghany, D. Raden, N. Nelson, and E. Racker. 1987. Polypeptide-dependent protein kinase from bakers' yeast. *Proc. Natl. Acad. Sci. USA*, 84, pp. 925–929.

63. Brandão, R.L., N.M. de Magalhães-Rocha, R. Alijo, J. Ramos, and J.M. Thevelein. 1994. Possible involvement of a phosphatidylinositol-type signaling pathway in glucose-induced activation of plasma membrane H(+)-ATPase and cellular proton extrusion in the yeast *Saccharomyces cerevisiae*. *Biochim. Biophys. Acta*, 1223, pp. 117–124.

64. Kaibuchi, K., A. Miyajima, K. Arai, and K. Matsumoto. 1986. Possible involvement of RAS-encoded proteins in glucose-induced inositolphospholipid turnover in *Saccharomyces cerevisiae*. *Proc. Natl. Acad. Sci. USA*, 83, pp. 8172–8176.

65. Coccetti, P., R. Tisi, E. Martegani, L. Souza Teixeira, R. Lopes Brandão, I. de Miranda Castro, and J.M. Thevelein. 1998. The PLC1 encoded phospholipase C in the yeast *Saccharomyces cerevisiae* is essential for glucose-induced phosphatidylinositol turnover and activation of plasma membrane H^+-ATPase. *Biochim. Biophys. Acta*, 1405, pp. 147–154.

66. Kraakman, L., K. Lemaire, P. Ma, A.W. Teunissen, M.C. Donaton, P. Van Dijck, J. Winderickx, J.H. de Winde, and J.M. Thevelein. 1999. A *Saccharomyces cerevisiae* G-protein coupled receptor, Gpr1, is specifically required for glucose activation of the cAMP pathway during the transition to growth on glucose. *Mol. Microbiol.*, 32, pp. 1002–1012.

67. Ansari, K., S. Martin, M. Farkasovsky, I.M. Ehbrecht, and H. Küntzel. 1999. Phospholipase C binds to the receptor-like GPR1 protein and controls pseudohyphal differentiation in Saccharomyces cerevisiae. *J. Biol. Chem.*, 274, pp. 30052–30058.

68. Ozcan, S., J. Dover, A.G. Rosenwald, S. Wölfl, and M. Johnston. 1996. Two glucose transporters in *Saccharomyces cerevisiae* are glucose sensors that generate a signal for induction of gene expression. *Proc. Natl. Acad. Sci. USA*, 93, pp. 12428–12432.

69. Ozcan, S., and M. Johnston. 1999. Function and regulation of yeast hexose transporters. *Microbiol. Mol. Biol. Rev.*, 63, pp. 554–569.

70. Souza, M.A., M.J. Trópia, and R.L. Brandão. 2001. New aspects of the glucose activation of the H(+)-ATPase in the yeast *Saccharomyces cerevisiae*. *Microbiology*, 147, pp. 2849–2855.
71. Kotyk, A., G. Lapathitis, and J. Horák. 2003. Critical findings on the activation cascade of yeast plasma membrane H^+-ATPase. *FEMS Microbiol. Lett.*, 226, pp. 175–180.
72. Nakajima-Shimada, J., S. Sakaguchi, F.I. Tsuji, Y. Anraku, and H. Iida. 2000. Ca^{2+} signal is generated only once in the mating pheromone response pathway in *Saccharomyces cerevisiae*. *Cell Struct. Funct.*, 25, pp. 125–131.
73. Tisi, R., S. Baldassa, F. Belotti, and E. Martegani. 2002. Phospholipase C is required for glucose-induced calcium influx in budding yeast. *FEBS Lett.*, 520, pp. 133–138.
74. Tökés-Füzesi, M., D.M. Bedwell, I. Repa, K. Sipos, B. Sümegi, A. Rab, and A. Miseta. 2002. Hexose phosphorylation and the putative calcium channel component Mid1p are required for the hexose-induced transient elevation of cytosolic calcium response in *Saccharomyces cerevisiae*. *Mol. Microbiol.*, 44, pp. 1299–1308.
75. Tisi, R., F. Belotti, S. Wera, J. Winderickx, J.M. Thevelein, and E. Martegani. 2004. Evidence for inositol triphosphate as a second messenger for glucose-induced calcium signalling in budding yeast. *Curr. Genet.*, 45, pp. 83–89.
76. Trópia, M.J., A.S. Cardoso, R. Tisi, L.G. Fietto, J.L. Fietto, E. Martegani, I.M. Castro, and R.L. Brandão. 2006. Calcium signaling and sugar-induced activation of plasma membrane H(+)-ATPase in *Saccharomyces cerevisiae* cells. *Biochem. Biophys. Res. Commun.*, 343, pp. 1234–1243.
77. Muller, E.M., E.G. Locke, and K.W. Cunningham. 2001. Differential regulation of two Ca(2+) influx systems by pheromone signaling in Saccharomyces cerevisiae. *Genetics*, 159, pp. 1527–1538.
78. Bouillet, L.E., A.S. Cardoso, E. Perovano, R.R. Pereira, E.M. Ribeiro, M.J. Trópia, L.G. Fietto, R. Tisi, E. Martegani, I.M. Castro, and R.L. Brandão. 2012. The involvement of calcium carriers and of the vacuole in the glucose-induced calcium signaling and activation of the plasma membrane H(+)-ATPase in *Saccharomyces cerevisiae cells*. *Cell Calcium*, 51, pp. 72–81.
79. Kellermayer, R., D.P. Aiello, A. Miseta, and D.M. Bedwell. 2003. Extracellular Ca(2+) sensing contributes to excess Ca(2+) accumulation and vacuolar fragmentation in a *pmr1*Δ mutant of *S. cerevisiae*. *J. Cell Sci.*, 116, pp. 1637–1646.

80. Groppi, S., F. Belotti, R.L. Brandão, E. Martegani, and R. Tisi. 2011. Glucose-induced calcium influx in budding yeast involves a novel calcium transport system and can activate calcineurin. *Cell Calcium*, 49, pp. 376–386.
81. Aiello, D.P., L. Fu, A. Miseta, and D.M. Bedwell. 2002. Intracellular glucose 1-phosphate and glucose 6-phosphate levels modulate Ca^{2+} homeostasis in *Saccharomyces cerevisiae*. *J. Biol. Chem.*, 277, pp. 45751–45758.
82. Fu, L., A. Miseta, D. Hunton, R.B. Marchase, and D.M. Bedwell. 2000. Loss of the major isoform of phosphoglucomutase results in altered calcium homeostasis in Saccharomyces cerevisiae. *J. Biol. Chem.*, 245, pp. 5431–5440.
83. Aiello, D.P., L. Fu, A. Miseta, K. Sipos, and D.M. Bedwell. 2004. The Ca^{2+} homeostasis defects in a *pgm2* Δ strain of *Saccharomyces cerevisiae* are caused by excessive vacuolar Ca^{2+}uptake mediated by the Ca^{2+}-ATPase Pmc1p. *J. Biol. Chem.*, 279, pp. 38495–38502.
84. Dlugai, S., S. Hippler, R. Wieczorke, and E. Boles. 2001. Glucose-dependent and-independent signalling functions of the yeast glucose sensor Snf3. *FEBS Lett.*, 505, pp. 389–392.
85. Ascher-Landsberg, J., T. Saunders, M. Elovitz, and M. Phillippe. 1999. The effects of 2-aminoethoxydiphenyl borate, a novel inositol 1,4, 5-trisphosphate receptor modulator on myometrial contractions. *Biochem. Biophys. Res. Commun.*, 264, pp. 979–982.
86. Palmer, C.P., X.L. Zhou, J. Lin, S.H. Loukin, C. Kung, and Y. Saimi. 2001. A TRP homolog in *Saccharomyces cerevisiae* forms an intracellular Ca(2+)-permeable channel in the yeast vacuolar membrane. *Proc. Natl. Acad. Sci. USA*, 98, pp. 7801–7805.
87. Miranda, M., K.E. Allen, J.P. Pardo, and C.W. Slayman. 2001. Stalk segment 5 of the yeast plasma membrane H^+-ATPase: Mutational evidence for a role in glucose regulation. *J. Biol. Chem.*, 276, pp. 22485–22490.
88. Miranda, M., J.P. Pardo, K.E. Allen, and C.W. Slayman. 2002. Stalk segment 5 of the yeast plasma membrane H(+)-ATPase. Labeling with a fluorescent maleimide reveals a conformational change during glucose activation. *J. Biol. Chem.*, 277, pp. 40981–40988.
89. Portillo, F., I.F. de Larrinoa, and R. Serrano. 1989. Deletion analysis of yeast plasma membrane H^+-ATPase and identification of a regulatory domain at the carboxyl-terminus. *FEBS Lett.*, 247, pp. 381–385.

90. Portillo, F., P. Eraso, and R. Serrano. 1991. Analysis of the regulatory domain of yeast plasma membrane H^+-ATPase by directed mutagenesis and intragenic suppression. *FEBS Lett.*, 287, pp. 71–74.

91. Serrano, R., C. Montesinos, M. Roldán, G. Garrido, C. Ferguson, K. Leonard, B.C. Monk, D.S. Perlin, and E.W. Weiler. 1991. Domains of yeast plasma membrane and ATPase-associated glycoprotein. *Biochim. Biophys. Acta*, 1062, pp. 157–164.

92. Lecchi, S., K.E. Allen, J.P. Pardo, A.B. Mason, and C.W. Slayman. 2005. Conformational changes of yeast plasma membrane H(+)-ATPase during activation by glucose: Role of threonine-912 in the carboxy-terminal tail. *Biochemistry*, 44, pp. 16624–16632.

93. Lecchi, S., C.J. Nelson, K.E. Allen, D.L. Swaney, K.L. Thompson, J.J. Coon, M.R. Sussman, and C.W. Slayman. 2007. Tandem phosphorylation of Ser-911 and Thr-912 at the C terminus of yeast plasma membrane H^+-ATPase leads to glucose-dependent activation. *J. Biol. Chem.*, 282, pp. 35471–35481.

94. Rubenstein, E.M., and M.C. Schmidt. 2007. Mechanisms regulating the protein kinases of *Saccharomyces cerevisiae. Eukaryot. Cell*, 6, pp. 571–583.

95. Eraso, P., and F. Portillo. 1994. Molecular mechanism of regulation of yeast plasma membrane H(+)-ATPase by glucose. Interaction between domains and identification of new regulatory sites. *J. Biol. Chem.*, 269, pp. 10393–10399.

96. Eraso, P., M.J. Mazón, and F. Portillo. 2006. Yeast protein kinase Ptk2 localizes at the plasma membrane and phosphorylates in vitro the C-terminal peptide of the H^+-ATPase. *Biochim. Biophys. Acta*, 1758, pp. 164–170.

97. de la Fuente, N., A.M. Maldonado, and F. Portillo. 1997. Yeast gene YOR137c is involved in the activation of the yeast plasma membrane H^+-ATPase by glucose. *FEBS Lett.*, 420, pp. 17–19.

98. Goossens, A., N. de La Fuente, J. Forment, R. Serrano, and F. Portillo. 2000. Regulation of yeast H(+)-ATPase by protein kinases belonging to a family dedicated to activation of plasma membrane transporters. *Mol. Cell. Biol.*, 20, pp. 7654–7661.

99. Gomes, K.N., S.M. Freitas, T.M. Pais, J.L. Fietto, A.H. Totola, R.M. Arantes, A. Martins, C. Lucas, D. Schuller, M. Casal, I.M. Castro, L.G. Fietto, and R.L. Brandão. 2005. Deficiency of Pkc1 activity affects glycerol metabolism in *Saccharomyces cerevisiae. FEMS Yeast Res.*, 5, pp. 767–776.

100. Salgado, A.P., D. Schuller, M. Casal, C. Leão, F.C. Leiper, D. Carling, L.G. Fietto, M.J. Trópia, I.M. Castro, and R.L. Brandão. 2002. Relationship between protein kinase C and derepression of different enzymes. *FEBS Lett.*, 532, pp. 324–332.

101. Schmitz, H.P., and J.J. Heinisch. 2003. Evolution, biochemistry and genetics of protein kinase C in fungi. *Curr. Genet.*, 43, pp. 245–254.

102. Pereira, M.B., R. Tisi, L.G. Fietto, A.S. Cardoso, M.M. França, F.M. Carvalho, M.J. Trópia, E. Martegani, I.M. Castro, and R.L. Brandão. 2008. Carbonyl cyanide m-chlorophenylhydrazone induced calcium signaling and activation of plasma membrane H(+)-ATPase in the yeast Saccharomyces cerevisiae. *FEMS Yeast Res.*, 8, pp. 622–630.

Index